Climate Change and International Shipping

Legal Aspects of Sustainable Development

Series Editor

David Freestone

This series publishes work on all aspects of the international legal dimensions of the concept of sustainable development. Its aim is to publish important works of scholarship on a range of relevant issues including conservation of natural resources, climate change, biodiversity loss and the role of international agreements, international organizations and state practice.

VOLUME 23

The titles published in this series are listed at *brill.com/lasd*

Climate Change and International Shipping

The Regulatory Framework for the Reduction of Greenhouse Gas Emissions

By

Yubing Shi

BRILL
NIJHOFF

LEIDEN | BOSTON

Library of Congress Cataloging-in-Publication Data

Names: Shi, Yubing, author.
Title: Climate change and international shipping : the regulatory framework for the reduction of greenhouse gas emissions / by Yubing Shi.
Description: Leiden ; Boston : Brill Nijhoff, 2017. | Series: Legal aspects of sustainable development ; volume 23 | Based on author's thesis (doctoral—Australian National centre for Ocean Resources and Security (ANCORS), University of Wollongong, 2014) issued under title: "The regulatory framework for the reduction of greenhouse gas emissions from international shipping. | Includes bibliographical references and index.
Identifiers: LCCN 2016036874 (print) | LCCN 2016041254 (ebook) | ISBN 9789004329300 (hardback : alk. paper) | ISBN 9789004329317 (E-book)
Subjects: LCSH: Greenhouse gas mitigation—Law and legislation. | Marine engines—Exhaust gas—Law and legislation.
Classification: LCC K3593.5.C37 S449 2017 (print) | LCC K3593.5.C37 (ebook) | DDC 344.05/738746—dc23
LC record available at https://lccn.loc.gov/2016036874

Typeface for the Latin, Greek, and Cyrillic scripts: "Brill". See and download: brill.com/brill-typeface.

ISSN 1875-0923
ISBN 978-90-04-32930-0 (hardback)
ISBN 978-90-04-32931-7 (e-book)

This book is printed on acid-free paper and produced in a sustainable manner.

Contents

Foreword

The process to regulate the reduction of greenhouse gas (GHG) emissions from international shipping is comprehensive and controversial. It involves political, historical, economic and technical considerations and cuts across international maritime law, international environmental law, and international trade law. It has been necessary to find political compromises and develop innovative regulatory efforts. The aim of this book was not only to examine the evolution and adequacy of the current regulatory framework for the reduction of GHG emissions from international shipping, but more so to provide options and recommendations for legal and institutional reforms to improve this framework. In attempting to take a holistic approach, this book synthesised applicable international environmental law principles and assessed the responses of the key stakeholders to the challenge of reducing GHG emissions. While both the international climate change regime under the United Nations Framework Convention on Climate Change process and the IMO through its Marine Environment Protection Committee have been grappling with this issue, this book also examines the *Paris Agreement* adopted in December 2015 and its implications on the international regulation of shipping GHG emissions. This book relies on primary sources from various international fora dated before April 2016.

Excerpts from this book have been published in refereed Journals. An earlier draft of section 7.5 has been published as 'Reducing greenhouse gas emissions from international shipping: Is it time to consider market-based measures?', *Marine Policy*, 2016, Vol. 64. Parts of Chapters 5 and 6 appeared in modified form in 'Greenhouse Gas Emissions from International Shipping: the Response from China's Shipping Industry to the Regulatory Initiatives of the International Maritime Organization', *International Journal of Marine and Coastal Law*, 2014, Vol. 29, No. 1; and 'Gigantic Shipbuilders under the IMO Mandate of GHG Emissions: With Special References to China, Japan and Korea', *Journal of East Asia & International Law*, 2014, Vol. 7, No. 2. Section 2.1 has been partially included in 'Are Greenhouse Gas Emissions from International Shipping a Type of Marine Pollution?', *Marine Pollution Bulletin*, 2016. Parts of Chapter 4 have been published as 'The Challenge of Reducing Greenhouse Gas Emissions from International Shipping: Assessing the International Maritime Organization's Regulatory Response', *Yearbook of International Environmental Law*, 2012, Vol. 23, No. 1.

Series Editor Preface

This is the twenty-third volume in the Brill Nijhoff series on *Legal Aspects of Sustainable Development* published under my General Editorship. The aim of this series is to publish works at the cutting edge of legal scholarship that address both the practical and the theoretical aspects of this important concept.

I am very pleased to be able to include this work by Dr Yubing Shi in this series. It is a revised and updated version of his PhD thesis at the University of Wollongong in Australia. In 2007, UN Secretary-General Ban Ki-moon called climate change the defining issue of our era. The passing of time has only reinforced this view. The good news is that after more than a decade of negotiations the Parties to the UN Framework Convention on Climate Change (UNFCCC) agreed in December 2015 in Paris to a new global agreement to address both mitigation of, and adaptation to, climate change. However the Paris Agreement does not expressly address two controversial issues: greenhouse gas emissions from aircraft and from ships. These have been left to negotiations within the existing regulatory bodies—the International Civil Aviation Organization (ICAO) and the International Maritime Organization (IMO) respectively. Although each accounts for a relatively small percentage of global emissions—both air traffic and seaborne traffic are still increasing exponentially. Data derived from satellite observations from 1992 to 2012 suggests that maritime traffic increased 300 percent in that 20 year period.

This very timely study puts the whole issue of maritime vessel emissions into perspective. Dr Shi examines in detail both the legal regime of the UNFCCC as well as the work of the IMO. Strides have been made at IMO in amending Annex VI of MARPOL 1973/78 to require reduction of greenhouse gases from vessels, but as this study shows, there are still major implementation issues.

I am delighted to include this volume in the series. Based on excellent research and scholarship, it addresses a highly practical aspect of our global effort to achieve sustainable development.

David Freestone
Washington DC

Acknowledgements

This book is substantially based on my doctoral thesis conducted at the Australian National Centre for Ocean Resources and Security (ANCORS), University of Wollongong, Australia. Through this process I was lucky to have Professor Robin Warner and Professor Warwick Gullett as my PhD supervisors. I would not have finished my thesis that addresses a multidisciplinary and controversial issue without their constant guidance, constructive criticisms and encouragement. Their valuable advice and comments assisted my understanding of some controversial issues relating to principles of international environmental law.

I am grateful for the support I received from many of the staff of ANCORS. In particular, I thank Professor Martin Tsamenyi, Professor Stuart Kaye and Professor Clive Schofield for their advice and encouragement in carrying out my research. To my fellow students at ANCORS, past and present, thanks for their companionship and friendship. I am also thankful for the discussions I had with Dr. Edmund Hughes from the IMO, Dr. James Harrison from the University of Edinburgh, and Professor Kuan-Hsiung Wang from National Taiwan Normal University.

Finally, I owe endless thanks to my wife, Bo Liu, my son, Harry, and my parents for their love, trust and sacrifice. I dedicate this work to them.

List of Figures

List of Tables

List of Acronyms

ADP	Ad Hoc Working Group on the Durban Platform for Enhanced Action
AISR	Australian International Shipping Register
AMS	American Mathematical Society
APA	Ad Hoc Working Group on the Paris Agreement
APSAI	Australian Peak Shippers Association Inc.
ASA	Australian Shipowners Association
ASF	Asian Shipowners Forum
AWG-KP	Ad Hoc Working Group on Further Commitments for Annex I Parties under the Kyoto Protocol
AWG-LCA	Ad Hoc Working Group on Long-term Cooperative Action
BAT	Best Available Technology
BAPA	Buenos Aires Plan of Action
BATNEEC	Best Available Technology Not Entailing Excessive Cost
BIMCO	Baltic and International Maritime Council
CANSI	China Association of the National Shipbuilding Industry
CBD	Convention on Biological Diversity
CBDR	Common but Differentiated Responsibility
CCS	China Classification Society
CDM	Clean Development Mechanism
CER	Certified Emission Reduction
CESA	Community of European Shipyards' Associations
CESS	Committee for Expertise of Shipbuilding Specifics
CIA	Central Intelligence Agency
CLIA	Cruise Lines International Association
CLRTAP	Geneva Convention on Long-Range Transboundary Air Pollution
CMA	Conference of the Party serving as the Meeting of the Parties to the Paris Agreement
CMP	Conference of the Party serving as the Meeting of the Parties to the Kyoto Protocol
COP	Conference of the Party
COSCO	China Ocean Shipping (Group) Company
CPRS	Carbon Pollution Reduction Scheme
CSA	China Shipowners Association
DSB	Daewoo Shipbuilding
CSC	Clean Shipping Coalition

DCME	Design, Construction, Manning or Equipment
DWT	Dead Weight Tonnage
EC	European Community
EEDI	Energy Efficiency Design Index
EEOI	Energy Efficiency Operational Indicator
EEZ	Exclusive Economic Zone
EIA	Environmental Impact Assessment
EIS	Efficiency Incentive Scheme
EMEP	Cooperative Programme for the Monitoring and Evaluation of the Long-range Transmission of Air Pollutants in Europe
ER	Emissions Registry
ERU	Emission Reduction Unit
ET	Emissions Trading
ETS	Emissions Trading System
EU	European Union
EU ETS	Union Emissions Trading Scheme
FAO	Food and Agriculture Organization of the United Nations
FOC	Flags of Convenience
FPSOS	Floating Production Storage and Offloading Facilities
FSUS	Floating Storage Units
GCF	Green Climate Fund
GDP	Gross Domestic Product
GESAMP	Group of Experts on Scientific Aspects of Marine Pollution
GEF	Global Environment Facility
GHG	Greenhouse Gas
GHG-WG	Working Group on GHG Emissions from Ships
GSAPS	Greek Shipowners Association for Passenger Ships
GT	Gross Tonnage
HCS	Hellenic Chamber of Shipping
HHI	Hyundai Heavy Industries
HSSA	Hellenic Shortsea Shipowners Association
IAC	InterAcademy Council
IACS	International Association of Classification Societies
IAPH	International Association of Ports and Harbors
IAPP	International Air Pollution Prevention Certificate
IBIA	International Bunker Industry Association
ICAO	International Civil Aviation Organization
ICCSA	Indian Coastal Conference Shipping Association
ICJ	International Court of Justice
ICS	International Chamber of Shipping

ICT	Information and Communications Technology
ICSU	International Council of Science
ILC	International Law Commission
IMarEst	Institute of Marine Engineering Science and Technology
IMCO	Inter-Governmental Maritime Consultative Organization
IMO	International Maritime Organization
INC	Intergovernmental Negotiating Committee
INDC	Intended Nationally Determined Contribution
INSA	Indian National Shipowners' Association
INTERCARGO	International Association of Dry Cargo Shipowners
INTERTANKO	International Association of Independent Tanker Owners
IPCC	Intergovernmental Panel on Climate Change
IPITCA	International Petroleum Industry Environmental Conservation Association
IPPIC	International Paint and Printing Ink Council
IPTA	International Parcel Tankers Association
ISF	International Shipping Federation
ISO	International Organization for Standardization
ITLOS	International Tribunal for the Law of the Seas
ITTC	International Towing Tank Conference
IUCN	International Union for Conservation of Nature
IUMI	International Union of Marine Insurance
JECKU	Japan, Europe, China, Korea and USA Shipbuilders' Association
JI	Joint Development
JWG	Joint Industry Working Group
KCCI	Korean Chamber of Commerce and Industry
KOSHIPA	Korea Offshore & Shipbuilding Association
KRS	Korean Register of Shipping
KSA	Korea Shipowners' Association
LDCS	Least Developed Countries
LIS	Leveraged Incentive Scheme
LLDCS	Land Locked Developing Countries
LNG	Liquefied Natural Gas
LOSC	United Nations Convention on the Law of the Sea
LPG	Liquefied Petroleum Gas
MBI	Market-Based Instrument
MBM	Market-Based Measure
MEA	Multilateral Environmental Agreement
MEPC	Marine Environment Protection Committee

MOP	Meeting of Parties
MOU	Memorandum of Understanding
MRV	Measurement, Reporting and Verification
MSC	Maritime Safety Committee
MUA	Maritime Union of Australia
NAS	National Academy of Science
NDC	Nationally Determined Contribution
NMFT	No More Favourable Treatment
NGO	Non-Governmental Organisation
OCIMF	Oil Companies International Marine Forum
ODS	Ozone Depleting Substance
OECD	Organisation for Economic Co-operation and Development
POP	Persistent Organic Pollutant
PSCO	Port State Control Officer
PSL	Port State Levy
PSSA	Particularly Sensitive Sea Area
RINA	Royal Institution of Naval Architects
RM	Rebate Mechanism
RO	Recognised Organisation
ro-ro	roll-on/roll-off
SAI	Shipyards Association of India
SAL	Shipping Australia Limited
SECT	Ship Efficiency and Credit Trading
SEEMP	Ship Energy Efficiency Management Plan
SEMP	Ship Efficiency Management Plan
SBI	Subsidiary Body for Implementation
SBSTA	Subsidiary Body on Scientific and Technological Advice
SHI	Samsung Heavy Industries
SIDS	Small Island Developing States
SIGTTO	Society of International Gas Tanker and Terminal Operators
SMS	Safety Management System
SSA	UK Shipbuilders and Ship Repairers Association
TOR	Terms of Reference
UCS	UK Chamber of Shipping
UGS	Union of Greek Shipowners
UNCED	United Nations Conference on Environment and Development
UNCTAD	United Nations Conference on Trade and Development
UNDP	United Nations Development Programme
UNECE	UN Economic Commission for Europe

UNEP	United Nations Environment Program
UNGA	United Nations General Assembly
US	United States of America
USCG	United States Coast Guard
VES	Vessel Efficiency System
VLCC	Very Large Crude Carriers
VOCs	Volatile Organic Compounds
WFP	United Nations World Food Programme
WMO	World Meteorological Organisation
WPCI	World Ports Climate Initiative
WSC	World Shipping Council
WTO	World Trade Organization
WWF	World Wide Fund for Nature

CHAPTER 1

Introduction

Climate change has been discussed broadly around the world and is recognised as a factor contributing to all global issues.[1] As an environmental, cultural and political phenomenon, climate change has been reshaping the way that people think about themselves, about their societies and about humanity's place on earth.[2] Greenhouse gas (GHG) emissions constitute the largest contribution to climate change,[3] and have thus attracted mounting attention from the international community as to how to effectively reduce GHG emissions on a global scale. One of the crucial global efforts is the international climate change regime, which comprises rules, norms, principles and procedures applicable to a range of activities.[4] International, regional and national regulations have been developed since the late 1970s to reduce GHG emissions.[5] Among them,

1 Milke Hulme, 'The Idea of Climate Change' (2010) 19(3) *GAIA: Ecological Perspectives for Science & Society* 171, 171. Hulme asserts that climate change has become an idea that now travels well beyond its origins in the natural sciences. Climate change takes on new meanings and serves new purposes, and has thus become 'the mother of all issues'. See also Susanne Moser, Heide Hackmann and Françoise Caillods, 'Global Environmental Change Changes Everything: Key Messages and Recommendations' in ISSC/UNESCO (ed), *World Social Science Report 2013: Changing Global Environments* (OECD Publishing and Unesco Publishing, 2013) 50. This report concludes that 'the social sciences must help to fundamentally reframe climate and global environmental change from a physical into a social problem'.

2 Hulme, above n. 1.

3 Intergovernmental Panel on Climate Change (IPCC), 'Fifth Assessment Report: Working Group I Report' (2013) <http://www.climatechange2013.org/images/uploads/WGIAR5-SPM_Approved27Sep2013.pdf> accessed 17 November 2013, Summary for Policymakers, p. 8. The summary for policymakers of the Working Group I Report asserts that 'the largest contribution to total radiative forcing [of climate change] is caused by the increase in the atmospheric concentration of CO_2 since 1750'.

4 Xinyuan Dai, 'Global Regime and National Change' (2010) 10(6) *Climate Policy* 622, 623. See also Patricia W. Birnie, Alan E. Boyle and Catherine Redgwell, *International Law and the Environment* (Oxford University Press, 3rd ed, 2009) 336.

5 See, e.g., *Convention on Long-Range Transboundary Air Pollution*, opened for signature 13 November 1979, 18 ILM 1442 (entered into force 16 March 1983); *Vienna Convention for the Protection of the Ozone Layer*, opened for signature 22 March 1985, 26 ILM 1529 (entered into force 22 September 1988); *Midwest Greenhouse Gas Reduction Accord*, signed on 15 November 2007, <http://www.c2es.org/us-states-regions/regional-climate-initiatives/mggra> accessed

 | DOI 10.1163/9789004329317_002

the *United Nations Framework Convention on Climate Change* (*UNFCCC*)[6] and its *Kyoto Protocol*[7] have provided the foundation for subsequent efforts to promote the international climate change regime. Various global efforts and outcomes, including the 2007 *Bali Road Map*,[8] 2010 *Cancun Agreements*,[9] 2011 *Durban Package*,[10] 2012 *Doha Climate Gateway*,[11] 2013 *Warsaw Outcomes*,[12] as well as the *Paris Agreement* adopted in December 2015,[13] have been shaping and will continue to shape the current international climate change regime.[14]

One shortcoming of the international climate change regime is that producers of GHG emissions from international shipping are exempt from liabilities under the *Kyoto Protocol*, notwithstanding that the contribution of GHG emissions from international shipping to climate change is significant and has been increasing.[15] Given the urgency of emission reduction and the global nature of the shipping industry, a global approach must be employed to regulate GHG emissions from shipping. The *UNFCCC* and the International Maritime

---

17 November 2013; *Clean Air Act of the United States of America*, 17 December 1963, 42 USC 7401–7626.

6 *United Nations Framework Convention on Climate Change*, opened for signature 9 May 1992, 31 ILM 848 (entered into force 21 March 1994) ('*UNFCCC*').

7 *Kyoto Protocol to the United Nations Framework Convention on Climate Change*, opened for signature 16 March 1998, 37 ILM 22 (entered into force 16 February 2005) ('*Kyoto Protocol*').

8 *Bali Action Plan*, Decision 1/CP.13, Report of the Conference of the Parties on its Thirteenth Session, Doc FCCC/CP/2007/6/Add.1 (14 March 2008).

9 *The Cancun Agreements*, Decisions 1–2/CMP.6, Report of the Conference of the Parties serving as the Meeting of the Parties to the Kyoto Protocol on its Sixth Session, FCCC/KP/CMP/2010/12/Add.1 (15 March 2011); Decision 1/CP.16, Report of the Conference of the Parties on its Sixteenth Session, FCCC/CP/2010/7/Add.1 (15 March 2011).

10 UNFCCC, *Durban: Towards Full Implementation of the UN Climate Change Convention* (2011) <http://unfccc.int/key_steps/durban_outcomes/items/6825.php> accessed 17 November 2013.

11 UNFCCC, *The Doha Climate Gateway* (2012) <http://unfccc.int/key_steps/doha_climate_gateway/items/7389.php> accessed 17 November 2013.

12 UNFCCC, *Warsaw Outcomes* (2013) <http://unfccc.int/key_steps/warsaw_outcomes/items/8006.php> accessed 19 April 2014.

13 *Paris Agreement to the United Nations Framework Convention on Climate Change* (2015), opened for signature 22 April 2016, FCCC/CP/2015/L.9 (not yet in force) ('*Paris Agreement*').

14 Although most of these outcomes are not legally binding, these achievements advanced the process of the global joint efforts effectively and to some extent could be deemed as 'a more elaborate and extended version of the 1992 UNFCCC'. Michael Grubb, 'Cancun: the Art of the Possible' (2011) 11(2) *Climate Policy* 847, 847. See also Navroz K. Dubash and Lavanya Rajamani, 'Beyond Copenhagen: Next Steps' (2010) 10(6) *Climate Policy* 593, 593.

15 The specific data on GHG emissions from international shipping is provided at 1.2.2.2 of this chapter.

Organization (IMO) have responded to this imperative and have commenced development of a regulatory framework.

GHG emissions from international shipping have been partially regulated by the IMO in the form of amendments to Annex VI to the *International Convention for the Prevention of Pollution from Ships* (*MARPOL 73/78*).[16] However, this regulation was adopted by a majority vote within the IMO rather than by a consensus. It is thus uncertain whether this regulation can be implemented uniformly by the global shipping industry. Meanwhile, it is important to identify the deficiencies existing in the current regulatory framework for this GHG issue and to provide measures for its improvement. The IMO is currently discussing the next step in addressing GHG emissions from shipping, and the *Paris Agreement* as a universal climate change agreement does not touch upon the reduction requirements for the international shipping sector but defers relevant matters to its subsequent conferences. Therefore, it is timely to examine the issues. This book responds to the need for an effective international regime to address GHG emissions from international shipping by exploring the application of international law principles. It analyses and assesses the responses from the UN, the IMO, the shipping industry, flag States and port States, and proposes legal, policy and institutional reforms to address gaps in the existing framework.

This introductory chapter provides a background to the global concern about GHG emissions from international shipping and identifies the central issue to be addressed by the book: how to improve the current regulatory framework in reducing GHG emissions from international shipping? The chapter is divided into three parts. The first part examines the relationship between GHG emissions and climate change and identifies the transboundary nature of GHG emissions. The second part analyses the sources and impacts of GHG

16 *International Convention for the Prevention of Pollution from Ships* (*MARPOL 73/78*), signed 2 November 1973, 12 ILM 1319, as amended by the 1978 Protocol to the 1973 Convention, 1341 UNTS 3, 17 ILM 546 (entered into force 2 October 1983). To date, *MARPOL 73/78* has adopted 6 annexes and their revisions, namely, Annex I. Oil (entered into force 2 October 1983), Annex II. Noxious Liquid Substances carried in Bulk (entered into force 6 April 1987), Annex III. Harmful Substances carried in Packaged Form (entered into force 1 July 1992), Annex IV. Sewage (entered into force 27 September 2003), Annex V. Garbage (entered into force 31 December 1988), and Annex VI. Air Pollution from Ships (entered into force 19 May 2005).

The reason why GHG emissions from international shipping have been 'partially' regulated lies in the fact that only certain types of ships engaged in international shipping have been regulated by amended Annex VI to *MARPOL 73/78*, and of the three routes within the IMO to regulate this GHG issue, only technical and operational measures have been employed whereas market-based measures are still under discussion.

emissions from international shipping. The third part explains the problem addressed by the book, presents the book objective, research questions and methods, and outlines the structure of the book.

1.1 Greenhouse Gas Emissions as a Regulatory Challenge

In a broad sense, GHGs consist of natural gases and anthropogenically produced gases. The former comprises water vapour (H_2O), carbon dioxide (CO_2), nitrous oxide (N_2O), methane (CH_4) and ozone (O_3), and the latter includes the halocarbons, and other chlorine and bromine—containing substances regulated under the *Montreal Protocol on Substances that Deplete the Ozone Layer*.[17] However, only seven types of GHGs are listed in the *Kyoto Protocol*, namely carbon dioxide (CO_2), methane (CH_4), nitrous oxide (N_2O), hydrofluorocarbons (HFCs), perfluorocarbons (PFCs), sulphur hexafluoride (SF_6) and nitrogen trifluoride (NF_3).[18] For the purposes of this book, the scope of GHGs is that defined in accordance with the *Kyoto Protocol*.

As a gaseous constituent in the atmosphere, GHGs absorb thermal infrared radiation, emitted by the Earth's surface, the atmosphere and clouds. Atmospheric radiation is emitted in all directions, including downward to the Earth's surface. In this way, GHGs trap heat within the surface-troposphere system. This is called the 'greenhouse effect'.[19] The earth's 'greenhouse effect' is what makes this planet suitable for life since without it the earth's surface would be much colder.[20] Therefore, GHGs are indispensable for the earth. However, apart from purely human-produced synthetic halocarbons, most GHGs have both natural and anthropogenic sources, and it is the latter—anthropogenically induced GHG emissions—that are 'extremely likely' to

17 Intergovernmental Panel on Climate Change (IPCC), 'Fourth Assessment Report' (2007) <http://www.ipcc.ch/publications_and_data/ar4/syr/en/contents.html> accessed 17 November 2013, Appendix Glossary; *Montreal Protocol on Substances that Deplete the Ozone Layer*, opened for signature 16 September 1987, 26 ILM 1550 (entered into force 1 January 1989) ('*Montreal Protocol*').

18 *Kyoto Protocol*, Annex A. The 1997 *Kyoto Protocol* only listed six types of GHGs, namely CO_2, CH_4, N_2O, HFCs, PFCs and SF_6, but a seventh type of GHG, NF_3,was added to the category in the Durban Climate Change Conference in 2011. NF_3 only applies from the beginning of the second commitment period (1 January 2013). *Doha Amendment to the Kyoto Protocol*, adopted 8 December 2012, Decision 1/CMP.8, C.N.718.2012.TREATIES-XXVII.7.c (not yet in force).

19 IPCC, above n. 17.

20 Ibid.

cause climate change.[21] The relationship between GHG emissions and climate change, together with the transboundary nature of GHG emissions, makes it a challenge to regulate GHG emissions under an international regulatory regime.

1.1.1 *Greenhouse Gas Emissions and Climate Change*

Excessive GHG emissions have been regarded as the main contribution to global climate change. However, this view has not been agreed by all. This section briefly reviews the debate on climate change and examines how GHG emissions contribute to climate change.

1.1.1.1 An Overview of Climate Change

Technically, climate change refers to 'a change in the state of the climate that can be identified by changes in the mean and/or the variability of its properties, and that persists for an extended period, typically decades or longer', which may result from either natural internal processes and 'external forcings' or anthropogenically—induced activities.[22] From the perspective of law, climate change has been defined as 'a change of climate which is attributed directly or indirectly to human activity that alters the composition of the global atmosphere and which is in addition to natural climate variability observed over comparable time periods'.[23] Climate change is a comparatively broad concept and the popular term 'global warming' only serves as a part of that.[24] The distinction between climate change and global warming, however, is often ignored or misinterpreted by the media.[25]

---

21 IPCC, above n. 3, 12.

22 IPCC, above n. 17. 'External forcing' refers to a forcing agent outside the climate system causing a change in the climate system, and some of its examples include volcanic eruptions, solar variations and anthropogenic changes in the composition of the atmosphere and land-use change.

23 *UNFCCC* art. 1(2).

24 The manifestations of climate change include but are not limited to: global warming, humidity, rainfall, wind, and severe weather events, although global warming also contributes to climate change. See IPCC, above n. 17.

25 See, e.g., Dorothy Parker, *Paul Sheehan, Ian Plimer, Heaven and Earth, and Is It Safe?* <http://themichaelduffyfiles.blogspot.com/2009/04/paul-sheehan-ian-plimer-heaven-and.html> accessed 10 May 2011; James Delingpole, *Meet the Man Who has Exposed the Great Climate Change Con Trick* (11 July 2009) <http://www.mannkal.org/downloads/environment/meetthemanwhohasexposedthegreatclimatechangecontrick.pdf> accessed 19 April 2014. In this literature, the authors often treat climate change and global warming as the same concept.

Beginning in the early 1970s, scientists came to understand the major trends occurring in many of the drivers of environmental change, including air pollution and climate change. The international community commenced efforts, through the *Stockholm Declaration*,[26] *Rio Declaration*,[27] and the *Johannesburg Declaration on Sustainable Development*,[28] to raise global consciousness on environmental protection, and also adopted international environmental conventions. Examples include the *1979 Convention on Long-Range Transboundary Air Pollution*,[29] the *1985 Convention for the Protection of the Ozone Layer*,[30] and the *1992 UNFCCC*.

The Intergovernmental Panel on Climate Change (IPCC) was established by the United Nations Environment Program (UNEP) and the World Meteorological Organization (WMO) in 1988 to provide knowledge and assessment of climate change.[31] On the basis of input from thousands of scientists throughout the world, IPCC has published five assessment reports (1990, 1995, 2001, 2007 and 2014). The Working Group I report of the fifth assessment report, which was released in September 2013, states that

> [w]arming of the climate system is unequivocal, and since the 1950s, many of the observed changes are unprecedented over decades to millennia. The atmosphere and ocean have warmed, the amounts of snow and ice have diminished, sea level has risen, and the concentrations of greenhouse gas have increased.[32]

Compared with previous IPCC reports, this latest IPCC report is unequivocal about the seriousness of anthropogenic climate change, and leaves 'fewer uncertainties about the serious consequences of inaction, in spite of the fact that there remain knowledge gaps and uncertainties in some areas of climate

---

26 *Stockholm Declaration on the Human Environment*, 11 ILM 1416 (16 June 1972).

27 *Rio Declaration on Environment and Development*, 31 ILM 874 (14 June 1992).

28 *Declaration on Sustainable Development and Plan of Implementation of the World Summit on Sustainable Development*, A/CONF.199/CRP/7 and A/CONF/L/6/Rev.2 (4 September 2002).

29 *Convention on Long-Range Transboundary Air Pollution*, opened for signature 13 November 1979, 18 ILM 1442 (entered into force 16 March 1983).

30 *Vienna Convention for the Protection of the Ozone Layer*, opened for signature 22 March 1985, 26 ILM 1529 (entered into force 22 September 1988).

31 IPCC, *Organization* <http://www.ipcc.ch/organization/organization.shtml> accessed 19 April 2014.

32 IPCC, above n. 3, 3.

science'.[33] Furthermore, this report points out that limiting climate change 'require[s] substantial and sustained reduction of greenhouse gas emissions'.[34]

The IPCC assessment reports are not universally welcomed. Various sustained criticisms exist. For example, Horner has asserted that 'environmentalism has served for decades as the best excuse to increase government control over actions [by most people in the society]' and that global warming may not be true in that: (a) heat has always benefited life when the earth has been warmer in the past; (b) only a small portion of greenhouse gases are anthropogenically induced; (c) some places of the world, Antarctica as an example, are getting colder; (d) 'the media only recently abandoned the "global cooling" scare'; and (e) 'global warming has not made hurricanes worse'.[35] As the coordinating lead author for the chapter on 'Climate scenario development' for the IPCC Third Assessment Report and contributing author for several other chapters, Hulme queried the rationality of reducing GHG emissions and its outcomes in practice.[36] To prove his opinions, Hulme put forward two ways of seeing climate change—treat it as an idea to be debated, adapted and used, or regard it as a physical phenomenon that can be observed, quantified and measured.[37] He preferred the first view in that it is more practical and a solution to climate change 'seemingly remains beyond our reach'.[38] The debate about global warming is certain to continue. However, the phenomenon of climate change has been generally accepted by most international organisations and scholars.[39] This book proceeds on the assumption that climate change

---

33 Dahe Qin, *Opening Remark at Working Group I—Twelfth Session* (23 September 2013) <http://www.ipcc.ch/meetings/session36/speeches/op_wg1_p12_Dahe_Qin.pdf> accessed 9 April 2014, p. 2.

34 IPCC, above n. 3, 14.

35 Christopher Horner, *The Politically Incorrect Guide to Global Warming and Environmentalism* (Regnery Publishing, 2007) preface.

36 Hulme, above n. 1, 171–172.

37 Mike Hulme, *Why We Disagree about Climate Change* (Cambridge University Press, 2009) 32.

38 Hulme, above n. 1, 172. Hulme asserts that reducing GHG emissions is 'beyond our reach' on the ground that global GHG emissions have accelerated rather than reduced after the adoption of the Kyoto Protocol. However, Hulme does not take the projected growth of international trade into account, which makes his argument less persuasive.

39 See, e.g., United Nations Environment Programme, 'Annual Report 2010' (2011) <www.unep.org/annualreport> accessed 10 May 2011; IPCC, above n. 3; ø. Buhaug et al., 'Second IMO GHG Study 2009' (International Maritime Organization (IMO), 2009); Philippe Sands, *Principles of International Environmental Law* (Cambridge University Press, 2nd ed, 2003) 358; Birnie et al., above n. 4, 335.

exists, that it is exacerbated by human activities, and that it warrants serious examination to identify effective mitigation strategies.

1.1.1.2 Contribution of Greenhouse Gas Emissions to Climate Change

Although many factors contribute to global climate change,[40] the preponderance of scientific evidence maintains that the principal cause is anthropogenically-induced GHG emissions. The IPCC Third Assessment Report stated that 'most of the observed warming over the last 50 years is *likely* to have been due to the increase in GHG concentrations'.[41] The IPCC Fourth Assessment Report reinforced this assessment, stating that 'most of the observed increase in global average temperatures since the mid-20th century is *very likely* due to the observed increase in anthropogenic GHG concentrations'.[42] The Working Group I report of the IPCC Fifth Assessment Report asserted that 'it is *extremely likely* that human influence has been the dominant cause of the observed warming since the mid-20th century'.[43] The scientific evidence provided by these IPCC reports has been strengthened and GHGs are regarded as the largest contribution to this phenomenon.[44]

The roles of different GHGs in global anthropogenic GHG emissions are illustrated in Figure 1.1. In the period 1970 to 2004, global GHG emissions from human activities increased by 70 per cent, and CO_2 increased by 80 per cent. It represented 77 per cent of total GHG emissions in 2004 indicating that it is the most important anthropogenic GHG.[45] However, CH_4 and N_2O accounted for only 14.3 per cent and 7.9 per cent of total GHG emissions respectively in 2004.

40 IPCC, above n. 3, 8–9. Theoretically both natural and manmade substances and processes that alter the Earth's energy budget, or in other words, change the energy balance of climate system, contribute to climate change.

41 Intergovernmental Panel on Climate Change (IPCC), 'Third Assessment Report' (2001) <http://www.grida.no/publications/other/ipcc_tar/> accessed 10 May 2011, Synthesis Report, p. 51.

42 IPCC, above n. 17, Synthesis Report, p. 39.

43 IPCC, above n. 3, 12.

44 Ibid. 3.

45 IPCC, above n. 17, Synthesis Report, p. 36.

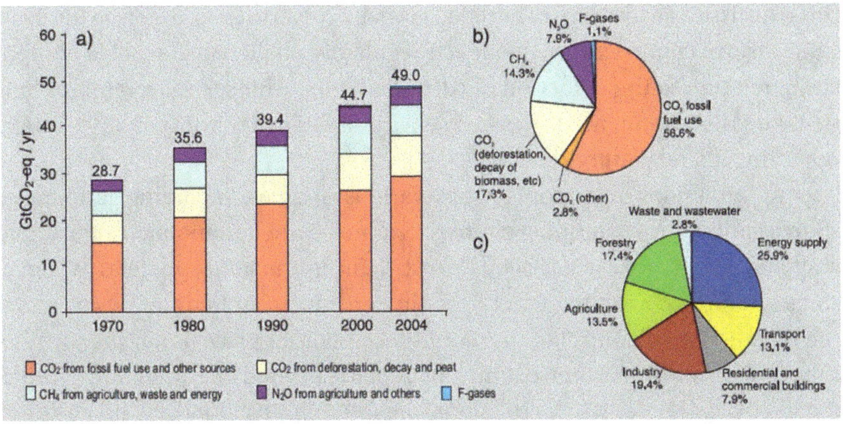

FIGURE 1.1 *Global anthropogenic GHG emissions.*[46]

1.1.2 *Transboundary Nature of Greenhouse Gas Emissions*

As far as climate change is concerned, oceans have been treated 'both as victims of the problem and as part of the solution.'[47] On the positive side, the oceans can transfer the heat between surface waters of the ocean and the lower atmosphere so as to adjust the global climate and weather; they also serve as a vital sink for absorbing authropogenic GHG emissions.[48] On the negative side, scientific data from all continents and most oceans has revealed that climate changes resulting from GHG emissions have endangered marine systems, leading to global marine-species redistribution and marine-biodiversity reduction in sensitive regions, ocean acidification, and other future risks.[49] The impacts from GHG emissions have provided further reasons to combat climate change. However, the transboundary nature of GHG emissions makes it a challenge to regulate. Given the transboundary nature of GHG emissions, the regulation of

46 Ibid. In Figure 1, a) Global annual emissions of anthropogenic GHGs from 1970 to 2004; b) Share of different anthropogenic GHGs in total emissions in 2004 in terms of CO_2-eq; c) Share of different sectors in total anthropogenic GHG emissions in 2004 in terms of CO_2-eq. (Forestry includes deforestation.).

47 David Freestone, 'Climate Change and the Oceans' (2009) 3(4) *Carbon & Climate Law Review* 383, 383.

48 Duncan E.J. Currie and Kateryna Wowk, 'Climate Change and CO_2 in the Oceans and Global Oceans Governance' (2009) 3(4) *Carbon & Climate Law Review* 387, 388.

49 Intergovernmental Panel on Climate Change (IPCC), 'Fifth Assessment Report: Working Group II Report' (2014) <http://ipcc-wg2.gov/AR5/images/uploads/IPCC_WG2AR5_SPM_Approved.pdf> accessed 19 April 2014, Summary for Policymakers, pp. 16–17.

this GHG issue needs to be conducted globally. Nevertheless, differing interests from various countries have made it a regulatory challenge to reach a consensus in relation to the reduction of GHG emissions. This section examines how GHG emissions influence the marine environment in a transboundary context, rendering it a regulatory challenge.

GHG emissions may come from various sources, including land-based sources and marine shipping sources. Examples are automobile exhaust and shipping discharges. As a gaseous constituent in the atmosphere, GHG emissions often travel with the wind from the territory of or in other places under the jurisdiction or control of one country, to another place under the jurisdiction or control of another country or a place beyond the limits of national jurisdiction. Hence, GHG emissions are often transboundary in nature.[50] The impacts of GHG emissions on the marine environment in a transboundary context include but are not limited to the following four aspects.[51]

Firstly, GHG emissions may gradually lead to the rise of ocean temperature so as to alter the dynamics of the marine environment. As a result of excessive emissions from GHGs, atmospheric CO_2 concentrations have increased from 280 to 380 ppm since the beginning of the industrial revolution, which is estimated to have led to a 0.74°C ± 0.18°C global temperature rise during the past 100 years.[52] It is 'virtually certain' that the upper ocean (0–700 metres depths) warmed from 1971 to 2010, and 'likely' that the ocean warmed from depths between 700 to 2000 metres over the period 1957–2009.[53] Under these circumstances, species distribution, polar systems, and global and regional weather patterns may be changed.[54] Some of the carbon stored in the form of methane hydrates from the seabed may ultimately be released once the deep ocean warms,[55] and the dynamics, structure and biodiversity of marine ecosystems is also likely to shift.[56]

50 See ch. 2, 2.3.

51 Ibid. According to the IPCC Fourth Assessment Report, the four aspects are generally the results of GHG emissions after quite a long period, and these approaches are obviously transboundary on the ground that they all involve the jurisdiction or control of several countries or places beyond the limits of national jurisdiction.

52 R. Hale et al., 'Predicted Levels of Future Ocean Acidification and Temperature Rise could Alter Community Structure and Biodiversity in Marine Benthic Communities' (2011) 120(5) *Oikos* 661, 661.

53 IPCC, above n. 3, 4–5.

54 Currie and Wowk, above n. 48, 389.

55 Ibid.

56 Hale et al., above n. 52.

Secondly, GHG emissions may result in sea-level rise and engender adverse impacts. Observations indicate that sea levels have risen by an average of 1.7 ± 0.3 mm per year since 1950, and that this rate increased to 3.3 ± 0.4 mm per year from 1993 to 2009, suggesting that the seal level rise is not only happening but that it is also accelerating.[57] The 2007 IPCC Fourth Assessment Report concluded that under global warming conditions caused by excessive GHG emissions and ignoring the contribution from melting sea ice, global sea level will rise by 18 to 59 centimetres (cm) in this century and reach 59 cm by 2099.[58] Even though, the IPCC's predictions on sea level rise are regarded as 'remarkably conservative' and 'wildly optimistic', largely owing to the IPCC's methodology of not taking into account the potential melting land ice.[59] The 2014 IPCC Fifth Assessment Report strengthened this trend and asserted that more than 95 per cent of the ocean area will experience sea level rise by the end of the 21st century.[60] It was further confirmed that global sea-level rise rates are 'very likely' accelerating.[61] Due to sea level rise, certain coastal hazards are more likely, such as flooding of coastal land, storm surges, erosion, destruction of infrastructure, settlements and facilities. Coastal residents may need to move so as to avoid larger losses and seek more secure shelter,[62] and some low-lying coastal States, for instance the Maldives and Tuvalu, are even facing the risk of disappearance.[63] In 2009 the Carteret Islanders of Papua New Guinea became the world's first entire community to be displaced by climate change.[64] Vanuatu communities have also been displaced from the Torres Islands as a

57 Robert J. Nicholls and Anny Cazenave, 'Sea-Level Rise and Its Impact on Coastal Zones' (2010) 328 (18 June 2010) *Science* 1517, 1517; See also A. Church John and J. White Neil, 'A 20th Century Acceleration in Global Sea-level Rise' (2006) 33(L01602) *Geophysical Research Letters* 1, 1.

58 IPCC, above n. 17, Synthesis Report, p. 45.

59 Clive Schofield, 'Shifting Limits? Sea Level Rise and Options to Secure Maritime Jurisdictional Claims' (2009) 3(4) *Carbon & Climate Law Review* 405, 406.

60 IPCC, above n. 3, 19.

61 Ibid. 18.

62 Peter Boehm, *Global Warning: Devastation of an Atoll* The Independent <http://www.commondreams.org/headlines06/0830-07.htm> accessed 18 August 2011. In this case, an entire coastal village in the north of Tegua Island in Vanuatu was relocated to higher ground in late 2005.

63 Currie and Wowk, above n. 48, 390.

64 Brian Merchant, *First Official Climate Change Refugees Evacuate Their Island Homes for Good* (8 May 2009) <http://www.treehugger.com/corporate-responsibility/first-official-climate-change-refugees-evacuate-their-island-homes-for-good.html> accessed 19 April 2014.

result of sea-level rise.[65] Additionally, sea level rise may influence maritime jurisdictional claims by coastal States due to the changes in their baselines.[66]

Thirdly, GHG emissions may cause ocean acidification and influence marine ecosystems negatively. Given the ocean's role as a carbon sink, growing CO_2 levels have led to enhanced absorption of CO_2 into the surface water of the ocean. As the CO_2 dissolves into the seawater and acts as a weak acid, the carbonates in the ocean are reduced.[67] This chemical process is known as ocean acidification. Acidification of the deep ocean may occur as one of the potential side effects of a process known as ocean fertilisation,[68] an activity designed to mitigate the deleterious effects of excess GHG emissions.[69] To date, the oceans have taken up one-third of all anthropogenically sourced CO_2.[70] The mean surface ocean pH has dropped 0.1 units from 8.2 to 8.1 since 1750, and this number is projected to drop an extra 0.3 to 0.4 units by the end of the century.[71]

The impacts of ocean acidification on marine ecosystems and biodiversity are profound. First, ocean acidification can affect the growth and viability of calcifying organisms, such as corals, bivalves, crustaceans and plankton, in that these marine organisms need carbonate to build their shells and skeletons while the decreased carbonate ion concentration makes this impossible.[72] Second, lower pH levels can alter the acid-base regulation, reproduction, respiration, metabolism and behaviour of some marine species.[73] Third, the photosynthesis necessary for some primary producers in the oceans can be shifted

---

65 IPCC, above n. 49, ch. 29, p. 6.

66 Schofield, above n. 59, 405.

67 Currie and Wowk, above n. 48, 391.

68 Rosemary Rayfuse, Mark G. Lawrence and Kristina M. Gjerde, 'Ocean Fertilisation and Climate Change: The Need to Regulate Emerging High Seas Uses' (2008) 23(2) *The International Journal of Marine and Coastal Law* 297, 298, 305–306. Generally the term 'ocean fertilisation' refers to the process of 'large-scale fertilising of the ocean with nutrients such as iron, nitrogen or phosphorus in an attempt to produce massive phytoplankton blooms which may assist in increasing absorption of CO_2 from the atmosphere'.

69 Robin Warner, 'Marine Snow Storms: Assessing the Environmental Risks of Ocean Fertilization' (2009) 3(4) *Carbon & Climate Law Review* 426, 427.

70 Ibid. 426.

71 James Orr et al., 'Anthropogenic Ocean Acidification over the Twenty-first Century and Its Impact on Calcifying Organisms' (2005) 437 *Nature* 681, 681. Measured on a logarithmic scale, pH is a dimensionless measure of the acidity of water or any solution. Given that pure water's pH is 7, acid solutions' pH is smaller than 7 while basic solutions' pH is larger than 7. IPCC, above n. 17, Appendix Glossary, p. 85.

72 Currie and Wowk, above n. 48, 391.

73 Cheryl Logan, 'A Review of Ocean Acidification and America's Response' (2010) 60(10) *BioScience* 819, 823.

due to incrementally dissolved CO_2.[74] Last but not least, GHG emissions may shift ranges and distribution of marine species. There is mounting evidence that species distribution and abundance could be affected by climate change,[75] and this is also the case for marine species. Due to increased ocean temperature and decreased pH resulting from excess GHG emissions, the ranges and distribution of marine species have been altered.[76]

Species' ranges are generally projected to shift towards higher latitudes and the spread of diseases, parasites, and non-native invasive species may be accelerated.[77] In marine areas where upwelling is important, any decrease in upwelling frequency or intensity could lead to reduced productivity.[78] With sea temperature rising, species distribution tends to favour those species that are better adapted to warmer, lower pH conditions, in which harmful algal blooms grow rampantly.[79] Major mortality of fish, crustaceans, and other organisms may occur in hypoxic zones.[80] Some ecologically rich zones, including the Great Barrier Reef and Queensland Wet Tropics, are projected to suffer a significant loss of biodiversity by 2020.[81] Besides the above direct impacts on marine species from exorbitant GHG emissions, some indirect effects are also obvious. For example, people may need to change their traditional fishing practices, and top predators such as sharks are more likely to be affected indirectly through prey and habitat changes.[82] Generally speaking, ocean acidification reduces the ocean's capacity to absorb humanmade CO_2, leads to

74 Ibid. 821.

75 See, e.g., Wim H. Van der Putten, Mirka Macel and Marcel E. Visser, 'Predicting Species Distribution and Abundance Responses to Climate Change: Why It is Essential to Include Biotic Interactions across Trophic Levels' (2010) 365 *Philosophical Transactions of the Royal Society B: Biological Sciences* 2025, 2025.

76 Currie and Wowk, above n. 48, 393.

77 Jennifer Hoffman, Ana Fonseca and Carlos Drews (eds), 'Cetaceans and Other Marine Biodiversity of the Eastern Tropical Pacific: Options for Adapting to Climate Change' (Report from a workshop held at San Jose, Costa Rica, 2009) <http://wwf.panda.org/?uNewsID=166824> accessed 15 May 2011, p. 13.

78 Ibid.

79 Ibid.

80 Ibid.

81 IPCC, above n. 17, Synthesis Report, p. 50.

82 Hoffman, Fonseca and Drews, above n. 77.

economic loss and engenders food security.[83] Limiting CO_2 emissions is 'the only' realistic mitigation option to address ocean acidification.[84]

1.2 Contribution of International Shipping to Greenhouse Gas Emissions

The contribution of GHG emissions to climate change has been underpinned by increasing scientific observations and analyses, and there has also been mounting evidence that international shipping contributes to GHG emissions and the consequential climate change impacts.[85] This section reviews the sources and impacts of GHG emissions from international shipping. It also provides an overview of international shipping and explains the contributions to its current GHG emissions.

1.2.1 *Introduction to International Shipping*

International shipping has been defined by the IMO as 'shipping between ports of different countries, as opposed to *domestic shipping*',[86] and excludes military and fishing vessels engaged on such voyages. For the purpose of this book, the definition by the IMO is adopted although military and fishing vessels can be on international voyages.[87] Consistent with the *2006 IPCC Guidelines for*

83 The International Geosphere-Biosphere Programme, The Intergovernmental Oceanographic Commission and The Scientific Committee on Oceanic Research, 'Ocean Acifidication Summary for Policymakers—Third Symposium on the Ocean in a High-CO_2 World' (International Geosphere-Biosphere Programme, Stockholm, Sweden, 2013) 1.

84 Ibid.

85 See, e.g., Veronika Eyring et al., 'Transport Impacts on Atmosphere and Climate: Shipping' (2010) 44(37) *Atmospheric Environment* 4735; Stathis Palassis, 'Climate Change and Shipping' in Robin Warner and Clive Schofield (eds), *Climate Change and the Oceans: Gauging the Legal and Policy Currents in the Asia Pacific and Beyond* (Edward Elgar Publishing Limited, 2012) 200.

86 Buhaug et al., above n. 39, 13. According to the *Second IMO GHG Study 2009*, 'domestic shipping' refers to 'shipping between ports of the same country, as opposed to *international shipping*', and excludes military and fishing vessels.

87 In 2012 the Food and Agriculture Organization (FAO) decided to address GHG emissions from fishing vessels engaged on international voyages due to the growing contribution of fishing vessels to marine environmental deterioration and global climate change. Theoretically the amended Annex VI to *MARPOL 73/78* applies to fishing vessels of 400 gross tonnage and above. However, in practice most fishing vessels are below 400 gross tonnage, which makes it necessary for the FAO or other competent international organisations to address GHG emissions from fishing vessels engaged on interna-

National Greenhouse Gas Inventories (*IPCC 2006 Guidelines*),[88] this definition also indicates that the same ship under an international voyage may frequently be engaged in both international and domestic shipping operations.[89] Indeed, this feature of international shipping constitutes the main barrier to including GHG emissions from international shipping in the State-based *Kyoto Protocol*.[90]

A number of differing definitions of 'ship' or 'vessel' exist in international law depending on the distinct purposes of treaties.[91] Based on the nature of the issue under discussion, this book adopts the definition of ship contained in

---

tional voyages. This issue will not be discussed in this book due to the different nature of the fishing industry. *Follow-Up to the Recommendations of the Twenty-Ninth Session of the Committee on Fisheries, Rome, Italy, 31 January–4 February 2011*, COFI 30th Session, FAO Doc COFI/2012/Inf.5 (9–13 July 2012) Agenda Item 8, xxx. Due to the complex sovereignty and international politics considerations involved, it is less likely that GHG emissions from military vessels engaged in international shipping could be regulated in the short term.

88 Intergovernmental Panel on Climate Change (IPCC), 'IPCC Guidelines for National Greenhouse Gas Inventories' (2006) <http://www.ipcc-nggip.iges.or.jp/public/2006gl/vol2.html> accessed 22 August 2011, Volume 2, Chapter 3, 3.5.1, p3.48. Under the IPCC 2006 Guidelines, international shipping is called 'international water-borne navigation' which may take place at sea, on inland lakes and waterways and in coastal waters. It includes journeys that depart in one country and arrive in a different country, but excludes such voyages by fishing vessels. International military navigation can be included as a separate sub-category of international shipping 'provided that the same definitional distinction is applied and data are available to support the definition'. Due to confidentiality issues, many inventory compilers may have difficulty obtaining data for the quantity of military fuel use. In this case, generally such voyages by military vessels are also excluded from the definition of international shipping.

89 Buhaug et al., above n. 39, 13.

90 See ch. 3, 3.2.2.1.

91 Mikhail Kashubsky, *Offshore Petroleum Security: Analysis of Offshore Security Threats, Target Attractiveness, and the International Legal Framework for the Protection and Security of Offshore Petroleum Installations* (PhD Thesis, University of Wollongong, 2011) 155–156. For example, the 2001 International Convention on Civil Liability for Bunker Oil Pollution Damage defines 'ship' as 'any seagoing vessel and seaborne craft, of any type whatsoever'. *IMO International Convention on Civil Liability for Bunker Oil Pollution Damage*, opened for signature 27 March 2001, 40 ILM 1493 (entered into force 21 November 2008) art. 1(1). The *1989 International Convention on Salvage* defines 'vessel' as 'any ship or craft, or any structure capable of navigation'. *International Convention on Salvage*, opened for signature 28 April 1989, 1953 UNTS 165 (entered into force 14 July 1996) art. 1(b). The *1924 International Convention for the Unification of Certain Rules relating to Bills of Lading for the Carriage of Goods by Sea* defines 'ship' as 'any vessel used for the carriage of goods by sea'. International Convention for the Unification of Certain Rules relating to Bills of

MARPOL 73/78. It defines 'ship' as 'a vessel of any type whatsoever operating in the marine environment and includes hydrofoil boats, air-cushion vehicles, submersibles, floating craft and fixed or floating platforms'.[92] This definition underscores different ships' operation in the marine environment, and is consistent with the purpose of international shipping.[93] Technically not all ships are used for international shipping due to their different operational capacity. Therefore it is easier to rely on this definition to identify which type of ships may be utilised to engage in international shipping.

Ship categories may vary depending on the different purposes of particular voyages, and they are often regulated or listed diversely under different treaties or legal documents. For instance, only seven types of ships were regulated by the revised Annex VI to *MARPOL 73/78* in 2011 under the new Energy Efficiency Design Index (EEDI) requirements.[94] However, the amendments of *MARPOL* Annex VI adopted in April 2014 extended these seven types of ships to 12 due to technological and regulatory improvements resulting in more types of vessels.[95] In the context of global climate change, ships are characterised by the types of cargo they are designed to carry. Ship categories may be classified as in Table 1.1. Table 1.1 lists the definitions of primary ship categories that have been used in the emissions inventory by the IMO.

Lading for the Carriage of Goods by Sea, opened for signature 25 August 1924, 120 LNTS 155 (entered into force 2 June 1931) art. 1(d).

92 *MARPOL 73/78* art. 2(4).

93 Kashubsky, above n. 91, 155; Michael Summerskill, *Oil Rigs: Law and Insurance* (Stevens & Sons, 1979) 13. Summerskill asserts that the term 'vessel' has 'a broader meaning than ship' and that 'the term "vessel" designates a variety of maritime craft, while the term "ship" is limited to a few species of the same genus'. However, some treaties, *LOSC* as an example, do not distinguish the two terms. The terms 'ship' and 'vessel' are used interchangeably in this book.

94 *MARPOL 73/78* Annex VI (2011 amendments) regs 20–21. In accordance with the new EEDI requirement, seven types of ships are listed, namely bulk carrier, gas tanker, tanker, container ship, general cargo ship, refrigerated cargo carrier and combination carrier.

95 *MARPOL 73/78* Annex VI (2014 amendments) regs 20–21. These added five types of ships are Liquefied Natural Gas (LNG) carriers, roll-on/roll-off (ro-ro) cargo ships (vehicle carriers), ro-ro cargo ships, ro-ro passenger ships and cruise passenger ships having non-conventional propulsion.

TABLE 1.1 *Definitions of the ship categories that have been used in the emissions inventory*[96]

Ship Type	Sub-type of Ships and their Definitions
Cargo Ships	**Crude Carriers:** include tankers which are intended for carrying crude oil. **Bulk Carriers:** ships designed to carry bulk goods such as grain, iron ore, coal and more. **General Cargo Carriers:** include a wide variety of cargo ships from small one-hold vessels to highly advanced multi-purpose vessels. Some of the ships are designed to carry containers as well as break-bulk cargos. Many of these ships are equipped with their own lifting gear. **Other Dry Carriers:** carriers of refrigerated cargo and other special dry cargo ships. **Products Tankers:** carry various types of refined petroleum products. **Chemical Tankers:** carry various types of industrial chemicals. **LPG Tankers:** specialised tankers for the carriage of liquefied petroleum gas and often also other products, for example ammonia. **LNG Tankers:** specialised tankers for the carriage of liquefied natural gas. **Other Tankers:** include a large number of bunker tankers and also those that carry a wide range of liquid niche products such as orange juice, bitumen, wine and water. **Container Ships:** built to carry containerised cargo and nothing else, i.e. fully cellular ships designed to carry containers both on deck and under deck. **Vehicle Ships:** designed to carry (new) cars, trucks and sometimes other special cargo on wheels. **Ro-Ro Ships:** ships that are loaded and discharged by driving the cargo on board on wheels.[97]

96 Buhaug et al., above n. 39, 15. Fishing vessels are removed from this table for the purpose of the book.

97 *MARPOL 73/78* Annex VI, Regulation 2, 3.33–3.36. Broadly speaking, Ro-Ro Ships consist of Ro-Ro Passenger Ships and Ro-Ro Cargo Ships, whereas the latter also include vehicle carriers, volume carriers and weight carriers.

TABLE 1.1 *Definitions of the ship categories* (cont.)

Ship Type	Sub-type of Ships and their Definitions
Other	**Ferries:** carry cars and passengers on regular schedules. This also includes overnight ferries. **Cruise Ships:** carry passengers on pleasure voyages. **Yachts:** large pleasure vessels. **Offshore:** encompasses a wide range of platform supply vessels and offshore support vessels. Drilling rigs are not included in this figure. **Service:** mainly tugs but also work-boats, dredgers, research vessels and more.

As a comparatively cost effective, clean and safe method of transportation,[98] international shipping offers an important means of moving goods internationally and enables other activities such as leisure cruising. International shipping is the backbone of global trade and a driving force of economic globalisation.[99] Many factors contribute to the development of international shipping. As an

98 International Maritime Organization (IMO), *Introduction to IMO* <http://www.imo.org/About/Pages/Default.aspx> accessed 17 May 2011; see also Rajiv Saxena, 'Overseas Shipping Made Cheaper' (2010) 42(7) *Industrial Engineer: IE* 24. But see Oceans Beyond Piracy, 'The Economics Cost of Somali Piracy 2012' (2013) <http://oceansbeyondpiracy.org/sites/default/files/attachments/View%20Full%20Report_1.pdf> accessed 22 November 2013, pp. 1–4. This 2012 report by Oceans Beyond Piracy reveals that maritime piracy costs the global economy between $5.7 and $6.1 billion in 2012, and has made international shipping costly. The increased cost includes increased 'per incident' costs, increased cost of armed guards, increased cost of increased speeds, consistent ratio of recurring costs vs. investments, ransoms and recovery, military operations, security equipment and guards, re-routing, increased speed, labour, prosecutions and imprisonment, insurance, and counter-piracy organisations.

99 International shipping carries around 80 per cent of global trade by volume. United Nations Conference on Trade and Development (UNCTAD), 'Review of Maritime Transport 2013' (2013) <http://unctad.org/en/PublicationsLibrary/rmt2013_en.pdf> accessed 14 April 2014, xi. See also Martin Stopford, *Martin Economics* (Routledge, 3rd ed, 2009) 3; G.P. Pamborides, *International Shipping Law: Legislation and Enforcement* (Kluwer Law International, 1999) 145. Pamborides asserts that 'shipping is too valuable to the world's economy to jeopardise'.

example, the evolution of ship propulsion has progressed from sailing ships to steam ships powered by coal and then to an almost universal use of diesel engines, significantly accelerating international trade.[100] Similarly, advances in telecommunication and information and communications technology (ICT) infrastructure,[101] reductions in trade barriers, and low energy costs have also contributed to the expansion of international shipping and seaborne trade.[102]

The world's merchant fleet is the main component of international shipping, and its development has been facilitated by world seaborne trade. The relationship between the world's merchant fleet and seaborne trade is displayed in Figure 1.2. With the rapid development of seaborne trade, the gross tonnage of the global merchant fleet has expanded significantly since 1985. In 2012 international seaborne trade by volume increased 4.3 per cent, nearly the same rate

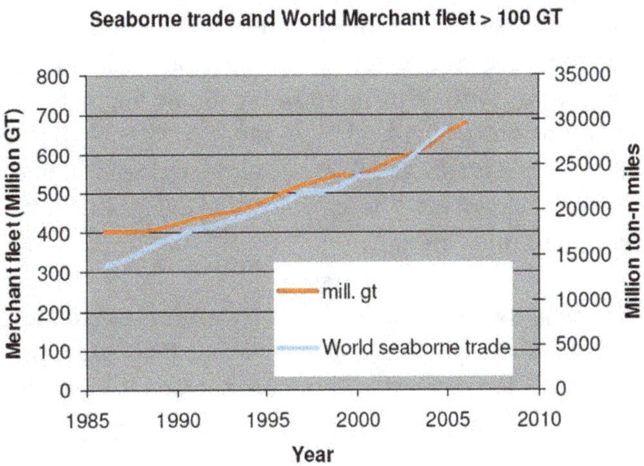

FIGURE 1.2 *World seaborne trade and merchant fleet (Fearnleys Review).*[103]

100 Sujith Kollamthodi et al., 'Greenhouse Gas Emissions from Shipping: Trends, Projections and Abatement Potential: Final Report' (The Shadow Committee on Climate Change, 2008) 3.

101 ICT infrastructure is an overall name used to describe all the computer and communications hardware and software used to manage clerical, administrative, and management tasks in organisations.

102 Kollamthodi et al., above n. 100.

103 Kollamthodi et al., above n. 100, 4.

as 2011; whereas in the same year the world economy decelerated with gross domestic product (GDP) increasing by 2.2 per cent, down from 2.8 per cent in 2011.[104] It appears that international shipping performed better than the global economy.[105] Therefore it is important to take the projected growing seaborne trade into account in the development of regulatory measures in tackling GHG emissions from international shipping.

1.2.2 *Greenhouse Gas Emissions from International Shipping*

The increase in fuel consumption associated with growing seaborne trade has led to a rise in atmospheric emissions from international shipping, which has attracted growing global attention.[106] This section examines the categories, statistics and impacts of GHG emissions from international shipping.

1.2.2.1 *Categories of Emissions from Ships*

In view of the nature of emissions from ships and using the definition of international shipping, it is not necessary to divide emissions categories into international and domestic emissions for the purposes of discussing them in this section.[107] GHG emissions from ships can be categorised differently based on differing criteria. This section briefly discusses two of these classifications.

Based on the sources of emissions, GHG emissions from ships can be classified into the following four categories:[108]

- *Emissions of exhaust gases*. As the main emissions from ships, exhaust gas emissions come from sources such as main engines, auxiliary engines, boilers and incinerators.[109] However, exhaust from incinerators is regarded as being a very small contribution to emissions and is often ignored.[110]

---

104 UNCTAD, above n. 99, 2,6.

105 Ibid. 6.

106 For example, the regulation of GHG emissions from international shipping has been widely discussed under the *UNFCCC* process and within the IMO. Meanwhile, the global shipping industry and various flag and port States have also responded actively to this GHG issue. The discussion of these issues is provided in the following chapters.

107 According to the above definition of international shipping, the same ship may frequently be engaged in both international and domestic shipping, which means that total emissions from international shipping may also include emissions from parts of domestic shipping. It is also not possible to split emissions into domestic emissions and international emissions technically.

108 Buhaug et al., above n. 39, 23.

109 Ibid.

110 Ibid.

- *Emissions of refrigerants.* As a necessity for the refrigeration and/or freezing of cargo and provisions and in air-conditioners, refrigerants generally emit through two main channels: leaks during operation and during maintenance of refrigeration and air-conditioning equipment, or during the dismantling process when the emissions are usually allocated to the country in which the ship was scrapped.[111]
- *Cargo emissions.* The emissions comprise various emissions and leakages, in particular leaks of refrigerant from refrigerated containers and trucks, and volatile compounds emissions (CH_4 and Non-methane Volatile Organic Compounds (NMVOCs)) from liquid cargoes.[112]
- *Other emissions.* This category includes emissions from testing and maintenance of fire-fighting equipment, and other equipment.[113]

According to their relations with combustion, emissions from ships can be divided into direct emissions and indirect emissions. Direct emissions are the results of combustion, whereas indirect emissions refer to those from non-combustible sources and fugitive emissions.[114]

Regarding direct emissions, the dominance of CO_2 emissions is obvious which can be seen clearly from Table 1.2. CO_2 is produced from vessels as a by-product of the oxidation of carbon in diesel fuel.[115] The second direct emission is from N_2O which arises from combustion but there has been little research into this type of emission in the context of the shipping sector because it is regarded as a relatively minor source of emissions.[116] The third direct type of emission is from CH_4 which is also a small source of emissions (see Table 1.2). Additionally, some GHG relevant substances also fit within this category, such as Volatile Organic Compounds (VOCs), which are emitted from fuel tankers during loading and unloading operations and the transport of crude oil.[117]

---

111 Ibid.

112 Ibid.

113 Ibid.

114 Kollamthodi et al., above n. 100, 5; Intergovernmental Panel on Climate Change (IPCC), above n. 88, Volume 2, Chapter 4, 4.1, p. 4.6. The IPCC defines 'fugitive emissions' as the 'intentional or unintentional release of GHGs during the extraction, processing and delivery of fossil fuels to the point of final use'.

115 Ibid.

116 Ibid.

117 Ibid.

Indirect emissions mainly consist of refrigerant gases HCFC-22, CFCs and halons from shipping.[118] Emissions of these gases are relatively small and hence have not been the focus of assessment. Further, these gases are not GHGs regulated under the *Kyoto Protocol* but are regulated under the *1985 Vienna Convention for the Protection of the Ozone Layer*, the *Montreal Protocol on Substances that Deplete the Ozone Layer*, and its successive adjustments.[119]

It is clear that exhaust gases are the primary source of ship emissions with CO_2 being the most important GHG emitted by ships. Other GHG emissions from ships are less important in terms of quantity and of global warming potential.[120] In view of the dominance of CO_2 in GHG emissions from international shipping, in practice the question of how to reduce GHG emissions from international shipping is often narrowly interpreted as how to reduce CO_2 emissions from international shipping.[121]

1.2.2.2 General Emissions Statistics from International Shipping

The emissions from shipping, especially international shipping, have been increasing in recent years at a high rate. In 2007, CO_2 emissions from international shipping reached 870 million tonnes (Table 1.2), which accounted for 2.7 per cent of the global emissions of CO_2 (Figure 1.3). However, in the same year, the contribution of international aviation to global total CO_2 emissions was only 1.9 per cent (Figure 1.3). In 2012, CO_2 emissions from

---

118 Ibid.

119 The 1987 Montreal Protocol was adjusted and/or amended in London 1990, Copenhagen 1992, Vienna 1995, Montreal 1997 and Beijing 1990 respectively, regulating HCFC-22, CFCs, halons and other relevant substances. Whereas the 1985 Vienna Convention for the Protection of the Ozone Layer is basically a framework and requires further actions by the parties.

120 These conclusions have been confirmed by the 'Second IMO GHG Study 2009'. See Buhaug et al., above n. 39, 1.

121 See, e.g., when the IMO started its work on the reduction of GHG emissions from ships in 1997, it adopted Resolution 8 on 'CO_2 emissions from ships' which requests the IMO to undertake a study on GHG emissions from ships and to consider feasible emissions-reduction strategies. International Maritime Organisation (IMO), 'Main Events in IMO's Work on Limitation and Reduction of Greenhouse Gas Emissions from International Shipping' (2011) <http://www.imo.org/MediaCentre/resources/Pages/Greenhouse%20gas%20emissions.aspx> accessed 22 November 2013, p. 3.

international shipping were 2.2 per cent of global CO_2 emissions.[122] For the period 2007–2012, the average CO_2 emissions from international shipping were 2.6 per cent of global CO_2 emissions.[123] There was a decline of CO_2 emissions from international shipping from 2.7 per cent of global CO_2 emissions in 2007 to 2.2 per cent of global CO_2 emissions in 2012 as a result of global economic crisis. However, maritime emissions are expected to increase significantly due to projected growth in demand for maritime transport services.[124] If no aggressive regulatory policies are introduced, CO_2 emissions from international shipping may grow by 150–250 per cent by 2050 compared with 2007.[125] Thesee statistics reveal that the increasing trend of GHG emissions from international shipping will be maintained in the long term, and should be recognised as a growing problem for scientists, industry and environmental policy makers.

TABLE 1.2 *Summary of GHG emissions from shipping* during 2007*

	International shipping (million tonnes)	Total shipping	
		million tonnes	CO_2 equivalent
CO_2	870	1050	1050
CH_4	Not determined*	0.24	6
N_2O	0.02	0.03	9
HFC	Not determined*	0.0004	≤6

SOURCE: SECOND IMO GHG STUDY 2009
* A SPLIT INTO DOMESTIC AND INTERNATIONAL EMISSIONS IS NOT POSSIBLE.

122 Smith, T.W.P. et al., 'Third IMO GHG Study' (International Maritime Organization (IMO), 2014) executive summary, Table 1.

123 Ibid.

124 Ibid. para. 5.1.

125 Buhaug et al., above n. 39, 1.

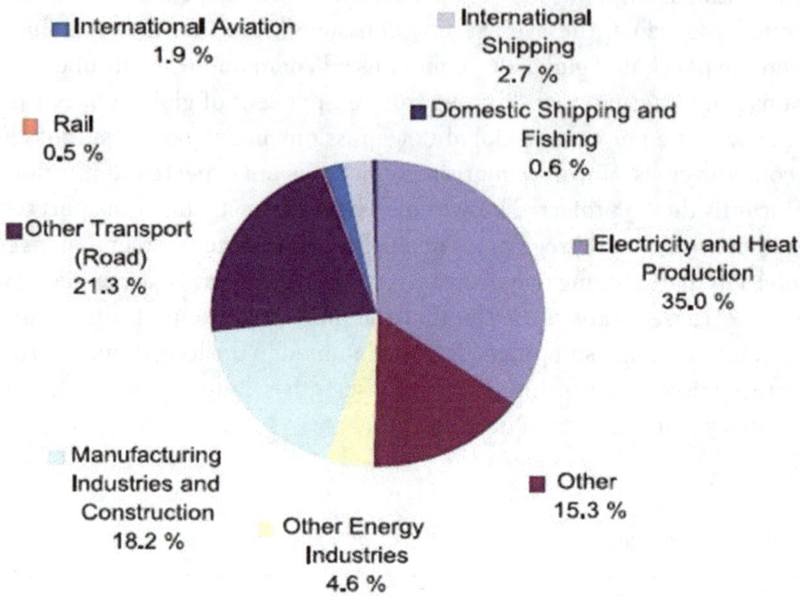

FIGURE 1.3 *Emissions of CO_2 from shipping compared with global total emissions.*
SOURCE: SECOND IMO GHG STUDY 2009[126]

1.2.2.3 Impacts of Greenhouse Gas Emissions from International Shipping

In addition to the impacts of GHG emissions on the marine environment in a transboundary context, discussed in Section 1.1.2, GHG emissions from international shipping also have some unique effects on the environment. These adverse effects involve atmospheric composition, human health and climate,[127] and are by their nature transboundary and chronically accumulative.

Firstly, most ships' emissions, excluding those in ports and in the vicinity of coastlines, are emitted in or transported to the marine boundary layer

126 The data provided in Table 1.2 and Figure 1.3 have been criticised because they do not take account of the global economic downturn since 2009. To provide a better foundation for IMO's future work, an update of the 2009 IMO Study of GHG Emissions Estimates for International Shipping was carried out and the final report was approved by the IMO in October 2014. *Report of the Expert Workshop on the Update of GHG Emissions Estimate for International Shipping* (*Update-EW*), note by the Secretariat, MEPC 65th Session, Agenda Item 5, IMO Doc MEPC 65/5/2 (4 March 2013) paras 8, 62; Smith et al., above n. 122.

127 See ibid. 124; Eyring et al., above n. 85, 4744–4759.

(MBL)[128] where they affect atmospheric composition negatively.[129] Generally a ship emits locally at relatively high concentrations.[130] Therefore, once GHG emissions from shipping are injected into the atmosphere, they mix with the ambient air and become diluted.[131] Meanwhile, these emissions are chemically transformed before new secondary species (such as ozone) are produced, and some of them are subsequently removed from the atmosphere by wet and dry deposition.[132] As a result, the atmosphere's composition will be affected.

Secondly, GHG emissions from international shipping may also indirectly affect human health by means of the formation of ground-level ozone and particulate matter.[133] Nearly 70 per cent of the emissions from oceangoing shipping take place within 400 km of the coastline along the main seaborne trade routes.[134] It can thus be deduced that most of the emissions from international shipping also occur within this distance from coastlines. If this is the case, these emissions may be transported hundreds of kilometres inland, bringing about air quality problems and impacting on human health.[135]

---

128 The marine boundary layer (MBL), also known as marine atmospheric boundary layer (MABL), is where the ocean and atmosphere exchange large amounts of heat, moisture, and momentum, primarily via turbulent transport. It is that part of the atmosphere that has direct contact and, hence, is directly influenced by the ocean. See Alvaro Semedo et al., 'Wave-Induced Wind in the Marine Boundary Layer' (2009) 66(8) *Journal of the Atmospheric Sciences* 2256, 2256.

129 Eyring et al., above n. 85, 4744–4745, 4752–4753.

130 Ibid. 4744.

131 Ibid.

132 Ibid. 4744–4745.

133 Ibid. 4752–4754.

134 James J. Corbett, Paul S. Fischbeck and Spyros N. Pandis, 'Global Nitrogen and Sulfur Inventories for Oceangoing Ships' (1999) 104(3) *Journal of Geophysical Research* 3457, 3465. Generally 'oceangoing shipping' refers to large cargo-carrying ships engaged in ocean-crossing trade. See Buhaug et al., above n. 39, 13.

135 Eyring et al., above n. 85, 4752–4754. See also Axel Michaelowa and Karsten Krause, 'International Maritime Transport and Climate Policy' (2000) 35(3) *Intereconomics* 127, 130–131. Michaelowa and Krause assert that climate change also has potential positive and negative impacts on international shipping. For instance, reduction of sea ice in certain areas due to global warming would save the shipping cost, enhanced public awareness of the greenhouse effect and the implementation of the Kyoto commitments in Annex I States would make more people choose seaborne transport; but meanwhile sea level rise would increase the cost for protecting current port infrastructure, and increased run-off and precipitation would cause higher sediment load in rivers.

Thirdly, ship emissions have an impact on climate—changing clouds and radiative forcing (RF).[136] As a residual product remaining at the end of the crude oil refining chain, heavy fuel oil (HFO) has been widely used by various ships due to its competitive price.[137] However, the by-products of combustion from HFO, including CO_2, black carbon (BC), nitrogen oxides (NO_x), sulphur dioxide (SO_2) and carbon monoxide (CO), produce significant impacts on climate by means of various physical and chemical interactions.[138]

As discussed earlier, GHG emissions from international shipping are also transboundary and borderless. This constitutes a feature of international shipping and makes it difficult to allocate these emissions to specific countries. This is illustrated by the following hypothetical scenario: A ship was built in South Korea, owned by Greeks but registered in Panama. It was chartered by Japanese and received an order to transport iron ore from Australia to China. The officers and crew were Filipinos. The iron ore was eventually processed to produce steel products for export to Germany. In this case, there will be debate about which country should be responsible for GHG emissions from these voyages.[139]

1.3 Statement of the Problem

The global community is experiencing rapid physical and technological changes, shaping every aspect of human life. Climate change is altering ecosystems and affecting communities around the globe. It has become one of the most prominent issues of global concern. There is mounting evidence that GHG emissions from international shipping contribute significantly to climate change.[140] Since limiting climate change requires 'substantial and sustained

136 Buhaug et al., above n. 39, 112. As a common metric to quantify climate impacts from different sources in units of W/m^2, RF refers to 'the change in the Earth—atmosphere energy balance since the pre-industrial period'.

137 P. Crist, 'Greenhouse Gas Emissions Reduction Potential from International Shipping' (OECD/ITF, 2009). In this report, Crist asserts that vessels engaged in coastal trips use either HFO or lighter marine distillate oil, and HFO accounts for about 77 per cent of total maritime transport fuel usage.

138 See Buhaug et al., above n. 39, 112; Eyring et al., above n. 85, 4766.

139 The nature of GHG emissions from international shipping was one of the main reasons why the SBSTA of the *UNFCCC* did not reach consensus on the allocation of GHG emissions from bunker fuels to specific countries in 1996. See ch. 3, 3.2.2.1.

140 See, e.g., IPCC, above n. 3; Buhaug et al., above n. 39; Eyring et al., above n. 85.

reductions' of GHG emissions,[141] the global community has turned its attention to mitigating these emissions. The 1992 *UNFCCC* and its *Kyoto Protocol* and *Paris Agreement* represent some of these achievements.

Given the growing contribution of GHG emissions from international shipping to global climate change, the international community has realised the importance of reducing shipping emissions and has made some regulatory efforts. To date international regulatory initiatives on addressing GHG emissions from international shipping have been conducted within two parallel regimes: the international climate change regime and the IMO GHG reductions regime. Under the international climate regulatory process, the *UNFCCC*, its *Kyoto Protocol* and *Paris Agreement*, as well as their various Conferences of the Parties (COPs), the COPs serving as the Meeting of the Parties to the Kyoto Protocol (CMPs) and the COPs serving as the Meeting of the Parties to the Paris Agreement (CMAs), have been discussing this GHG issue since 1995. Within the IMO, Party members have been discussing and negotiating the approaches to regulating the GHG emissions issue since 1997 when the IMO adopted Resolution 8 on 'CO_2 emissions from ships', which requested the IMO to undertake a study on GHG emissions from ships and consider feasible CO_2 reduction strategies.

No substantial outcomes on the reduction of GHG emissions from international shipping have been achieved under the *UNFCCC* process. Within the IMO the regulatory framework for the reduction of GHG emissions from international shipping is in the preliminary stages of its development although other vessel source pollution has been comprehensively regulated by the IMO. On 15 July 2011, the IMO adopted amendments to Annex VI to *MARPOL 73/78* from technical and operational perspectives. This revision makes mandatory the Energy Efficiency Design Index (EEDI) for new ships and the Ship Energy Efficiency Management Plan (SEEMP) for all ships, representing 'the first ever mandatory global greenhouse gas reduction regime' for the international shipping industry.[142] However, it is still arguable whether this regulation can lead to an absolute GHG emissions reduction,[143] and whether market-based measures (MBMs) should be adopted for furthering the reduction of GHG

---

141 IPCC, above n. 3, 14.

142 International Maritime Organization (IMO), *Mandatory Energy Efficiency Measures for International Shipping Adopted at IMO Environment Meeting* (15 July 2011) <http://www.imo.org/MediaCentre/PressBriefings/Pages/42-mepc-ghg.aspx> accessed 25 August 2011.

143 See, e.g., an IMO-commissioned assessment study in 2011 indicates that based on the 2010 CO_2 emissions level, it is almost impossible to achieve absolute emission reduction from 2010 to 2050 using the EEDI and SEEMP alone. Zabi Bazari and Tore Longva, '*Assessment*

emissions from international shipping. These controversies still remain even after further amendments of Annex VI to *MARPOL 73/78* were adopted in April 2014.

The goal of the international community in tackling climate change is to limit to 2 or 1.5 degrees Celsius the increase in the global average temperature by 2100.[144] However, a report in 2013 by the Asian Development Bank reveals that an increase of two degrees Celsius by 2050 is 'almost unavoidable'.[145] With the exponential growth in international seaborne trade, GHG emissions from international shipping will continue to have adverse impacts on the environment, human health and climate change. As such, a study suggests that compared with 1990, an approximately 50:50 chance of avoiding two degrees demands a 70–80 per cent reduction in emissions from energy by 2050.[146] Accordingly the scale of reduction from shipping GHG emissions to meet the two degrees target is unprecedented, and it is significant to expedite the global regulation of this issue. The 2011 amendments to Annex VI to *MARPOL 73/78* are the first mandatory reduction regime on this GHG issue, but they were not adopted by consensus and only entered into force on 1 January 2013. It is thus not easy to achieve the 70–80 per cent reduction target as suggested in this study. Accordingly, how to construct a sound regulatory framework for the reduction of GHG emissions from international shipping becomes the key to addressing this problem, which involves balancing the interests of different

of IMO Mandated Energy Efficiency Measures for International Shipping' (IMO DOC MEPC 63/INF.2, Annex, 31 October 2011) executive summary, p. 8.

144 The 2 degrees Celsius goal was first put forward by the G-8 in 2009, and later agreed in the Copenhagen Accord. In 2010 this goal was formally incorporated into the *UNFCCC* process, and Article 2 of the *Paris Agreement* reaffirms this goal and pursues efforts to limit the temperature increase to 1.5 degrees Celsius above pre-industrial levels. Nevertheless, the specific reduction targets and time frame for achieving this goal have not yet been agreed under the *UNFCCC* process. See Lavanya Rajamani, 'The Cancun Climate Agreements: Reading the Text, Subtext and Tea Leaves' (2011) 60(2) *The International and Comparative Law Quarterly* 499, 501; *Paris Agreement* art. 2(1)(a).

145 Michael Westphal, Gordon Hughes and Jorn Brommelhorster (eds), *Economics of Climate Change in East Asia* (Asian Development Bank, 2013) executive summary, xvi.

146 Kevin Anderson and Alice Bows, 'Executing a Scharnow Turn: Reconciling Shipping Emissions with International Commitments on Climate Change' (2012) 3(6) *Carbon Management* 615, 622. See also A. Bows-Larkin et al., 'High Seas, High Stakes: High Seas Project Final Report' (Tyndall Centre for Climate Change Research, University of Manchester, 2014) <http://www.mace.manchester.ac.uk/media/eps/schoolofmechanicalaerospaceandcivilengineering/research/centres/tyndall/pdf/High_Seas_High_Stakes_High_Seas_Project_Final_Report.pdf> accessed 16 April 2016.

stakeholders, including the UN, the IMO, the shipping industry, and various flag and port States.

1.4 Scope of the Book and Chapter Outline

This book aims to provide an analysis of international law principles relevant to the reduction of GHG emissions from international shipping, and the legal and policy responses to this issue from the key stakeholders: the UN, the IMO, the shipping industry and various flag and port States. This analysis underpins the subsequent suggestions regarding the ways to improve the current legal, policy and institutional framework for the reduction of GHG emissions from international shipping so as to mitigate these emissions more effectively and efficiently. To achieve this goal, the structure of the analysis proceeds as follows:

(1) Examining the sources of GHG emissions from international shipping and how they affect the climate, the marine environment and human health;
(2) Examining relevant international law principles and how they apply to the reduction of GHG emissions from international shipping;
(3) Analysing the legal, policy and institutional responses by the UN, IMO, the shipping industry, and flag and port States;
(4) Identifying gaps existing in the current regulatory framework; and
(5) Proposing gap-filling options for improving the current regulatory framework for the reduction of GHG emissions from international shipping.

The underlining argument of this book is that there are deficiencies in the current legal, policy and institutional frameworks governing GHG emissions from international shipping. The author asserts that legal and institutional reforms should be conducted to improve the regulatory framework for the reduction of GHG emissions from international shipping.

The discussion is presented in eight chapters. Chapter 1, the introductory chapter, introduces the relationship between climate change and GHG emissions, in particular GHG emissions from international shipping. This provides a background for addressing GHG emissions from international shipping. The introductory chapter also summarises the statement of the problem, objective of the research, and the book structure. Chapter 2 examines international law principles regarding international environmental responsibility and explores the possibility of applying these principles to the reduction of GHG emissions from international shipping. These principles are liability for transboundary harm, the precautionary principle, common but differentiated responsibility, no more favourable treatment, and the polluter-pays principle. The application of these principles underpins the responses from the UN, IMO, the shipping industry, and flag and port States, as well as the final gap-filling options.

The UN, IMO, the shipping industry, flag States and port States are the main stakeholders of this GHG emissions issue. Chapters 3 to 6 analyse and assess the legal, policy and institutional responses from these stakeholders in order to understand the current regulatory framework for this issue as well as to identify the gaps existing in this framework. It is hypothesised that the differing regulatory interests of these stakeholders have prevented them from making significant efforts to address this issue and achieving more progress.

Chapter 7 identifies the gaps existing in the current regulatory framework for this GHG emissions issue, and explores gap-filling options based on international law principles and responses from main stakeholders as analysed in previous chapters. The final chapter synthesises the results of the research and concludes with a summary of findings and recommendations from previous chapters. Table 1.3 provides the line of argument that will be followed by this book.

TABLE 1.3 *Book chapters and main line of argument*

Chapter	Main line of argument
One **Introduction**	Introduction to book topic, an overview of research objective, questions and methods, and main line of argument.
Two **International environmental law responsibility and its application to the issue of GHG emissions from international shipping**	This chapter examines the possibility of applying international law principles to the reduction of GHG emissions from international shipping. These principles are: liability for transboundary harm, the precautionary principle, common but differentiated responsibility, no more favourable treatment, and the polluter-pays principle. *It is argued that these principles are applicable to this GHG emissions issue.*
Three **UN response to the issue of GHG emissions from international shipping**	The UN's legal and institutional response to the GHG emissions issue, in particular its initiatives under the global climate change regime, is analysed. *It is suggested that the UN needs to make more efforts in tackling this GHG emissions issue.*

Chapter	Main line of argument
Four **IMO response to the issue of GHG emissions from international shipping**	The IMO's GHG mandate and the IMO GHG emissions regime, in particular the outcomes achieved within the MEPC, are discussed. *It is argued that the IMO's regulatory initiatives represent an advance in addressing this GHG issue but also create challenges for their implementation.*
Five **Response from the shipping industry to the issue of GHG emissions from international shipping**	This chapter examines the response from international and regional shipping organisations to this GHG emissions issue, and then discusses the responses from the shipping industries in *UNFCCC* Annex I States and non-Annex I States respectively. Case studies suggest that *national shipping industries took divergent views on this GHG emissions issue due to their differing regulatory interests.*
Six **Response from flag States and port States to the issue of GHG emissions from international shipping**	Based on the analyses of flag State control and port State control, this chapter examines the responses from main flag States and port States, as well as global and regional port State organisations, to this GHG emissions issue. Case studies suggest that *in comparison with UNFCCC Annex I flag States, non-Annex I flag States have more diverse responses towards this GHG issue due to their differing regulatory interests.*
Seven **The future development of legal and institutional frameworks to reduce GHG emissions from international shipping**	This chapter identifies the gaps existing in the current regulatory framework for this GHG emissions issue, and explores gap-filling options based on international law principles and responses from main stakeholders as analysed in previous chapters.
Eight **Conclusion**	The responses from main stakeholders to this GHG emissions issue are synthesised, which combined with the application of relevant international law principles, underpin the recommendations for furthering the reduction of GHG emissions from international shipping. Suggestions for steps forward are also provided.

CHAPTER 2

International Environmental Law Responsibility and its Application to the Issue of Greenhouse Gas Emissions from International Shipping

2.1 Introduction

Customary international law, general principles of law, and normative instruments have shaped and advanced the development of international environmental law, and with its evolution new norms and principles have emerged to meet new challenges.[1] One of these challenges is how to regulate greenhouse gas (GHG) emissions from international shipping. As the regulatory framework for GHG emissions from international shipping is still in the preliminary stages of its development, the application of the current or new principles of international environmental law to this issue will provide theoretical support for the further development of this framework.

This chapter examines the key principles and rules of international environmental law as reflected in treaties, binding acts of international organisations, State practice and soft law commitments, and applies them to the problem of GHG emissions from international shipping. This chapter is set out in six parts. The first part discusses the concept of 'pollution' and its relationship with GHG emissions from international shipping. The second part identifies the jurisdiction over this problem. The third part explores the environmental liability for transboundary harm caused by GHG emissions from international shipping. The fourth part examines the precautionary principle as it applies to the issue. The fifth part reviews the evolution and implications of the principle of 'Common but Differentiated Responsibility' (CBDR) and the 'No More Favourable Treatment' (NMFT) principle, and examines how to apply these principles to the problem under review. The last part seeks to identify the optimal allocation of responsibility among the relevant stakeholders in GHG emissions from international shipping in accordance with the 'Polluter-Pays' principle.

To better understand the nature of GHG emissions from international shipping and relate the problem to current treaties, the relationship between this

---

1 Alexandre Charles Kiss and Dinah Shelton, *International Environmental Law* (Transnational Publishers, 3rd ed., 2004) 175.

 | DOI 10.1163/9789004329317_003

GHG emissions issue and 'pollution' will first be examined. Various law of the sea obligations will apply if the GHG emissions from international shipping come under the definition of 'pollution' in the *1982 United Nations Convention on the Law of the Sea* (LOSC).[2]

2.1.1 *The Concept of 'Pollution'*

There is no uniform definition of 'pollution' in international law.[3] The term 'pollution' is used with different meanings depending on differing contexts and purposes.[4] For the purpose of this book, a definition of 'marine pollution' or 'pollution of the marine environment' is examined. Treaty definitions of 'pollution', in particular 'marine pollution', have expanded over time. Among various definitions, two typically reflect a change of views over time by the international community. One example is the narrow definition of 'marine pollution' initially adopted by the Joint Group of Experts on Scientific Aspects of Marine Pollution (GESAMP) in 1969. Under the GESAMP definition, 'marine pollution' means

> the introduction by man, directly or indirectly, of substances into the marine environment (including estuaries) *resulting in* such deleterious effects as *harm to living resources*, hazards to human health, hindrance to *marine activities including fishing*, impairment of quality for use of sea water and reduction of amenities.[5] [emphasis added]

2 *United Nations Convention on the Law of the Sea*, opened for signature 10 December 1982, 1833 UNTS 3 (entered into force 16 November 1994) ('*LOSC*').

3 See, e.g., V.S. Russell, 'Pollution: Concept and Definition' (1974) 6(3) *Biological Conservation* 157, 157; Timothy J. Sullivan, 'Environment and Enforcement: Regulation and the Social Definition of Pollution' (1984) 12(2) *Ecology Law Quarterly* 423, 423.

4 Patricia W. Birnie, Alan E. Boyle and Catherine Redgwell, *International Law and the Environment* (Oxford University Press, 3rd ed., 2009) 189; R.B. Clark, *Marine Pollution* (Oxford University Press, 5th ed., 2001) 8–9. Clark asserts that the word 'pollution' may be utilised broadly to refer to 'the environmental damage caused by wastes discharged into the sea ("inputs")', 'the occurrence of wastes in the sea ("contamination")', or 'the wastes themselves ("pollution")'. However, 'pollution' often means 'the wastes themselves' in the context of marine environment. There is also no generally accepted definition of pollution in municipal law. This issue is further discussed in next section.

5 Qing-nan Meng, *Land-based Marine Pollution: International Law Development* (Graham & Trotman, 1987) 4; Joint Group of Experts on Scientific Aspects of Marine Pollution (GESAMP), 'Report of the First Session (London, UN Doc.GESAMP I/11, 1969)' (1969) 5.

This definition was adopted by the 1972 Stockholm United Nations Conference on the Human Environment and the *1976 Barcelona Convention for the Protection of the Mediterranean Sea against Pollution* but added the words 'or energy' after the word 'substances'.[6] The *1974 Paris Convention on Prevention of Marine Pollution from Land-based Sources* developed this definition by expanding the scope of harms to 'marine ecosystems and other legitimate uses of the sea'.[7] Generally the definitions of pollution in the above conventions encompass a comparatively narrow scope of harms to the marine environment.

Subsequently, a broader definition of pollution was adopted by treaties such as the *1979 Geneva Convention on Long-Range Transboundary Air Pollution* (*CLRTAP*),[8] and the *1982 LOSC*.[9] Under this later definition, 'pollution (of the marine environment)' refers to

> the introduction by man, directly or indirectly, of substances or energy into the marine environment, including estuaries, which *results or is likely to result in* such deleterious effects as *harm to living resources and marine life*, hazards to human health, hindrance to *marine activities, including fishing and other legitimate uses of the sea*, impairment of quality for use of sea water and reduction of amenities.[10] [emphasis added]

Through the comparison of the above italicised parts, we can find that the second definition of pollution represents at least two improvements on the first one. In the first place, the definition in the *LOSC* also includes the risk of harm to ecosystems, endangered species and other forms of marine life while the first one only refers to actual harms. This approach is more consistent with the precautionary principle,[11] and can be deemed as an advance on the ground that it can better protect the environment or human health from

---

6 *Convention for the Protection of the Mediterranean Sea against Pollution*, signed 16 February 1976, 15 ILM 300 (entered into force 12 February 1978) art. 2(a). See also, Daud Hassan, *Protecting the Marine Environment from Land Based Sources of Pollution* (Ashagate, 2006) 14.

7 *Convention for the Prevention of Marine Pollution from Land-based Sources*, opened for signature 4 June 1974, 13 ILM 352 (entered into force 6 May 1978) art. 1(1).

8 *Convention on Long-Range Transboundary Air Pollution*, opened for signature 13 November 1979, 18 ILM 1442 (entered into force 16 March 1983) art. 1(a) ('*CLRTAP*').

9 *LOSC* art. 1(4).

10 Ibid.

11 See below 2.4. The precautionary principle, also referred to as 'the precautionary approach', exhorts decision-makers, 'where there are threats of serious or irreversible damage', not to use 'lack of full scientific certainty…as a reason for postponing cost-effective

potential damage. In the second place, concerning the adverse effect of pollution, the second definition focuses on environmental conservation broadly while the first one is more anthropocentric, stressing the 'impact on resources or amenities useful to man' narrowly.[12] Generally the *LOSC* definition predominates in definitions favoured by the Organisation for Economic Cooperation and Development (OECD) and the International Law Association (ILA) in that it 'presents a much more clear environmental perspective'.[13] However, some scholars advocate that there are only 'slight amendments' between the two definitions.[14]

Several implications of the LOSC definition of marine pollution are notable. Firstly, 'introduction by man' indicates that pollution occurs only due to human activities. Secondly, 'directly or indirectly' 'into the marine environment' refers to the marine environment including all maritime zones (high seas, exclusive economic zone, continental shelf, contiguous zone, territorial sea and internal waters), water column, seabed and subsoil. Thirdly, the expression 'substances or energy' encompasses solid, liquid, gaseous materials objects, noise, vibrations, heat and radiation.[15] However, this scope may be adjusted and potential new pollutants may be added with advancing technology and amendments to international treaties.

Fourthly, the expression 'deleterious effects' indicates that the threshold for pollution is that human activity leads to 'significant' environmental impact, such as endangering human health or resources.[16] Based on the International Law Commission (ILC) First Report on the Legal Regime for Allocation of Loss in Case of Transboundary Harm Arising out of Hazardous Activities, 'significant' harm may be judged from two factors: one is that it is "more than *de*

---

measures to prevent environmental degradation'. See also *Rio Declaration on Environment and Development*, 31 ILM 874 (14 June 1992) principle 15 ('*Rio Declaration*').

12 Birnie, Boyle and Redgwell, above n. 4, 188.

13 Ibid 189.

14 See, e.g., Meng, above n. 5.

15 Kiss and Shelton, above n. 1, 176.

16 Ibid 177. But see Birnie, Boyle and Redgwell, above n. 4, 186–188. Birnie, Boyle and Redgwell assert that it is very controversial to determine the threshold at which harm to the environment becomes a breach of obligation. Many treaties and cases impose 'significant' or 'serious or irreversible damage' to qualify reference to deleterious effects, while none of the relevant civil-liability conventions requires environmental harm to be serious or significant. Thus, they criticise that this difference may allow the utility of the activity to outweigh the seriousness of the harm (for instance, caused by pollution) and have the effect of converting an obligation to prevent harm (an absolute obligation) into an obligation of diligence or into a constraint on abuse of rights.

minimis, 'negligible', 'detectable', or 'appreciable' but need not be at the level of 'serious' or 'substantial'"; the other is that it "must lead to real detrimental effects on such aspects as human health, industry, property, the environment or agriculture in other states, measured by factual and objective standards".[17] Since 'pollution' falls into a category of environmental harm, it is arguable that if a type of transbounary harm is 'more than detectable' and has caused actual detrimental effects, this harm meets the threshold of being a type of pollution even though this harm is not 'serious' or 'substantial'.

The LOSC definition of marine pollution adopts a traditional approach, which relates pollution to 'a certain level of seriousness either in volume or in the context of their location'.[18] While this approach has been widely adopted by various treaties, the *1996 Protocol to the Convention on the Prevention of Marine Pollution by Dumping of Wastes and Other Matter, 1972* (*1996 Protocol to London Dumping Convention*)[19] provided an alternative. This opposite approach is called the 'reverse listing' where all waste dumping is deemed as pollution unless it can be proved harmless.[20] It appears that the 1996 Protocol provides more stringent criteria on pollution. However, this Protocol adapts the LOSC definition of pollution by replacing the term 'substances or energy' with 'wastes or other matter'.[21] Therefore, it is vital to judge whether something is a 'waste or other matter' before putting it under the category of pollution in this treaty. The *Convention on the Prevention of Marine Pollution by Dumping of Wastes and Other Matter, 1972* (*London Dumping Convention*) provides that,

---

17 Pemmaraju Sreenivasa Rao, '*First Report on the Legal Regime for Allocation of Loss in Case of Transboundary Harm Arising out of Hazardous Activities*, International Law Commission, 55th Session, 5 May–6 June and 7 July–8 August, 2003, UN Doc A/CN.4/531 (21 March 2003)' (2003).

18 Birnie, Boyle and Redgwell, above n. 4, 189.

19 *Protocol to the Convention on the Prevention of Marine Pollution by Dumping of Wastes and Other Matter*, opened for signature 7 November 1996, EMuT 972:96/D (entered into force 24 March 2006) ('*1996 Protocol to London Dumping Convention*').

20 Birnie, Boyle and Redgwell, above n. 4, 189.

21 *1996 Protocol to London Dumping Convention* art. 1(10). This article provides that, " 'pollution' means the introduction, directly or indirectly, by human activity, of wastes or other matter into the sea which results or is likely to result in such deleterious effects as harm to living resources and marine ecosystems, hazards to human health, hindrance to marine activities, including fishing and other legitimate uses of the sea, impairment of quality for use of sea water and reduction of amenities."

> The Contracting Parties pledge themselves to promote, within the competent specialised agencies and other international bodies, measures to protect the marine environment against *pollution caused by*:
> (c) *wastes generated in the course of operation of vessels*, aircraft platforms and other man-made structures at sea.[22] [emphasis added]

This provision indicates that wastes generated from shipping operations could cause pollution. In other words, these wastes could be regarded as pollution under the *London Dumping Convention.*

Finally, 'deleterious effects' should result from these 'substances or energy'. This cause-effect relationship, however, is sometimes difficult to measure in practice and relies heavily on scientific evidence.[23] It is arguable that these five factors could be utilised to judge whether a substance or energy is a type of marine pollution, if the *LOSC* definition of marine pollution is set as a criterion.

2.1.2 *'Pollution' and Greenhouse Gas Emissions from International Shipping*

As discussed in Chapter 1, GHG emissions from international shipping mainly include CO_2, CH_4, N_2O and HFC with CO_2 as the most important GHG. The question of whether GHG emissions from international shipping are a type of pollution is controversial and fiercely debated. It is also important to identify the nature of shipping GHG emissions as GHG emissions, being a type of pollution, may trigger the application of many pollution-related treaties to this issue. This section examines this issue from two perspectives, namely whether shipping GHG emissions meet the treaty definition of marine pollution, and using a comparative analysis of national legislations on the nature of shipping GHG emissions.

2.1.2.1 Legal Analysis of Treaty Definition of Pollution

The five factors drawn from the *LOSC* definition of pollution as discussed in the previous section can be summarised into three questions in the context of GHG emissions from international shipping. They are: (1) whether GHG emissions from international shipping are anthropogenic? (2) Whether they are 'substances or energy' or 'wastes or other matter' being brought into the

---

22 *Convention on the Prevention of Marine Pollution by Dumping of Wastes and Other Matter*, opened for signature 29 December 1972, 18 ILM 510 (entered into force 30 August 1975) ('*London Dumping Convention*') art. XII.

23 For example, to identify the adverse effects from GHG emissions by international shipping is often difficult. This will be examined further in the following sections.

marine environment? And, (3) whether they lead to 'deleterious effects'? GHG emissions from international shipping can be treated as a type of pollution if they meet the three criteria incorporated in the three questions.

Firstly, are GHG emissions from international shipping all anthropogenic? According to the analysis in Chapter 1, GHGs consist of natural gases and anthropogenic sources of emissions and GHG emissions from international shipping include emissions of exhaust gases, emissions of refrigerants, cargo emissions and other emissions. Emissions of exhaust gases mainly come from engines, boilers and incinerators, and cargo emissions result from leakages of refrigerant and volatile compounds emissions from liquid cargo. Such emissions mainly come from engines, refrigerants and other equipment. Therefore it is axiomatic that GHG emissions from international shipping are human-induced.

Secondly, are GHG emissions from international shipping 'substances or energy' or 'wastes or other matter' being brought into the marine environment? Based on the above definition of 'substances or energy', gaseous materials including GHG emissions from international shipping are within this category. Indeed, GHG emissions from ships have been deemed to be a kind of 'substance', both theoretically and practically. Annex VI to *International Convention for the Prevention of Pollution from Ships* (*MARPOL 73/78*) provides that '[e]*mission* means any release of substances subject to control by this Annex from ships into the atmosphere or sea'.[24] The Australian Government and some Australian States and Territories have regulated carbon-based products in onshore underground storage areas in order to facilitate, promote and encourage the storage of GHG substances in geological formations.[25] Examples include *Greenhouse Gas Geological Sequestration Act 2008* (Victoria) section 1, *Greenhouse Gas Storage Act 2009* (Queensland) section 3, *Petroleum and Geothermal Energy Act 2000* (South Australia) section 3(a), *Barrow Island Act 2003* (West Australia), *Carbon Capture and Storage Act* (Commonwealth) section 3. GHG emissions are treated as 'substances' under *MARPOL 73/78* and the above national legislations.

Whether GHG emissions from international shipping are 'wastes or other matter' as defined under the *London Dumping Convention* is not so straightforward. The *London Dumping Convention* prohibits the dumping of all 'wastes and other matter' listed in Annex I and requires a prior special

24 *International Convention for the Prevention of Pollution from Ships* (*MARPOL 73/78*), signed 2 November 1973, 12 ILM 1319, as amended by the 1978 Protocol to the 1973 Convention, 1341 UNTS 3, 17 ILM 546 (entered into force 2 October 1983) annex VI reg 2(7).

25 Nicola Durrant, 'Carbon Capture and Storage Laws in Australia: Project Facilitation or a Precautionary Approach?' (2010) 18(4) *Environmental Liability Journal* 148, 155.

permit for the dumping of wastes listed in Annex II.[26] However, as the most important shipping GHG emission, CO_2 is not specifically referred to in either Annex I or Annex II. It was argued that CO_2 would fall under the 'industrial waste' category in Annex I if it is produced from a 'manufacturing or processing operation'.[27] Based on this understanding, CO_2 derived from fossil fuels has been regarded as an 'industrial waste' by the Scientific Group of the *London Dumping Convention*, as well as by the United Kingdom Government.[28] Furthermore, CO_2 streams from CO_2 capture processes for sequestration 'may be considered for dumping being mindful of the Objectives and General Obligations of the [1996] Protocol set out in articles 2 and 3'.[29] While shipping CO_2 may derive from a ship's 'manufacturing or processing operation', it is arguable that shipping CO_2 may fall within the category of 'industrial waste' and thus make it a type of pollution under the *London Dumping Convention*. The amended *1996 Protocol to London Dumping Convention* only treats CO_2 streams from CO_2 capture processes for sequestration as a type of dumping. Nevertheless, the fact that shipping CO_2 may dissolve into the seawater and be absorbed into the seabed[30] reveals that shipping CO_2 is of a similar nature to CO_2 from CO_2 capture processes for sequestration. Thus shipping CO_2 may also be treated as dumping, or pollution under the *London Dumping Convention*.

Thirdly, do GHG emissions from international shipping lead to 'deleterious effects'? As discussed earlier, the environmental harm caused by pollution needs to be 'more than detectable', but it need not be 'serious' or 'substantial'. GHGs are emitted during the whole voyage of a vessel. The amount of discharge depends on many factors such as engine and ship design, cargo volume, and shipping speed.[31] These features make GHG emissions from international

26 *London Dumping Convention* art. IV(1).

27 Yvette Carr, 'The International Legal Issues Relating to the Facilitation of Sub-seabed CO_2 Sequestration Projects in Australia' (2007) 14 *Australian International Law Journal* 137, 143; Ray Purdy and Richard Macrory, *Geological Carbon Sequestration: Critical Legal Issues* (January 2004) Tyndall Centre for Climate Change Research <http://www.tyndall.ac.uk/sites/default/files/wp45.pdf> accessed 1 May 2014, p. 21.

28 Purdy and Macrory, above n. 27. However, this view has not achieved consensus among various countries.

29 *1996 Protocol to London Dumping Convention* (as amended in 2006) annex I, 1.8.

30 Duncan E.J. Currie and Kateryna Wowk, 'Climate Change and CO_2 in the Oceans and Global Oceans Governance' (2009) 3(4) *Carbon & Climate Law Review* 387, 391.

31 These can be inferred according to the categories of GHG emissions from international shipping as discussed in Chapter 1, 1.2.2.2. There are mainly four categories of GHG emissions from international shipping, namely emissions of exhaust gases (from sources

shipping cumulative, and indicate that not all such emissions bring about 'significant' environmental impact so as to be 'more than detectable'. However, in practice, the problem of how to measure whether GHG emissions are 'significant' is not straightforward. It needs to be assessed on a case-by-case basis. What is significant also depends on the specific context of each case, such as 'the nature of the harm in question, the risk it poses, the location of the harm in relation to natural features and human activity', and 'the particular capabilities of the state in question', and these factors may vary over time.[32] The inevitable subjective elements incorporated in these factors suggest that the judgement of 'deleterious effects' caused by shipping GHG emissions can never be totally objective.

It may be concluded that theoretically GHG emissions from international shipping meet the main characteristics reflected from the treaty definition of pollution and hence could be regarded as a type of pollution provided that these emissions engender 'deleterious effects' or lead to 'significant' environmental impact, or they could be treated as 'wastes' under the *London Dumping Convention*. Therefore, it is arguable that GHG emissions from international shipping are by their nature a type of 'conditional' pollution. In other words, GHG emissions from ships can be treated as pollution under certain circumstances. This view is also consistent with the regulatory practice within the International Maritime Organization (IMO). In July 2011 the reduction of GHG emissions from international shipping was regulated in the form of amendments of Annex VI to *MARPOL 73/78*, which relates to air 'pollution' from ships. However, this regulation does not clearly stipulate GHG emissions as air pollution, and it is still open to debate whether GHG emissions from ships are a type of pollution.

2.1.2.2 National Legislation on the Legal Nature of GHG Emissions

Given that GHG emissions from ships are a type of 'conditional' pollution, different countries have adopted national legislation on the basis that GHG emissions, including those from international shipping, are pollutants. Some countries listed in Annex I to the *United Nations Framework Convention on*

---

such as engines, auxiliary engines, boilers and incinerators), emissions of refrigerants, cargo emissions and others.

32 Rebecca M. Bratspies and Russell A. Miller (eds), *Transboundary Harm in International Law: Lessons from the Trail Smelter Arbitration* (Cambridge University Press, 2006) 112. See also Birnie, Boyle and Redgwell, above n. 4, 142; Kiss and Shelton, above n. 1, 177.

Climate Change (*UNFCCC*) have regulated GHG emissions as pollutants,[33] whereas many non-Annex I countries to the *UNFCCC* leave GHG emissions unregulated.

The United States of America (US) is one of those countries that have regulated GHG emissions as air pollutants. The *Clean Air Act of the US* provides that,

> the Administration [Environmental Protection Agency (EPA)] shall conduct a basic engineering research and technology program to develop, evaluate, and demonstrate nonregulatory strategies and technologies for air pollution prevention.... Such program shall include the following elements:
> (1) Improvements in nonregulatory strategies and technologies for preventing or reducing multiple *air pollutants, including sulphur oxides, nitrogen oxides, heavy metals, PM-10 (particular matter), carbon monoxide, and carbon dioxide*, from stationary sources, including fossil fuel power plants.[34] [emphasis added]

As the most important GHG, carbon dioxide (CO_2) is regulated as an air pollutant in this Act. Although the above provision only regulates CO_2 from stationary sources, it appears that the legal nature of CO_2 as air pollution will not change when the CO_2 emissions are from mobile sources (e.g., ships). This can also be seen from the definition of 'air pollutant'. The *Clean Air Act of the US* defines 'air pollutant' as 'any air pollution agent or combination of such agents, including any physical, chemical, biological radioactive (including source material, special nuclear material, and by-product material) substance or matter which is emitted into or otherwise enters the ambient air'.[35] This definition indicates that whether CO_2 is from stationary or mobile sources will not change its legal nature. In 2009 the Environmental Protection Agency of the US announced that CO_2 and five other GHGs threaten public health and the environment, and thus should be treated as 'dangerous pollutants'.[36] In 2005, the New Jersey State Department of Environmental Protection, based

33 *United Nations Framework Convention on Climate Change*, opened for signature 9 May 1992, 31 ILM 848 (entered into force 21 March 1994) ('*UNFCCC*').

34 *Clean Air Act of the United States of America*, Pub L No 108–201, Stat, 42 USC §7401 et seq. (1970, as amended in 1977 and 1990) Sec 103(g)(1).

35 Ibid. Sec 7602(g).

36 Nicolas Loris, *EPA Formally Declares CO_2 a Dangerous Pollutant* (7 December 2009) <http://blog.heritage.org/2009/12/07/epa-formally-declares-co2-a-dangerous-pollutant/> accessed 1 April 2016.

on the authorisation conferred on by the *Air Pollution Control Act of New Jersey*, regulated CO_2 as a pollutant,[37] thereby allowing State regulators to cap CO_2 emissions in tackling climate change.

Case law has also played an important role in pushing and shaping this expansion of the pollution concept. In *Massachusetts v. Environmental Protection Agency* (EPA),[38] the US Supreme Court held that the EPA has the authority to regulate GHG emissions as a response to petitions filed by environmental groups and the California Attorney General. Some environmental groups petitioned the EPA to take actions to reduce marine emissions, and the California Attorney General requested that the EPA regulate GHG emissions from oceangoing vessels.[39] The Supreme Court's decision not only addressed similar petitions from other entities,[40] but also accelerated the regulatory process of the US on GHG emissions.

GHG emissions have not been explicitly regulated as air pollutants in Australia, but they have been treated as pollutants in some proposed schemes. The Carbon Pollution Reduction Scheme (CPRS) was a cap-and-trade emission trading scheme proposed by the Rudd Government to the Australian Parliament in 2009.[41] The aim of this scheme is to reduce GHG emissions through adding a price to emit carbon. It can be regarded as the predecessor of the Carbon Tax scheme in Australia.[42] Although the CPRS failed for lack

37 Centre for Climate and Energy Solutions, *New Jersey Classifies Carbon Dioxide as Air Contaminant* (18 October 2005) <http://www.c2es.org/us-states-regions/news/2005/new-jersey-classifies-carbon-dioxide-air-contaminant> accessed 1 April 2016.

38 *Massachusetts v. Environmental Protection Agency*, 549 U.S. 497 (2007).

39 Timothy Nast, 'The Reponse of the International Shipping Industry to Global Climate Change' (2013) 44(1) *Journal of Maritime Law and Commence* 29, 32.

40 Ibid. 33.

41 Parliament of Australia, *Carbon Pollution Reduction Scheme* (22 October 2010) <http://www.aph.gov.au/About_Parliament/Parliamentary_Departments/Parliamentary_Library/Browse_by_Topic/ClimateChange/Governance/Domestic/national/cprs> accessed 1 May 2014.

42 *Australian Clean Energy Bill 2011 (Explanatory Memorandum)*, adopted by the Parliament of the Commonwealth of Australia and House of Representatives (2010–2011), Policy Context, p. 12. Australia's carbon tax scheme, also called carbon pricing mechanism, is incorporated into the *Australian Clean Energy Bill 2011*. This scheme commenced on 1 July 2012 with a price that would be fixed for the first three years, and it was expected that on 1 July 2015 the carbon price would transition to a fully flexible price under an emissions trading scheme when the price would be determined by the market. However, this scheme was finally abolished by the Abbott Government on 17 July 2014.

of public support,[43] the title of the scheme reveals that GHG emissions were regarded as a type of pollution by Australian policy makers.

Compared with the US and Australia, China, as the largest developing country, has not regulated or limited GHG emissions in its domestic legislation. The definition and scope of air pollutants are not provided for in Chinese regulation such as its *Environmental Protection Law*,[44] and the *Air Pollution Prevention and Control Law*.[45] It is anticipated that GHGs will not be regulated as a type of pollution in Chinese legislation in the short term. It was argued that the regulation of GHGs (mainly CO_2) as air pollutants would slow down Chinese economic development and trigger the application of more international treaty obligations.[46] As a non-Annex I State to the *UNFCCC*, China does not have compulsory emissions reduction targets which justify China's deregulation of GHG emissions. Similar to China, other large developing countries such as India, Brazil and South Africa also have not regulated GHG emissions in their national legislation.

It is concluded that GHG emissions from international shipping can be regarded as a type of 'conditional' pollution. This theoretical assertion provides considerable scope for various countries to adopt differing legislative choices on the legal nature of GHG emissions. In practice some developed countries have regulated GHG emissions as pollutants while most developing countries have not regulated GHG emissions, which is consistent with their respective obligations under the *Kyoto Protocol* to the *UNFCCC*.[47]

See Lenore Tayor, *Australia Kills Off Carbon Tax* (17 July 2014) <http://www.theguardian.com/environment/2014/jul/17/australia-kills-off-carbon-tax> accessed 17 July 2014.

43 Parliament of Australia, above n. 41.

44 *Environmental Protection Law of the People's Republic of China*, adopted at the 11th Meeting of the Standing Committee of the Seventh National People's Congress on 26 December 1989, and amended on 24 April 2014.

45 *Air Pollution Prevention and Control Law of the People's Republic of China*, adopted in 1987 and amended in 1995, 2000 and 2015 respectively, entered into force on 1 January 2016.

46 李志文 [Li Zhiwen], '《船舶温室气体减排国际立法的新发展及其启示》 [New Development of International Regulation in Reducing Greenhouse Gas Emissions from Ships and Enlightenments to China]' (2012) 152(6) 法商研究 *Journal of Studies in Law and Business* 141, 145.

47 *Kyoto Protocol to the United Nations Framework Convention on Climate Change*, opened for signature 16 March 1998, 37 ILM 22 (entered into force 16 February 2005).

2.2 Jurisdiction over Greenhouse Gas Emissions from International Shipping

In international law, responsibility occurs when the legal interest of one subject of the law is invaded by another legal person.[48] To determine international environmental law responsibility and how it applies to specific areas, a good understanding of the principles of State jurisdiction is fundamental. Given that GHG emissions from international shipping can be regarded as a type of 'conditional' pollution, many treaties relating to marine environmental pollution, including the *1982 LOSC* and *MARPOL 73/78*, will apply to this GHG issue. The next part discusses the concept of jurisdiction, and examines State jurisdiction over GHG emissions from international shipping from three perspectives, namely flag State jurisdiction, coastal State jurisdiction and port State jurisdiction.[49]

2.2.1 *The Concept of 'Jurisdiction'*

As a fundamental concept of international law, State jurisdiction over a particular event is a key to analysing many international disputes. In the context of marine pollution regulation, 'jurisdiction' refers to 'the competence of states to prescribe and enforce legislation against vessels engaged in pollution'.[50] This definition indicates that there are two types of jurisdiction. One is 'prescriptive' or 'legislative' jurisdiction, which empowers a State to 'enact or promulgate substantive pollution control standards'.[51] These standards, especially those applying in zones beyond the internal waters and territorial sea of a State, should comply with 'generally accepted' international standards and are often consistent with 'internationally agreed' standards.[52] They are generally not beyond internationally accepted standards in order to ensure the freedom of navigation.[53] In the case of GHG emissions from international shipping, the IMO is the international organisation which drafts such standards. The second type of jurisdiction is 'enforcement' jurisdiction, which authorises a State

---

48 Ian Brownlie, *Principles of Public International Law* (Oxford University Press, 7th ed, 2008) 433.

49 Flag State control and the issue of 'Flag of Convenience', and port State control, are examined separately in Chapter 6 of this book.

50 Alan Khee-Jin Tan, *Vessel-Source Marine Pollution: the Law and Politics of International Regulation* (Cambridge University Press, 2006) 176.

51 Ibid.

52 *LOSC* arts. 211(2), 212(1). See also ibid. Tan asserts that under certain circumstances, international law may endorse the prescription of national standards.

53 See *LOSC* arts. 17, 58.

to 'prevent or punish the actual violation of the relevant standards'.[54] Some scholars have raised another so called 'adjudicative' or 'judicial' jurisdiction, referring to 'the power of national courts or tribunals to adjudicate prosecutions against a vessel or a person for transgressions of prescribed standards'.[55] In this book 'enforcement' jurisdiction is interpreted as encompassing the 'adjudicative' or 'judicial' authority of States consistent with the interpretation of jurisdiction adopted in international agreements such as the *LOSC* and IMO instruments[56] and the approach that has been adopted by some scholars.[57]

A flag, coastal or port State of a particular vessel has different prescriptive or enforcement jurisdiction.[58] Historically, in order to gain an equitable balance between coastal and navigational interests, the jurisdiction reallocation between different State actors was eventually formed with the joint effort of the whole international community.[59] Among the international achievements, the *LOSC* and *MARPOL 73/78* form the current jurisdiction regime in the context of vessel-source pollution.[60] Jurisdiction under the *LOSC* and *MARPOL 73/78* regime, however, should also apply to the GHG emissions from international shipping due to the nature of such emissions being a kind of pollution. Thus, jurisdiction over GHG emissions from international shipping is examined from the perspective of flag, coastal and port States respectively in the following parts of this section.

54 Tan, above n. 50, 176.

55 Ibid. See also R.R. Churchill and A.V. Lowe, *The Law of the Sea* (Manchester University Press, 3rd ed., 1999) 344. Churchill and Lowe subdivide enforcement jurisdiction into the competence to arrest (arrest jurisdiction) and the competence of courts to deal with alleged breaches of the law (judicial jurisdiction).

56 For instance, in section 6 (enforcement) of *LOSC*, art. 217(4) and art. 218(1) stipulate that the flag State and port State may institute proceedings under certain circumstances, which are of the nature of adjudicative or judicial jurisdiction. See also *LOSC* art. 62.

57 See, e.g., Malcolm N. Shaw, *International Law* (Cambridge University Press, 6th ed., 2008) 572; Tan, above n. 50, 177.

58 Churchill and Lowe, above n. 55, 344.

59 Tan, above n. 50, 177; Churchill and Lowe, above n. 344–353. Through three United Nations Law of the Sea Conferences, coastal and port State jurisdiction expanded; flag State jurisdiction diminished but still remained prominent.

60 See, e.g., Churchill and Lowe, above n. 55, 344–352; Tan, above n. 50, 184–222; Donald Rothwell and Tim Stephens, *The International Law of the Sea* (Hart Publishing, 2010) 353–358.

2.2.2 *Flag State Jurisdiction*

A flag State refers to 'the State whose nationality a particular vessel has',[61] or in other words, 'the State in which the vessel is registered or whose flag it is entitled to [fly]'.[62] In customary law, the flag State enjoys the primary jurisdiction over the ship flying its flag, and it is the only subject which has jurisdiction to enforce regulations applicable to ships on the high seas.[63] This form of jurisdiction is reflected in the provisions of *MARPOL 73/78* and the *1982 LOSC*.

Under the *MARPOL 73/78* regime, the flag State enjoys both prescriptive and enforcement jurisdiction. Regarding prescriptive jurisdiction, flag States are required to adopt laws to ensure that the regulatory provisions of *MARPOL 73/78* are applied to ships on their registries.[64] Any violation of *MARPOL 73/78* is to be prohibited wherever it occurs, and sanctions shall be established under the law of the flag State.[65] Additionally, the penalties specified under flag State law shall be adequate in severity to discourage violations of *MARPOL 73/78*, and shall be equally severe irrespective of where the violations occur.[66] As for enforcement jurisdiction, flag States have three obligations under *MARPOL 73/78*. Firstly, flag States are obliged to institute proceedings against any of their ships suspected of having violated *MARPOL 73/78*.[67] In order to facilitate flag State prosecution of such offences, all parties to *MARPOL 73/78* are required to report incidents at sea involving harmful substances,[68] no matter where the offence is committed. Secondly, flag States shall act appropriately to either inspect, investigate, or detect the ship on suspected violation of *MARPOL 73/78* when informed of suspected violations by other parties, or impose penalties when such violations have been proved.[69] Thirdly, flag States are to conduct

61 Churchill and Lowe, above n. 55, 344.

62 Birnie, Boyle and Redgwell, above n. 4, 400.

63 Ibid 401. For instance, in the *Lotus Case* (*Lotus Case* (*France v Turkey*) (1927) PCIJ Series A, No. 10), the Permanent Court of International Justice cited the principle that 'no state may exercise any kind of jurisdiction over foreign ships on the high seas', which indicated that foreign ships should not be arrested or detained while they are on the high seas. However, the flag State still exercises its jurisdiction over the ships flying its flag no matter where it is operating.

64 *MARPOL 73/78* art. 3.

65 *MARPOL 73/78* art. 4(1).

66 *MARPOL 73/78* art. 4(4).

67 *MARPOL 73/78* art. 4(1).

68 *MARPOL 73/78* arts. 4(2), 6(3).

69 *MARPOL 73/78* art. 8.

surveys, to issue or authorise other parties to issue certificates, to ensure the compliance of their ships with the convention.[70]

It is primarily the responsibility of flag States to regulate the issue of GHG emissions from international shipping. For instance, flag States that have ratified amendments of Annex VI to *MARPOL 73/78* shall incorporate the energy efficiency requirements for ships as specified in amended Annex VI into their domestic legislation, and, if violations occur, institute proceedings in relation to such offences. When informed of a suspected violation of *MARPOL 73/78*, the flag State is obliged to cooperate with relevant parties in detecting, inspecting or investigating the violation. Furthermore, flag States that are parties to *MARPOL* Annex VI shall conduct regular surveys, issue or empower other parties to issue the International Energy Efficiency Certificate (IEE Certificate) so as to comply with IMO standards.[71]

The jurisdictional competence of flag States under *LOSC* is consistent with *MARPOL 73/78*. Flag States are obliged to pass pollution control laws for their ships under *LOSC* provisions.[72] Article 211(2) requires flag States to adopt laws and regulations to prevent marine pollution that 'at least have the same effect' as that of generally accepted international rules and standards.[73] Since *MARPOL 73/78* has been regarded by some scholars as representing 'generally accepted international rules and standards [in the context of regulation of vessel pollution]', it was argued that Article 211 of the *LOSC* has made *MARPOL 73/78* and 'other relevant international standards' 'an obligatory minimum'.[74] Another similar view is that *MARPOL 73/78* and all of its annexes 'which have entered into force, and have attracted high participation' could be treated as 'generally accepted international standards'.[75] However, how to judge 'high participation' remains unclear. It is arguable that both views should not apply to the 2011 amendments of Annex VI to *MARPOL 73/78*. This is because the 2011

70 *MARPOL 73/78* art. 5.

71 *MARPOL 73/78 Annex VI* regs 5–9.

72 See, e.g., *LOSC* art. 94(1). This article reads that, 'Every [flag] State shall effectively exercise its jurisdiction and control in administrative, technical and social matters over ships flying its flag [on the high seas].'

73 *LOSC* art. 211(2). This article stipulates that,

'States shall adopt laws and regulations for the prevention, reduction and control of pollution of the marine environment from vessels flying their flag or of their registry. Such laws and regulations shall at least have the same effect as that of generally accepted international rules and standards established through the competent international organization or general diplomatic conference.'

74 Birnie, Boyle and Redgwell, above n. 4, 413.

75 Rothwell and Stephens, above n. 60, 355.

amendments to *MARPOL* Annex VI were adopted by a majority vote rather than a consensus, and some major shipping nations, such as China, Brazil, Kuwait, and Saudi Arabia, voted against the amendments.[76] The participation was still high (49 out of 59 parties to Annex VI voted yes), and the 2011 amendments entered into force on 1 January 2013. However, it is less persuasive for the 2011 amendments of Annex VI to *MARPOL 73/78* to be 'generally accepted' although the amendments may fall within the 'internationally agreed rules' as specified in Article 212(1) of the *LOSC*.[77] Based on Article 31 of the *Vienna Convention on the Law of Treaties* in 1969,[78] Article 212(1) of the *LOSC* might be interpreted as meaning that rules adopted at the international level could be treated as 'internationally agreed rules'.[79] Nevertheless, Article 212(1) of the *LOSC* only requested States to 'take into account internationally agreed rules, standards and recommended practices and procedures' when they adopt laws and regulations to prevent, reduce and control marine pollution from or through the atmosphere. This provision imposes a very weak obligation on flag States.[80]

Concerning the enforcement jurisdiction of flag States, *LOSC* requires flag States to take necessary measures for the implementation and enforcement of international rules and standards.[81] These measures include the investigation of pollution offences, inspection, certification, and instituting proceedings under certain circumstances.[82] In fact, these measures are exactly what *MARPOL 73/78* demands and are thus 'nothing novel in principle'.[83]

---

76 *Report of the Marine Environment Protection Committee on Its Sixty-Second Session*, MEPC 62nd Session, Agenda Item 24, IMO Doc MEPC 62/24 (26 July 2011) para. 6.110.

77 James Harrison, 'Recent Developments and Continuing Challenges in the Regulation of Greenhouse Gas Emissions from International Shipping' (2012) *University of Edinburgh Research Paper Series* <http://ssrn.com/abstract=2037038> accessed 6 June 2014, pp. 20, 22–23. The rules incorporated in Articles 211(1) and 212(1) of the LOSC are so-called 'rules of reference', which require parties to the LOSC to comply with rules and standards as specified in other international instruments.

78 *Vienna Convention on the Law of Treaties*, opened for signature 23 May 1969, 8 ILM 679 (entered into force 27 January 1980) art. 31(1) ('*1969 Vienna Treaty Convention*'). This article provides that,

'A treaty shall be interpreted in good faith in accordance with the ordinary meaning to be given to the terms of the treaty in their context and in the light of its object and purpose.'

79 Harrison, above n. 77, 23.

80 Ibid.

81 *LOSC* art. 217.

82 Ibid.

83 Birnie, Boyle and Redgwell, above n. 4, 413. Birnie, Boyle and Redgwell assert that Article 217 of the LOSC just 'fully accords with existing customary and conventional law' and is thus 'nothing novel'. But see Rothwell and Stephens, above n. 60. Rothwell and Stephens

Therefore, flag State jurisdiction over GHG emissions from international shipping under the *MARPOL 73/78* regime is almost the same as that under the *LOSC* framework. Flag States are primarily responsible for the regulation and control of GHG emissions from their ships. However, in practice some flag States may not exercise their entire jurisdiction in terms of GHG emissions from their own fleets, and empirical survey suggests that flag States impose lower fines than port States with regard to the average fines for violating *MARPOL* standards.[84] There are many reasons why flag States lack incentives to exercise such jurisdiction. For instance, GHG emissions from international shipping are often outside the territory of the flag State and may only imperil the environment of others, so the incentives for a flag State to enforce may be low.[85] Further, such enforcement is often costly. The principal, or indeed sole, interest of many flag States is often to obtain economic benefits by means of registration fees or taxes from ships registered there due to the existence of 'flags of convenience'(FOC),[86] where registration is the 'only substantial connection' with the flag State.[87]

2.2.3 *Coastal State Jurisdiction*

Churchill and Lowe provide an apposite definition of coastal State, based on which a coastal State is 'the State in one of whose maritime zones a particular vessel lies'.[88] Different from flag States, coastal States have incentives to impose severe restrictions upon ships navigating within their maritime zones. In general, the pollution caused by ships, including GHG emissions from international

assert that 'the LOSC does not alter the capacity of flag States to take action to enforce pollution control standards', but it 'transforms the customary law capacity into a positive obligation', which should be an advance.

84 Ho-Sam Bang, 'Recommendations for Policies on Port State Control and Port State Jurisdiction' (2013) 44(1) *Journal of Maritime Law and Commence* 115, 127.

85 Michael Faure and Ying Song (eds), *China and International Environmental Liability: Legal Remedies for Transboundary Pollution* (Edward Elgar Publishing, 2008) 87.

86 Ibid 85. To date there is no uniform definition of FOC. See, e.g., Egiyan defines the FOC as 'national flags of those States in which shipowners register their ships so as to avoid: (a) financial obligations; and (b) the nature and conditions of shipping were their vessels registered in their own countries'. G.S. Egiyan, 'Flag of Convenience' or 'Open Registration' of Ships' (1990) 14(2) *Marine Policy* 106, 107; Griffin defines the FOC as 'flags of certain countries whose laws make it easy and attractive for ships owned by foreign nationals or companies to fly these flags'. Andrew Griffin, 'MARPOL 73/78 and Vessel Pollution: A Glass Half Full or Half Empty?' (1994) 1(2) *Indiana Journal of Global Legal Studies* 489, 506.

87 Birnie, Boyle and Redgwell, above n. 4, 360.

88 Churchill and Lowe, above n. 55, 344.

shipping, often makes the coastal States the victim of such damage. The damage frequently occurs either in the exclusive economic zone or in the territorial sea of the coastal State, although such damage is cumulative and global in nature in the context of GHG emissions from ships. However, the claims for stricter jurisdiction by coastal States are restricted by the *LOSC* in order to maintain navigational rights and an equitable balance between coastal States and foreign flag States. Coastal States' jurisdiction to regulate and enforce their laws against vessels depends on their sovereignty or sovereign rights over maritime zones contiguous to their coasts, and the *LOSC* provides the framework for dealing with this issue. Hence this part examines coastal State jurisdiction based on these maritime zones under the *LOSC*, and reviews the regulations from *MARPOL 73/78* and other treaties.

In internal waters, such as ports, coastal States enjoy full legislative and enforcement jurisdiction. Generally they are free to apply national laws and determine conditions of entry for foreign vessels when such ports are open, but appropriate publicity and communication with the IMO is necessary.[89] On the one hand, internal waters form part of the territory of coastal States thus the coastal State has full territorial sovereignty over these waters.[90] On the other hand, this arrangement was also recognised by *MARPOL 73/78* and utilised by the United States and other countries in passing stringent national legislation applicable to foreign shipping.[91] As the first State to ban all single-hull oil tankers from entering its ports, the US did not wait for agreement in the IMO, and this approach was then adopted by the European Union in a similar ban following the sinking of the *Prestige* (oil spill incident) in 2002.[92] Accordingly, in the context of GHG emissions from international shipping, coastal States may in principle legislate and enforce their own national requirements on emissions, and apply them to their internal waters and ports as a condition for the entry of foreign vessels.

In contrast to internal waters, the legislative and enforcement jurisdiction of coastal States in the territorial sea is not unlimited. Concerning legislative jurisdiction, the coastal State enjoys sovereignty, and may apply its national laws on environmental protection to its territorial sea. Furthermore, international treaties on dumping or pollution from ships accord three rights to coastal States in the territorial sea, namely the designation of Emission Control

---

89 *LOSC* art. 211(3).

90 See *LOSC* art. 211(3). The only exception is for vessels in distress, which have a right to take refuge in the nearest port.

91 *MARPOL 73/78* art. 5(3).

92 Birnie, Boyle and Redgwell, above n. 4, 414.

Areas,[93] the designation and control of navigation routes for safety and environmental purposes,[94] and the prohibition of pollution discharges.[95] However, such rights should not hamper the exercise of innocent passage of foreign ships,[96] and such laws and regulations should not include matters related to the design, construction, manning or equipment of foreign ships (CDEM standards) 'unless they are giving effect to generally accepted international rules or standards'.[97] As for enforcement by coastal States in their territorial sea, the *LOSC* stipulates such measures as inspection, proceedings and detention of foreign vessels for coastal States under certain circumstances.[98] These measures, however, are regarded as 'rarely used' for anti-pollution purposes due to their possible hampering of innocent passage of foreign ships.[99]

Regarding the issue of GHG emissions from international shipping, the coastal State's jurisdiction in its territorial sea may be more in the nature of prescriptive rather than enforcement jurisdiction. Coastal States may adopt

---

93 *MARPOL 73/78* Annex VI regs 2.8, 13, 14. Regulation 2.8 reads that:

'*Emission Control Area* means an area where the adoption of special mandatory measures for emissions from ships is required to prevent, reduce and control air pollution from NO_x or SO_x and particulate matter or all three types of emissions and their attendant adverse impacts on human health and environment.'

Regulation 13 stipulates that an 'Emission Control Area shall be any sea area, including any port area, designated by the Organization'. This means that an Emission Control Area could also be located in the exclusive economic zone or other maritime zones of a coastal State.

94 *LOSC* art. 22.

95 *MARPOL 73/78* art. 4(2); *LOSC* art. 21(1)(f); *Convention on the Prevention of Marine Pollution by Dumping of Wastes and Other Matter*, opened for signature 29 December 1972, 18 ILM 510 (entered into force 30 August 1975) art. 4(3) ('*London Dumping Convention*'). See also Birnie, Boyle and Redgwell, above n. 4, 414.

96 *LOSC* art. 24(1), 211(4).

97 *LOSC* art. 21(2).

98 *LOSC* art. 220(2). According to this article, where there are clear grounds for believing that a vessel navigating in the territorial sea of the coastal States has violated laws and regulations consistent with international standards, then the coastal State may undertake a physical inspection of the vessel relating to the violation and may institute proceedings, including the detention of the vessel.

99 See *LOSC* art. 19(2). According to this article, only pollution that is 'wilful and serious' and contrary to the LOSC will deprive a vessel in passage of its innocent character, which is rare. However, these anti-pollution measures pose serious danger to navigational freedom and will generally hamper the innocent passage of foreign ships. In view of this concern, often the preferable solution will be to rely on port States for anti-pollution enforcement purpose. See also Birnie, Boyle and Redgwell, above n. 4, 417; Faure and Song, above n. 85, 93.

their national standards on GHG emissions from ships, such as operational requirements in the territorial sea.[100] They may also propose the establishment of GHG Emission Control Areas in their territorial sea to the IMO, as provided in *MARPOL* Annex VI for the purpose of reducing NO_x and SO_x emissions from shipping.

The establishment of the Exclusive Economic Zone (EEZ) is regarded by many commentators as the 'most significant reform' of the *1982 LOSC*.[101] The EEZ confers on coastal States sovereign rights over living and non-living resources, and jurisdiction relating to the protection and preservation of the marine environment.[102] Under the *LOSC*, coastal States may regulate pollution from seabed installations and dumping within the EEZ, but this prescriptive jurisdiction is limited to the application of international rules and standards, namely IMO rules and standards.[103] In other words, this national legislation should neither be 'less demanding' nor 'more stringent' than IMO rules and standards.[104] A State can only adopt stricter rules when such rules are regulating the IMO designated Particularly Sensitive Sea Areas (PSSAs) or ice-covered areas which are within a coastal State's EEZ.[105] Regarding enforcement jurisdiction, only when there are 'clear grounds' for believing that a vessel has committed a violation in the EEZ and such violation threatens substantial damage to the coastal State, may the coastal State 'require the vessel to give information regarding its identity and port of registry'.[106] The coastal State may exercise its power over EEZ pollution control only when the vessel is still navigating in its

---

100 An operational requirement/measure is one of the three methods considered so far within the IMO for regulating GHG emissions from ships (the other two are technical measures and market-based measures) and this method has been adopted by the IMO in the form of the Ship Energy Efficiency Management Plan (SEEMP) in 2011 amendments of Annex VI to *MARPOL 73/78*. It includes the requirements during the course of getting on board, checking of certificate and documents and inspection of or on other pollution prevention measures or facilities. For example, the coastal State may regulate the speed limit of foreign vessels in its territorial sea so as to reduce the GHG emissions and ensure better safety.

101 Rothwell and Stephens, above n. 60, 356. As a new maritime zone introduced by the 1982 *LOSC*, the EEZ extends to 200 nautical miles from the territorial sea baseline which to a significant extent expands the sovereign rights of a coastal State.

102 *LOSC* art. 56.

103 *LOSC* arts. 208, 210, 211(5)(6).

104 Rothwell and Stephens, above n. 60, 356.

105 Ibid. 357.

106 *LOSC* art. 220(3)(5).

EEZ and this power is limited to a request for information.[107] Therefore, this enforcement is not favoured by the coastal State in that it cannot effectively prevent, stop or penalise possible violation of relevant international regulations by the foreign vessel.[108] Accordingly, in the context of GHG emissions from international shipping, what a coastal State may do is to incorporate IMO regulations, currently the energy efficiency requirements on ships under amendments to Annex VI of *MARPOL 73/78*, into its domestic environmental protection regime so that these regulations can be applied in its EEZ.

No State has territorial jurisdiction in the high seas.[109] In these waters, the flag State has exclusive jurisdiction over its ships when they produce pollution including GHG emissions. The coastal State is not permitted to take measures unless it is threatened by the damage resulting from 'pollution or threat of pollution following upon a maritime casualty or acts relating to such casualty'.[110] Nevertheless, there is no evidence of such enforcement for GHG emissions.

2.2.4 *Port State Jurisdiction*

A port State refers to 'the State in one of whose ports a particular vessel lies'.[111] In contrast to the limited jurisdiction of coastal States, port States have substantial jurisdiction over pollution within their jurisdiction. This has been deemed as a 'corrective measure to remedy the inadequacy of flag State jurisdiction'.[112] As a port is situated in a State's internal waters, a port State may legislate for the prevention, reduction and control of marine pollution, including GHG emissions from international shipping, as a condition for the entry of foreign vessels into its ports or internal waters.[113] However, it is the port State's enforcement jurisdiction that attracts more attention since it might prove to be an effective deterrent against ships polluting any part of the sea because they are likely to face investigation or the institution of proceedings in the port State. In a broad sense, the enforcement jurisdiction of the port State includes both enforcement or administrative jurisdiction and judicial jurisdiction. Administrative jurisdiction is often called port State control and primarily involves the

107 *LOSC* art. 220(3).

108 Faure and Song, above n. 85, 94. Under the circumstances, the only remedy for the coastal State is probably to inform the flag States or the next port State of the possible violation, so that these States may take actions to investigate or institute proceedings.

109 *LOSC* art. 89.

110 *LOSC* art. 221.

111 Churchill and Lowe, above n. 55, 344.

112 Faure and Song, above n. 85, 98.

113 *LOSC* art. 211(3).

inspection and certification by a port, whereas judicial jurisdiction of the port State involves the prosecution of offences committed in its ports or coastal State maritime zones, or outside the internal waters, territorial sea or EEZ of the port State.[114]

Under *MARPOL 73/78* the inspection and certification rules serve as the basis for the enforcement jurisdiction of port States. Different certificates are designed and required by the IMO as various standards to measure corresponding aspects of pollution from ships.[115] For example, the IEE Certificate is related to reducing GHG emissions from international shipping.[116] According to *MARPOL* Annex VI, such certificates will only be issued to ships which meet the Ship Energy Efficiency Management Plan (SEEMP), and for new ships, the Energy Efficiency Design Index (EEDI) is also mandatory.[117]

As for inspection, two points are notable. Firstly, inspection should be limited to verifying that there is a valid certificate on board. If there are clear grounds for believing that the condition of the ship, or its equipment, does not correspond substantially with the particulars of the certificate, or there is no valid certificate, the port State must ensure that the ship does not sail until it can proceed to sea without presenting an unreasonable threat of harm to the marine environment.[118] If inspection detects violation of *MARPOL 73/78* or its Annexes, the port State shall forward a report to the flag State so that the appropriate action may be taken.[119] Secondly, port States must apply *MARPOL 73/78* standards to all ships calling at their ports in that *MARPOL* adopts 'no more favourable treatment' with respect to the ships of non-Parties to the

114 *LOSC* art. 218; See also Bang, above n. 84, 119. The difference between port State control and port State jurisdiction, and regional Memoranda of Understanding (MOUs) on port State control are examined in Chapter 6.

115 But these certificates shall be issued or endorsed either by the Administration (the flag State) or any organization duly authorized by it. In every case, the Administration assumes full responsibility for the certificate. See, e.g., *MARPOL* Annex VI reg 6(5).

116 *MARPOL* 73/78 Annex VI (2011 amendments) reg 6(4)(5).

117 *MARPOL* 73/78 Annex VI (2011 amendments) regs 6,7. See also International Maritime Organization (IMO), *Mandatory Energy Efficiency Measures for International Shipping Adopted at IMO Environment Meeting* (15 July 2011) <http://www.imo.org/MediaCentre/PressBriefings/Pages/42-mepc-GHG.aspx> accessed 31 October 2011.

118 *MARPOL73/78* art. 5(2).

119 *MARPOL73/78* art. 6(5).

convention.[120] Furthermore, a type of Memorandum of Understanding (MOU) relating to port State inspection has been developed to coordinate regional port State control.[121] Port State MOUs were designed to ensure that ships do not evade *MARPOL 73/78* rules through calling at ports where the inspection regime is lax. To date there are nine MOUs on port State control which have covered most of the regions of the world.[122] If we relate the issue of GHG emissions from international shipping to port State jurisdiction, the port State will inspect the vessel to verify whether the IEE Certificate is on board and whether there is any violation of *MARPOL* Annex VI even though the flag State of the ship is not a party to the convention.

Article 218 of the *LOSC* gives port States a discretionary power to investigate and prosecute discharge violations wherever they have taken place.[123] Regarding violations within the coastal zones of another State,[124] the port State can only act by request from the State concerned.[125] As to violations on the high seas, the port State may prosecute directly and independently in the public interest.[126] In this sense, port State jurisdiction has been viewed as a kind of 'universal jurisdiction'.[127] Therefore, if there is any violation of IMO regulations, including those relating to GHG emissions from international shipping (amendments of Annex VI to *MARPOL 73/78* or any other regulations), the investigation or prosecution by the port State could provide a kind of correction. However, in practice port States seldom exercise their judicial jurisdiction

120 *MARPOL73/78* art. 5(4). This article reads that,

> 'With respect to the ship of non-Parties to the Convention, Parties shall apply the requirements of the present Convention as may be necessary to ensure that no more favourable treatment is given to such ships.'

121 Rothwell and Stephens, above n. 60, 354.

122 Ibid. These nine MOUs are Paris MOU, Latin American MOU, Tokyo MOU, Caribbean MOU, Mediterranean MOU, Indian Ocean MOU, West and Central African MOU, Black Sea MOU and Riyadh MOU.

123 *LOSC* art. 218(1).

124 *LOSC* art. 218(2). 'Coastal zones' refers to the internal waters, territorial sea or EEZ of another State.

125 Ibid. 'The States concerned' may be that State (violation occurs in its coastal zones), the flag State, a State damaged or threatened by the discharge violation, or the violation has caused or is likely to cause pollution in the coastal zones of the State instituting the proceedings.

126 Ibid. Although the port State's jurisdiction under this article is independent and no request from the flag State is necessary, the flag State does enjoy a right of pre-emption, which enables it to insist on taking control of any prosecution. See *LOSC* art. 228(1).

127 Birnie, Boyle and Redgwell, above n. 4, 422.

to prosecute on the ground that instituting legal proceedings can be costly.[128] Some States, Netherlands and South Korea as examples, have not prosecuted any foreign ship but only utilise administrative penalties such as detention or charging inspection fees.[129]

2.3 Environmental Liability for Transboundary Harm Caused by Greenhouse Gas Emissions from International Shipping

As observed by Sands, two opposite fundamental objectives guided the development of the rules of international environmental law, namely 'that states have sovereign rights over their natural resources' and 'that states must not cause damage to the environment'.[130] The latter objective involves environmental liability for transboundary harm since the 'environment' not only includes areas within national jurisdiction, but also transboundary contexts and areas beyond national jurisdiction.[131] As one of the central tenets of international environmental law, the rules of liability on transboundary harm have been widely applied and developed.[132] This part examines these rules and explores their application to the issue of GHG emissions from international shipping.

128 Bang, above n. 84, 126. Bang asserts that a very small number of *MARPOL* violations have been prosecuted by a few port States, but there is no evidence of such prosecution on illegal discharges.

129 Ibid. 127.

130 Philippe Sands, *Principles of International Environmental Law* (Cambridge University Press, 2nd ed., 2003) 235. These objectives are set out in Principle 21 of the Stockholm Declaration, which provides that:

'States have, in accordance with the Charter of the United Nations and the principles of international law, the sovereign right to exploit their own resources pursuant to their own environmental policies, and the responsibility to ensure that activities within their jurisdiction or control do not cause damage to the environment of other States or of areas beyond the limits of national jurisdiction.'

Stockholm Declaration on the Human Environment, 11 ILM 1416 (16 June 1972) principle 21 ('*Stockholm Declaration*').

131 Ibid.

132 Cases involving the rules of transboundary harm include but are not limited to: *Trail Smelter Case* (*United States of America v Canada*) (*Reports of International Arbitral Awards*) (1938 & 1941) 3 UN RIAA 1905; *Nuclear Tests Case* (*Australia v France*) (*Interim Protection*) (1974) ICJ Reports 253 ; *Lac Lanoux Arbitration* (*France v Spain*) (1957) 12 UN RIAA 285; *Corfu Channel Case* (*U.K. v Albania*) (1949) ICJ Rep. 4;*The MOX Plant Case* (*Ireland v United Kingdom*) (2001) 47 ILM 405; *ITLOS*, Order of 3 December 2001 on Provisional Measures; *Pulp Mills on the River Uruguay* (*Argentina v Uruguay*) (2006) ICJ Reports.

2.3.1 *An Overview of Transboundary Harm*

As a broader concept than transboundary pollution,[133] 'transboundary harm' generally refers to 'harm caused in the territory of or in other places under the jurisdiction or control of a State other than the State of origin, whether or not the States concerned share a common border'.[134] The areas damaged by transboundary harm may be either within a jurisdiction or beyond national jurisdiction. The general duty to prevent and to minimise the risk of transboundary harm is derived from the fundamental principle *sic utere tuo ut alienum non laedas* or 'principle of good neighbourliness'. It has been underpinned by State practice, judicial decisions, multilateral environmental agreements, and the work of the International Law Commission (ILC).[135] In particular, Principles 2, 18 and 19 of the *1992 Rio Declaration on Environment and Development* (*Rio Declaration*) provide specific principles applicable to transboundary harm and environmental risks.[136] Two transboundary harm rules could be drawn from these principles. They are:

> (1) States have a duty to prevent, reduce, and control transboundary pollution and environmental harm resulting from activities within their jurisdiction or control; and
> (2) States also have a duty to cooperate in mitigating transboundary environmental risks and emergencies, through notification, consultation, negotiation, and in appropriate cases, environmental impact assessment.[137]

---

133 Birnie, Boyle and Redgwell, above n. 4, 188. This difference can be seen clearly from Articles 1 and 2 of 1985 Vienna Convention for the Protection of the Ozone Layer (*Vienna Convention for the Protection of the Ozone Layer*, opened for signature 22 March 1985, 26 ILM 1529 (entered into force 22 September 1988)('*Vienna Ozone Convention*')) and 1992 *UNFCCC* on the definition of 'adverse effects', and Article 1 of the 1979 Convention on Long-Range Transboundary Air Pollution (*Convention on Long-Range Transboundary Air Pollution*, opened for signature 13 November 1979, 18 ILM 1442 (entered into force 16 March 1983)) ('*CLRTAP*') on the definition of 'pollution'.

134 *Draft Articles on Prevention of Transboundary Harm from Hazardous Activities*, ILC Report GAOR A/56/10 (2001) art. 2(c) ('*Draft Articles*'). Regarding this definition, 'State of origin' means the State in the territory or otherwise under the jurisdiction or control of which the activities referred to in article 1 are planned or are carried out; 'States concerned' means the State of origin and the State likely to be affected.

135 Birnie, Boyle and Redgwell, above n. 4, 137.

136 Ibid.

137 Ibid.

The two rules have been applied in many international judicial decisions.[138] The two rules have the status of customary international law, but in certain respects these rules can also be treated as general principles of law.[139] The rules on transboundary harm have been gradually formed and developed since the 1930s. This process, according to different elements reflected in the treaties or cases, can be summarised into three stages.

The first stage can be traced back to the *Trail Smelter* arbitration—the origins of a rule on transboundary harm and also the first international environmental law decision in the world.[140] The *Trail Smelter* dispute covered a period of 13 years from 1928 to 1941, and is usually the only case cited in which general principles of international law on State liability was applied to address issues involving transboundary damage.[141] One of the main contributions from the *Trail Smelter* arbitration could be summarised as the well-known '*Trail Smelter* principles',[142] which include: (1) each state has a duty to prevent transboundary harm;[143] and (2) the polluter-pays principle, which asserts that 'the polluting

---

138 For example, the first rule was applied in *Trail Smelter Case* and the *Corfu Channel Case. Trail Smelter Case*, below n. 140; *Corfu Channel Case* (*U.K. v Albania*) (1949) ICJ Rep. 4. The second rule was applied in *Lac Lanoux Arbitration case. Lac Lanoux Arbitration* (*France v Spain*) (1957) 12 UN RIAA 285.

139 Birnie, Boyle and Redgwell, above n. 4, 137. Customary international law and general principles of law are two sources of international law regulated by Article 38 of the *Statute of the International Court of Justice. Charter of the United Nations and Statute of the International Court of Justice*, signed 26 June 1945, 59 STAT 1031 (entered into force 24 October 1945). The two sources have different elements and implications. See Brownlie, above n. 48, 6–12, 16–18.

140 *Trail Smelter Case* (*United States of America v Canada*) (*Reports of International Arbitral Awards*) (1938 & 1941) 3 UN RIAA 1905 ('*Trail Smelter* (1941)'). The Trail Smelter arbitration of 1938 and 1941 was a landmark decision about a dispute over environmental degradation between the United States and Canada. A tribunal was set up by Canada and the United States to resolve a dispute over damages to US citizens and property in the State of Washington caused by a smelter on the Canadian side of the border. The tribunal decided that Canada had to pay the United States for damages, and further that it was obliged to abate the pollution. The second Trail Smelter dispute in 2003 on the contamination of the Upper Columbia River in Washington State of the United States by Trail Smelter in Canada, however, is not discussed in this chapter.

141 Hanqin Xue, *Transboundary Damage in International Law* (Cambridge University Press, 2003) 269.

142 Bratspies and Miller, above n. 32, 3.

143 *Trail Smelter* (1941), above n. 140. This rule was reflected in the famous conclusion made by the tribunal that,

'Under the principles of international law... no state has the right to use or permit the use of territory in such a manner as to cause injury by fumes in or to the territory of

State should pay compensation for the transboundary harm it has caused.'[144] These principles have been widely accepted as rules of customary international law and applied or cited by judges in some of the subsequent cases such as the *Lac Lanoux Arbitration* and the *Nuclear Tests Case*.[145]

Furthermore, the rules on transboundary harm are reflected in some treaties. For instance, the *1951 International Plant Protection Convention* recognised the need to prevent the introduction and spread of plant pests and diseases across national boundaries.[146] The *1963 Nuclear Test Ban Treaty* prohibits nuclear tests if the explosion would cause radioactive debris 'to be present outside the territory limits of the state under whose jurisdiction or control such explosion is conducted.'[147] The *1968 African Conservation Convention* provides that the States Parties shall cooperate 'whenever any national measure is likely

---

another or the properties or persons therein, when the case is of serious consequence and the injury is established by clear and convincing evidence.'

144 Bratspies and Miller, above n. 32. In this case the Tribunal's decision holds that a state should be strictly liable for damages arising from activities by a private corporation operating within the state's jurisdiction. See also ibid.

The *Trail Smelter case* can be deemed as the application of the polluter-pays principle, which from the author's point of view could be understood that the State (Canada) is actually also a 'polluter' in this case. According to a European Community Council recommendation of November 7, 1974, 'polluter' refers to 'someone who directly or indirectly damages the environment or who creates conditions leading to such damage'. The smelter was within the jurisdiction of the Canadian government and managed and taxed by the latter, so the Canadian government had the due diligence obligation to make it not produce transboundary air pollution. In this case, however, the Canadian government 'create[d] conditions leading to such damage'.

145 *Lac Lanoux Arbitration* (*France v Spain*) (1957) 12 UN RIAA 285. This case involves the proposed diversion of an international river by France (the upstream state), and the arbitral tribunal finally affirmed that a state (France) has an obligation not to exercise its rights to the extent of ignoring the rights of another (Spain, the downstream state).

Nuclear Tests Case (*Australia v France*) (*Interim Protection*) (1974) ICJ Reports 253. This case involves Australia's claims on the possible environmental damage from France's atmospheric nuclear tests in the Pacific. Judge de Castro stated: 'If it is admitted as a general rule that there is a right to demand prohibition of the emission by neighbouring properties of noxious fumes, the consequences must be drawn, by an obvious analogy, that the Applicant is entitled to ask the Court to uphold its claim that France should put an end to the deposit of radio-active fall-out on its territory.'

146 *International Plant Protection Convention*, opened for signature 6 December 1951, 150 UNTS 67 (entered into force 3 April 1952) preamble.

147 *Treaty Banning Nuclear Weapon Tests in the Atmosphere, in Outer Space and under Water*, opened for signature 5 August 1963, 480 UNTS 43 (entered into force 10 October 1963) art. I(1)(b).

to affect the natural resources of any other State.'[148] In summary, one common feature that both cases and treaties share is that they only deal with transboundary harm to other states.

The second stage commenced in 1972 when the *Stockholm Declaration* was adopted. The concept of transboundary harm during this period was expanded from mere relations between two States to relations which also include those between one State and global commons areas, namely the areas beyond the limits of national jurisdiction. Examples of global commons areas are the high seas and the airspace above them, the deep sea-bed, outer space, the Moon and other celestial bodies.[149] These changes were reflected in Principle 21 of *Stockholm Declaration*, Principle 2 of *Rio Declaration*, Article 3 of the *Convention on Biological Diversity* (CBD),[150] Preamble of the UNFCCC, and other conventions and UN documents.[151]

Among these international instruments, Principle 21 of the *Stockholm Declaration* has two relevant implications. Firstly, States have a due diligence obligation to regulate all public and private activities within their jurisdiction and control so as to prevent and control the transboundary harm to other States or areas outside the limits of their jurisdiction.[152] This affirmed and

---

148 *African Convention on the Conservation of Nature and Natural Resource*, opened for signature 15 September 1968, 1001 UNTS 4 (entered into force 9 October 1969) art. XVI(1)(b).

149 Birnie, Boyle and Redgwell, above n. 4, 145. Some relevant treaties include 1967 Outer Space Treaty (*Treaty on Principles Governing the Activities of States in the Exploration and Use of Outer Space, including the Moon and Other Celestial Bodies*, opened for signature 27 January 1967, 610 UNTS 205 (entered into force 10 October 1967)), 1979 Moon Treaty (*Agreement Governing the Activities of States on the Moon and Other Celestial Bodies*, opened for signature 5 December 1979, 18 ILM 1434 (entered into force 11 July 1984)), 1972 London Dumping Convention (*Convention on the Prevention of Marine Pollution by Dumping of Wastes and Other Matter*, opened for signature 29 December 1972, 18 ILM 510 (entered into force 30 August 1975)), and 1982 LOSC art. 145, 209.

150 *Convention on Biological Diversity*, opened for signature 5 June 1992, 31 ILM 818 (entered into force 29 December 1993) ('*CBD*').

151 For example, the United Nations General Assembly emphasised that State parties 'must not produce significant harmful effects in zones situated outside their national jurisdiction' during the course of the exploration, exploitation and development of their natural resources. See, *Cooperation between States in the Field of the Environment*, UNGA Res 2995 XXVII (15 December 1972).

152 Kiss and Shelton, above n. 1, 189. See also Birnie, Boyle and Redgwell, above n. 4, 147–150. Birnie asserts that 'due diligence' addresses two issues: one is that it 'requires the introduction of policies, legislation, and administrative controls applicable to public and private conduct which are capable of preventing or minimizing the risk of transboundary

improved the '*Trail Smelter* principles'. Secondly, States should apply the same rules not only within their jurisdiction (for example, land territory, territorial sea, continental shelf, and exclusive economic zone) but also to activities and processes under their control, such as 'ships, airplanes and spacecraft having the nationality of the State, missions to Antarctica, troops stationed in foreign territories, and any occupied or dependent territories'.[153] The incorporation of 'global commons areas' into the protection regime against transboundary harm by Principle 21 was a significant advancement which suggests that the State obligation to prevent, reduce and control environmental harm was 'no longer solely bilateral in character';[154] instead it benefits all humankind.

The third stage began with the adoption of *LOSC*. However, some of the subsequent treaties or UN documents still fall into the category of the second stage.[155] The significance of *LOSC* concerning the prevention of transboundary harm mainly lies in the shift of emphasis from a negative obligation to prevent transboundary harm to a positive commitment to preserve and protect the environment.[156] To that end, two changes were made. Firstly, *LOSC* transforms the 'responsibility' into a 'duty' under Article 193[157] which probably indicates more moral commitment whereas the 'responsibility', which was used in

---

harm to other States or the global environment'; the other is that it 'entails an evolving standard of technology and regulation' since internationally agreed 'ecostandards' can be easily detailed and precise, as in the annexes to *MARPOL* 73/78, and usually be easily updated, often using soft-law instruments or decisions of the parties.

153 Kiss and Shelton, above n. 1, 189–190.

154 Birnie, Boyle and Redgwell, above n. 4, 145.

155 For instance, essentially the 1992 *CBD* and the *Rio Declaration* adhere to the features on the prevention of transboundary harm reflected in the *Stockholm Declaration*.

156 Sands, above n. 130, 244; Birnie, Boyle and Redgwell, above n. 4, 151.

'Negative obligation' was reflected in previous treaties or regulatory documents where this state obligation was only mentioned and imposed by some limitations. For example, the second part of Rio Principle 2 only mentions the avoidance of the fact of harm, rather than the conduct of the state in bringing it about or failing to prevent it; and the first part of Rio Principle 2 probably indicates that economic activities outweigh the seriousness of the possible environmental harm.

157 *LOSC* art. 193. Article 193 reads that:

'States have the sovereign right to exploit their natural resources pursuant to their environmental policies and in accordance with their *duty* to protect and preserve the marine environment.'

But Sands asserts that 'it is unclear what was intended by the change'. Sands, above n. 130, 244.

previous treaties, is generally a condition of being responsible. Secondly, Article 194(2) utilises strong language to indicate such commitment. It provides that,

> States *shall* take *all measures necessary* to ensure that activities under their jurisdiction or control *are so conducted as not to cause damage* by pollution *to other States and their environment*, and that pollution arising from incidents or activities under their jurisdiction or control *does not spread beyond the areas* where they exercise sovereign rights in accordance with [the] Convention.[158] [emphasis added]

Following the LOSC, the rules on the prevention of transboundary harm continued to develop through subsequent treaties and the International Court of Justice (ICJ) rulings, such as the 1985 ASEAN Convention[159] and the ICJ's Advisory Opinion on the Legality of the Threat or Use of Nuclear Weapons.[160]

The liability regime on transboundary harm also developed with the evolution of rules on transboundary harm. The International Law Commission (ILC) has been working on transboundary environmental harm since 1978 with the 'improbable' title of 'Liability for Injurious Consequences of Acts Not Prohibited by International Law'.[161] In 1996 the ILC released draft articles and commentary,[162] which initially put forward the three-element damage

---

158 *LOSC* art. 194(2).

159 *Association of South Ease Asian Nations Agreement on the Conservation of Nature and Natural Resources*, opened for signature 9 July 1985, 15 EPL 64 (not yet in force) art. 20(1). Article 20(1) recognises the second element of Principle 21 of Stockholm Declaration and Principle 2 of Rio Declaration as a 'general accepted principle of international law'. It reads that,

'Contracting Parties have in accordance with generally accepted principles of international law the responsibility of ensuring that activities under their jurisdiction or control do not cause damage to the environment or the natural resources under the jurisdiction of other Contracting Parties or of areas beyond the limits of national jurisdiction.'

160 *Legality of the Threat or Use of Nuclear Weapons (Advisory Opinion)* (1996) ICJ Reports 241, para. 29. In this advisory opinion, the ICJ stated that:

'The existence of the general obligation of States to ensure that activities within their jurisdiction and control respect the environment of other States or of areas beyond national control is now part of the corpus of international law relating to the environment.' This statement could be interpreted as a confirmation of the role of the prevention of transboundary harm as a rule of customary international law.

161 Birnie, Boyle and Redgwell, above n. 4, 146; See also *II Yearbook International Law Commission* (1980) Pt1, 160, 138–139.

162 *Report of the Working Group on International Liability*, in International Law Commission Report (1996) GAOR A/51/10, Annex 1, 235.

structure made up of prevention, cooperation and strict liability.[163] However, more important contributions from the ILC are the *2001 Draft Articles on the Prevention of Transboundary Harm from Hazardous Activities* (*Draft Articles*) and the *2006 Draft Principles on the Allocation of Loss in the Case of Transboundary Harm Arising Out of Hazardous Activities* (*Draft Principles*).[164]

The *2001 Draft Articles* apply to all activities within the jurisdiction or control of States which involve a risk of causing significant transboundary harm,[165] and the 'risk' covers both unlikely but disastrous accidents and highly probable but smaller scale harm.[166] Since the harm or the risk of harm has to be 'significant', a thorough determination of 'significance' is thus important.

The *2006 Draft Principles* are basically an international standard of liability involving both compensation for damage and the procedures and remedies. As the core principle, Principle 6(1) sets out the objective of prompt, adequate and effective compensation by means of competent judicial and administrative bodies of the State.[167] Concerning the allocation of loss arrangement

---

163 *Report of the Working Group on International Liability*, annex I, C, Ch I, art. 4 (Prevention), art. 5 (Liability), and art. 6 (Cooperation).

Article 4 reads that, 'States shall take all appropriate measures to prevent or minimize the risk of significant transboundary harm and, if such harm has occurred, to minimize its effects.' This article, together with article 6, provides the basic foundation for the articles on prevention. The obligation of States to take preventive or minimization measures is one of due diligence.

Article 5 stipulates that, 'In accordance with the present articles, liability arises from significant transboundary harm caused by an activity referred to in article 1 and shall give rise to compensation or other relief.' This principle of liability and reparation is a necessary corollary and complement to article 4. That article obliges States to prevent or minimize the risk from activities that are not prohibited by international law. Article 5, on the other hand, establishes an obligation to provide compensation or other relief whenever significant transboundary harm occurs.

Article 6 specifies that, 'States concerned shall cooperate in good faith and as necessary seek the assistance of any international organization in preventing or minimizing the risk of significant transboundary harm and, if such harm has occurred, in minimizing its effects both in affected States and in States of origin.' This kind of all-round cooperation is essential in designing and implementing effective policies to prevent or minimize the risk of causing significant transboundary harm.

164 *Draft Principles on the Allocation of Loss in the Case of Transboundary Harm Arising Out of Hazardous Activities*, Annex, UN Doc A/RES/61/36 (18 December 2006) ('*Draft Principles*').

165 *Draft Articles* art. 1.

166 *Draft Articles* art. 2(a).

167 *Draft Principles* art. 6(1). Art 6(1) reads that:

'States shall provide their domestic judicial and administrative bodies with the necessary jurisdiction and competence and ensure that these bodies have prompt, adequate and effective remedies available in the event of transboundary damage caused by

in the *2006 Draft Principles*, three features are notable. Firstly, it adopts strict liability[168] in treaties and in national law instead of proof of fault.[169] The ILC commentary indicates that this choice has taken the inherent risks of hazardous activities into consideration and that it would be unjust and inappropriate to apply proof of fault once accidents occur.[170] Secondly, it imposes liability for damage on the operator and/or other person or entity,[171] which is drawn from the existing civil liability and compensation schemes and seems more flexible.[172] Thirdly, it offers more options for supplementary compensation from the industry and/or State in case the financial resources of the operator are insufficient to cover the damage suffered due to an incident.[173]

In addition to *MARPOL 73/78*, the *1999 Basel Liability Protocol*, the *2001 Bunker Convention*, and above treaties, some other treaties on specific areas, such as oil pollution from ships, have been adopted to cope with possible damage from transboundary harm. Typical examples are the 1969/1971 regime (*1969 CLC*,[174]

---

hazardous activities located within their territory or otherwise under their jurisdiction or control.'

168 'Strict liability' means that a person is liable for any harm he causes even if he is not at fault or negligent. The rationale is that it is technically difficult to prove fault for the victims in the environmental context due to the complex process of tracing the formation of harm. This approach aims to better provide compensation for victims and reduce potential harm. David Weisbach, 'Negligence, Strict Liability, and Responsibility for Climate Change' (2012) 97(2) *Iowa Law Review* 521, 554–555.

169 *Draft Principles* art. 4(2). See also Birnie, Boyle and Redgwell, above n. 4, 320.

170 *International Law Commission Report* (2004) Commentary to Principle 4, 15–17.

171 *Draft Principles* art. 4(2)(3).

172 Take the *2001 Bunker Convention* and the *1999 Basel Liability Protocol* as examples: under the *2001 Bunker Convention*, the shipowner, charterer, manager and operator are jointly and severally liable; while the *1999 Basel Liability Protocol* makes the generators, exporters, importers and disposers all potentially liable at different stages of the wastes' journey to its eventual destination. See *IMO International Convention on Civil Liability for Bunker Oil Pollution Damage*, opened for signature 27 March 2001, 40 ILM 1493 (entered into force 21 November 2008) ('*2001 Bunker Convention*') art. 3,7; *Protocol on Liability and Compensation for Damage Resulting from Transboundary Movements of Hazardous Wastes and Their Disposal*, opened for signature 10 December 1999, EMuT 989:22/B (not yet in force) art. 4,5,9('*1999 Basel Liability Protocol*'). See also Birnie, Boyle and Redgwell, above n. 4, 320.

173 *Draft Principles* art. 7.

174 *International Convention on Civil Liability for Oil Pollution Damage*, opened for signature 29 November 1969, 973 UNTS 3 (entered into force 19 June 1975) ('*1969 CLC*').

and the *1971 Fund Convention*[175]), the 1992 regime (the *1992 CLC*,[176] and the *1992 Fund Convention*[177]), and the *1996 HNS Convention*,[178] which will be further discussed in the polluter-pays principle section of this chapter.

2.3.2 *The Application of Transboundary Harm Rules to the Issue of Greenhouse Gas Emissions from International Shipping*

As already noted, GHG emissions from international shipping can be regarded as a type of 'conditional' pollution. Given that harm is a broader term than pollution, are GHG emissions from international shipping a kind of transboundary harm? And if so, how can the rules on the prevention of transboundary harm apply to this GHG emissions issue? This section provides an analysis of these issues.

First of all, can GHG emissions from international shipping produce transboundary harm? To address this question, we need to examine the terms 'harm' and 'transboundary' respectively.

'Harm' means adverse effects caused to persons, property or the environment.[179] As illustrated in Chapter 1, GHG emissions from international shipping may result in many deleterious effects on atmospheric composition, marine ecosystems, human health and climate. These effects, however, may not be considered 'serious' or 'substantial' due to the cumulative nature of GHG emissions. Some serious consequences, such as the inundation of some islands as a result of sea level rise,[180] or the extinction of some marine species due to ocean acidification, may be caused by many factors over quite a long term where GHG emissions from international shipping are only part of the

175 *International Convention on the Establishment of an International Fund for Compensation for Oil Pollution Damage*, opened for signature 18 December 1971, 11 ILM 284 (entered into force 16 October 1978) ('*1971 Fund Convention*').

176 *International Convention on Civil Liability for Oil Pollution Damage*, opened for signature 27 November 1992, IMO LEG/CONF.9.15 (entered into force 30 March 1996) ('*1992 CLC*').

177 *International Convention on the Establishment of an International Fund for Compensation for Oil Pollution Damage*, opened for signature 27 November 1992, 87 UKTSCm 3433 (entered into force 30 May 1996) ('*1992 Fund Convention*').

178 *International Convention on Liability and Compensation for Damage in Connection with the Carriage of Hazardous and Noxious Substances by Sea*, opened for signature 3 May 1996, 25 ILM 1406 (not yet in force) ('*1996 HNS Convention*').

179 *Draft Articles* art. 2(b).

180 For example, in late 2005 an entire coastal village in the north of Tegua Island in Vanuatu was relocated to higher ground; in 2009 the Carteret Islanders of Papua New Guinea became the world's first entire community to be displaced by climate change. See ch. 1, 1.1.2.

cause and only add to the quantum of harm. On the other hand, the harm could be 'more than detectable' in a comparatively short period. For instance, excessive GHG emissions from international shipping may contribute to shifting ranges and distribution of species which will have direct impacts on fish stocks and can be easily noticed. Researchers have found that in harbour cities ship emissions, including GHG emissions and other emissions, are often a dominant source of urban pollution.[181] Therefore, as discussed earlier, the harm resulting from GHG emissions from international shipping may be 'more than detectable', but whether it is 'serious' or 'substantial' needs to be examined on a case-by-case basis. This element, based on the requirements from the *2001 Draft Articles* and other international instruments, meets the threshold for 'significant' harm under certain circumstances.

The harm caused by GHG emissions from international shipping may also be 'transboundary'. The definition of 'transboundary harm' in the *2001 Draft Articles* indicates that the 'boundary' refers to territorial boundaries, jurisdictional boundaries or control boundaries of the State.[182] In the case of international shipping, the 'State of origin' is generally the flag State of the ship,[183] and the 'State likely to be affected' may be the coastal State, port State or a third State.[184] Since ships always sail between ports of different countries in the context of international shipping,[185] the harms caused by GHG emissions from international shipping to other areas may fall into four main scenarios.

The first scenario is that GHG emissions from international shipping cause significant harm to the high seas and the deep seabed, or the international airspace which is the airspace above the high seas and exclusive economic

181 Veronika Eyring et al., 'Transport Impacts on Atmosphere and Climate: Shipping' (2010) 44(37) *Atmospheric Environment* 4735, 4753.

182 *Draft Articles* art. 2(c). This article stipulates that,

'Transboundary harm' means harm caused in the *territory* of or in other places under the *jurisdiction* or *control* of a State other than the State of origin, whether or not the States concerned share a common border.

This sentence is a summary for the above article, stressing that there are actually three types of 'boundaries'.

183 *Draft Articles* art. 2(d). This article reads:

'State of origin' means the State in the territory or otherwise under the jurisdiction or control of which the activities referred to in article 1 are planned or are carried out.

184 *Draft Articles* art. 2(e). This article provides that:

'State likely to be affected' means the State or States in the territory of which there is the risk of significant transboundary harm or which have jurisdiction or control over any other place where there is such as risk.

185 See ø. Buhaug et al., 'Second IMO GHG Study 2009' (International Maritime Organization (IMO), 2009) 13.

zones (EEZ) of a coastal State. In this case, the harms occur between the flag State of the ship and the global commons areas, or in other words, between one national jurisdiction and the areas beyond the limits of national jurisdiction. The second scenario is that GHG emissions from international shipping impose adverse effects on the EEZ, territorial sea (or archipelagic waters), continental shelf, the land territory, the atmosphere above the land and territorial sea (or archipelagic waters) of a State. In these circumstances, the harms are inflicted to the coastal States or the port States (or the archipelagic States) by the flag State of the ship, and this scenario involves harms between two national jurisdictions. The third scenario is that GHG emissions from international shipping result in substantial harm to a third State, such as its land territory or territorial sea or airspace under its sovereignty, via the territorial sea or land of another coastal State or port State (a second State). In this scenario, the coastal State or port State (the second State) has jurisdiction over GHG emissions from ships of the flag State although they are not under an obligation to assume such jurisdiction.[186] In this case, the flag State will be responsible for the harm it causes to the third State directly. Clearly this scenario also involves harm between two national jurisdictions. In addition, there is another possibility, or the fourth scenario. GHG emissions from one ship produce harm to another ship or platform of another State (irrespective of whether they are on the high seas or anywhere on the sea) during their international voyages. This scenario involves harm between two national jurisdictions (two flag States). However, it rarely occurs in practice.

It is clear that all four scenarios fall into the category of harm between two national jurisdictions (Scenario two, three and four) or from one national jurisdiction to the areas beyond national jurisdiction (Scenario one). This observation underpins the transboundary nature of GHG emissions from international shipping, which was discussed earlier in this section.

The above discussion of the terms 'harm' and 'transboundary' in the context of GHG emissions from international shipping make possible the application of the rules on the prevention of transboundary harm to this specific issue. However, how can the rules on the prevention of transboundary harm apply

---

186 *LOSC* art. 211(4). This article stipulates that,

'Coastal States *may*, in the exercise of their sovereignty within their territory sea, adopt laws and regulations for the prevention, reduction and control of marine *pollution* from foreign vessels, including vessels exercising the right of innocent passage…'

Since GHG emissions from international shipping can be treated as a type of 'conditional' pollution as discussed at the beginning of this chapter, this article applies to this scenario.

to this GHG emissions issue? A brief examination of the four scenarios in relation to possible transboundary harm is now provided to justify the application of the two transboundary harm rules to GHG emissions from international shipping.

In the first scenario, harm occurs between one national jurisdiction and the areas beyond national jurisdiction. As discussed earlier, the flag State of the ship has primary jurisdiction when GHG emissions from international shipping cause significant harm to the high seas or international airspace.[187] Coastal States are not permitted to take action unless they are threatened by the damage (transboundary harm),[188] which generally does not apply in this scenario. In this case, in accordance with the two rules on transboundary harm, it is mainly the flag State that has a duty to prevent, reduce and control the possible transboundary harm from the GHG emissions of its ship. To achieve this goal, the flag State is required to adopt national laws on the reduction of shipping GHG emissions, which should at least have the same effect as that of generally accepted international rules and standards established through the IMO,[189] taking into account the amended Annex VI to *MARPOL 73/78* irrespective of whether they have ratified the amendments.[190] When a violation occurs, the flag State shall impose administrative penalties or institute proceedings in relation to such offences. Moreover, flag States shall conduct

---

187 However, the *1982 LOSC* also empowers the port State enforcement jurisdiction with respect to violations committed beyond its territorial sea (including the global commons areas) by a ship flying a foreign flag, where the flag State may be reluctant to do so, and/or where the coastal State is unable or incompetent to act. See *LOSC* art. 218.

188 *LOSC* art. 221.

189 *LOSC* art. 211(2). Currently there are no specific generally-accepted IMO rules relating GHG emissions from international shipping.

190 See *LOSC* art. 212(1). This provision provides that,

'States shall adopt laws and regulations to prevent, reduce and control pollution of the marine environment from or through the atmosphere, applicable to the air space under their sovereignty and to vessels flying their flag or vessels or aircraft of their registry, taking into account internationally agreed rules, standards and recommended practices and procedures and the safety of air navigation.'

When applying this provision to the adoption of amended Annex VI to *MARPOL 73/78*, it can be argued that this amendment is 'internationally agreed' rather than 'generally accepted' as indicated in Article 211(2) of the LOSC. This is because the amendment was adopted by a majority vote within the IMO in July 2011 when some large shipping nations, such as China, India, Brazil, Chile, Kuwait and Saudi Arabia, voted against this adoption. This lack of consensus makes the amendment less likely to be a 'generally accepted' rule, but the broad participation of many countries still enable it to be an 'internationally agreed' amendment. See Harrison, above n. 77, 21–23.

regular surveys, issue or empower other parties to issue the IEE Certificate to ships flying their flags. Another aspect is the duty of cooperation by States concerned. When informed of suspected violations of *MARPOL 73/78* in areas beyond national jurisdiction, the flag State is obliged to cooperate with relevant parties, such as port States or sometimes coastal States, in detecting, inspecting or investigating such violations.

The second scenario deals with the harm between the flag State and a coastal State or port State. In accordance with the two rules on transboundary harm, two requirements apply to both parties. First of all, the flag State has the prescriptive and enforcement jurisdiction and responsibility to prevent, reduce and control transboundary harm resulting from GHG emissions from the ship entitled to fly its flag. This duty is exactly the same as that in the first scenario. Meanwhile, the coastal State may adopt national laws on the reduction of GHG emissions from international shipping applicable to either its territorial sea or its EEZ. However, such legislation should not hamper the right of innocent passage of foreign vessels in its territorial sea and should be consistent with Annex VI to *MARPOL 73/78* in its EEZ if the State is a party to Annex VI. In their internal waters and ports, both the coastal State and the port State are free to make and enforce national laws dealing with such emissions. More significantly, port States may investigate and prosecute discharge violations wherever they have taken place.[191] In the second place, once damage or risk of damage occurs, the States concerned, including the coastal State, port State, or other parties, shall notify the vessel or the flag State of the violation of Annex VI to *MARPOL 73/78* or the *LOSC* if applicable, so that measures can be taken to reduce the possible loss.

The third scenario involves harm between the flag State and a third State. The discussion for the second scenario applies if the third State is also a coastal State or a port State when the GHG emissions from international shipping are transferred to its territorial sea or other maritime zones via another coastal State or port State (the second State). However, if the third State is a landlocked State and the harm is caused to its land or the atmosphere above its land, the duties of the flag State are still the same as those in the first and second scenarios while the rights and obligations of the third State are very similar to those of the United States in the *Trail Smelter* case.

The fourth scenario involves harm between two flag States. When such harm occurs, it is mainly the two flag States that deal with the issue. The prescriptive and enforcement jurisdiction and obligations of the flag State illustrated in the first scenario will then apply.

---

191 *LOSC* art. 218(1).

Another relevant issue is the allocation of liabilities for the transboundary harm caused by GHG emissions from international shipping. Nowadays it is common that the vessel is registered in one State and managed by an operating company registered in another State, the crew is multinational and the beneficial owner is in another State. In these circumstances, when GHG emissions from international shipping cause transboundary harm, more jurisdictions will be involved besides the 'State of origin' and the 'State likely to be affected'. This issue will be further discussed in the section on the polluter-pays principle of this chapter.

2.4 The Precautionary Principle and Its Application to the Issue of Greenhouse Gas Emissions from International Shipping

The precautionary principle has emerged in environmental law and policy in response to the need for an effective method to deal with risks and uncertainties associated with activities with the potential to affect the environment. It has been generally accepted as a 'basic rule' that governs activities especially those affecting the ocean environment.[192] This principle prescribes a general rule which has been translated into specific responsibilities for the proponents of certain maritime activities to meet before they are undertaken, so as to mitigate the adverse effects of these activities on the marine environment.[193] This part first examines the concept of the precautionary principle from the perspectives of its evolution, legal status and implementation. Then, it explores the application of this principle to the issue of GHG emissions from international shipping.

2.4.1 *An Overview of the Precautionary Principle*

The precautionary principle, also called the precautionary approach or precautionary measures,[194] is closely related to other international environmental law principles including the obligation to prevent transboundary harm, the preventive principle and the polluter-pays principle. Firstly, the duty to

192 J.M. Van Dyke, 'Applying the Precautionary Principle to Ocean Shipments of Radioactive Materials' (1996) 27(4) *Ocean Development and International Law* 379, 379.

193 Ibid.

194 The term 'precautionary approach' is preferred by the US and many global agreements adopt it or 'precautionary measures', while the 'precautionary principle' is favoured by European treaties and European Community law. However, these differences are often regarded as less significant. See Birnie, Boyle and Redgwell, above n. 4, 155.

prevent transboundary harm requires that each State has a duty to prevent significant harm to other States or to common spaces. In other words, the State has the obligation of 'diligent prevention and control of foreseeable risks', which to some extent justifies the adoption of a precautionary approach.[195] Secondly, the polluter-pays principle, which is analysed in the following section, is closely associated with the precautionary principle.[196] The precautionary principle imposes an environmental duty of care to prevent potential harm through seeking 'collective environmental responsibility'.[197] Similarly, under the polluter-pays principle, not only present polluters but also potential polluters are responsible for their actions. It appears that both principles adopt a 'forward-looking approach'.[198] Finally, the precautionary principle has been developed on the basis of the preventive principle. However, prevention aims to avoid an 'identifiable threat' which has been scientifically proven, whereas precaution underscores avoiding 'uncertain outcomes which may, or may not, be harmful'.[199] In general, the polluter-pays and preventive principles deal with known situations and risks while the precautionary principle addresses the scientific uncertainty of issues.[200]

The precautionary principle emerged from the early concept of *vorsorge* (foresight, taking care) adopted by the former West Germany in its environmental management in the 1960s. It evolved into the *vorsorgeprinzip* (precautionary or foresight principle) to resolve the environmental issues faced by Germany and other European countries in the 1970s.[201] The principle was first employed internationally in the 1984 Conference on Protection of the North Sea.[202] It was later endorsed by the *1987 Montreal Protocol* and the 1990 *Bergen*

195 Ibid. 153.

196 Warwick Gullett, 'Environmental Protection and the Precautionary Principle: a Response to Scientific Uncertainty in Environmental Management' (1997) 14(1) *Environmental and Planning Law Journal* 52, 55.

197 Ibid. 54.

198 Minna Pyhålå, Anne Brusendorff and Hanna Paulomåki, 'The Precautionary Principle' in Malgosia Fitzmaurice, David M. Ong and Panos Merkouris (eds), *Research Handbook on International Environmental Law* (2010) 203, 204.

199 Warwick Gullett, 'The Precautionary Principle in Australia: Policy, Law and Potential Precautionary EIAs' (2000) 11(2) *Risk: Health, Safety & Environment* 93, 98.

200 Pyhålå, Brusendorff and Paulomåki, above n. 198, 205.

201 Elena McCarthy, 'Ocean Noise, Scientific Uncertainty, and the Paradox of the Precautionary Principle' (2007) 10(3) *Journal of International Wildlife Law & Policy* 233, 233; Pyhålå, Brusendorff and Paulomåki, above n. 198, 205.

202 McCarthy, above n. 201, 233; Birnie, Boyle and Redgwell, above n. 4, 154. The adoption of the 1984 Bremen Ministerial Declaration and the 1987 London Declaration of the First

Ministerial Declaration on Sustainable Development.[203] Since then, the precautionary principle has been incorporated into 'almost all' international agreements and declarations related to environmental protection.[204] These include the *1992 Rio Declaration, Agenda 21, 1992* UNFCCC, *1992* CBD, and *1996 Protocol to London Dumping Convention.*[205] Furthermore, many States, including both developed and developing States, have incorporated the precautionary principle in their domestic environmental policy and law.[206]

---

and Second International Conferences on the Protection of the North Sea indicates the early utilisation of the precautionary principle. Specifically, article D3 of the 1984 Bremen Ministerial Declaration says that, 'Precautionary measures for air quality control by reduction of emissions at source should also be determined for the protection of the North Sea, based on the best available technology'. Article 7 of the 1987 London Declaration purports that, 'in order to protect the North Sea from possibly damaging effects of the most dangerous substances, a precautionary approach is necessary which may require action to control inputs of such substances even before a causal link has been established by absolutely clear scientific evidence'.

203 *Montreal Protocol on Substances that Deplete the Ozone Layer*, opened for signature 16 September 1987, 26 ILM 1550 (entered into force 1 January 1989) preamble ('*Montreal Protocol*').

It states that, '[parties to the Montreal Protocol] determined to protect the ozone layer by taking precautionary measures to control equitably total global emissions of substances that deplete it, with the ultimate objective of their elimination on the basis of developments in scientific knowledge, taking into account technical and economic considerations'.

Bergen Ministerial Declaration on Sustainable Development, 20 EPL 200 (15 May 1990) principle 7.

It advocates that, 'in order to achieve sustainable development, policies must be based on the precautionary principle. Environmental measures must anticipate, prevent and attack the causes of environmental degradation. Where there are threats of serious or irreversible damage, lack of full scientific certainty should not be used as a reason for postponing measures to prevent environmental degradation'.

Kiss and Shelton note that the 1990 Bergen Ministerial Declaration on Sustainable Development was 'the first international instrument to treat the principle as one of general application and linked to sustainable development'.

See Kiss and Shelton, above n. 1, 269–207.

204 Kiss and Shelton, above n. 1, 207.

205 *Rio Declaration* prin 15; *Agenda 21* ch. 17, para. 17.1; UNFCCC art. 3.3; CBD preamble; *1996 Protocol to London Dumping Convention* art. 3(1).

206 For example, in Australia the precautionary principle has been explicitly included in many fisheries laws and employed in a number of cases. Examples are the *Fisheries Management Act 1991* (Cth), *Dixon v Australian Fisheries Management Authority* (AFMA) [2000], *Arno Blank v AFMA* [2000], and *Latitude Fisheries Pty Ltd and Anor v AFMA* [2000].

There have been a number of discussions on the concept of precautionary principle and how it should be interpreted and implemented.[207] Arguably Principle 15 of the *Rio Declaration* is the most accepted formulation of the precautionary principle.[208] Principle 15 stipulates that:

> Where there are threats of serious or irreversible damage, lack of full scientific certainty shall not be used as a reason for postponing cost-effective measures to prevent environmental degradation.[209]

This formulation provides constraints or guidance on the decision-making process relating to a proposed action, and can be regarded as an advance when compared to the traditional preventive principle.[210] As such, Fisher

---

See Warwick Gullett, Chris Paterson and Elizabeth Fisher, 'Substantive Precautionary Decision-Making: the Australian Fisheries Management Authority's "Lawful Pursuit" of the Precautionary Principle' (2001) 7(2) *The Australian Journal of Natural Resources Law and Policy* 95, 110–114. In India and Pakistan, the precautionary principle is treated as a principle of international law and invoked by their Supreme Courts. Birnie, Boyle and Redgwell, above n. 4, 159.

207 See, e.g., Timothy O'Riordan and James Cameron, *Interpreting the Precautionary Principle* (Earthscan Publications, 1994); Julian Morris, *Rethinking Risk and the Precautionary Principle* (Butterworth-Heinemann, 2000); Poul Harremoës, *The Precautionary Principle in the 20th Century: Late Lessons from Early Warnings* (Earthscan Publications, 2002); Simon Marr, *The Precautionary Principle in the Law of the Sea: Modern Decision Making in International Law* (Martinus Nijhoff, 2003); Barney Dickson, Rosie Cooney and Ebscohost, *Biodiversity and the Precautionary Principle: Risk Uncertainty and Practice in Conservation and Sustainable Use* (Earthscan, 2005); Jacqueline Peel, *The Precautionary Principle in Practice: Environmental Decision-making and Scientific Uncertainty* (Federation Press, 2005).

208 Gullett, above n. 196.

209 *Rio Declaration* pin 15.

210 But this approach has also been criticised for its weakness, and is treated as 'deliberation-guiding' rather than 'action-guiding'. It is argued that the 'deliberation-guiding' approach as adopted in Principle 15 of the Rio Declaration is 'less stringent' than the 'action-guiding' approach as adopted by the 1989 report of the Nordic Council's International Conference on the Pollution of the Seas. This report reads that,

'The need for an effective precautionary approach, with that important principle intended to safeguard the marine ecosystem by, amongst other things, eliminating and preventing pollution emissions where there is reason to believe that damage or harmful effects are likely to be caused, even where there is inadequate or inconclusive scientific evidence to prove a causal link between emissions and effects.'

Dickson classifies the formulations of the precautionary principle into two versions: one is 'action-guiding' version of the principle calling for action against the practice that

treats the precautionary principle as 'a legal principle that is concerned with decision-making under scientific uncertainty in the context of risk regulation'.[211] Fisher asserts that risk regulation consists of such three activities as standard setting, the application of those standards, and enforcement, while the precautionary principle mainly involves standard setting.[212] While Fisher underscores the standard setting stage of the decision-making process, Gullett pays more attention to the outcome of applying the precautionary principle. Gullett takes the view that this principle should 'at minimum' be interpreted as 'requiring the adoption of sound environmental practices and the reduction of emissions of pollutants at source'.[213] The application of this principle 'normally involves accepting a known risk of environmental harm to guard against an uncertain environmental outcome'.[214]

The precautionary principle has received widespread support theoretically and practically although it has been subject to significant and sustained criticism for its subjective criteria or vague wording.[215] It is arguable that the precautionary principle can be an important environmental law principle and a rule of customary international law.[216] In practice, this principle

---

may cause damage; the other is 'deliberation-guiding' version which stipulates that lack of evidence shall not be used as a reason for postponing action against potentially harmful activities.

Barnabas Dickson, 'The Precautionary Principle in CITES: A Critical Assessment' (1999) 39(2) *Natural Resources Journal* 211, 213–214; *Nordic Council's International Conference on Pollution of the Seas*, Greenpeace 18, Annex 2 (1990) 27.

211 Elizabeth Fisher, "Precaution, Precaution Everywhere: Developing a Common Understanding of the Precautionary Principle in the European Community" (2002) 9(1) *Maastricht journal of European and comparative law* 21, 9.

212 Elizabeth Fisher, 'Is the Precautionary Principle Justiciable?' (2001) 13(3) *Journal of Environmental Law* 315, 317.

213 Gullett, above n. 196, 58.

214 Ibid.

215 See, e.g., Marr, above n. 207, 21. Marr asserts that the precautionary principle relies heavily on subjective criteria to trigger environmental action; Morris, above n. 207, 7–15. Morris claims that problems of this principle include the fallacy that the merest possibility of catastrophe should justify action, the precautionary principle is unnecessary, demands for a reversal of the burden of proof are disingenuous, the standard of proof is infinitely high, a duty to take action to prevent harm would be too broad, examining the full range of alternatives would be infinitely costly, and so on.

216 See, e.g., Sands, above n. 130, 279; Pyhälä, Brusendorff and Paulomäki, above n. 198, 210. Sands asserts that current State practice supports the view that the precautionary principle 'reflects a principle of customary law', and Pyhälä, Brusendorff and Paulomäki argue that this principle can be considered as a principle of customary international law 'at

has been invoked by many cases in the International Court of Justice (ICJ),[217] the International Tribunal for the Law of the Seas (ITLOS),[218] and the World Trade Organization (WTO).[219] However, these international bodies have not taken an explicit position as to whether the precautionary principle is a binding principle of customary international law although some judges have referred to it in individual judgments.[220] For example, in the 1999 *Southern Bluefin Tuna (Australia and New Zealand v. Japan)* case,[221] Judge Shearer commented that, 'the measures ordered by the Tribunal are rightly based upon considerations deriving from a precautionary approach'.[222] As noted earlier, this principle has also been incorporated in the statute law and case law of many countries.

Regarding the structure of the precautionary principle, the following four basic elements are generally found in most formulations.[223] They are: a threat of harm, a lack of scientific certainty or evidence, no proved causation between cause and effect, and the existence of the duty to act.[224] Then, in cases where there is reason to believe harm may occur but it cannot be proven scientifically, what precautionary responses or which precautionary measures should be taken by policy makers? Based on current research, the following three steps might be employed in applying the precautionary principle.

---

least from a regional perspective'. But Birnie, Boyle and Redgwell purport that it is 'far from evident' that the precautionary principle has the normative character of a rule of law. Birnie, Boyle and Redgwell, above n. 4, 161.

217 Examples are the 1995 Nuclear Tests case and the 1997 Gabcikovo-Nagymaros Case. See *Request for an Examination of the Situation in Accodance with Paragraph 63 of the Court's Judgment of 20 December 1974 in the Nuclear Tests (New Zealand v France)* (1995) ICJ Reports 288; *Gabcikovo-Nagymaros Case (Hungary v Slovakia)* (1997) ICJ Reports 7.

218 Examples are the 1999 Southern Bluefin Tuna case, and the 2001 MOX Plant Case. See *Southern Bluefin Tuna Case (Australia & New Zealand v Japan)* (1999) Case Nos. 3 and 4 ITLOS; *The MOX Plant Case (Ireland v United Kingdom)* (2001) 47 ILM 405; ITLOS, Order of 3 December 2001 on Provisional Measures.

219 Example is the 1988 Hormones case involving beef hormones and genetically modified organisms which has been settled by the WTO.

220 Pyhålå, Brusendorff and Paulomåki, above n. 198, 208.

221 *Southern Bluefin Tuna Case (Australia & New Zealand v Japan)* (1999) Case Nos. 3 and 4 ITLOS.

222 See Pyhålå, Brusendorff and Paulomåki, above n. 198, 222.

223 Peter L. deFur and Michelle Kaszuba, 'Implementing the Precautionary Principle' (2002) 288(1–2) *The Science of The Total Environment* 155, 157. L.deFur and Kaszuba used 7 cases to illustrate the 4-element structure of the precautionary principle, especially the possible situations for 'uncertainty'.

224 Ibid.

The first step is to identify current options for precautionary measures, or tools for incorporating the precautionary principle. This principle calls for a response in the face of scientific uncertainty. However, the selection of appropriate precautionary measures should take into account the differing ecological, cultural, political and economic interests and conditions of different countries.[225] Furthermore, preventative measures should be taken so as to fulfil the purpose of the precautionary principle.[226] These preventive measures and tools may include research, an Environmental Impact Assessment (EIA), alternative assessment,[227] ecosystem management, and mitigating all reasonably foreseeable damage.[228] Of these measures, EIA has been given a key role in that the precautionary principle can be integrated in the EIA process.[229]

The second step is to locate appropriate precautionary responses or measures. One unifying feature of the precautionary principle is its reversal of the burden of proof. There can be a range of precautionary responses based on differing requirements for the burden of proof. Traditionally the opponents of an activity will be permitted to conduct the activity unless there is proof of likely and unacceptable harm, whereas the precautionary principle requires the opponents of an activity to prove that the proposed activity will not adversely affect the environment before they are permitted to proceed.[230] As such, Gullett put forward four operational approaches to implement precaution, which range from strongest precautionary strength to weakest precautionary strength as follows:

- Completely reverse the burden of proof to require the proponent to meet a high evidentiary standard pointing to harmlessness before the activity—or modified activity—may be permitted;

---

225 Pyhålå, Brusendorff and Paulomåki, above n. 198, 217.

226 Ibid.

227 A definition of alternative assessment is provided in the next section. See below n. 235.

228 See, e.g., Van Dyke, above n. 192, 381–383; Anne Steinemann, 'Improving Alternatives for Environmental Impact Assessment' (2001) 21(1) *Environmental Impact Assessment Review* 3, 4–10.

229 See, e.g., Warwick Gullett, 'Environmental Impact Assessment and the Precautionary Principle: Legislating Caution in Environmental Protection' (1998) 5(3) (Sept 1998) *Australian Journal of Environmental Management* 146, 148–154; Gullett, above n. 199, 116–123. Gullett asserts that the precautionary principle can be integrated in the EIA process through three steps, namely threshold for operation of EIA, content of EIA, and substantive influence on decision-making.

230 Gullett, above n. 196, 59; Sands, above n. 130, 273.

- Approve the activity, contingent on a low "acceptability" level of uncertainty (determined in a manner similar to cost-benefit analyses or risk assessments);
- Approve the activity but require the proponent to use best available technology (BAT) or best available technology not entailing excessive cost (BATNEEC) and conduct stringent post-decision monitoring;
- Apply precautionary measures pursuant to the doctrine of "no regrets".[231]

These four approaches may be utilised to provide appropriate precautionary measures for a proposed activity. It can be inferred that the requirements on how much is known about a possible outcome increase when the precautionary strength increases. In other words, the complete reversal of the burden of proof requires the least knowledge of the outcome of a proposed activity. A medium strength formulation of the principle can be found in the 2000 *Convention on the Conservation and Management of Highly Migratory Fish Stocks in the Western and Central Pacific Ocean*. The parties agreed that,

> For new or exploratory fisheries, members of the Commission *shall adopt as soon as possible cautious conservation and management measures*, including, *inter alia*, catch limits and effort limits. *Such measures shall remain in force until there are sufficient data to allow assessment of the impact of the fisheries on the long-term sustainability of the stocks*, whereupon conservation and management measures based on that assessment shall be implemented. The latter measures shall, if appropriate, allow for the gradual development of the fisheries.[232] [emphasis added]

The burden of proof is reversed and it becomes necessary for the proponent (the Commission) to conduct impact assessment with sufficient data prior to the termination of precautionary conservation and management measures. In this case the proposed fishing activity is approved due to its low 'acceptability' level of uncertainty.

231 Gullett, above n. 196, 60. The 'no regrets' doctrine permits regulatory action even when there are uncertain consequences of taking such action. This doctrine is applicable as long as there will be other benefits of taking such action. In contrast, the precautionary principle involves uncertain consequences of inaction. See also Ronnie Harding and Elizabeth Fisher, *Perspectives on the Precautionary Principle* (Federation Press, 1999) 41.

232 *Convention on the Conservation and Management of Highly Migratory Fish Stocks in the Western and Central Pacific Ocean*, opened for signature 5 September 2000, 40 ILM 278 (entered into force 19 June 2004) art. 6(5).

The last step is to optimise the selected precautionary responses or measures taking relevant factors into consideration. Principle 15 of the *Rio Declaration* requires precautionary measures to be 'cost-effective'. The economic feasibility of proposed activities can be ensured through conducting cost-benefit analysis.[233] However, as noted earlier, environmental effectiveness should not be ignored while seeking cost-effectiveness of any proposed activity. It is thus important to seek a balance between cost-effectiveness and environmental effectiveness of proposed precautionary measures.

2.4.2 *The Applicability of the Precautionary Principle to the Issue of Greenhouse Gas Emissions from International Shipping*

This section will address two questions: is the precautionary principle applicable to the issue of GHG emissions from international shipping? If so, how should it be implemented in the context of GHG emissions from international shipping?

Two factors justify the application of the precautionary principle to GHG emissions from international shipping. First, the four elements constituting the formulation of the precautionary principle can also be found in this GHG emissions issue. As discussed in Chapter 1, five IPCC Assessment Reports have recognised the existence of harm or potential harm brought by GHG emissions from various sources, including those from international shipping. These harms include observed sea level rise, global warming and extreme weather. However, these harms cannot be proven scientifically, or in other words, there are uncertainties as to the outcome of proposed activities (i.e., international shipping). These IPCC Assessment Reports still utilise the terms 'likely' (IPCC Third Assessment Report), 'very likely' (IPCC Fourth Assessment Report), or 'extremely likely' (IPCC Fifth Assessment Report) to explore the causation between GHG emissions and their adverse effects. As climate change is a global issue, theoretically each State has a duty to act in reducing GHG emissions from ships. It is thus reasonable for the precautionary principle to apply to this issue.

Second, from an international law perspective, the precautionary principle should be applicable to GHG emissions from international shipping. Article 3 of the *UNFCCC* provides that '[t]he Parties should take precautionary measures to anticipate, prevent or minimize the causes of climate change and mitigate its adverse effects'. In essence, this principle represents the fundamental consensus of the international community in tackling global climate

233 Kristin Kuntz-Duriseti, 'Evaluating the Economic Value of the Precautionary Principle: Using Cost Benefit Analysis to Place a Value on Precaution' (2004) 7(4) *Environmental Science & Policy* 291, 292.

change. Currently the *UNFCCC* and the International Maritime Organization (IMO) are working on the regulation of GHG emissions from ships. While the IMO received its mandate to regulate the GHG issue from Article 2(2) of the *Kyoto Protocol* to the *UNFCCC*, it is reasonable that the precautionary principle embodied in the *UNFCCC* should also be applied to the regulation of GHG emissions from ships either under the IMO regime or through the *UNFCCC* process.

The precautionary principle could be implemented in relation to GHG emissions from international shipping through three steps as examined in the previous section. The first step is to identify available precautionary responses or tools for incorporating the precautionary principle relating to GHG emissions from ships. Currently there are three routes the IMO has taken to regulate GHG emissions from ships, namely technical measures, operational measures, and market-based measures (MBMs).[234] These measures can be regarded as precautionary responses which could be utilised to curb the negative impacts resulting from the proposed activity (international shipping). Technical and operational measures have been introduced in the form of amendments to Annex VI of *MARPOL 73/78*, so the remaining precautionary measures are either to enhance the current technical and operational measures, or to employ MBMs.

Alternative assessment and EIA might be utilised as tools for incorporating the precautionary principle. Alternative assessment is also referred to as options analysis, facility planning, source reduction planning, and pollution prevention planning.[235] It is a frequently used method to examine alternatives for achieving a specific purpose and selecting the one with the least potential impact on human health and environment.[236] Alternative assessment has been applied widely as a central aspect of the EIA process.[237]It indicates a shift from 'problem-based' environmental policy to 'solution-based' policy.[238] In the context of shipping GHG emissions, the EIA and alternative assessment may be used to identify the environmental impacts of international shipping, or possible alternatives to current practices in international shipping. Additionally, there could be other precautionary responses or duties for shipping States. For instance, they may have a duty to notify the possible risks to coastal States or port States before their ships arrive, they may have a duty to consult, to develop

234 These measures are examined in details in Chapter 4.

235 Joel A. Tickner and Ken Geiser, 'The Precautionary Principle Stimulus for Solutions and Alternatives-based Environmental Policy' (2004) 24(7–8) *Environmental Impact Assessment Review* 801, 803.

236 Pyhälä, Brusendorff and Paulomäki, above n. 198, 218.

237 Tickner and Geiser, above n. 235, 803.

238 Ibid. 801.

alternative techniques or other methods to mitigate all reasonably foreseeable damage.[239] The carbon tax package adopted by the Australian government in 2012 is an example of this type of measure.[240]

The second step is to select appropriate precautionary measures to address shipping GHG emissions. As the main impacts of GHG emissions from international shipping have been generally recognised by IPCC Assessment Reports and IMO GHG Studies, it is reasonable to infer that there is substantial knowledge about the possible outcomes of the proposed activity—international shipping. GHG emissions from international shipping lead to negative environmental impacts, but this is a cumulative process and international shipping serves as the most important means of transportation for international trade.[241] Therefore, based on the four-approach theory proposed by Gullett, less precaution will be needed to prevent negative impacts brought about by shipping GHG emissions. While the first approach, which reverses the burden of proof, would require shipowners or ship operators to prove the harmlessness of international shipping, the fourth approach does not provide any obligations on shipowners or ship operators. Therefore, the middle two approaches will be more suitable than the other two options. Regarding the acceptability level of uncertainty relating to impacts of shipping GHG emissions, this uncertainty is generally acceptable for most people. International shipping, as an important means of transportation, cannot be prohibited or replaced by other means of transportation due to the possible higher negative impacts from other alternatives. On this basis, the second approach that approves the activity based on a low acceptability level of uncertainty may not apply in the context of international shipping. It is thus arguable that the third approach should be adopted. That is, approve the activity (international shipping) but require the proponent (shipowners or ship operators) to use BAT or BATNEEC and conduct stringent post-decision monitoring. Currently this technical measure has been adopted by the IMO.

239 See Van Dyke, above n. 192, 382–383.

240 The carbon tax package adopted by the Australian government establishes a carbon pricing mechanism which commenced on 1 July 2012 with a price that will be fixed for the first three years, and on 1 July 2015 the mechanism will transition to an emissions trading scheme with the price determined by the market. See *Australian Clean Energy Bill 2011 (Explanatory Memorandum)*, Policy Context 12.

241 United Nations Conference on Trade and Development (UNCTAD), 'Review of Maritime Transport 2012' (2012) <http://unctad.org/en/PublicationsLibrary/rmt2012_en.pdf> accessed 30 July 2014, p. xiii. International shipping carries around 80 per cent of global trade by volume and over 70 per cent by value.

The last step is to achieve a balance between cost-effectiveness and environmental effectiveness of proposed precautionary measures. This issue has been raised by some countries during the discussions and negotiations of technical measures within the IMO. While it is difficult to achieve cost-effectiveness through upgrading the technical threshold for shipbuilding, it is feasible to achieve a balance between cost-effectiveness and environmental effectiveness.[242]

2.5 'Common but Differentiated Responsibility' and 'No More Favourable Treatment'

Whether the principles of 'Common but Differentiated Responsibility' (CBDR) or 'No More Favourable Treatment' (NMFT) should be applied to the GHG emissions issue has become a focal point in the debate since the IMO received its mandate to regulate GHG emissions from international shipping from Article 2(2) of the *Kyoto Protocol* in 1997. While developed States insist that the NMFT principle should apply as it typically does for all shipping conventions adopted under the auspices of the IMO, developing States argue that the CBDR principle should override it as IMO's mandate for this regulatory issue comes from the *Kyoto Protocol*, which endorses the CBDR principle. This dispute has impeded the process of international regulation by the IMO and has imposed challenges on future implementation of the adopted energy efficiency measures within the IMO.[243] This part first examines the principles of CBDR and NMFT, and then discusses whether both principles should be applied to the regulation of GHG emissions from international shipping, and explores the approaches for achieving this application.

242 See ch. 7, 7.5.2.2. In Chapter 7, cost-effectiveness and environmental effectiveness is treated as one of the criteria for selecting MBMs for reducing GHG emissions from international shipping.

243 This can be illustrated by two aspects of the debate. First, from the year of 1998 when the IMO got the mandate from the Kyoto Protocol to regulate GHG emissions from international shipping, such emissions were not regulated until July 2011. Second, regarding the IMO mandatory energy efficiency measures adopted in July 2011, consensus was not reached within the IMO which imposes challenges for the future enforcement of these measures. The CBDR principle is not fully incorporated in the adopted energy efficiency measures. See Md. Saiful Karim, 'IMO Mandatory Energy Efficiency Measures for International Shipping: The First Mandatory Global Greenhouse Gas Reduction Instrument for an International Industry' (2011) 7(1) *Macquarie Journal of International and Comparative Environmental Law* 111, 113.

2.5.1 *Common but Differentiated Responsibility*

As a nascent principle of international environmental law, the CBDR principle has received considerable attention from the international community. As far as its origins are concerned, there are generally four different opinions. Harris asserts that the CBDR principle originated from the principle of the 'common heritage of mankind',[244] which has evolved into another relevant principle of 'common concern of mankind'. This latter principle was first raised in the UN General Assembly Resolution 43/53 in 1988, where climate change is recognised as 'a common concern of mankind since climate is an essential condition which sustains life on earth'.[245] In order to resolve such 'common concern' properly, States should be allocated responsibilities. As a response to the question of which States bear the greatest responsibility for climate change, the principle of CBDR came into being.[246] Sands purports that the principle of CBDR evolved from the application of equity in general international law based on which the special needs of developing countries should be taken into account.[247] Cullet takes the view that the differentiated treatment, as the key part of the CBDR principle, could be traced back to the older principle of economic differentiation adopted in agreements on international trade and economic development.[248] This view, however, reflects more the development of North-South relations, or the relations between developed States and developing States, which shaped the content of the principle of CBDR.[249] Some other scholars treat the 1972 Stockholm Conference on the Human Environment as the origin of the principle of CBDR.[250] This view is based on the fact that

244 Paul Harris, 'Common but Differentiated Responsibility: the Kyoto Protocol and United States Policy' (1999) 27(7) *N.Y.U. Environmental Law Journal* 27, 28.This principle was adopted by many international and regional treaties, such as the 1982 LOSC, 1959 Antarctic Treaty and 1946 International Convention for the Regulation of Whaling.

245 *Protection of Global Climate for Present and Future Generations of Mankind*, GA/Res 43/53, 43rd sess, 70th plen mtg, UN Doc A/RES/43/53 (6 December 1988) art. 1.

246 See Harris, above n. 244, 28–29.

247 Sands, above n. 130, 285.

248 Philippe Cullet, 'Differential Treatment in International Law: Towards a New Paradigm of Inter-state Relations' (1999) 10(3) *European Journal of International Law* 549, 577–578.

249 Ibid. 565–578. During this process, especially with the establishment of the new international economic order in the 1970s, developing countries shifted from 'full cooperation with the North' to 'trying to impose on developed countries a new set of principles and rules of international law'. Their claims in the context of climate change set the foundation for the CBDR principle.

250 See, e.g., Nina E. Bafundo, 'Compliance with the Ozone Treaty: Weak States and the Principle of Common but Differentiated Responsibility' (2006) 21(3) *American University*

during this conference the concept of sustainable development was first raised and the different development priorities of developed countries and developing countries were identified.[251] Thus, the Stockholm conference represented 'the first time that an international consensus had been reached, at least in theory, on applying CBDR and differentiated standards to international environmental problems'.[252] However, the current content and interpretation of the CBDR principle derives something from all these sources, and has been evolving as international relations and politics change.

The CBDR principle was implicit in the *1987 Montreal Protocol*.[253] It was first explicitly formulated in Principle 7 of the *Rio Declaration*, which provides:

> States shall co-operate in a spirit of global partnership to conserve, protect and restore the health and integrity of the Earth's ecosystem. *In view of the different contributions to global environmental degradation, States have common but differentiated responsibilities.* The *developed countries acknowledge the responsibility* that they bear in the international pursuit of sustainable development in view of the pressures their societies place on the global environment and of the technologies and financial resources they command.[254] [emphasis added]

This elaboration, although criticised by both developed States and developing States,[255] has been widely accepted and endorsed in many conventions and

International Law Review 461, 468; Michael Weisslitz, 'Rethinking the Equitable Principle of Common but Differentiated Responsibility: Differential versus Absolute Norms of Compliance and Contribution in the Global Climate Change Context' (2002) 13(2) *Colorado Journal of International Environmental Law and Policy* 473, 479.

251 Weisslitz, above n. 250.

252 Ibid. 480.

253 *Montreal Protocol* art. 5. This protocol requires both developed countries and developing countries to work together to reduce controlled substances, but provides developing countries with a 10-year grace period. This practice is consistent with the principle of CBDR.

254 *Rio Declaration* pin 7.

255 Developed States did not want to be legally responsible for their past and current contributions to environmental degradation, while many developing States were not satisfied with its euphemistic expression on the liability of developed States. This dissatisfaction can be seen from the proposed text of the G77 Group of developing States, which provides that,

'..The major cause of the continuing deterioration of the global environment is the unsustainable patterns of production and consumption, particularly in developed countries... In view of their main historical and current responsibility for global

treaties, including the *1992 CBD*,[256] *1992 UNFCCC* and its *Kyoto Protocol* and *Paris Agreement*.[257] Based on this formulation, the CBDR principle consists of two elements. One is the establishment of the common responsibility of States to protect the global environment. The other is the acknowledgement by all States that differentiated responsibilities should be allocated to different States due to their different contributions to a particular environmental problem and their differing capacities to take remedial measures.[258] In other words, the CBDR principle requires both developed and developing States to contribute to addressing environmental problems, but developed States bear greater responsibility.

2.5.1.1 Common Responsibility

As mentioned above, the notion of 'common responsibility' evolved from the principle of 'common heritage of mankind', or 'common concern of humankind'. UN General Assembly Resolutions and many conventions, including the *1992 CBD* and *UNFCCC*, have recognised biological diversity and climate change as 'matters of common concern to humankind'.[259] Addressing these environmental problems is 'not solely a matter of domestic jurisdiction of each individual State'. Rather, all States, including developing States, are required to 'participate actively in the formation and implementation of international law

---

environmental degradation and their capability to address this common concern, developed countries shall provide adequate, new and additional financial resources and environmentally sound technologies on preferential and concessional terms to developing countries to enable them to achieve sustainable development'.

It is clear that the proposal by the developing States was 'much more direct on the point of responsibility' of developed States. Meanwhile the US issued an interpretative statement, stressing that Principle 7 does not 'imply a recognition … of any international obligations … or any diminution in the responsibility of developing countries'.

See *Proposal Submitted on behalf of the Group 77*, UN Doc.A/CONF.151/PC/WG.III/L.20/REV.1 (1992); *Report of the United Nations Conference on Environment and Development*, UN Doc.A/CONF.151/26 (1992) 20; Duncan French, 'Developing States and International Environmental Law: the Importance of Differentiated Responsibilities' (2000) (49) *International & Comparative Law Quarterly* 35, 36–37.

256 *CBD* art. 20(4).

257 *UNFCCC* art. 3–4; *Kyoto Protocol* art. 10; *Paris Agreement* preamble, arts. 2(2), 4(3)(4)(19).

258 Sands, above n. 130, 286.

259 Lavanya Rajamani, 'The Principle of Common but Differentiated Responsibility and the Balance of Commitments under the Climate Regime' (2000) 9(2) *Review of European Community & International Environmental Law* 120, 121.

for sustainable development'.[260] This notion, however, is rooted in the nature of the Earth and the spirit of solidarity.[261] This is underpinned clearly by the preamble of the *Rio Declaration*, which provides

> With the goal of establishing a new and equitable global partnership through the creation of new levels of co-operation among States, key sectors of societies and people.
> Recognizing the integral and interdependent nature of the Earth, our home.[262]

To gain the above 'equitable global partnership' so as to protect 'our home', *Agenda 21* also put forward similar objectives, and urged States to

> promote and support the *effective participation of all countries concerned, in particular developing countries*, in the negotiation, implementation, review and governance of international agreements or instruments, including appropriate provision of technical and financial assistance and other available mechanisms for this purpose, as well as the use of differential obligations where appropriate.[263] [emphasis added]

Certain means have been adopted by international institutions to facilitate the participation of developing States in jointly addressing international environmental problems, such as the establishment of global environmental protection funds (examples are the Global Environment Facility (GEF) and the Green Climate Fund (GCF)), technical and financial assistance, although the support is still 'insufficient for the task' and their effectiveness is to be improved.[264] However, these measures also suggest that 'common responsibilities can never be separated from differentiated responsibilities'.[265]

260 Yoshiro Matsui, "Some Aspects of the Principle of 'Common but Differentiated Responsibilities'" (2002) 2(2) *International Environmental Agreements: Politics, Law and Economics* 151, 153.

261 Ibid. 154; Rajamani, above n. 259.

262 *Rio Declaration* preamble.

263 *Agenda 21*, A/CONF.151/26 (Vols. I, II, III) (13 June 1992) para. 39.3 (c) ('*Agenda 21*').

264 Matsui, above n. 260, 154.

265 Ibid.

2.5.1.2 Differentiated Responsibility

As the other element of the principle of CBDR, differentiated responsibility means the allocation of differentiated environmental standards to developed States and developing States based on a range of factors. These factors, according to Sands, may include 'special needs and circumstances, future economic development of developing countries, and historic contributions to causing an environmental problem'.[266] This approach was widely endorsed by many UN documents and treaties. Examples include the *1972 Stockholm Declaration*,[267] *1974 Charter of Economic Rights and Duties of States*,[268] *1992 Rio Declaration*,[269] *1982 LOSC*,[270] *1987 Montreal Protocol*,[271] *1992 CBD*,[272] *1992 UNFCCC* and its *Kyoto Protocol* and *Paris Agreement*.[273]

In practice, differentiated responsibility leads to 'different legal obligations' so as to more effectively realise 'substantive equality'.[274] To achieve this goal, differentiated obligations are adopted by various international treaties and documents, and different techniques are employed to implement them.[275] The first approach is to establish differentiated standards. This approach applies not only between developed States and developing States but also between developed States.[276] Taking the *1992 UNFCCC* and its *Kyoto Protocol* as an example, Articles 4(1) and 4(2) of the *UNFCCC* stipulated that only Annex I States (mainly developed States) bear the responsibility of returning their GHG emission levels to 1990 levels by 2000. Article 11(2) of *Kyoto Protocol* puts forward the 'appropriate burden sharing' mechanism to guide the future negotiation of financial

---

266 Sands, above n. 130, 287.

267 *Stockholm Declaration* prin 23.

268 *Charter of Economic Rights and Duties of States*, UNGA Res.3281(XXIX), 29th Sess, Agenda Item 48, A/RES/29/3281 (12 December 1974) art. 30.

269 *Rio Declaration* prins 6, 11.

270 *LOSC* art. 194(2).

271 *Montreal Protocol* art. 5.

272 *CBD* art. 20(4).

273 *UNFCCC* art. 3, 4; *Kyoto Protocol* art. 10; *Paris Agreement* preamble, arts. 2(2), 4(3)(4)(19).

274 Sands, above n. 130, 289; Philippe Cullet, 'Common but Differentiated Responsibilities' in Malgosia Fitzmaurice, David M. Ong and Panos Merkouris (eds), *Research Handbook on International Environmental Law* (2010) 161.

275 Sands classifies these different techniques into 'grace' periods delaying implementation and less stringent commitments. See Sands, above n. 130, 289. But Matsui sorts these measures by using the concept of 'double standards' with one differentiating substantive rights and obligations, and the other differentiating the timing of the application of substantive provisions. See Matsui, above n. 260, 156–158.

276 French, above n. 255, 40.

commitments between developed States while its Annex B lists the individual reduction commitment of different developed States.[277] The second approach is to regulate 'grace' period to delay implementation by certain developing States. In this regard the *1987 Montreal Protocol* gave developing States 10 years' grace period for implementing regulated control measures.[278] The third approach is to have flexible and equitable requirements on different States as to the implementation of adopted measures, taking specific needs and special situations of developing States into consideration. Article 3 of the *UNFCCC* provides:

> In their actions to achieve the objective of the convention and to implement its provisions, the Parties shall be guided, inter alia, by the following: [1] ... the benefit of present and future generations of humankind, *on the basis of equity and in accordance with their common but differentiated responsibilities and respective capabilities* ... [2] The *specific needs and special circumstances of developing country Parties*, especially those that are *particularly vulnerable* to the adverse effects of climate change ...[279] [emphasis added]

Apart from these approaches, the transfer of financial and technological resources to developing States is also frequently utilised as a method of differentiating responsibilities between developed States and developing States. This kind of transfer, however, may take various forms including official assistance channels, the setting of different funds,[280] private means, and the Clean Development Mechanism (CDM) created by the *Kyoto Protocol*. These techniques are generally incorporated into multilateral environmental agreements (MEAs) by means of 'substantive provisions'. In other words, MEAs generally implement the differentiated responsibility, or the CBDR principle, by way of 'substantive provisions' instead of referring explicitly to such terms.[281]

277 Ibid.

278 *Montreal Protocol* art. 5(1).

279 *UNFCCC* art. 3.

280 Examples include the UNEP Environmental Fund, the World Heritage Fund, and the Multilateral Fund under the Montreal Protocol.

281 See Steinar Andresen and Ellen Hey, 'The Effectiveness and Legitimacy of International Environmental Institutions' (2005) 5(3) *International Environmental Agreements: Politics, Law and Economics* 211, 216. On 20 October 2011, at the 2011 Biennial Ingram Lecture organised by University of New South Wales, Professor Ellen Hey delivered a presentation entitled 'The Principle of CBDR and International Environmental Law'. She asserted that except for the *UNFCCC*, MEAs generally do not refer explicitly to the principle of CBDR instead they implement it by way of substantive provisions.

These different techniques employed to incorporate differentiated responsibility indicate that there are differing interpretations of the meaning of 'differentiation'. Many developing countries tend to interpret the 'differentiated responsibility' as different central obligations where developing countries are excluded from binding obligations such as GHG emissions reductions.[282] However, there has been less room for such interpretation during the negotiations of the international climate change regime. In particular, the 2010 *Cancun Agreements*[283] adopted at the 16th Conference of Parties (COP) to the *UNFCCC* reveals 'a shift towards greater parallelism between developed and developing countries' as to requirements relating to mitigation actions or targets and international Measurement, Reporting and Verification (MRV).[284] The 2015 *Paris Agreement* adopts a 'self-differentiation' model through creating the concept of 'Nationally Determined Contributions (NDCs)'.[285] Given the growing divergence in the interpretation of differentiation, Rajamani put forward a broad interpretation of differentiated responsibility. Based on current international environmental agreements, she asserts that differentiated responsibility consists of three categories, namely: differentiated central obligations, differentiated implementation arrangements, and the granting of assistance, including financial and technological assistance.[286]

---

282 See, e.g., *Report of the Working Group on Energy Efficiency Measures for Ships*, MEPC 61st Session, IMO Doc MEPC 61/WP.10 (30 September 2010) para. 4.31. In this meeting, while China insisted on the incorporation of the CBDR principle in regulating GHG emissions from international shipping, China proposed that '[t]he application of EEDI should be mandatory to developed countries and voluntary to developing countries'. See also Lavanya Rajamani, 'The Climate Regime in Evolution: The Disagreements that Survive the Cancun Agreements' (2011) 5(2) *Carbon & Climate Law Review* 136, 145. India underscored in one of its submissions to the UNFCCC that 'mitigation actions of developing countries will be voluntary' and they 'should under no circumstances be seen as taking on internationally legally binding commitments by these countries'.

283 *The Cancun Agreements*, Decisions 1–2/CMP.6, Report of the Conference of the Parties serving as the Meeting of the Parties to the Kyoto Protocol on its Sixth Session, FCCC/KP/CMP/2010/12/Add.1 (15 March 2011); Decision 1/CP.16, Report of the Conference of the Parties on its Sixteenth Session, FCCC/CP/2010/7/Add.1 (15 March 2011).

284 Rajamani, above n. 282, 144. See also ch. 3, 3.2.2.

285 NDCs refer to the reduction targets that each country intends to achieve under the *Paris Agreement*, and this contribution should reflect its common but differentiated responsibilities and respective capabilities, in the light of different national circumstances. See ch. 3, 3.2.2.3.

286 Lavanya Rajamani, *Differential Treatment in International Environmental Law* (Oxford University Press, 2006) 191.

The basis for differentiated responsibility has been analysed and supported by many commentators.[287] Two key justifications that have been invoked by international treaties and scholarly commentary are the historical responsibility of the North for current environmental degradation and their present capability to remedy such problems. The first justification is related to the differing contributions of States to environmental problems. From this perspective, the principle of CBDR can be deemed as the application of the polluter-pays principle.[288] As the main GHG emitter contributing to current environmental problems, developed States are the main polluters and thus should be responsible for this issue. The second justification involves different capabilities of States. This can theoretically be underpinned by the principle of equity or the concept of environmental justice. It has been generally accepted that justice is 'a compulsory part of international environmental law'.[289] Due to imbalanced historical and present distribution of resources and power, it can be argued that a form of distributive justice should be realised.[290] Therefore, differentiated responsibility can be viewed as a kind of 'entitlement' by developing States or obligation by developed States instead of on the basis of 'need' or 'compassionate measures'.[291]

Other justifications are based on the different priorities of developed and developing States,[292] 'international cooperation' or 'solidarity' being a feature of current international environmental law where developed States are obliged to bear more responsibilities,[293] and the utilisation of differentiated

---

287 See, e.g., Cullet, above n. 248; French, above n. 255; Rajamani, above n. 286; Matsui, above n. 260.

288 See, e.g., Matsui, above n. 260, 155; Rajamani, above n. 259, 122.

289 Cullet, above n. 274, 162.

290 Ibid.

291 S.R. Choudhary, 'Common but Differentiated Responsibility in International Environmental Law from Stockholm to Rio' in Konrad Ginther, Erik Denters and Paul J.I.M. de Waart (eds), *Sustainable Development and Good Governance* (1995) 322, 334. But Matsui treats this differentiated responsibility of developed States as the one with a moral or political nature. See Matsui, above n. 260, 155.

292 Generally protecting global environment or adapting to the consequences thereof has become one of the priorities of developed States whereas developing States pay more attention to their economies instead of environmental protection. See *Stockholm Declaration* 1.4; French, above n. 255, 52.

293 Ibid. 55. French attributes the basis of this 'international cooperation' to the Rio Declaration. The preamble of the Rio Declaration reads, '[w]ith the goal of establishing a new and equitable global partnership through the creation of new levels of cooperation among States, key sectors of societies and people'.

treatment as an incentive for developing States to participate in multilateral environmental agreements.[294]

2.5.1.3 Legal Status and Application

The inclusion of the CBDR principle in various treaties and UN documents has successfully increased the participation of developing States in international environmental regulation and management and achieved consensus in some areas especially within the climate regime. Its legal status, however, is still open to debate. Generally, it is regarded as a principle of international environmental law.[295] Nevertheless, due to different understandings of the term 'principle', it has been accorded different degrees of status by scholars—some have classified it as 'merely aspirational' while others regard it as 'legally binding'.[296] Currently its status in customary international law terms is not defined.[297]

The preamble of the *Kyoto Protocol* emphasises that 'in pursuit of the ultimate objective of the [*UNFCCC*] Convention as stated in its Article 2', States' actions should be 'guided by Article 3 of the [*UNFCCC*] Convention'.[298] Article 2 of the *UNFCCC* requires that the goal of GHG emission reduction should be conducted in a way to 'enable economic development to proceed in a sustainable manner', which implies the necessity of differentiated treatment if the requirement of 'sustainable development' and the context of this article is

---

Birnie, Boyle and Redgwell, above n. 4, 135. Birnie et al assert that solidarity is 'a key element' of the CBDR principle, which is reflected by chapter 17.2 of Agenda 21. It provides that, '... shall be commensurate with their technological and financial capacities and priorities in allocating resources for development needs and ultimately depends on technology transfer and financial resources required and made available to them'. Thus, through making obligations 'conditional' developing States can thereby impose pressure on developed States.

294 French, above n. 255, 56.

295 *UNFCCC* art. 3. The CBDR was regulated under the title of 'Principles' in article 3. See also Cullet, above n. 274, 161; Harrison, above n. 77, 7.

296 Lavanya Rajamani, 'The Nature, Promise, and Limits of Differentiated Treatment' (2005) 16(1) *Yearbook of International Environmental Law* 81, 102.

297 For example, Kiss regards it as one of the seven principles under sustainable development; Birnie et al., discuss it under the principles of global environmental responsibility. But Sands treats it as an independent general principle. Cullet is of the opinion that it is 'one of the important principles of international environmental law', however, its 'binding nature remains disputed'. See Kiss and Shelton, above n. 1, 218; Birnie, Boyle and Redgwell, above n. 4, 128–137; Sands, above n. 130, 285–289; Cullet, above n. 274, 161.

298 *Kyoto Protocol* preamble.

taken into account.[299] Article 3 of the *UNFCCC* treats the CBDR as one of the principles of the convention, and the preamble of the *UNFCCC* also acknowledges this principle. Therefore, through the incorporation of the CBDR principle into various conventions, especially the *UNFCCC* and its *Kyoto Protocol* and *Paris Agreement*, the cornerstone role of this principle within the global climate change regime has been generally accepted. However, the implication of the differentiation element of this principle has been evolving.

2.5.2 *No More Favourable Treatment*

2.5.2.1 An Overview

No more favourable treatment (NMFT), also called the equal treatment for all ships principle, or universal treatment principle, refers to 'port States enforcing applicable standards in a uniform manner to all ships in their ports, regardless of flag'.[300] Under the *IMO Convention*,[301] Article 1(b) describes the 'removal of discriminatory action' as one of the purposes of the IMO, and Article 3 treats the 'normal processes of international shipping business' as a recommended way to deal with shipping-related matters. Indeed, these two Articles provide a legal basis for the NMFT principle. The term NMFT was included in *MARPOL 73/78* and applies to all annexes to that Convention.

Article 5(4) of *MARPOL 73/78* stipulates that,

> With respect to the ship of non-Parties to the Convention, Parties shall apply the requirements of the present Convention as may be necessary to ensure that *no more favourable treatment* is given to such ships.[302] [emphasis added]

Article II of the 1978 Protocol to 1974 *International Convention for the Safety of Life at Sea* (*SOLAS*) also has a similar provision, which reads:

---

299 One element of sustainable development is to integrate environmental protection into economic and other development. This integration approach may also underpin the adoption of differentiated legal commitments based on the differentiated historical responsibility of States and their different capacity to respond to environmental requirements. Sands, above n. 130, 263.

300 Buhaug et al., above n. 185, 20.

301 *Convention on the Inter-Governmental Maritime Consultative Organization*, opened for signature 6 March 1948, 289 UNTS 3 (entered into force 17 March 1958), amended and renamed as *Convention on the International Maritime Organization*, opened for signature 14 November 1975, 9 UTS 61 (entered into force 22 May 1982) ('*IMO Convention*').

302 *MARPOL 73/78* art. 5(4).

> 3. With respect to *the ships of non-Parties to the Convention and the present Protocol,* the Parties to the present Protocol shall apply the requirements of the Convention and the present Protocol as may be necessary *to ensure that no more favourable treatment is given to such ships.*[303] [emphasis added]

Article X of the 1978 *International Convention on Standards of Training, Certification and Watchkeeping for Seafarers* (STCW) provides:

> This Article shall be applied as may be necessary to ensure that *no more favourable treatment is given to ships entitled to fly the flag of a non-Party* than is given to ships entitled to fly the flag of a Party.[304] [emphasis added]

It appears that the NMFT principle only applies to Article X of this convention. However, the fact that this Article addresses the issue of 'control' indicates the application of this principle to the whole convention since the main purpose of this convention is to deal with the control of foreign ships while in the ports of a State.[305]

To date this principle has been consistently applied without exception to all 53 IMO treaty instruments currently in existence.[306]

2.5.2.2 Legal Status and Application

The NMFT principle has been widely applied to treaties adopted by the IMO. Nevertheless, it is only a customary rule applicable within the IMO regime. The application of this principle is one of the key features of IMO's efforts in exercising uniform standards around the world, and it has assisted the IMO to fulfil the regulatory purposes of these treaties. First, the introduction of the NMFT principle has been proven to be an effective means and incentive for non-participating States to become contracting parties to an IMO treaty.[307]

---

303 *International Convention for the Safety of Life at Sea,* opened for signature 1 November 1974, 1184 UNTS 2 (entered into force 25 May 1980) art II ('*SOLAS*').

304 *International Convention on Standards of Training, Certification and Watchkeeping for Seafarers,* opened for signature 7 July 1978, 1361 UNTS 2 (entered into force 28 April 1984), as amended by the 1995 Protocol, 1969 UNTS (entered into force 1 February 1997) art. X, para. 5 ('*STCW*').

305 G.P. Pamborides, *International Shipping Law: Legislation and Enforcement* (Kluwer Law International, 1999) 107–108.

306 International Maritime Organisation (IMO), *Status of Conventions: List of IMO Conventions and Their Amendments* (16 May 2013) <http://www.imo.org/About/Conventions/StatusOfConventions/Pages/Default.aspx> accessed 1 January 2014.

307 Pamborides, above n. 305, 108.

Under this principle, port State control will impose the standards as indicated in an IMO treaty on all ships calling at a port of a contracting party. In this way, it becomes more difficult for a state to avoid compliance with a convention adopted under the auspices of the IMO. Second, the application of the NMFT principle may relieve concern over the existence of the 'flag of convenience' (FOC) phenomenon.[308] Under the FOC, a ship may change its flag easily to have a non-Annex I State nationality if differentiated responsibility applies and ships flying the flags of non-Annex I States would then enjoy less stringent treatment.[309] If this were to be the case, the regulatory efforts by the IMO to address many maritime issues would be ineffective. Furthermore, 75 per cent of the world shipping tonnage, by deadweight, of all merchant ships on international voyages is registered in developing States.[310] Therefore, it would be 'ineffective' for the IMO to act by means of regulating only 25 per cent of the world's shipping tonnage if the NMFT principle were not in place.[311]

It is worth noting that the IMO has limited the application of the NMFT principle to IMO regulated treaties. For instance, the preamble of amended Annex VI to *MARPOL 73/78* in 2011 provides:

> Recognizing also that adoption of the amendments to Annex VI *in no way prejudges the negotiations held in other international fora*, such as *the United Nations Framework Convention on Climate Change* (*UNFCCC*), *nor affect the positions of the countries that participate in such negotiation*,[312] [emphasis added]

This statement reveals the IMO's view on the application scope of its NMFT principle. That is, the IMO's regulations on GHG emissions from ships, including the revised *MARPOL* Annex VI, are independent from those reached within the *UNFCCC-Kyoto Protocol* regime. The application of the NMFT principle to shipping GHG emissions issue should not be regarded as a precedent which may be applicable to the international climate change regime. Nevertheless, as an obligation under the *Kyoto Protocol*, the IMO still needs to report its progress on the GHG emissions issue to the *UNFCCC*'s SBSTA on a regular basis.

---

308 Definitions of FOC is provided at footnote 86.

309 Haifeng Wang, 'GHG Emissions from the International Goods Movement by Ships and the Adaptation Funding Distribution' in Zongwei Luo (ed), *Green Finance and Sustainability: Environmentally-Aware Business Models and Technologies* (Business Science Reference, 2011) 274, 283.

310 Buhaug et al., above n. 185, 21.

311 Ibid.

312 *MARPOL Annex VI* (*2011 amendments*) preamble.

2.5.3 *Application of Both 'Common but Differentiated Responsibility' and 'No More Favourable Treatment' Principles to the Issue of Greenhouse Gas Emissions from International Shipping*

The CBDR principle and the NMFT principle are distinct in terms of their content and their scope of application. Generally the CBDR principle applies to the global climate change regime, whereas the NMFT principle applies to all IMO treaties. Regulating GHG emissions from international shipping involves both global climate change and the IMO. In this case, how the two principles should be applied to this GHG emissions issue is controversial.

2.5.3.1 Applicability of Two Regulatory Principles

Three divergent views exist as to the applicability of the CBDR and NMFT principles to the regulation of GHG emissions from international shipping. One view is that only the CBDR principle should be applied to this GHG emissions issue on the ground that the IMO received its mandate to regulate GHG emissions from international shipping from Article 2(2) of the *Kyoto Protocol* to the *UNFCCC*. This view has been supported by many developing countries[313] and some scholars.[314] This interpretation of the IMO's mandate justifies the application of the CBDR principle, which runs through the *UNFCCC* and its *Kyoto Protocol*, to this issue. Article 2(2) of the *Kyoto Protocol* provides,

> *The parties included in Annex I* shall pursue limitation or reduction of emissions of greenhouse gases not controlled by the Montreal Protocol from aviation and *marine bunker fuels, working through* the International

313 See, e.g., China, India, Brazil, Saudi Arabia and some other developing countries all hold that the IMO's mandate comes from the Kyoto Protocol. *Report of the Marine Environment Protection Committee on its 59th Session*, Statement by the Delegation of China on GHG Issues, IMO Doc MEPC 59/24/Add.1 Annex 13 (2009) para. 1; *Report of the Marine Environment Protection Committee on Its Sixty-First Session*, MEPC 61st Session, Agenda Item 24, IMO Doc MEPC 61/24 (6 October 2010) annex 3.

314 See, e.g., Wang, above n. 309, 275; A. Miola, M. Marra and B. Ciuffo, 'Designing A Climate Change Policy for the International Maritime Transport Sector: Market-Based Measures and Technological Options for Global and Regional Policy Actions' (2011) 39(9) *Energy Policy* 5490, 5492; Derya Aydin Okur, *The Challenge of Regulating Greenhouse Gas Emissions from International Shipping and the Complicated Principle of 'Common but Differentiated Responsibilities'* (2012) <http://web.deu.edu.tr/hukuk/dergiler/dergimiz13-1/2-deryaaydinokur.pdf> accessed 1 January 2014, p. 28; Jodie Moffat, 'Arranging Deckchairs on the Titanic: Climate Change, Greenhouse Gas Emissions and International Shipping' (2010) 24(2) *Australian and New Zealand Maritime Law Journal* 104, 105.

> Civil Aviation Organization and *the International Maritime Organization*, respectively.[315] [emphasis added]

Marine bunker fuel, also called degraded residue heavy fuel oil, is the main fuel used by ships on international voyages.[316] Therefore, it is arguable that the above article can be interpreted as meaning that the IMO has a mandate from the *Kyoto Protocol* to regulate GHG emissions from international shipping. Furthermore, this provision may also be interpreted as meaning that only Annex I States (developed States) are under the commitment to conduct the emissions reductions.[317] However, this view has been opposed by the Sub-Division for Legal Affairs of the IMO. It asserted that the IMO did not receive its GHG mandate from the *Kyoto Protocol*, and this provision should not be interpreted as meaning that non-Annex I States are exempt from any obligations. Rather, it should be interpreted that the reduction of such emissions is 'a task which is properly within the purview of IMO', and 'only Annex I countries should be involved in the negotiations within IMO'.[318] Article 31 of the *1969 Vienna Convention on Treaties* stipulates that,

> 1. A treaty shall be interpreted *in good faith in accordance with the ordinary meaning* to be given to the *terms of the treaty in their context* and in the light of its object and purpose.
> 2. The context for purpose of the interpretation of a treaty shall comprise, in addition to the text, including its preamble and annexes...[319] [emphasis added]

In accordance with these rules, in particular based on the ordinary meaning of the terms and the context of negotiating the *Kyoto Protocol*, Article 2(2) of the *Kyoto Protocol* may be understood as meaning that only Annex I States are obliged to make reductions in international shipping, which is consistent with the rest of the *Kyoto Protocol* where the CBDR principle has been fully incorporated. However, this interpretation will only be logical if the IMO receives its

315 *Kyoto Protocol* art. 2.2.

316 Md. Saiful Karim and Shawkat Alam, 'Climate Change and Reduction of Emissions of Greenhouse Gases from Ships: An Appraisal' (2011) 1(1) *Asian Journal of International Law* 131, 131.

317 See, e.g., W.B. Fitzgerald, O.J.A. Howitt and I.J. Smith, 'Greenhouse Gas Emissions from the International Maritime Transport of New Zealand's Imports and Exports' (2011) 39(3) *Energy Policy* 1521, 1523; Moffat, above n. 314, 104.

318 *Legal Aspects of the Organization's Work on Greenhouse Gas Emissions in the Context of the Kyoto Protocol*, note by the Secretariat, IMO Doc MEPC 58/4/20 (1 August 2008).

319 *1969 Vienna Treaty Convention* art. 31 (1)(2).

GHG mandate from this provision. The IMO has denied this possibility and its documents indicate that no consensus was achieved as to the interpretation of this provision after the adoption of the *Kyoto Protocol*.[320] Therefore, it seems that the first view would be more acceptable only when it is recognised that the IMO receives its GHG mandate from the *Kyoto Protocol*.

The second view supports the sole application of the NMFT principle to this GHG emissions issue. This view is held by the Sub-Division for Legal Affairs of the IMO and some scholars.[321] The IMO has been the main international institution working on the regulation of GHG emissions from international shipping since 1997. Therefore, there is little doubt that the NMFT principle is applicable to this issue. As noted earlier, this is because the NMFT principle has been consistently applied to all IMO treaties and has become a customary practice within the IMO regime. Furthermore, the reduction of GHG emissions from ships has been partially regulated in the form of amendments of Annex VI to *MARPOL 73/78* in 2011 and 2014 respectively. Since the NMFT principle is explicitly stipulated in Article 5(4) of *MARPOL 73/78*, it follows that the regime of GHG emission reductions from ships in Annex VI is subject to this article. In other words, the principle of NMFT applies to GHG emissions from international shipping in this context. In order to exclude the application of the CBDR principle, the proponents of this view assert that the IMO derives its global mandate from the *IMO Convention*, the *LOSC* and IMO Regulation 8, but not from Article 2(2) of the *Kyoto Protocol*.

The third view recognises the application of both the CBDR and the NMFT principles to this GHG emissions issue, but insists that this might only be achieved through market-based mechanisms.[322] However, the proponents of

---

320 See, e.g., *Report of the Marine Environment Protection Committee on Its Forty-Ninth Session*, MEPC 49th Session, Agenda Item 22, IMO Doc MEPC 49/22 (8 August 2003) para. 4.9. At the 49th MEPC meeting in 2003, the Committee agreed that the regulation of GHG emissions from international shipping should be based on the NMFT principle rather than the CBDR principle. However, no consensus was achieved as to this agreement.

321 See, e.g., International Maritime Organisation (IMO), 'Main Events in IMO's Work on Limitation and Reduction of Greenhouse Gas Emissions from International Shipping' (2011) <http://www.imo.org/MediaCentre/resources/Pages/Greenhouse%20gas%20emissions.aspx> accessed 14 June 2014, p. 28; Sebastian Oberthür, 'Institutional Interaction to Address Greenhouse Gas Emissions from International Transport: ICAO, IMO and the Kyoto Protocol' (2003) 3(3) *Climate Policy* 191, 195.

322 See, e.g., Karim and Alam, above n. 316, 144–147; European Federation for Transport and Environment, *Bunker Fuels and the Kyoto Protocol: How ICAO and the IMO Failed the Climate Change Test* <http://www.transportenvironment.org/sites/te/files/media/2009_06_aviation_shipping_icao_imo_history.pdf> accessed 1 January 2014.

this view have only explored the approaches of applying both principles to this issue, and some of them also assert that the IMO's mandate to regulate shipping GHG emissions does not derive from Article 2(2) of the *Kyoto Protocol*.[323]

To date none of these three views have been generally accepted by most countries and the shipping industry. Theoretically speaking, it appears that identifying the origin of the IMO's mandate to regulate GHG emissions from international shipping is a key to addressing this debate. Identifying the generally accepted origin of the IMO's mandate could determine which principles may apply to the regulation of this issue. Generally if an international agreement gives the IMO a specific mandate, it would appear reasonable that the principles reflected in that agreement should also apply to the regulation of the GHG issue by the IMO.

As discussed earlier, the proponents of the first view attribute the IMO's mandate in regulating GHG emissions from ships solely to the *Kyoto Protocol*. Article 2(2) of the *Kyoto Protocol* requests the Annex I States of the *UNFCCC* to 'work through the IMO' to limit or reduce their GHG emissions from ships. Whether this provision gives the IMO the exclusive mandate to regulate this GHG issue is open to debate and dependent on various interpretations of the term 'work through'. However, it 'establishes a formal link to the IMO' by authorising the IMO to regulate this GHG issue,[324] and implies that the IMO should 'take the lead' on this issue.[325] Furthermore, the acceptance of this mandate by the IMO is consistent with the *IMO Convention*.[326] Since then the IMO has reported its progress in regulating the GHG issue to the *UNFCCC*'s Subsidiary Body on Scientific and Technological Advice (SBSTA) on a regular basis, which could be regarded as one of its obligations in fulfilling this mandate.[327] Therefore, it is argued that it is not reasonable to assert that the IMO's mandate has nothing to do with the *Kyoto Protocol*.[328]

323 Karim and Alam, above n. 316, 147.

324 Bernd Hackmann, 'Analysis of the Governance Architecture to Regulate GHG Emissions from International Shipping' (2012) 12(1) *International Environmental Agreements: Politics, Law and Economics* 85, 90.

325 Harrison, above n. 77, 1.

326 *IMO Convention* art. 68. This provision stipulates that the IMO may take over functions or obligations within its scope from any other international organizations by means of international agreements.

327 But Hackmann asserts that this cooperation between the UNFCCC and the IMO is 'reciprocal exchange of information and a reciprocal participation in relevant meetings', and both institutions are independent in their decisions. Hackmann, above n. 324, 95.

328 See, e.g., Karim and Alam assert that the IMO's mandate 'is not subject to the UNFCCC or to its Kyoto Protocol'. Karim and Alam, above n. 316, 147–148.

The proponents of the second view assert that the IMO derives its global mandate from the *IMO Convention*, the *LOSC* and IMO Resolution 8, but not from Article 2(2) of the *Kyoto Protocol*. According to this view, Articles 1(a) and 64 of the *IMO Convention* provide the IMO with a global mandate and global competence 'in the field of shipping and the effect of shipping on the marine environment',[329] in particular in relation to 'technical matters of all kinds affecting shipping engaged in international trade'.[330] Articles 211(1) and 212(3) of the *LOSC* request States Parties to 'establish global rules, standards, and recommended practices and procedures' to prevent, reduce and control atmospheric and vessel-source marine pollution. In particular, these actions shall be conducted through diplomatic conferences or a competent international organization (the IMO). Therefore, the *LOSC* defines flag, coastal and port State jurisdiction, while the IMO specifies how member State jurisdiction should be exercised to meet IMO safety and shipping anti-pollution regulations.[331] Furthermore, Resolution 8 on 'CO_2 emissions from ships' was adopted by the *MARPOL* Conference of the Parties in 1997. This resolution requested the IMO to start its work on the reduction of GHG emissions from ships and has therefore been regarded as a key legal document underpinning subsequent regulatory efforts by the IMO. In addition, those who take the view that attributes the IMO's mandate to these three sources exclude Article 2(2) of the *Kyoto Protocol* as a source of the IMO's mandate relating to the GHG issue. It is asserted that there has been no precedent for any IMO treaty instruments adopting a common but differentiated responsibility approach similar to that incorporated in the *Kyoto Protocol*.[332] This argument, however, runs counter to the legal basis for the first view on the IMO's mandate to regulate the GHG emissions issue. Generally an organisation which receives and accepts a mandate under an international agreement cannot question principles incorporated in that agreement simply based on its own previous practice which is incompatible with such principles.[333]

---

329 *IMO Convention* art. 64.

330 *IMO Convention* art. 1(a).

331 *Implications of the United Nations Convention on the Law of the Sea for the International Maritime Organization*, IMO Doc LEG/MISC.6 (10 September 2008) 13.

332 IMO, above n. 321, 28.

333 *1969 Vienna Treaty Convention* art. 31; *IMO Convention* art. 41. In accordance with Article 31 of the *Vienna Convention on the Law of Treaties*, a provision of a treaty shall be interpreted based on the context, the object and purpose of the treaty and the context includes the text, the preamble and annexes. On this basis, the provision that provides a mandate for an organisation should be subject to the principle incorporated in that treaty, which also enables the application of the principle to the organisation. For example, Article 41 of the

From an international law perspective, the above two views on the origin of the IMO's GHG mandate both have their legal bases. There is no clear hierarchy between the *Kyoto Protocol* and *IMO Convention* and the *LOSC* on the issue and it is open to debate which rules should prevail if there is a conflict between these treaties.[334] For this reason, it might be appropriate to strike a compromise between the two views. It is clear that the *IMO Convention* and the *LOSC* provide the IMO with general competence to regulate GHG emissions from ships, while the *Kyoto Protocol* gives the IMO a specific mandate to regulate this matter. The two interpretations of the IMO's mandate are thus consistent and the IMO can utilise both these competences to regulate GHG emissions from international shipping. An important implication of this compromise interpretation is that principles incorporated in the *Kyoto Protocol* and the *IMO Convention* will also apply to the regulation of the GHG issue, namely, the CBDR and the NMFT principles. It therefore appears that the third view is more logical. However, the approaches that have so far been proposed for applying these two principles to this issue as expressed in the third view can be improved. This is examined in the next section.

2.5.3.2 Approaches to Applying the Two Principles

It is a challenge to determine how to incorporate both the CBDR and NMFT principles into the IMO's regulation of shipping GHG emissions. To address this issue, two assumptions can be made. One is that the CBDR principle is State-based whereas the NMFT principle is ship-based, so there is no irreconcilable conflict between them. The other is that common responsibility and differentiated responsibility are two core elements of the CBDR principle and common responsibility has been incorporated into this issue via the NMFT principle,[335] so the key to applying the CBDR principle is effective incorporation of differentiated responsibility. There are different interpretations of the implications of the CBDR principle, in particular the meaning of 'differentiated responsibility'. Due to the complexity of the issue of GHG emissions from shipping, the adoption of a broad interpretation of differentiated treatment would be practical. As discussed earlier, Rajamani put forward a broad interpretation of

---

IMO Convention provides that the MEPC, 'when exercising the functions conferred upon it by or under any international convention or other instrument, shall conform to the relevant provisions of the convention or instrument in question.'

334 Since the *Kyoto Protocol*, *IMO Convention* and the *LOSC* vary extensively in their context and scope, Article 30 of the *Vienna Convention on the Law of Treaties* cannot apply to addressing conflicts between these treaties.

335 *MARPOL 73/78* art. 5(4).

differentiated responsibility which consists of three categories, namely differentiated central obligations, differentiated implementation arrangements, and the granting of assistance, including financial and technological assistance.[336] If this broad interpretation of differentiation is employed, the CBDR principle could be applied to the GHG issue in different ways depending on the nature of various measures for addressing this issue.

There are three routes for regulating shipping GHG emissions that have been considered within the IMO: technical measures, operational measures, and MBMs. In terms of technical and operational measures adopted by the IMO in 2011, strengthening effective transfer of technologies and financial assistance from developed countries to developing countries in relation to these technical and operational measures would constitute an application of the CBDR principle to this issue as indicated in the third category of differentiated responsibility.[337] As a requirement of the NMFT principle, port States exercise uniform control over all ships calling at their ports through participation in various MOUs on Port State Control.[338] For this reason, it would be difficult to implement differentiated central obligations as indicated in the first category of differentiated responsibility with regard to the issue of GHG emissions from shipping.[339] Indeed this category of differentiated responsibility is often claimed by developing countries as the main form of the CBDR principle.[340] Meanwhile, the difficulty in applying the first category

336 Rajamani, above n. 286, 191.

337 Regulation 23 of the amended *MARPOL* Annex VI in 2011 stipulates the transfer of technology and financial assistance; however, this regulation is still very weak and thus needs to be strengthened.

338 Port State Control refers to 'the inspection of foreign ships in national ports to verify that the condition of the ship and its equipment comply with the requirements of international regulations and that the ship is manned and operated in compliance with these rules'. With the support of the IMO, to date various regional port State control organizations and agreements on Port State Control, namely the MOUs, have been signed to cover all of the world's oceans. International Maritime Organization (IMO), *Port State Control* <http://www.imo.org/blast/mainframe.asp?topic_id=159> accessed 1 January 2014.

339 Some States proposed that combining both principles could be achieved by differentiating commitments for developed and developing countries based on certain routes of shipping without relying on the nationality of ships. However, due to various regional MOUs on port State control, in practice this proposal is not feasible. Miola, Marra and Ciuffo, above n. 314, 5492.

340 For instance, when China and India have mentioned the application of the CBDR principle to this issue, they have generally explained that only developed countries should commit themselves to compulsory GHG emission reductions from international shipping, while energy-efficiency measures should be voluntary for developing countries. *Report of*

of differentiated responsibility to technical and operational GHG-reduction measures is also underpinned by the existence of Flag-of-Convenience (FOC) States. It may be hypothesised that developing flag States are exempt from complying with IMO GHG-reduction regulations, as implied by the first category of the CBDR principle. In this circumstance, shipowners from developed countries would probably opt for flagging their ships under these FOC States to avoid the stringent regulations and increased cost in their own States flowing from compliance with these regulations. As of 1 January 2013, ships registered in developing countries (excluding transition economies) accounted for 75.49 per cent of the world fleet by deadweight tonnage (dwt), which if combined with the FOC would render these GHG-reduction measures barely effective.[341] Theoretically, it would be feasible to phase-in application of the CBDR principle to this GHG issue as indicated in the second category of differentiated responsibility.[342] Indeed during the discussions within the IMO, some developing countries proposed this approach to postpone the application of regulations to developing countries.[343] However, due to the concern for the FOC and the urgency of addressing this issue against the backdrop of global climate change, this option was not adopted by the IMO.

More options are available to incorporate the two principles with respect to MBMs. One possibility is to apply the CBDR principle to the issue by allocating differentiated central obligations to developed countries and developing

the Marine Environment Protection Committee on Its Sixty-First Session, MEPC 61st Session, Agenda Item 24, IMO Doc MEPC 61/24 (6 October 2010) 1–3.

341 United Nations Conference on Trade and Development (UNCTAD), 'Review of Maritime Transport 2013' (2013) <http://unctad.org/en/PublicationsLibrary/rmt2013_en.pdf> accessed 1 January 2014, p. 57.

342 See Miola, Marra and Ciuffo, above n. 314, 5492. For example, it was proposed that a three-phased approach could be employed to address this GHG issue, namely the set-up of a scheme for voluntary participation by the countries and ports as the first step, a scheme that covers all traffic in the ports of *UNFCCC* Annex I countries as the second step, and finally this scheme would be extended to cover all countries on a global level.

343 *Comments on the Proposed Mandatory Energy Efficiency Regulations*, submitted by China, Saudi Arabia and South Africa, MEPC 62nd Session, Agenda Item 5, IMO Doc MEPC 62/5/10 (5 May 2011) para. 14. In this document, the co-sponsors proposed a draft text which provided that, 'the regulations of EEDI and SEEMP shall apply to ships of developing countries five years after the date of their entry into force'; or 'shall be phased in over a period of eight years for ships built for developing countries and during the period of phasing in, developing countries shall only apply 50% of the required EEDI reduction rate'.

countries so as to ensure 'no net incidence on developing countries',[344] as indicated in the first category of differentiated responsibility. Currently some proposed MBMs incorporating both principles have been submitted to the IMO for further discussion.[345] This approach has also been supported by the *UNFCCC* Secretariat,[346] as well as by other countries and international organisations.[347] Since these MBM proposals also apply the NMFT principle, the effectiveness of these measures is unlikely to be influenced by FOC States. A detailed assessment of current MBM proposals and the selection of MBMs for addressing GHG emissions from international shipping is provided in Chapters 4 and 7 of this book.

2.6 The Polluter-Pays Principle and Its Application to the Issue of Greenhouse Gas Emissions from International Shipping

Preventing and controlling pollution is a costly process where a significant financial investment in human and material resources and infrastructure or a large amount of compensation for victims may be required.[348] But who is going to pay these bills? The polluter-pays principle provides some rules for addressing this problem. Despite the view from many States that it is only

344 *Ensuring No Net Incidence on Developing Countries from A Global Maritime Market-Based Mechanism*, submitted by World Wide Fund for Nature (WWF), MEPC 63rd Session, Agenda Item 5, IMO Doc MEPC 63/5/6 (22 December 2011).

345 See, e.g., the Rebate Mechanism proposed by the International Union for Conservation of Nature (IUCN) has incorporated the CBDR and NMFT principles.

346 *Report of the Marine Environment Protection Committee on Its Sixty-First Session*, MEPC 61st Session, Agenda Item 24, IMO Doc MEPC 61/24 (6 October 2010) annex 6, p. 2. At the 61st MEPC meeting, the *UNFCCC* Secretariat made a statement, which asserts that '[w]e have to commit ourselves to work on a solution which respects both principles, and allows each treaty regime to retain the integrity of its principles and practices'.

347 See, e.g., Malaysia, Ethiopia and WWF support the adoption of both principles in addressing the GHG issue. *Report of the Marine Environment Protection Committee on Its Sixtieth Session*, MEPC 60th Session, Agenda Item 22, IMO Doc MEPC 60/22 (12 April 2010) annex 4, p. 10; *Report of the Marine Environment Protection Committee on Its Sixty-First Session*, MEPC 61st Session, Agenda Item 24, IMO Doc MEPC 61/24 (6 October 2010) annex 3, p. 11; *Ensuring No Net Incidence on Developing Countries from A Global Maritime Market-Based Mechanism*, submitted by World Wide Fund for Nature (WWF), MEPC 63rd Session, Agenda Item 5, IMO Doc MEPC 63/5/6 (22 December 2011).

348 See Kenneth A. MacInnis, 'The Polluter Pays Principle: Preventing Ship-Source Pollution in the Arctic' in Aldo E. Chircop et al. (eds), *The Regulation of International Shipping: International and Comparative Perspectives* (2012) 143, 143.

applicable at the domestic level rather than at the international level,[349] the polluter-pays principle has been evolving and is widely applied in various international instruments. This part first examines the polluter-pays principle as to its evolution, content and implementation, and then discusses how it can be applied to the issue of GHG emissions from international shipping.

2.6.1 *An Overview of the Polluter-Pays Principle*

The polluter-pays principle refers to the requirement that 'the costs of pollution should be borne by the person responsible for causing the pollution'.[350] Although its meaning and application are still open to debate,[351] the polluter-pays principle has been recognised worldwide and is referred to in both national legislation and international declarations and agreements.[352] The origin of this principle can be traced back to 1972 when the Organisation for Economic Co-operation and Development (OECD) formally propounded it as a means of coping with environmental problems.[353] The formulation of this principle was first contained in the *1972 OECD Guiding Principles*,[354] which provide:

349 Sands, above n. 130, 281.

350 Ibid. 279.

351 Ibid. 280.

352 James A. Tobey and Henri Smets, 'The Polluter-Pays Principle in the Context of Agriculture and the Environment' (1996) 19(1) *WORLD ECONOMY* 63, 63.

353 Concerning the origin of the polluter-pays principle, there are some different views.

Some scholars trace it back to the *1972 OECD Guiding Principles* which recommend the adoption of the polluter-pays principle to allocate costs of pollution prevention and control measures. *Recommendation of the Council on Guiding Principles concerning International Economic Aspects of Environmental Policies* (26 May 1972) OECD Recommendation C (72) 128 art. 1 A. See, e.g., Kiss and Shelton, above n. 1, 213; Priscilla Schwartz, 'The Polluter-Pays Principle' in Malgosia Fitzmaurice, David M. Ong and Panos Merkouris (eds), *Research Handbook on International Environmental Law* (2010) 243, 244.

Other scholars, Sands for example, asserts that the polluter-pays principle in treaty law can be traced back to some of the first instruments setting minimum rules on civil liability for damage by hazardous activities, and the earliest one is the *1960 Paris Convention* which provides that the operator of the nuclear installation, whether a private entity or the state, is strictly liable for injury to or loss of life of any person and damage to or loss of property. Sands, above n. 130, 281; See also *OECD Convention on Third Party Liability in the Field of Nuclear Energy*, opened for signature 29 July 1960, 956 UNTS 251 (entered into force 1 April 1968) art. 3 (1).

354 *Environment and Economics: Guiding Principles Concerning International Economic Aspects of Environmental Policies*, OECD Doc. No.C(72)128, 1972 WL 24710 (26 May 1972).

> The principle to be used for allocating costs of pollution prevention and control measures to encourage rational use of scarce environmental resources and to avoid distortions in international trade and investment is the so-called "Polluter-Pays-Principle". This Principle means that the polluter should *bear the expenses of carrying out the above mentioned measures decided by public authorities* to ensure that the environment is in an acceptable state. In other words, the cost of these measures should be reflected in the cost of goods and services which cause pollution in production and/or consumption. Such measures *should not be accompanied by subsidies* that would create significant distortions in international trade and investment.[355] [emphasis added]

This formulation of the polluter-pays principle specifies the costs and suggests the basic rules for the implementation of this principle. However, environmental damage is excluded from such costs.[356] To make the principle better suit changing situations, the OECD adopted further recommendations in 1974, 1989 and 1991 respectively, supplementing the content of the Guidelines on the implementation of the principle and its exceptions as well.[357] Similarly, the European Community (EC) also adopted the polluter-pays principle in its various recommendations and acts of EC secondary legislation.[358]

The polluter-pays principle has not only been applied 'in a geographic region subject to uniform environmental law' such as the OECD and EC,[359] it has also been widely endorsed in a number of international instruments. Sands asserts that the polluter-pays principle in international treaty law originated from early treaties on civil liability for damages from hazardous activities,[360] such as the

355 Ibid annex para. 4.

356 Sands, above n. 130, 281.

357 For instance, in 1974 a new OECD recommendation called on its member States to observe the polluter-pays principle uniformly, and defined 'polluter' as 'someone who directly or indirectly damages the environment or who creates conditions leading to such damage'; in 1989 the OECD Recommendation on the Application of the Polluter-Pays Principle to Accidental Pollution implied that the operator of a hazardous installation should bear relevant cost, provided guidance on 'reasonable measures', and listed certain exceptions to the principle; in 1991 a final recommendation urged OECD member States to treat 'economic instruments' as a means of implementing this principle. See ibid.; Schwartz, above n. 353, 244; Kiss and Shelton, above n. 1, 215.

358 Sands, above n. 130, 283–284.

359 Kiss and Shelton, above n. 1, 215.

360 Sands, above n. 130.

1963 IAEA Liability Convention,[361] *1969 CLC*,[362] and *1971 Oil Fund Convention*.[363] In these treaties, the ideas on the polluter-pays principle were reflected by providing that the damage resulting from hazardous activities should be borne by the shipping industry and oil cargo interests,[364] although the polluter-pays principle was not explicitly invoked. In 1992, Principle 16 of the *Rio Declaration* explicitly raised the polluter-pays principle to the global level, providing that:

> National authorities should endeavour to promote the *internalization of environmental costs* and the use of economic instruments, taking into account the approach that *the polluter should, in principle, bear the cost of pollution*, with due regard to the public interest and without distorting international trade and investment.[365] [emphasis added]

The formulations of the polluter-pays principle by the OECD and Principle 16 of the *Rio Declaration* indicate that the purpose of this principle is to internalise the economic costs of pollution control and prevent governments from subsidising these environmental costs. Due to the global participation and profound significance of the 1992 Rio Conference on Environment and Development (UNCED), the polluter-pays principle was for the first time recognised globally as an environmental policy in 1992.[366] But essentially this principle was still not legally binding due to its not having achieved the status of 'the normative character of a rule of law'.[367] After UNCED it was endorsed by more international instruments.[368] Examples include *1992 Convention for the Protection of the Marine Environment of the North-East Atlantic* (OSPAR

361 *IAEA Vienna Convention on Civil Liability for Nuclear Damage*, opened for signature 29 May 1963, 1063 UNTS 265 (entered into force 12 November 1977).

362 *International Convention on Civil Liability for Oil Pollution Damage*, opened for signature 29 November 1969, 973 UNTS 3 (entered into force 19 June 1975).

363 *International Convention on the Establishment of an International Fund for Compensation for Oil Pollution Damage*, opened for signature 18 December 1971, 11 ILM 284 (entered into force 16 October 1978).

364 Sands, above n. 130, 281.

365 *Rio Declaration* prin 16.

366 Birnie, Boyle and Redgwell, above n. 4, 322.

367 Ibid.

368 Before the Rio Conference, the polluter-pays principle was endorsed by the 1990 OPRC in its preamble that the polluter-pays principle is 'a general principle of international environmental law'. See *International Convention on Oil Pollution Preparedness, Response and Cooperation*, opened for signature 30 November 1990 (entered into force 13 May 1995) preamble ('*OPRC*').

Convention),[369] *1992 Convention on the Protection and Use of Transboundary Watercourses and International Lakes* (*Watercourses Convention*),[370] *1992 Convention on the Transboundary Effects of Industrial Accidents* (*Industrial Accidents Convention*),[371] and *1996 Protocol to the Contracting Parties to the Convention on the Prevention of Marine Pollution by Dumping of Wastes and Other Matter* (*1996 Protocol to London Dumping Convention*).[372]

The polluter-pays principle is treated as 'a rule of economic, juridical and political good sense'.[373] It seeks to address the issues of 'cost allocation' and 'cost internalisation'.[374] 'Cost allocation' of this principle resolves the question of 'who pays' for the pollution prevention and control, whereas its 'cost internalisation' answers the question of 'how much should be paid'.[375] As noted earlier, cost internalisation is a concept from economics. With this concept, the polluter-pays principle aims to improve economic efficiency by 'internalising external environmental costs of production and consumption into

369 *Convention for the Protection of the Marine Environment of the North-East Atlantic*, opened for signature 22 September 1992, 32 ILM 1068 (entered into force 25 March 1998) art. 2 ('*OSPAR Convention*'). This article reads that 'The Contracting Parties shall apply: … the polluter pays principle, by virtue of which the costs of pollution prevention, control and reduction measures are to be borne by the polluter.'

370 *Convention on the Protection and Use of Transboundary Watercourses and International Lakes*, opened for signature 17 March 1992, 31 ILM 1312 (entered into force 6 October 1996) art. 2 ('*Watercourses Convention*'). This article stipulates that 'the Parties shall be guided by the following principles: … The polluter-pays principle, by virtue of which costs of pollution prevention, control and reduction measures shall be borne by the polluter.'

371 *Convention on the Transboundary Effects of Industrial Accidents*, opened for signature 17 March 1992, 31 ILM 1330 (entered into force 19 April 2000) preamble ('*Industrial Accidents Convention*'). The preamble states that '[t]aking account of the polluter-pays principle as a general principle of international environmental law.'

372 *Protocol to the Contracting Parties to the Convention on the Prevention of Marine Pollution by Dumping of Wastes and Other Matter*, opened for signature 7 November 1996, 36 ILM 1 (entered into force 24 March 2006) art. 3 ('*1996 Protocol to London Dumping Convention*'). This article reads that 'Taking into account the approach that the polluter should, in principle, bear the cost of pollution, each Contracting Party shall endeavour to promote practices whereby those it has authorized to engage in dumping or incineration at sea bear the cost of meeting the pollution prevention and control requirements for the authorized activities, having due regard to the public interest.'

373 Organisation for Economic Co-operation and Development (OECD), *The Polluter Pays Principle* (OECD Publishing, 1975) 25.

374 Tobey and Smets, above n. 352, 64.

375 Ibid.

market prices'.[376] Another relevant question is 'how to pay', which is sometimes interpreted as the implementation of the polluter-pays principle.

Since the meaning of the polluter-pays principle varies in different contexts,[377] currently there are no indisputable answers to the above three questions. However, some of the interpretations or options under discussion are provided below in Table 2.1. Firstly, who pays for the pollution? Or, who is the polluter? In contrast to the rigid definition provided by the OECD, in practice the concept of the polluter varies depending on different categories in different contexts. According to Schwartz, at least three categories may apply based on different criteria.[378] A list of these categories and types of persons is summarised in Table 2.1.

TABLE 2.1 *Different categories of polluters*[379]

Category Criteria of Polluters	Types of Persons as Polluters
Personality	States, corporations, industries, individuals
Nature and effects of conduct or activity	Any activity that contributes to the deterioration of the environment, including natural resource use for economic or social purposes and attaching liability to direct or indirect environmental consequences. Examples: the handling or disposal of waste; the use and management of water resources; enjoyment of environmental quality as in use of recreational facilities ...

376 Ibid.

377 Jonathan Remy Nash, "Too Much Market? Conflict between Tradable Pollution Allowances and the 'Polluter Pays' Principle" (2000) 24(2) *Harward Environmental Law Review* 465, 472–473.

378 Schwartz, above n. 353, 247–248.

379 Ibid.

TABLE 2.1 *Different categories of polluters* (cont.)

Category Criteria of Polluters	Types of Persons as Polluters
Scope of responsibility	Subject: individual or collective, partial or total, actual or potential Measures: aid, technology transfer, or emission reduction programmes provided by developed States to developing States (application of the CBDR)

These categories of polluters, however, may not always be responsible for the pollution they cause. As indicated in Table 2.1, sometimes the polluter may be only partially responsible. So in practice a case-by-case examination for determining the particular polluter's responsibility for the pollution should always be undertaken.

Secondly, what should the polluter pay for? Or in other words, what is the cost? Similarly to the first question, different regional or international instruments list various types of cost to be borne or paid as a result of incorporating the polluter-pays principle. Some of these views are listed in Table 2.2.

TABLE 2.2 *Different cost bearing arrangements by regional or international instruments*[380]

Regional or International Instruments	Types of Cost Bearing
OECD 1972, 1974	(1) Cost related to measures needed to prevent, control and reduce pollution; (2) Cost of administrative measures by the authorities in response to pollution, including those implementing anti-pollution policies, developing anti-pollution technologies and grants for modernising out-of-date plants.

380 This table is developed based on the following sources: MacInnis, above n. 348, 148; Schwartz, above n. 353, 248–249.

Regional or International Instruments	Types of Cost Bearing
Principle 16 of the Rio Declaration Agenda 21 OECD 1974, 1989, 1991 Environmental Liability Directive 2004	Cost of pollution and 'environmental cost', including: cost related to pollution regulation, environmental protection and management. Cost of achieving prescribed environmental quality, preventing or remedying environmental damage, preventing accidental pollution, and the clean-up or reinstatement of the environment after an accident, the 'cost of exceptional measures' needed to protect human health and the environment.
Landfill Directive 1999	Social cost, remaining external cost of investment on technology, the cost incurred when a ban is placed on polluting activity, indemnity cost, operational cost, including the cost of present and future expenditure and loss of profit, even if not accessible in monetary terms.
Antarctic Treaty Secretariat 2005	(1) Cost of abatement, compensation and reparation; (2) Cost of promoting best environmental practice, best available technology, and the cost of a prompt and effective response to environmental emergencies.
A General Summary	*(1) Costs of pollution control by governments;* *(2) Emergency response and clean-up costs; and* *(3) Compensation to victims of pollution.*

Through examining different cost-bearing arrangements in regional and international instruments, it may be inferred that based on the polluter-pays principle, a polluter often needs to pay three types of cost. They are costs of pollution control by governments, emergency response and clean-up costs, and compensation to victims of pollution.[381] While in a particular case, there could be so many types of cost for the polluter to bear, generally the polluter is

381 MacInnis, above n. 348, 148.

the first one to pay such cost. However, it will not invalidate the polluter-pays principle if the polluter passes on such costs to the consumer.[382] As argued by Kiss, 'if the polluter holds a right to pollute, the victims must pay for cessation or reduction of the activity'.[383] The mechanism of the polluter-pays principle is underpinned by both economic and social theories.

Thirdly, how to pay? Or, how to implement the polluter-pays principle? To answer this question, the OECD provides seven options, namely direct controls, taxes, payments, subsidies, various incentives (tax benefits, accelerated amortisation, credit facilities), the auction of pollution rights and charges.[384] However, Schwartz puts forward four categories of implementation methods under the polluter-pays principle. The main content is listed in Table 2.3.

TABLE 2.3 *Different implementation methods of the polluter-pays principle*[385]

Category	Sub-category	Examples
Regulatory Regime	Command and control	Fuel sulphur emission limits in *MARPOL* Annex VI
	Self-regulation	Market-based instruments, e.g., environmental fees, tradable permit, liability rules
	Voluntary initiatives	Energy Efficiency Operational Indicator (EEOI) developed by the IMO
Economic Strategies	Internalisation	Cost incorporated with the external effects
	Incentives	ET, CDM, JI within the Kyoto Protocol; Carbon tax
	Initiatives	Employ best available techniques

382 OECD, above n. 373, 27.
383 Kiss and Shelton, above n. 1, 214.
384 OECD, above n. 373, 28.
385 This table is developed partially based on the following source: Schwartz, above n. 353, 249–255.

Category	Sub-category	Examples
	Innovations	Advocate investment by adopting new measures, technology designs or environmentally friendly products
Liability Regime	Fault-based liability	Deterrence, redress, reparation, and restitution of States
	Strict liability	Operator non-fault liability
Cooperative Regime	Bilateral, regional and global cooperation	managing transboundary risks/harm

Generally the above methods may be utilised in a particular case independently or jointly so as to improve performance in combating pollution. However, for the purpose of this book, the assessment of the regulatory regime is the focus.

As the term 'in principle' used in Principle 16 of the *Rio Declaration* may imply, the polluter-pays principle allows certain exceptions from its strict application. Generally speaking, States should follow the polluter-pays principle 'except when it would be socially, economically, or environmentally unreasonable to do so'.[386] This policy arrangement is consistent with the concept of 'equitable internalisation' in that they both take into account the different responsibilities or capacities among different States when such policies apply.[387] This feature, however, also indicates its 'soft' law nature.

386 Sanford E. Gaines, 'The Polluter-Pays Principle: from Economic Equity to Environmental Ethos' (1991) 26(3) *Texas International Law Journal* 463, 477.

387 Nash, above n. 377, 476–477. 'Equitable internalisation' refers to the allocation of 'abatement costs and the costs of residual pollution among polluters and between polluters and victims'.

2.6.2 *The Application of the Polluter-Pays Principle to the Issue of Greenhouse Gas Emissions from International Shipping*

Although endorsed by various regional and international instruments, the polluter-pays principle has not gained 'the same degree of support or attention' in recent years as the preventive and precautionary principles,[388] or the CBDR. As asserted by Gaines, the polluter-pays principle cannot address all environmental problems.[389] Similarly the application of the polluter-pays principle to the issue of GHG emissions from international shipping is not straightforward.

As discussed in 2.3 of this chapter, transboundary harm caused by GHG emissions of ships to other areas may include four scenarios. In each scenario, the harm may lead to adverse effects to persons, property or the environment. According to the polluter-pays principle, generally the polluter should bear all the costs that such emissions may generate. The polluter may include the flag State,[390] ship owners and operators,[391] or in some cases, individuals who should be responsible for such damage. However, in practice ship owners and ship operators are generally regarded as the polluters of GHG emissions from ships due to their direct contribution to these emissions. The recently-adopted technical and operational measures by the IMO clearly reflect this rule. Through implementing the EEDI and SEEMP, shipowners primarily pay for the higher shipbuilding cost whereas ship operators pay the cost relating to implementing stringent operational requirements. Flag States may also be responsible for transboundary harm resulting from such emissions under certain circumstances, which may include regulation and negotiation related costs as can be seen from the *Trail Smelter* case.

Compared with the identification of the polluter of GHG emissions from international shipping, the cost bearing and implementation methods of the polluter-pays principle are more complicated. International shipping refers to 'shipping between ports of different countries, as opposed to *domestic shipping*'.[392] Such voyages may consist of domestic voyages (shipping within

---

388 Sands, above n. 130, 280.

389 Gaines, above n. 386, 487.

390 In international shipping, the flag State may be responsible for the transboundary damage caused by the emissions from the ships flying its flag due to its role of exercising its jurisdiction.

391 Ship operators generally include ship managers and charterers. They should be held liable for their choice to employ the services of a substandard vessel. See Pamborides, above n. 305, 145.

392 Buhaug et al., above n. 185, 13. According to the *Second IMO GHG Study 2009*, 'domestic shipping' refers to 'shipping between ports of the same country, as opposed to *international shipping*', and excludes military and fishing vessels.

the maritime zones of a State, including the internal water, territory sea, and EEZ) and international voyages (shipping outside the maritime zones of a State) of the flag State. Regarding international voyages, in practice two approaches of dealing with transboundary harm from ships are in place based on the different injuries. When such injuries are 'slight and infrequent', the polluter may be more willing to 'absorb them without increasing its level of control'.[393] This offer is generally accepted by the victims in that such harm may not be easily recognised due to its cumulative nature and the litigation or arbitration cost may be higher than the value of such a claim.[394] However, when the possible compensation arising from the injuries is significant, the polluter may prefer to avoid or reduce such costs through legal means.[395] In this respect, the polluter-pays principle generally applies. These approaches are argued to be 'appropriate' if examined from the economic, environmental and social ethics perspective.[396] When harm occurs during the domestic voyage of the flag State's vessel, the polluter-pays principle may not be applicable. Instead, the traditional legal principle that 'injuries incidental to lawful activities will not be compensated' may apply.[397] This is because of the fact that the affected party, as a member from the same State, probably benefits indirectly from the shipping activities. In this context, depending on different situations, the flag State may apply relevant domestic legislation to this issue, into which the polluter-pays principle may not be incorporated. It appears that a uniform cost-bearing mechanism in relation to GHG emissions from ships is necessary to be established globally so as to address this divergence in current shipping practice.

The previous section concludes that the costs may include costs of pollution control by governments, emergency response and clean-up costs, and compensation to victims of pollution. Due to the cumulative nature of GHG emissions from ships, the cost relating to shipping GHG emissions may only include the first and third of these costs. While pollution costs can be calculated through identifying measures that have been taken by governments, the identification of victims is complicated. This is mainly due to the nature of this issue being a part of the global climate change regime. Often while the

---

393 Gaines, above n. 386, 492. Gaines refers to general environmental harm when analysing these two possibilities, which in the view of the present author could also be applied to the environmental harm brought about by international shipping.

394 Ibid.

395 Ibid.

396 Ibid.

397 Ibid.

polluter is discharging GHG emissions from ships, they also suffer from its adverse effects either directly or indirectly.[398] In this sense, polluters are often victims of such pollution. Therefore, the application of the polluter-pays principle, or specifically the implementation of allocation of costs, to the issue has to be put in a global context through adopting globally uniform measures. If this assertion is related to the MBMs that are currently under discussion within the IMO, MBMs which involve the global emissions reduction of different sectors may better reflect the polluter-pays principle.[399]

Among the four categories of the implementation methods mentioned above, the liability regime is less useful in this context due to the difficulty of identifying specific polluters and victims. The cooperative regime is necessary but could be integrated into other categories whenever it applies. Economic strategies are useful, which can be clearly seen from the three mechanisms established under the *Kyoto Protocol*.[400] As good examples of economic strategies, emissions trading (ET), the clean development mechanism (CDM) and joint development (JI) have been achieving success. For the purpose of this book, the development of a regulatory regime for the issue under discussion is the focus and there is still ample room for further steps.

Concerning the method of 'command and control', the 2011 amendments of Annex VI to *MARPOL 73/78* regulate the mandatory EEDI for new ships and SEEMP for all ships. These measures will have profound influences on the reduction of GHG emissions from ships.[401] As for voluntary initiatives, a wide range of discussions were held within the IMO and as a result, many measures, such as Energy Efficiency Operational Indicator (EEOI), have been suggested by the IMO as voluntary measures for all States. The third type of regulatory regime, self-regulation, often called a market-based instrument (MBI) or MBMs, has aroused intense debate within the IMO. The adoption of

398 See Vesselin Popovski and Kieran G. Mundy, 'Defining Climate-Change Victims' (2012) 7(1) *Sustainability Science* 5, 5. Some small island States have announced that they would take those States contributing to the most emissions to international legal proceedings. For example, in 2002 Tuvalu (later joined by Kiribati and Maldives) announced that it was taking Australia to the International Court of Justice for the damages Australia has caused via its climate policy. However, Koivurova asserts that this approach would not work. Timo Koivurova, 'International Legal Avenues to Address the Plight of Victims of Climate Change: Problems and Prospects' (2007) 22(2) *Journal of Environmental Law and Litigation* 267, 277, 298.

399 See ch. 4, 4.3.4; ch. 7, 7.5.2.2.

400 The three mechanisms under the *Kyoto Protocol* are clean development mechanism (CDM), emissions trading (ET) and joint implementation (JI).

401 See ch. 4, 4.3.3.3.

MBMs complies with the polluter-pays principle. The environmental fees (contributions), being one type of MBMs,[402] can be taken as an example. An environmental fee is generally imposed on a unit of pollution thus providing the polluter with an incentive to reduce the amount of pollution in order to avoid heavy fees.[403] The setting of suitable fee rates, or in other words, the calculation of internalised cost, is crucial. If a fee is set too low compared with the cost needed for the reduction of one unit of pollution, the polluter may prefer to pay and continue polluting.[404] One example of the environmental fee is a fund for GHG emissions from international shipping (GHG Fund). Basically the contributions to the GHG Fund are paid per tonne of bunker fuel by the polluter as the cost for preventive measures in this context, and are allocated to possible affected parties or victims whenever it applies. This approach complies with the polluter-pays principle in this regard. Lastly it is arguable that equitable consideration for the implementation of the polluter-pays principle cannot be ignored in the context of GHG emissions from international shipping. This is because equitable consideration not only imposes flexibility on the implementation of the polluter-pays principle, but also resonates with the CBDR principle. This approach may better encourage the participation of developing States in global emissions reduction from international shipping.

2.7 Conclusion

This chapter serves as the theoretical foundation for the book especially the gap analysis and gap-filling options relating to the current regulatory framework for the reduction of GHG emissions from international shipping to be raised in the following chapters. It was first argued that GHG emissions from international shipping, in particular CO_2, are a type of 'conditional' pollution. Therefore, many marine pollution-related treaties apply to this GHG emissions issue. This argument and the principles relating to flag State, coastal State and port State jurisdiction also underpin the application of international environmental law principles to GHG emissions from international shipping.

---

402 Three main types of MBIs include environmental fees, tradable permit (allowance) schemes, and liability rules. *Scientific Study on International Shipping and Market-Based Instruments*, IMO Doc. MEPC 60/INF.21 (15 January 2010) annex, para. 2.3; see also ch. 4, 4.3.4.2.1.

403 Ibid. para. 2.4.

404 Ibid.

It was further argued that GHG emissions from international shipping might lead to transboundary harm under four scenarios. On this basis, the duties associated with transboundary harm would apply in the context of GHG emissions from international shipping. These include a flag State's primary prescriptive and enforcement jurisdiction and responsibility to prevent, reduce and control transboundary harm resulting from GHG emissions from the ships entitled to fly its flag. To achieve this goal, flag States need to adopt national legislation on the reduction of such emissions, taking into account the amended Annex VI to *MARPOL 73/78* irrespective of whether they have ratified this amendment. Flag States need to conduct regular surveys, issue or empower other parties to issue the IEE Certificate to ships flying their flags, as well as impose administrative penalties or institute proceedings in relation to offences. Furthermore, coastal States and port States also have a duty to cooperate in mitigating transboundary environmental risks arising from excessive GHG emissions from international shipping.

GHG emissions from ships have been recognised as harmful, but there is not yet scientific proof that they have caused specific impacts. The application of the precautionary principle in this context would justify the action of States in taking proactive steps to tackle shipping GHG emissions. In contrast to the precautionary principle, the polluter-pays principle aims to address three relevant questions, namely: Who is the polluter? What should the polluter pay for? And, how to pay? It has been argued in this chapter that in the context of the GHG emissions issue, the polluter should include ship owners, ship operators and flag States under certain circumstances. The cost should be put in a global context through adopting uniform measures, whereas the means of payment could include various technical and operational measures and MBMs. In particular, MBMs which involve the global emissions reduction of different sectors may better reflect the polluter-pays principle.

Whether the CBDR and NMFT principle should be applied to GHG emissions from international shipping is a controversial issue. This chapter argues that both the CBDR and the NMFT principles have solid ground for their application to this GHG emissions issue, and it is nearly impossible to exclude either of them in this regard. In this context it was argued that the *IMO Convention* and the *LOSC* provide the IMO with general competence to regulate GHG emissions from ships, while the *Kyoto Protocol* gives the IMO a specific mandate to regulate this matter. It is thus reasonable for both principles to apply to this GHG emissions issue. It was further argued that depending on the nature of regulatory measures, the CBDR and NMFT principles could be incorporated into the regulation of GHG emissions from ships in different ways.

The application of these selected international law principles to the regulation of GHG emissions from international shipping has several implications. It reveals that the GHG emissions issue is reflective of, or subject to, the underlying principles of international environmental law. These principles should thus be taken into account in the developing regulatory regime of GHG emissions from ships. In addition, the development of this regime also has resulted in new implications for these principles, and impacts on their evolution. This interaction can be seen from the interpretation of the CBDR and NMFT principles in the context of shipping GHG emissions.

CHAPTER 3

The UN Response to the Issue of Greenhouse Gas Emissions from International Shipping

3.1 Introduction

Greenhouse gas (GHG) emissions from international shipping contribute to global warming and climate change, while international regulation on shipping emissions is subject to the evolution of the international climate change regime.[1] Climate change did not become an issue of global concern until it was brought to the attention of the UN. In 1987 a report entitled *Our Common Future* was discussed in the UN General Assembly, attracting worldwide attention to the global issues of development and environment. In this report, 'climate change' was mentioned nine times as a fast-growing global threat.[2] It was also in this meeting that the scientific community formally brought the climate change issue to the political agenda under the auspices of the World Meteorological Organization (WMO) and the United Nations Environment Programme (UNEP), specialised agencies of the UN.[3] One year

1 The international climate change regime, often called the *UNFCCC-Kyoto Protocol* regime, mainly refers to a series of multilateral environmental agreements (MEAs), their related protocols and soft law in relation to climate change. See Patricia W. Birnie, Alan E. Boyle and Catherine Redgwell, *International Law and the Environment* (Oxford University Press, 3rd ed, 2009) 84. The international regulation on the reduction of GHG emissions from international shipping is primarily the mandate of the IMO as indicated in Article 2(2) of the *Kyoto Protocol*. On this basis, the regulation by the IMO on this GHG emissions issue should be subject to the international climate change regime, or in other words, the UNFCCC-Kyoto Protocol regime. This issue is discussed at 3.3 of this chapter.

2 The term 'climate change' was mentioned nine times in the text and twice in the footnotes of the report. Item 32, Chapter 1 of the report reads that, 'it is true globally for such threats as climate change, ozone depletion, and species loss, [and the] risks increase faster than do our abilities to manage them.' Item 11, chapter 7 of the report reads that, "[t]he environmental risks and uncertainties of a high energy future are also disturbing and give rise to several reservations ... the serious probability of climate change generated by the 'greenhouse effect' of gases emitted to the atmosphere, the most important of which is carbon dioxide (CO_2) produced from the combustion of fossil fuels." World Commission on Environment and Development (WCED), *Our Common Future* (Oxford University Press, 1987) 35, 146–147.

3 Bert Bolin, *A History of the Science and Politics of Climate Change: the Role of the Intergovernmental Panel on Climate Change* (Cambridge University Press, 2007) 40.

 | DOI 10.1163/9789004329317_004

later, the WMO and the UNEP established the Intergovernmental Panel on Climate Change (IPCC). In 1992 the *United Nations Framework Convention on Climate Change* (*UNFCCC*)[4] was adopted at the Rio United Nations Conference on Environment and Development (UNCED), and its *Kyoto Protocol* and *Paris Agreement* was then adopted in 1997 and 2015 respectively.[5] The *UNFCCC*, its *Kyoto Protocol* and *Paris Agreement*, together with the agreements or declarations adopted in their Conferences of Parties (COPs), the COPs serving as the Meeting of the Parties to the Kyoto Protocol (CMPs), and the COPs serving as the Meeting of the Parties to the Paris Agreement (CMAs), constitute the core elements of the current climate change regime. They have significantly shaped the direction of international regulation on the reduction of GHG emissions from international shipping.

This chapter examines the responses from the UN to the issue of GHG emissions from ships, aiming to identify the areas that need to be improved to facilitate and improve the global regulation of GHG emissions from international shipping. This chapter consists of two main parts. The first part introduces the UN's institutional responses to the issue, in particular the responses from the IPCC, as well as the interaction among the IPCC, UNEP, and other UN agencies. The second part examines the international legal framework on climate change from two perspectives: analysis of two conventions on the prevention of atmospheric pollution prior to the *1992 UNFCCC* and a critical review of the *UNFCCC*, its *Kyoto Protocol* and agreements produced in their COPs and CMPs.

3.2 The UN Institutional Responses

Climate change is an urgent environmental problem of a global nature, which makes it difficult for individual States to develop an effective national regulatory response. To cope with this issue, the UN, including its various agencies, has responded actively. This part reviews the work conducted by the IPCC in combating climate change, and the contributions from the UNEP, WMO, and other UN institutions. Due to their key roles in regulating GHG emissions from

4 *United Nations Framework Convention on Climate Change*, opened for signature 9 May 1992, 31 ILM 848 (entered into force 21 March 1994) ('*UNFCCC*').

5 *Kyoto Protocol to the United Nations Framework Convention on Climate Change*, opened for signature 16 March 1998, 37 ILM 22 (entered into force 16 February 2005) ('*Kyoto Protocol*'); *Paris Agreement to the United Nations Framework Convention on Climate Change* (2015), opened for signature 22 April 2016, FCCC/CP/2015/L.9 (not yet in force) ('*Paris Agreement*').

international shipping, the responses by the IMO, a UN specialised agency, are examined in Chapter 4.

3.2.1 *The UN and the IPCC*

The late 1970s and the 1980s witnessed a growing debate among scientists and policy makers on the risks associated with human-induced climate change. The need for independent, scientific and technical advice became apparent to inform decision-making on this important and complex issue. The first efforts were made by the United States (US) although this initiation was triggered by the energy crisis in the 1970s.[6] The US government treated climate change as 'a threat to humankind', and its National Academy of Science (NAS) conducted an assessment on possible future human-induced changes of climate in 1977.[7] The inclusion of this issue in the political arena of the US government in 1978 encouraged more research in relation to climate change.[8] However, it was only in 1980 that the International Council of Science (ICSU), UNEP and WMO jointly developed a first international assessment on climate change, although this effort proved to be 'not very successful'.[9] To cope with this challenge, in 1988 the 43rd Session of the United Nations General Assembly (UNGA) adopted a resolution entitled 'Protection of the global climate for present and future generations of mankind'. The resolution endorsed the action by UNEP and WMO in jointly establishing the IPCC, and requested the IPCC to prepare a comprehensive review and recommendations on all aspects of climate change and its impacts, with a view to formulating realistic response strategies.[10] Therefore, the IPCC was set up by the WMO and UNEP as an effort by the UN to provide the governments of the world with a reliable scientific view on climate change. As discussed earlier, the report named *Our Common Future* triggered this process within the UN. The IPCC is intended to serve as a link between the scientific

---

6 In 1978, the Carter administration of the US intended to use domestic coal to solve the energy crisis, which brought the issue of climate change into the political agenda for the first time. Nicolas Nierenberg, Walter R. Tschinkel and Victoria J. Tschinkel, 'Early Climate Change Consensus at the National Academy: The Origins and Making of Changing Climate' (2010) 40(3) *Historical Studies in the Natural Sciences* 318, 319.

7 Bolin, above n. 3, 33.

8 Examples are the report by the JASON defense advisory panel chaired by Gordon MacDonald in 1979 and a report by an ad hoc National Academy of Sciences (NAS) with Jule G. Charney as the lead author in the same year. Nierenberg, Tschinkel and Tschinkel, above n. 6.

9 Bolin, above n. 3, 35.

10 *Protection of Global Climate for Present and Future Generations of Mankind*, GA/Res 43/53, 43rd sess, 70th plen mtg, UN Doc A/RES/43/53 (6 December 1988) art. 5, 10.

community and political institutions, and thus promote the construction and improvement of the international climate change regime.

Under the auspices of the UN, the IPCC's structure and working mechanisms have been improving. Currently the IPCC has three Working Groups and a Task Force on National Greenhouse Gas Inventories. Working Group I deals with 'the Physical Science Basis of Climate Change', Working Group II with 'Climate Change Impacts, Adaptation and Vulnerability' and Working Group III with 'Mitigation of Climate Change'. The Task Force on National Greenhouse Gas Inventories aims to develop and refine a methodology for the calculation and reporting of national GHG emissions and removals. It meets in Plenary at the level of Representatives of Governments, and is assisted by Technical Support Units (TSU) hosted and financially supported by the Government of the developed country co-chair of that Working Group/Task Force. Other departments within the IPCC include the IPCC Bureau, IPCC Secretariat and IPCC Executive Committee.[11] This structure is illustrated in Figure 3.1.

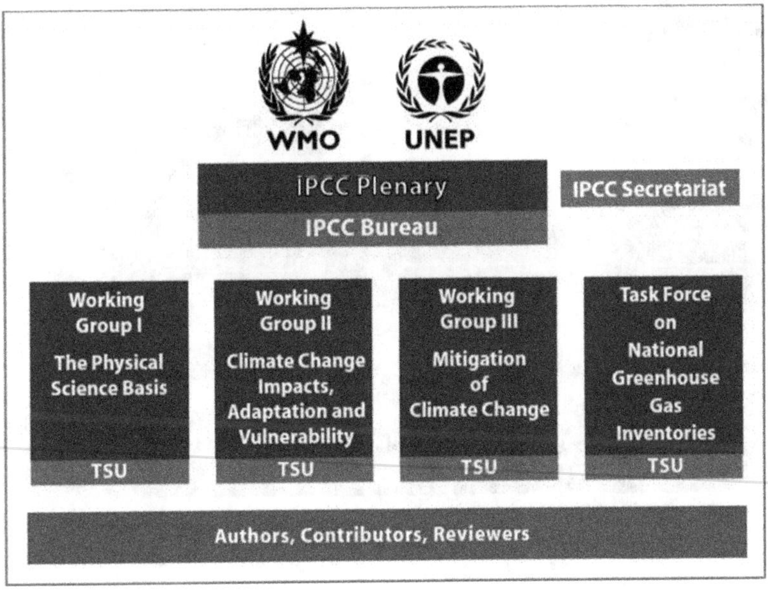

FIGURE 3.1 *Structure of the IPCC.*[12]

11 Intergovernmental Panel on Climate Change (IPCC), *Structure* <http://www.IPCC.ch/organization/organization_structure.shtml> accessed 22 August 2012.

12 Ibid.

Due to its scientific and intergovernmental nature, the IPCC's work is to be 'policy-relevant and yet policy-neutral, never policy-prescriptive'.[13] Apart from that, scientific integrity, objectivity, openness and transparency are other principles that the IPCC should apply.[14] Generally the IPCC provides reports for the information of policy-makers within governments. To ensure the incorporation of the principles mentioned above into its reports, the IPCC has to follow strict procedures. In 2010, as a response to the request by the IPCC Chair and the Secretary-General of the UN, the InterAcademy Council (IAC) reviewed the IPCC's processes and procedures and put forward some proposals for improvement which were partly adopted by the IPCC. Figure 3.2 describes how the IPCC reports are currently produced.

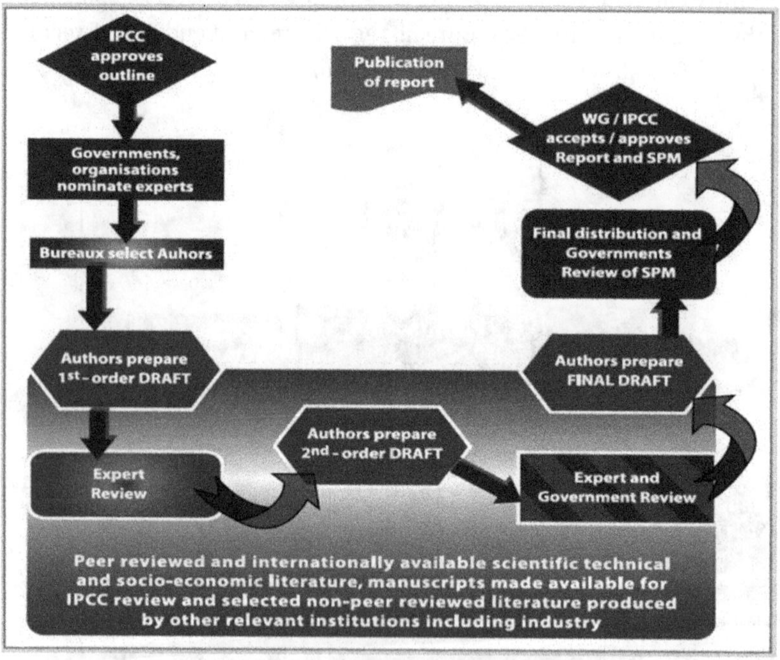

FIGURE 3.2 *The procedure of drafting and reviewing reports by the IPCC.*[15]

13 Intergovernmental Panel on Climate Change (IPCC), *Organization* <http://www.IPCC.ch/organization/organization.shtml> accessed 17 July 2012.

14 World Meteorological Organization (WMO), *Intergovernmental Panel on Climate Change (IPCC)*<http://www.WMO.int/pages/themes/climate/international_IPCC.php> accessed 22 August 2012.

15 Intergovernmental Panel on Climate Change (IPCC), *Principles and Procedures* <http://www.IPCC.ch/organization/organization_procedures.shtml> accessed 22 August 2012.

To date the IPCC has issued five Assessment Reports contributing to the combating of climate change around the world. The IPCC First Assessment Report of 1990 revealed the significance of climate change as a natural and political issue, and thus played a 'decisive' role in the adoption of the *UNFCCC*. The 'Legal measures' paper submitted by the Response Strategies Working Group of the IPCC laid the foundation for the drafting and adoption of the *UNFCCC*.[16] Apart from that, the IPCC has remained the most important source of scientific, technical and socio-economic information for the *UNFCCC*, after the entry into force of the Convention, through its Special Reports, Technical Papers and Methodology Reports. Since 1991 the IPCC has supported the *UNFCCC* by preparing Methodology Reports for National GHG Inventories.[17] Thus, the relationship between the *UNFCCC* and the IPCC is deemed as 'a model for interaction between science and decision-makers'.[18] The IPCC Second Assessment Report of 1995 provided key input to the further development of the *UNFCCC*, in particular the adoption of its *Kyoto Protocol*. The IPCC Third Assessment Report of 2001 and the IPCC Fourth Assessment Report of 2007 further confirmed the contribution of GHG emissions to climate change and global warming. In September 2013 and early 2014, the IPCC released the reports of its three working groups, and a synthesis report was released on 2 November 2014.[19] These reports further strengthen the scientific evidence of anthropogenic climate change, and leave 'fewer uncertainties about the serious consequences of inaction'.[20]

It is clear that the UN helped to establish the IPCC and monitor its sound development. The IPCC, conversely, underpins the efforts of the UN in persuading countries around the world to recognise and combat climate change jointly. One of these achievements is the *UNFCCC* and its *Kyoto Protocol*, which determines the direction of global regulating GHG emissions from international shipping through giving the IMO a GHG mandate, setting the reduction

---

16 Jill Barrett, 'The Negotiation and Drafting of the Climate Change Convention' in Robin Churchill and David Freestone (eds), *International Law and Global Climate Change* (1991) 183, 184–187.

17 Intergovernmental Panel on Climate Change (IPCC), *Understanding Climate Change: 22 Years of IPCC Assessment* <http://www.IPCC.ch> accessed 17 July 2012.

18 Ibid.

19 Intergovernmental Panel on Climate Change (IPCC), 'Fifth Assessment Report (AR5)' (2014) <http://www.IPCC.ch/> accessed 18 June 2014.

20 Dahe Qin, *Opening Remark at Working Group I—Twelfth Session* (23 September 2013) <http://www.IPCC.ch/meetings/session36/speeches/op_wg1_p12_Dahe_Qin.pdf> accessed 18 June 2014, p. 2.

targets for UNFCCC Annex I States, and discussing regulatory principles for GHG emissions from ships.

3.2.2 *Other Institutions and Their Interaction*

In addition to the IPCC, some other UN subsidiary bodies or agencies also contribute to combating of climate change. As the 'voice for the environment within the United Nations system' established in 1972,[21] UNEP established a Climate Change sub-program. In this program, UNEP works with countries, particularly developing countries, to raise public awareness of the Earth's changing climate, strengthen countries' ability to adapt to climate change, and integrate climate change responses into their national development processes.[22] Essentially UNEP is assigned 'a catalytic and coordinating role' in the management of the climate change issue within the UN system.[23] The World Meteorological Organization (WMO) is regarded as the UN system's 'authoritative voice on the state and behaviour of the Earth's atmosphere'.[24] It exercises important functions such as weather and climate observation and monitoring, understanding of climate processes, the development of clear, precise and user-targeted information and other services for policy makers.[25] The UN Economic Commission for Europe (UNECE) also contributed significantly to the adoption of the *1979 Convention on Long-Range Transboundary Air Pollution* (CLRTAP).[26] The above work makes a substantial contribution to international efforts in fighting climate change. However, this work essentially implements the outcomes within the *UNFCCC-Kyoto Protocol* regime rather than regulating climate change. The scope of these institutions seldom includes the GHG emissions from international shipping.

Other institutions, the Committee on Trade and Environment (CTE) within the World Trade Organization (WTO) as an example,[27] contribute little to the issue of climate change. Although the CTE deals with the relationship between

---

21 United Nations Environment Programme (UNEP), *About UNEP* <http://www.UNEP.org> accessed 18 July 2012.

22 Ibid.

23 See Robin Churchill and David Freestone (eds), *International Law and Global Climate Change* (Graham & Trotman/M. Nijhoff, 1991) 167.

24 World Meteorological Organization (WMO), *WMO and Climate Change* <http://www.WMO.int/pages/themes/WMO_climatechange_en.html> accessed 18 July 2012.

25 Ibid.

26 *Convention on Long-Range Transboundary Air Pollution*, opened for signature 13 November 1979, 18 ILM 1442 (entered into force 16 March 1983).

27 The forerunner to the CTE was the Group on Environmental Measures and International Trade (GEMI) established in 1971 but it did not meet until 1992. In 1994 the CTE replaced

Multilateral Environmental Agreements (MEAs) and the international trading regime,[28] as of 30 July 2012, no conflict relating to an MEA, or a matter directly involving climate change, has been submitted to a dispute settlement panel within the WTO.[29] This is probably because of the limited authority of the CTE, which is confined to making recommendations rather than making decisions. Nevertheless, litigation seeking climate change mitigation or adaptation has been initiated in many countries such as Australia and the United States.[30] Examples are the *Anvil Hill Project Watch Association v Minister for the Environment and Water Resources* (2007) in Australia and the *Massachusetts v Environmental Protection Agency* (2007) in the USA.[31] Thus, it might be inferred that global issues need international responses, but national responses might

---

the GEMI under the *Marrakech Agreement* while the WTO took over the 1947 General Agreement on Tariffs and Trade (GATT).

28 The CTE has the mandate to 'identify the relationship between trade measures and environmental measures', and 'make recommendations on changes that might be necessary to the multilateral trading system both to enhance positive interaction between trade and environmental measures and avoid protectionist trade measures'. This mandate comes from the 'Decision on Trade and Environment on 14 April 1994'. See Farhana Yamin and Joanna Depledge, *The International Climate Change Regime: A Guide to Rules, Institutions and Procedures* (Cambridge University Press, 2004) 531–532.

29 World Trade Organization (WTO), *Chronological List of Disputes Cases* <http://www.WTO.org/english/tratop_e/dispu_e/dispu_status_e.htm> accessed 22 August 2012.

30 Jacqueline Peel, 'Issues in Climate Change Litigation' (2011) 5(1) *Carbon & Climate Law Review* 15, 15.

31 *Anvil Hill Project Watch Association Inc v Minister for the Environment and Water Resources* (2007) FCA 1481. In this case Centennial Coal proposed to build a large open-cut coal mine in NSW and received State approval under Part 3A of the *Environmental Planning and Assessment Act 1979* (*NSW*), while the Anvil Hill Project Watch Association, as a local community association, sued the Minister to review the government's decision in that the proposed mine is to produce up to 10.5 million tons of coal per annum and operate for 12 years and thus have a significant environmental impact. However, the Minister argued that the estimated annual emissions from burning coal harvested from the mine would constitute only 0.04 per cent of global GHG emissions. Justice Stone dismissed the application for review on the ground that the likelihood and extent of adverse impact on matters protected under the *Environment Protection and Biodiversity Conservation Act 1999* (*Cth*) (*EPBC Act*) was not significant enough to trigger the application of the *EPBC Act*.

Massachusetts v Environmental Protection Agency 549 U.S. 497 (2007), 127 S.Ct. 1438. This case was held in the US Supreme Court in which 12 states and several cities of the US brought suit against the Environmental Protection Agency (EPA), aiming to push the federal agency to regulate carbon dioxide and other GHGs as pollutants. The Court held that the EPA has the authority to regulate GHG emissions as a response to petitions filed by environmental groups and the California Attorney General. The majority opinion of

also be effective under certain circumstances.[32] In the context of GHG emissions from international shipping, national or regional initiatives in tackling this issue might be of significance in advocating or pushing the international negotiation process. For instance, the European Union (EU) has taken unilateral measures to deal with GHG emissions from aviation and has planned to take similar measures to tackle GHG emissions from ships.[33] These measures might impose some pressure on the IMO in regulating this issue in a timely manner,[34] or provide the IMO with some approaches for reference. This issue is examined in more detail in Chapter 7.

Within the UN system, these agencies interact with each other in jointly combating climate change, and thus promote the efficiency of such work. This interaction can take different forms. The establishment of the IPCC is an example where the UNEP and WMO collaborated closely in the 1980s. Moreover, the UNEP has cooperated with other UN agencies actively in addressing climate

---

the justices commented that GHGs meet the definition of air pollutant regulated under the *Clean Air Act*.

32 One of the motivations for a country to regulate a global issue like climate change is that climate change is a 'multiscalar' environmental problem with both global impacts and local impacts. The climate change impact caused by an activity might be insignificant globally but could be 'measurable and significant' within the country. Peel, above n. 30, 17.

33 The EU has been working to include aviation and maritime carbon taxes in the EU Emission Trading System (EU ETS), and the aviation tax entered into force on 1 January 2012 (EU Directive 2008/10/101/EC) which applies to all airlines that fly in and out of the EU. In December 2012 the EU suspended this policy due to improved performance by the International Civil Aviation Organization (ICAO), or perhaps because of strong opposition from many countries, including the US, Russia, China and India. In the same year, the EU published a consultation document seeking the views on how best to reduce GHG emissions from ships so as to finally include GHG emissions from international shipping in an EU ETS. See, e.g., Elena Ares, *EU ETS and Aviation* (23 May 2012) <www.parliament.uk/briefing-papers/SN05533.pdf> accessed 24 August 2012; Aoife O'Leary, David Holyoake and Marta Ballesteros, 'Legal Implications of EU Action on GHG Emissions from the International Maritime Sector' (2011) 5–6; Will Nichols, *EU Launches Attempt to Deliver Shipping Emissions Trading Scheme* (24 January 2012) <http://www.businessgreen.com/bg/news/2140997/eu-launches-attempt-deliver-shipping-emissions-trading-scheme> accessed 1 January 2014.

34 Both the IMO and the ICAO received their GHG mandates from Article 2(2) of the *Kyoto Protocol* in the same year, so any significant regulatory progress occurred in one institution would possibly encourage the other institution to take further steps. Furthermore, the possible inclusion of shipping GHG emissions into an EU ETS would impair the regulatory authority of the IMO in this regard. See Sebastian Oberthür, 'Institutional Interaction to Address Greenhouse Gas Emissions from International Transport: ICAO, IMO and the Kyoto Protocol' (2003) 3(3) *Climate Policy* 191, 202.

change internationally, such as the *UNFCCC* Secretariat, the IPCC Secretariat and the World Bank.[35] Further, both the UNEP and WMO's work is shaped by the negotiations process of the *UNFCCC*, its *Kyoto Protocol* and *Paris Agreement*.[36] Other types of UN institutions include the COPs, CMPs and CMAs established under the *UNFCCC* process. They have been pushing the negotiations process of the international climate change regime through organising rounds of conferences for their State Parties. In particular, the Subsidiary Body on Scientific and Technological Advice (SBSTA) and the Ad Hoc Working Group on Long-term Cooperative Action (AWG-LCA) under the *UNFCCC* had been working on GHG emissions from international bunker fuels before 2012. The Ad Hoc Working Group on the Durban Platform for Enhanced Action (ADP) established in 2011 had been working on negotiating the *Paris Agreement* that was adopted in 2015 and is expected to enter into force from 2020. Currently the Ad Hoc Working Group on the Paris Agreement (APA) has been established under the *Paris Agreement* to conduct the work mandated by the Agreement in conjunction with other bodies under the *UNFCCC*.

The *UNFCCC* cooperates with the IMO through reciprocal exchange of information and reciprocal participation in relevant meetings.[37] However, it is open to debate as to the regulatory roles of the *UNFCCC* and the IMO, in particular whether the IMO should be the exclusive international organisation responsible for the regulation of GHG emissions from international shipping.[38] Another form of institutional collaboration exists in the adoption of similar or common definitions through which the UN institutions might provide a common basis for regulation. One example of such collaboration lies in the adoption of a definition for 'air pollution'. The definition of 'air pollution' in Article 1(a) of the *CLRTAP* was generally adopted by the subsequent UN Conventions as the definition of marine pollution under the *United Nations*

---

35 Yamin and Depledge, above n. 28, 533–534, 539–540.

36 United Nations Environment Programme (UNEP), *Climate Change* <http://www.UNEP.org/gc/gc26/factsheet/pdfs/Climate_change.pdf> accessed 24 August 2012, p. 1; World Meteorological Organization (WMO), *WMO at UNFCCC/COP Sessions* <http://www.WMO.int/pages/prog/wcp/cop17/background_en.html> accessed 24 August 2012.

37 Bernd Hackmann, 'Analysis of the Governance Architecture to Regulate GHG Emissions from International Shipping' (2012) 12(1) *International Environmental Agreements: Politics, Law and Economics* 85, 95.

38 There is no hierarchy between the two institutions (UNFCCC and the IMO) in regulating GHG emissions from international shipping, and both institutions have been involved in the regulation of this GHG emissions issue. Therefore, currently different interpretations exist. The IMO's mandate has been discussed in Chapter 2 (2.5.3.1), and the IMO's role in regulating GHG emissions from ships is examined in Chapters 4 (4.2) and 7 (7.4.3.2, 7.5.5).

Convention on the Law of the Sea (LOSC)[39] and the definition of emission under Annex VI to the *International Convention for the Prevention of Pollution from Ships* (MARPOL 73/78).[40]

Aside from the UN institutional collaboration, a certain degree of institutional conflict or 'fragmentation' also exists in international environmental governance,[41] as well as the climate change regime. This fragmentation, however, is regarded as the main factor leading to slow development of the regulation by the IMO on GHG emissions from international shipping.[42] The impact of institutional fragmentation on the reduction of shipping emissions, as well as possible options for its improvement, is examined in Chapter 7 of this book.

The UN's institutional responses to the GHG emissions issue, or in other words, climate change, have implications for the reduction of GHG emissions from international shipping. The establishment of the IPCC links the scientific community and political institutions. As a growing source of GHG emissions contributing to climate change, emissions from international shipping have also been recognised by the IPCC in its Assessment Reports.[43] Other institutions, such as the UNEP and WMO, raise the awareness of the Earth's

---

39 *United Nations Convention on the Law of the Sea*, opened for signature 10 December 1982, 1833 UNTS 3 (entered into force 16 November 1994) art. 1(4) ('*LOSC*').

40 See *CLTRAP* art. 1(a); LOSC art. 1(4); *International Convention for the Prevention of Pollution from Ships* (*MARPOL 73/78*), signed 2 November 1973, 12 ILM 1319, as amended by the 1978 Protocol to the 1973 Convention, 1341 UNTS 3, 17 ILM 546 (entered into force 2 October 1983) annex VI, art. 2(7) ('*MARPOL 73/78*'). See also ch. 2, 2.1.1.

41 Karen N. Scott, 'International Environmental Governance: Managing Fragmentation through Institutional Connection' (2011) 12(1) *Melbourne Journal of International Law* 177, 179–182.

42 Hackmann, above n. 37, 1. But, some scholars have asserted that this kind of fragmentation could be considered 'a strength rather than a weakness'. See, e.g., O.R. Young, *The Institutional Dimensions of Environmental Change: Fit, Interplay, and Scale* (MA: MIT Press, 2002); Steinar Andresen, 'The Effectiveness of UN Environmental Institutions' (2007) 7(4) *International Environmental Agreements: Politics, Law and Economics* 317;. T. Gehring and S. Oberthür, 'Interplay: Exploring Institutional Interaction' in Oran R. Young, Leslie A. King and Heike Schroeder (eds), *Institutions and Environmental Change: Principal Findings, Applications, and Research Frontiers* (MA: MIT Press, 2008).

43 See, e.g., Intergovernmental Panel on Climate Change (IPCC), 'Fourth Assessment Report' (2007) <http://www.IPCC.ch/publications_and_data/ar4/syr/en/contents.html> accessed 27 August 2012, p. 36; Intergovernmental Panel on Climate Change (IPCC), 'Fifth Assessment Report: Working Group III Report' (2014) <http://report.mitigation2014.org/spm/IPCC_wg3_ar5_summary-for-policymakers_approved.pdf> accessed 18 June 2014, p. 8.

climate change, provide technical knowledge on combating climate change and implement the outcomes within the international climate change regime. Their work, although not specialised in the reduction of shipping emissions, is indispensable in uniting people from both developed countries and developing countries. As one of the main institutions coping with GHG emissions from ships, the *UNFCCC* and its *Kyoto Protocol* and *Paris Agreement*, as well as its COPs, CMPs, CMAs, SBSTA, AWG-LCA, ADP and APA, provide crucial platforms for different countries to discuss and negotiate the reduction of such emissions. Furthermore, given that international regulation of shipping GHG emissions within the UN institutions is a lengthy and complex process, any regulatory or enforcement initiative or unilateral action by individual States or the EU might facilitate or improve the global regulation of the GHG emissions issue under the auspices of these UN institutions.[44] Therefore, any initiatives made by individual States or regional organisations to reduce shipping GHG emissions should be studied and promoted if applicable.

3.3 International Legal Framework on Climate Change

In a broad sense, the international legal framework on climate change covers various global and regional treaties and non-binding political agreements to combat climate change by States or through intergovernmental organisations. As a relatively narrow concept, the international climate change regime usually refers to the *1992 UNFCCC* and its *Kyoto Protocol* and *Paris Agreement*, as well as its COPs, CMPs and CMAs.[45] As such, the climate change regime was formally established in 1992 when the *UNFCCC* was adopted; whereas the broader international legal framework on climate change also comprises the previous regional and international efforts in regulating atmospheric pollution. This part first reviews the UN's efforts in tackling air pollution from the perspectives of the *1979 Convention on Long-Range Transboundary Air*

44 See Oberthür, above n. 34.

45 This definition of the international climate change regime has been supported by many scholars. See, e.g., Birnie, Boyle and Redgwell, above n. 1, 356; Yamin and Depledge, above n. 28, 24–29; Ronald D. Brunner, 'Science and the Climate Change Regime' (2001) 34(1) *Policy Sciences* 1, 1; Sebastian Oberthür, 'The Climate Change Regime: Interactions with ICAO, IMO, and the EU Burden-Sharing Agreement' in Sebastian Oberthür and Thomas Gehring (eds), *Institutional Interaction in Global Environmental Governance* (The MIT Press, 2006) 53, 54.

Pollution (CLRTAP),[46] and *the 1985 Convention for the Protection of the Ozone Layer* (*Vienna Convention*).[47] Based on the analysis of the UNFCCC and its *Kyoto Protocol*, this part then examines the key outcomes of the *UNFCCC-Kyoto Protocol* regime during its series of negotiations with a particular emphasis on the newly-adopted *Paris Agreement*. The analysis of these outcomes indicates that international regulation on the reduction of GHG emissions from international shipping is subject to the evolution and direction of the UN climate change regime.

3.2.1 *The Prevention of Atmospheric Pollution*

The issue of atmospheric pollution is generally discussed in the academic literature separately from climate change.[48] Thus atmospheric pollution was excluded from the climate change regime. However, it could be a part of the international legal framework on climate change and also one aspect of the international legal framework on the reduction of GHG emissions from international shipping. At least three factors lead to this conclusion. First, the international legal framework on climate change is a concept broader than the climate change regime. It encompasses not only current conventions regulating climate change, but also the formation of the key regulatory tool, the 'framework treaty', which was initially adopted by two conventions on the prevention of atmospheric pollution: the *1979 Convention on Long-Range Transboundary Air Pollution* (CLRTAP), and the *1985 Convention for the Protection of the Ozone Layer* (*Vienna Convention*). The two conventions were regarded as the first 'framework treaties' to address atmospheric pollution, and this approach was later followed by the UNFCCC and its *Kyoto Protocol*. Second, climate change and atmospheric pollution are 'interlinked'.[49] Certain types of atmospheric pollution, transboundary air pollution as an example, and the depletion of the ozone layer, have been proven to contribute to global warming and climate change.[50] Effective international regulation on climate change could reduce atmospheric pollution. Third, as discussed in Chapter 1, GHGs can be a type of 'conditional' pollution, and the

---

46 *Convention on Long-Range Transboundary Air Pollution*, opened for signature 13 November 1979, 18 ILM 1442 (entered into force 16 March 1983) ('*CLRTAP*').

47 *Vienna Convention for the Protection of the Ozone Layer*, opened for signature 22 March 1985, 26 ILM 1529 (entered into force 22 September 1988) ('*Vienna Convention*').

48 See, e.g., Philippe Sands, *Principles of International Environmental Law* (Cambridge University Press, 2nd ed, 2003) 322–356; Alexandre Charles Kiss and Dinah Shelton, *International Environmental Law* (Transnational Publishers, 3rd ed, 2004) 562–579.

49 Birnie, Boyle and Redgwell, above n. 1, 336.

50 See ibid. 336.

broad definition of GHGs includes those resulting in atmospheric pollution.[51] For example, the release of GHGs including chlorofluorocarbons (CFCs), halons, and other chlorine-based substances may lead to the destruction of the ozone layer.[52] Thus, the issue of atmospheric pollution becomes an indispensable part of the international legal framework on climate change, as well as GHG emissions from international shipping.

3.2.1.1 The *1979 Convention on Long-Range Transboundary Air Pollution*

As early as the *Trail Smelter* case in 1941, transboundary air pollution has been a matter of international concern. The *1982 LOSC* is considered to be 'the first binding rules of a global nature' on atmospheric pollution.[53] Its Articles 212 and 222 grant States legislative and enforcement responsibilities regarding air pollution. However, the role of the *LOSC* in combating climate change is generally less mentioned than the *1979 Convention on Long-Range Transboundary Air Pollution* (*CLRTAP*). This is probably because the *CLRTAP* is a 'framework treaty' that relates it to climate change. As 'the first international legally binding instrument' dealing with regional air pollution,[54] the *1979 CLRTAP* was adopted by the UNECE, signed by all European States, the US and Canada. It established a regional framework to combat transboundary air pollution. The main characteristics of the convention consist of two aspects. It provides for a 'soft commitment' by all parties that they should 'endeavour to limit and, as far as possible, gradually reduce and prevent air pollution including long-range transboundary air pollution'.[55] Thus it establishes a general obligation on parties to limit their emissions of air pollutants, although there was no specific target or timetable for such a limit.[56] For example, it does not limit such pollution to a given level in certain years. The other important feature lies in its broad definition of 'air pollution', which leaves room for further regulation by means of protocols. Article 1(a) of the convention provides:

> 'air pollution' means the introduction by man, directly or indirectly, of substances or energy into the air resulting in deleterious effects of such a

51 See ch. 1, 1.1.

52 Birnie, Boyle and Redgwell, above n. 1, 336.

53 Kiss and Shelton, above n. 48, 564.

54 United Nations Economic Commission for Europe (UNECE), *The 1979 Geneva Convention on Long-rang Transboundary Air Pollution* <http://www.UNECE.org/env/lrtap/lrtap_h1.html> accessed 2 August 2012.

55 *CLRTAP* art. 2; Sands, above n. 48, 325.

56 Sands, above n. 48, 325.

> nature as to endanger human health, harm living resources and ecosystems and material property and impair or interfere with amenities and other legitimate uses of the environment, and 'air pollutants' shall be construed accordingly.[57]

In this definition, 'substances or energy' directly or indirectly introduced by man is a broad expression and covers both GHG emissions and ozone depleting substances.[58] It enables various types of air pollutants to be regulated by the protocols of the *1979 CLRTAP*. To date the *CLRTAP* has eight protocols which have set specific targets for reduction of air pollutants, ranging from sulphur emissions,[59] Nitrogen Oxides,[60] Volatile Organic Compounds (VOCs),[61] heavy metals,[62] and Persistent Organic Pollutants (POPs).[63] For instance, its 1985 Protocol required its parties to reduce the sulphur emissions or their transboundary fluxes by 30 per cent by 1993, using 1980 levels as the basis for calculation of reductions.[64] Additionally, the convention established a 'Cooperative Programme for the Monitoring and Evaluation of the Long-range Transmission of Air Pollutants in Europe' (EMEP), an Executive Body and Secretariat to monitor the air pollutants and develop relevant procedures. The *1979 CLRTAP* has its strengths and weaknesses. It is weak due to its nature of being a 'framework treaty'. It is strong in providing such framework for future cooperation and regime development of more effective measures against pollution.[65] The approach of combining a framework treaty followed by protocols was

---

57 *CLRTAP* art. 1(a). This definition of air pollution was later generally adopted by the *LOSC* and *MARPOL* Annex VI, which may be regarded as a kind of collaboration within UN institutions. See ch. 3, 3.2.2.

58 Sands, above n. 48, 325.

59 The 1985 Protocol on the Reduction of Sulphur Emissions or their Transboundary Fluxes by at least 30 per cent (entered into force 2 September 1987); the 1994 Protocol on Further Reduction of Sulphur Emissions (entered into force 5 August 1998).

60 The 1988 Protocol concerning the Control of Nitrogen Oxides or their Transboundary Fluxes (entered into force 14 February 1991).

61 The 1991 Protocol concerning the Control of Emissions of Volatile Organic Compounds or their Transboundary Fluxes (entered into force 29 September 1997).

62 The 1998 Protocol on Heavy Metals (entered into force 29 December 2003).

63 The 1998 Protocol on Persistent Organic Pollutants (POPs) (entered into force 23 October 2003).

64 The 1985 Protocol on the Reduction of Sulphur Emissions or their Transboundary Fluxes by at least 30 per cent art. 2.

65 See Birnie, Boyle and Redgwell, above n. 1, 345; Kiss and Shelton, above n. 48, 565–568.

adopted by subsequent environmental treaties to combat climate change and ozone depletion.[66]

In the context of GHG emissions from international shipping, the framework treaty model reflected in the *1979 CLRTAP* was adopted by both the *1992 UNFCCC* and its *Kyoto Protocol*, and *MARPOL 73/78*. The *CLRTAP* broad definition of 'air pollution' also covers GHG emissions from international shipping. However, its definition of 'long-range transboundary air pollution' specifies that the distance between the polluter and the victim should generally make it impossible 'to distinguish the contribution of individual emission sources or groups of sources'.[67] In reality, this article excludes the application of this treaty to the issue of GHG emissions from shipping. The distance between the ship which emits GHGs and the victims from such emissions should be long enough that the ship cannot be identified, so that the treaty may be applicable to GHG emissions from ships.[68] Yet, it would be meaningless whether the treaty could be applied in this context as such, since the ship might avoid liability against its emissions, which is also inconsistent with the polluter pays principle.[69] Additionally, the definition of 'air pollution' by the treaty indicates that the pollution should be actual and has 'result[ed] in deleterious effects'.[70] This expression did not include the risk of pollution or damage, which was based on the knowledge at that time. As time went on, the precautionary principle was incorporated into the Protocols of the *CLRTAP*.[71]

3.2.1.2 The *1985 Convention for the Protection of the Ozone Layer*

As a layer in the Earth's atmosphere containing high concentrations of ozone (O_3), the ozone layer is thought to prevent people from harmful exposure to

---

66 Examples are the *1992 UNFCCC* and its protocols and the *1985 Convention for the Protection of the Ozone Layer* and its protocols. They are examined in the following sections.

67 *CLRTAP* art. 1(b). This article reads that,

> "Long-range transboundary air pollution' means air pollution whose physical origin is situated wholly or in part within the area under the national jurisdiction of one State and which has adverse effects in the area under the jurisdiction of another State at such a distance that it is not generally possible to distinguish the contribution of individual emission sources or groups of sources."

68 See Kiss and Shelton, above n. 48, 564.

69 See ch. 2, 2.6.

70 *CLRTAP* art. 1(a).

71 Henrik Selin and Noelle Eckley, 'Science, Politics, and Persistent Organic Pollutants: The Role of Scientific Assessments in International Environmental Co-operation' (2003) 3(1) *International Environmental Agreements* 17, 27. See also ch. 2, 2.4.

ultraviolet radiation from the sun and adjust the temperature structure of the earth. Since the 1960s, losses in the ozone layer over the Antarctic, the Arctic, Australia and some other areas have been observed.[72] As a response to this issue, an international ozone regime was established under the auspices of the UNEP. Currently the regime consists of the *1985 Convention for the Protection of the Ozone Layer* (*Vienna Convention*), the *1987 Montreal Protocol on Substances that Deplete the Ozone Layer* (*Montreal Protocol*),[73] and four amendments to the Protocol, adopted in London (1990),[74] Copenhagen (1992),[75] Montreal (1997)[76] and Beijing (1999).[77]

Similar to the *1979 CLRTAP*, the 1985 *Vienna Convention* is essentially a framework treaty. It does not set any targets or timetable for action, but requires its parties to 'take appropriate measures' to cooperate in four respects. Article 2(2) of the convention lists these obligations:

> To this end the Parties shall . . . :
> (a) Co-operate by means of systematic observations, research and information exchange in order to better understand and assess the effects of human activities on the ozone layer and the effects on human health and the environment from modification of the ozone layer;
> (b) Adopt appropriate legislative or administrative measures and co-operate in harmonizing appropriate policies to control, limit, reduce or prevent human activities under their jurisdiction or control should it be found that these activities have or are likely to have adverse effects resulting from modification or likely modification of the ozone layer;

---

72 Sands, above n. 48, 343.

73 *Montreal Protocol on Substances that Deplete the Ozone Layer*, opened for signature 16 September 1987, 26 ILM 1550 (entered into force 1 January 1989) ('*Montreal Protocol*').

74 *The 1990 Amendment to the Montreal Protocol on Substances that Deplete the Ozone Layer* (*London*), opened for signature 29 June 1990, 30 ILM 537 (entered into force 10 August 1992) ('*The 1990 Amendment*').

75 *The 1992 Amendment to the Montreal Protocol on Substances that Deplete the Ozone Layer* (*Copenhagen*), opened for signature 25 November 1992, 32 ILM 874 (entered into force 14 June 1994) ('*The 1992 Amendment*').

76 *The 1997 Amendment to the Montreal Protocol on Substances that Deplete the Ozone Layer* (*Montreal*), opened for signature 25 September 1997, UNEP/OzL.Pro.9/12 (entered into force 10 November 1999) ('*The 1997 Amendment*').

77 *The 1999 Amendment to the Montreal Protocol on Substances that Deplete the Ozone Layer* (*Beijing*), opened for signature 17 December 1999, UNEP/OzL.Pro.11/10 (entered into force 25 February 2002) ('*The 1999 Amendment*').

> (c) Co-operate in the formulation of agreed measures, procedures and standards for the implementation of this Convention, with a view to the adoption of protocols and annexes;
> (d) Co-operate with competent international bodies to implement effectively this Convention and protocols to which they are party.[78]

These obligations are general, and are implemented by parties at their discretion based on relevant scientific and technical considerations, taking their capabilities into consideration.[79] The above Article 2(a)(b), however, incorporates the precautionary principle in that it requests its State Parties to take actions once 'these activities have or are likely to have adverse effects'. Compared with the *1979 CLRTAP*, the *1985 Vienna Convention* was a major advance in this regard, and was thus regarded as 'one of the first' to recognise and adopt the precautionary approach.[80] This approach was also applied in its *1987 Montreal Protocol* and its amendments[81] and the IMO negotiation process on the reduction of GHG emissions from international shipping.[82] The spirit of cooperation indicated in the above article is another important feature of the convention, which was strengthened in its *1987 Montreal Protocol*.[83] The cooperation among States, in particular between developed States and developing States, is regarded as one of the important reasons for the success of the *1985 Vienna Convention* and its *Montreal Protocol*,[84] and arguably it was the absence of this cooperation that resulted in the slow progress of global climate change

78 *Vienna Convention* art. 2(2).

79 *Vienna Convention* art. 2(2)(4).

80 Birnie, Boyle and Redgwell, above n. 1, 351.
The precautionary approach was also explicitly invoked in the Preamble of the *Vienna Convention*. The Preamble of the *Vienna Convention* notes that, '[m]indful also of the precautionary measures for the protection of the ozone layer which have already been taken by the national and international levels.' See also ch. 2, 2.4.

81 The preamble of the *Montreal Protocol* notes that, '[d]etermined to protect the ozone layer by taking precautionary measures to control equitably total global emissions of substances that deplete it.' This expression goes beyond the limitation of the *1985 Vienna Convention* to precautionary measures that already existed. The Montreal Protocol and its amendments stipulate specific legal obligations and timetables for such requirements with the absence of scientific evidence, which could be deemed as the application of the precautionary principle in this context.

82 See ch. 2, 2.4.

83 See, e.g., *Montreal Protocol* arts. 9, 10.

84 Bryan A. Green, 'Lessons from the Montreal Protocol: Guidance for the Next International Climate Change Agreement' (2009) 39(1) *Environmental Law* 253, 259.

negotiations. The acquisition of alternative technology as stipulated in Article 4 was regarded as 'most unsatisfactory' on the ground that the parties only need to cooperate in accordance with their domestic laws and regulations.[85] However, the definition it provides on 'adverse effects' clearly indicates that ozone depletion might lead to climate change,[86] which relates the convention to the international climate change regime.

The *1987 Montreal Protocol* was regarded as a 'landmark international environmental agreement' due to the new regulatory techniques, institutional arrangements, and innovative financial mechanisms that it adopted.[87] As a Protocol to the *1985 Vienna Convention*, it sets specific targets for reducing and eliminating consumption and production of ozone depleting substances. These substances were listed and regulated by the Protocol and its four amendments.[88] As these substances, including O_3, are types of GHGs in a broad sense, the adoption of the Protocol paved the way for the negotiation and adoption of the *UNFCCC* and its *Kyoto Protocol*. The adoption of the Common but Differentiated Responsibility (CBDR) principle[89] is another important feature of the Protocol. Article 5(1) of the Protocol offers developing country parties a grace period of ten years beyond the dates set for phase-out of the controlled substances regulated under Article 2 of the Protocol.[90]

---

85 Birnie, Boyle and Redgwell, above n. 1, 350. Article 4(2) of the *Vienna Convention* provides that, '[t]he parties shall cooperate, consistent with their national laws, regulations and practices and taking into account in particular the needs of the developing countries, in promoting, directly or through competent international bodies, the development and transfer of technology and knowledge.'

86 Article 1(2) of the *Vienna Convention* reads that,

"'Adverse effects' means changes in the physical environment or biota, including changes in climate, which have significant deleterious effects on human health or on the composition, resilience and productivity of natural and managed ecosystems, or on materials useful to mankind."

87 Sands, above n. 48, 345–346. UNEP, *Key Achievements of the Montreal Protocol To Date* <http://ozone.UNEP.org/Publications/MP_Key_Achievements-E.pdf>accessed29August 2012. According to the statistics by the UNEP, as of the end of 2009, the Parties to the *Montreal Protocol* had phased out the consumption of 98 per cent of all the chemicals controlled by the Protocol. In this sense, the *Montreal Protocol* was successful in achieving its goals of phasing out the ODSs. See Green, above n. 84, 259.

88 Totally there are 5 Annexes ranging from Annex A, Annex B, Annex C, Annex D, and Annex E listing different groups of controlled substances under this regime.

89 See ch. 2, 2.5.

90 Article 5(1) of the Protocol provides,

'Any Party that is a developing country and whose annual calculated level of consumption of the controlled substances is less than 0.3 kilograms per capita on the date of the entry

Article 5(2) and 5(3) aims to facilitate access to 'environmentally safe alternative substances and technology' by developing countries, and promises to offer them 'subsidies, aid, credits, guarantees or insurance programmes'.[91] These regulations attract participation from developing countries including India and China. Meanwhile, the Protocol's institutional arrangements for the regular meetings of parties (MOPs), and the 'Financial Mechanism' including the establishment of a Multilateral Fund, have also gained wide support.[92] These methods have been relied on extensively in the subsequent treaties on climate change.

Three approaches adopted by the *Montreal Protocol* might be valuable to the issue of GHG emissions from international shipping. One is its flexible arrangement, which has helped both cooperation between various States and participation from developing States.[93] For instance, parties are required to review and modify the provisions of the Protocol in the future as a new economic or environmental situation or technology appears.[94] Three categories of exemptions of certain Ozone Depleting Substances (ODSs) for certain uses were provided,[95] which might alleviate concerns from relevant States due to their special situations.[96] Additionally, rules on trade of ODSs are not rigid. The trade on production allowances among party States is allowed, which makes it possible for low-producing States to meet their domestic needs.[97] In contrast to the above flexible agreement, the IMO explicitly sets the principle of No More Favourable Treatment (NMFT) in its *MARPOL 73/78*,[98] which makes it less attractive for developing States to participate in the reduction of GHG emissions from ships. When comparing the successful practice under the *1987 Montreal Protocol* with the slow progress within the IMO in regulating

---

into force of the Protocol for it, or any time thereafter within ten years of the date of entry into force of the Protocol shall, in order to meet its basic domestic needs, be entitled to delay its compliance with the control measures set out in paragraphs 1 to 4 of Article 2 by ten years after that specified in those paragraphs.'

91 *Montreal Protocol* art. 5(2)(3).

92 See, e.g., Sands, above n. 48, 354–357; Kiss and Shelton, above n. 48, 575–579.

93 Green, above n. 84, 262–265.

94 *Montreal Protocol* art. 6.

95 The three categories of exemptions include 'the critical uses of methyl bromide, essential uses for all other [ODSs], and laboratory and analytical uses'. UNEP Ozone Secretariat, *Exemption Information* <http://ozone.UNEP.org/Exemption_Information/> accessed 28 August 2012.

96 Green, above n. 84, 263.

97 *Montreal Protocol* art. 2(5).

98 *MARPOL 73/78* art. 5(4); see also ch. 2, 2.5.

the shipping GHG emissions issue, it might be inferred that certain kinds of flexibility are necessary to facilitate reaching consensus in the IMO on future action to regulate shipping GHG emissions. Indeed, the IMO had demonstrated flexibility in its newly-adopted Energy Efficiency measures for reducing shipping emissions in that it gave new ships a six and a half year delay in applying new shipbuilding standards, although this benefit applies to ships registered in both developing and developed States.[99] It is arguable that more flexibility is needed for adopting market-based measures (MBMs) to further reduce GHG emissions from international shipping.[100]

Another innovative approach adopted by the *Montreal Protocol* lies in its well-designed application of the CBDR principle. First, it links the obligations by developing State parties with the effective implementation of financial cooperation and the transfer of technology from developed State parties.[101] Article 5(5) of the amended Protocol in 1990 provides:

> Developing the capacity to fulfil the obligations of the Parties operating under paragraph 1 of this Article to comply with the control measures set out in Articles 2A to 2E and Article 2I, and any control measures in Articles 2F to 2H that are decided pursuant to paragraph 1 *bis* of this Article, and *their implementation by those same Parties will depend upon the effective implementation of the financial co-operation as provided by Article 10 and the transfer of technology* as provided by Article 10A.[102]

As discussed in Chapter 2, the financial cooperation and transfer of technology included in the above article is a kind of 'differentiated treatment', and constitutes a form of the CBDR principle. The Protocol is innovative in setting the implementation of financial cooperation and technology transfer as the conditions for the developing States to fulfil their obligations under the Protocol. This approach was later adopted by the *UNFCCC*.[103] Second, the provisions of

---

99 John Vidal, *Maritime Countries Agree First Ever Shipping Emissions Regulation* <http://www.guardian.co.uk/environment/2011/jul/18/maritime-countries-shipping-emissions-regulation> accessed 29 August 2012; see also ch. 4, 4.3.2.

100 See ch. 4, 4.3.4, ch. 7, 7.5. There are three routes within the IMO in regulating GHG emissions from international shipping, namely technical measures, operational measures, and MBMs. Currently technical and operational measures have been adopted by the IMO, while the MBMs are still under discussion.

101 Sands, above n. 48, 354.

102 *The 1990 Amendment* art. 5(5).

103 *UNFCCC* art. 4(7).

the Multilateral Fund required that the financial assistance would only cover the incremental costs undertaken by developing States in fulfilling their obligations under the Protocol.[104] As such, both developing and developed States found this an appropriate measure to relieve their concerns: developing States received the assistance that they were lacking while developed States were guaranteed that their assistance would be utilised for the purpose of reducing ODSs.

Currently the mechanisms on financial cooperation and technology transfer are in place under the *UNFCCC*,[105] as well as various funds. It is important to maintain and improve these mechanisms to incorporate the successful practice of the *Montreal Protocol* into the reduction of GHG emissions from shipping. In particular, it might be necessary to apply both the CBDR and the NMFT principles in a creative manner. Due to the growing status of developing States in the international economy and politics, it is important to take into account the interests from this group of nations in the development of global regulation of shipping GHG emissions. To that end, the adoption of the CBDR principle might take different forms, such as certain kinds of MBMs, so as to be more flexible and attractive for most States. It is arguable that the obligation or commitment to reduce GHG emissions by developing countries should *only be activated* once the financial assistance and technology transfer as agreed by both parties are effectively implemented by developed countries. This approach, however, is stronger than those indicated within the *Montreal Protocol* and the *1992 UNFCCC*.[106] Similar restrictions on the utilisation of the funds by the Protocol might also be applied through a GHG Fund. The key to attracting the participation from developing States lies in a sound arrangement on the utilisation and allocation of these benefits as can be seen from the *Montreal Protocol*. These issues are further discussed in Chapters 4 and 7.

The third successful approach adopted by the *Montreal Protocol* lies in its dealing with non-parties by means of trade sanctions which effectively attracted the involvement and cooperation from the industry. Article 4 of the Protocol as revised in 1990 and 1997 requires each party to ban the import and export of controlled substances or products containing such substances from and to non-parties. Due to their consistency with Article 20(b) of the *General*

104 Green, above n. 84, 266.

105 *UNFCCC* art. 4(7).

106 *UNFCCC* art. 4(7).

Agreement on Tariffs and Trade (GATT) and the WTO,[107] these trade measures could be used as 'sticks' while financial assistance serves as a 'carrot'.[108] More importantly, the Protocol encouraged the utilisation of 'environmentally safe alternative substances'.[109] In this way it successfully ensured the industry that a worldwide reduction of ODSs would eventually eliminate the market for ODSs, and it would be secure and profitable from a long term perspective to invest in the research and development of alternatives to ODSs.[110] Thus the industry actively participated in the exploration of new alternatives, as well as the transfer of ozone safe technology, as these technologies were mostly owned by the private sector rather than the governments of developed States.

While ozone depletion and climate change are both issues of a global nature, these trade measures, or methods of dealing with non-parties, particularly in the way they attracted participation from industry, may also have a significant role to play in accelerating the reduction of shipping emissions. To avoid the non-parties' competitive advantage gained from possible future regulations by the IMO, the functions of the port State need to be strengthened since any ship's entry into a port State is subject to the jurisdiction of that State. In this case, the port State might strengthen its legislation and enforcement of vessel entry into its port, and this arrangement could be made through improving current Memoranda of Understanding (MOUs) in the port State control regime.[111] However, trade-related measures might not be applied directly in this context. Furthermore, in view of the fact that most ships around the world are owned by private shipping companies, it is important to examine their needs and get them involved in the reduction of shipping emissions. Chapter 5 discusses this issue in more detail.

107 There are a lot of discussions on the possible conflicts or necessary coordination between trade measures and environmental concerns, but some scholars assert that trade measures are justified and legitimate in this regard. See, e.g., Birnie, Boyle and Redgwell, above n. 1, 353; Scott N. Carlson, 'The Montreal Protocol's Environmental Subsidies and GATT: A Needed Reconciliation' (1994) 29(2) *Texas International Law Journal* 211, 229; Zhongxiang Zhang, 'Multilateral Trade Measures in a Post-2012 Climate Change Regime? What Can Be Taken from the Montreal Protocol and the WTO?' (2009) 37(12) *Energy Policy* 5105, 5105.

108 Zhang, above n. 107.

109 *Montreal Protocol* art. 5(2), 9(1).

110 Green, above n. 84, 267.

111 See ch. 6, 6.5.2, ch. 7, 7.4.4.

3.2.2 *The* UNFCCC-Kyoto Protocol *Regime*

It is generally accepted that the international climate change regime, also referred to as the *UNFCCC-Kyoto Protocol* regime,[112] was established in 1992 when the UNFCCC was adopted, culminated in 1997 when the *Kyoto Protocol* was signed, and is currently under development.[113] There are different views on the development stages of this regime. Some scholars classify the regime into five periods,[114] while others divide it into four periods.[115] Based on the recent development of the regime, the evolution of the *UNFCCC-Kyoto Protocol* regime is divided into three stages as illustrated in Table 3.1. While the early international responses to the issue of climate change have been discussed in the first part of this chapter, this section mainly examines the establishment of the regime and the post-Kyoto efforts by the international community with a focus on the relevance this has on the international regulation of GHG emissions from international shipping. The *Paris Agreement* as the new outcome of this regime and its implications on shipping GHG emissions are analysed in a separate subsection.

112 D.M. Ong, 'International Legal Efforts to Address Human-induced Global Climate Change' in M. Fitzmaurice, D.M. Ong and Panos Merkouris (eds), *Research Handbook on International Environmental Law* (Edward Elgar Publishing Limited, 2010) 450, 451.

113 See, e.g., Yamin and Depledge, above n. 28, 22–29; Kevin A. Baumert, 'Participation of Developing Countries in the International Climate Change Regime: Lessons for the Future' (2006) 38(2) *The George Washington International Law Review* 365; Daniel Bodansky, 'The History of the Global Climate Change Regime' in Urs Luterbacher and Detlef F. Sprinz (eds), *International relations and global climate change* (Cambridge, MA: MIT Press, 2001) 23; Lavanya Rajamani, 'The Cancun Climate Agreements: Reading the Text, Subtext and Tea Leaves' (2011) 60(2) *The International and Comparative Law Quarterly* 499, 499.

114 See, e.g., Bodansky, above n. 113, 23–24. Bodansky divided the period till 1997 into five periods, namely the foundational period, the agenda-setting phrase from 1985 to 1988, a pre-negotiation period from 1988 to 1990, the formal intergovernmental negotiations phase of the UNFCCC, Post-Rio Developments and the Negotiation of the *Kyoto Protocol*.

115 See, e.g., Yamin and Depledge, above n. 28, 22–29. Yamin and Depledge classify the regime into three stages, namely the emergence of the climate change regime, entry into force of the UNFCCC and the Berlin Mandate, the post-Kyoto era, and the post-Marrakesh era.

TABLE 3.1 *Evolution of the international climate change regime*[116]

Stage of Development	Major Outcomes	Meeting Date
Early International Responses	UNGA Resolution 43/53	6 Dec 1988
	The IPCC First Assessment Report	1990
	Ministerial Declaration	29 Oct–7 Nov 1990
	UNGA Resolution 45/212	21 Dec 1990
Establishment of the Regime	UNFCCC	3–14 Jun 1992
	Berlin Mandate	28 Mar–7 Apr 1995
	Geneva Ministerial Declaration	8–19 Jul 1996

116 This table is compiled by the author based on the following sources: UNFCCC, *Meetings* <http://UNFCCC.int/meetings/items/6240.php> accessed 19 June 2014; UNFCCC, *A Brief Overview of Decisions* <http://UNFCCC.int/documentation/decisions/items/2964.php> accessed 19 June 2014; Yamin and Depledge, above n. 28, 22–29.

COP/CMP or Organiser	Key Elements/Contributions
UN	1. Climate change as a 'common concern of mankind'; 2. Endorsed the action in establishing the IPCC and requested it to prepare a review & recommendation on climate change.
IPCC	Global mean temperature likely to increase by about 0.3°C per decade, under business-as-usual emission scenario.
Second World Climate Change Conference	1. Countries need to stabilise GHG emissions, developed States should establish emissions targets and/or national programs or strategies; 2. Called for negotiations on a framework convention on climate change.
UN	1. Establishment of INC; 2. INC to host the negotiating and drafting of the UNFCCC.
UNCED	1. Defined an ultimate objective and principles; 2. Divided countries into Annex I, Annex II and non-Annex I, and specified general commitments to different Parties respectively; **3. Included general obligations by all parties to promote and cooperate in the reduction of GHG emissions from the transport sector. (Art. 4(1)(c)).**
UNFCCC COP 1	1. Assessed specific commitments for Annex I Parties under the UNFCCC as 'not adequate'; 2. Launched negotiations on 'a protocol or another legal instrument' to be concluded by COP 3; **3. Requested its SBSTA and the SBI to address the allocation and control of emissions from international aviation and shipping.**
COP 2	1. Endorsed the 1995 IPCC Second Assessment Report; called for accelerating negotiations on a legally binding Protocol or another legal instrument; **2. SBSTA negotiated allocation and control of international bunker fuels.**

TABLE 3.1 *Evolution of the International Climate Change Regime* (cont.)

Stage of Development	Major Outcomes	Meeting Date
	Kyoto Protocol (KP)	1–10 Dec 1997
Development of the Regime (Post-Kyoto Era)	Buenos Aires Plan of Action (BAPA)	2–13 Nov 1998
	22 Decisions	25 Oct–5 Nov 1999
		13–24 Nov 2000
	Bonn Agreement	16–27 Jul 2001
	Marrakesh Accords	29 Oct–9 Nov 2001
	Delhi Ministerial Declaration	23 Oct–1 Nov 2002

COP/CMP or Organiser	Key Elements/Contributions
COP 3	1. Provided all parties with general commitments, and Annex I parties with individual emission targets; set flexibility mechanisms: CDM, JI, ET; Reporting and Review, Compliance system; **2. Delegated IMO the mandate to regulate GHG emissions from international shipping (Art. 2(2));** **3. Urged SBSTA to elaborate further on the inclusion of international bunker emissions in the overall GHG inventories of Parties.**
COP 4	Included 7 decisions focusing on strengthening the financial mechanism, technology transfer, adverse effects of climate change/implementation of response measures, activities implemented jointly, flexibility mechanisms, and preparation for COP/CMP. Many of these decisions or actions are to be finished before COP 6.
COP 5	Implementation of the BAPA; adoption of the guidelines for the preparation of national communications by Annex I States; capacity building, transfer of technology, and flexible mechanisms (no major conclusions).
COP 6	The Hague conference serves as Part I of COP 6.
COP 6–2	1. Adopted the IPCC Third Assessment Report of 2001; 2. Provided core elements for the implementation of the BAPA; established new Special Climate Change Fund, the Kyoto Protocol Adaptation Fund.
COP 7	Based on the Bonn Agreement, set out detailed rules, procedures, technical guidelines and work programmes, which actually completed the work of BAPA; brought to an end the post-Kyoto cycle of policy-making launched by the BAPA.
COP 8	Reiterated the need to build on the outcomes of the World Summit.

TABLE 3.1 *Evolution of the International Climate Change Regime* (cont.)

Stage of Development	Major Outcomes	Meeting Date
		1–12 Dec 2003
		6–17 Dec 2004
	Montreal Action Plan	28 Nov–9 Dec 2005
		6–17 Nov 2006
	Bali Road Map (Bali Action Plan)	3–14 Dec 2007
		1–12 Dec 2008

COP/CMP or Organiser	Key Elements/Contributions
COP 9	The Milan conference adopted decisions on the institutions and procedures of the Kyoto Protocol and the implementation of the UNFCCC.
COP 10	The Buenos Aires conference completed unfinished business from the Marrakesh Accords, reassessed the building blocks of the process, and discussed future policies.
COP 11 /CMP 1	Discussed capacity building, transfer of technologies, adverse efforts of climate change, etc.; **launched negotiations on the next phase of the KP with the establishment of AWG-KP.**
COP 12 /CMP 2	Accepted Belarus as an Annex B Party under the KP.
COP 13 /CMP 3	1. Endorsed the IPCC Fourth Assessment Report of 2007; 2. Bali Road Map is made up of a set of decisions that represent the work to be done under various negotiating 'tracks'; 3. As a part of the Bali Road Map, the Bali Action Plan pointed out a process to enable the full, effective and sustained implementation of the Convention through long-term cooperative action up to and beyond 2012. It includes five categories: shared vision, mitigation, adaptation, technology and financing; 4. **AWG-LCA was established to conduct this process, and is responsible for addressing shipping emissions' reduction under the sub-item 1b(iv) addressing cooperative sectoral approaches and sector-specific actions.**
COP 14 /CMP 4	The Poznan conference launched the Adaptation Fund under the KP.

TABLE 3.1 *Evolution of the International Climate Change Regime* (cont.)

Stage of Development	Major Outcomes	Meeting Date
	Copenhagen Accord	7–18 Dec 2009
	Cancun Agreements	29 Nov–10 Dec 2010
	Durban Package	28 Nov–9 Dec 2011

COP/CMP or Organiser	Key Elements/Contributions
COP 15 / CMP 5	1. Raised climate change policy to 'the highest political level'; 2. Committed developed States to USD30 billion fast-starting financing for adaptation and mitigation in developing States for the period 2010–2012; 3. Decided to establish the Copenhagen Green Climate Fund, and a Technology Mechanism.
COP 16 / CMP 6	Integrated many of the elements of the Copenhagen Accord: 1. Took note of the mitigation targets and actions communicated by States, provided for transparency in their implementation; established new mechanisms for **the measurement, reporting and verification (MRV)** of mitigation efforts and support **for both developed and developing States;** 2. Established an Adaptation Framework, a Technology Mechanism and a Green Climate Fund; 3. Created a framework for addressing deforestation in developing States.
COP 17 / CMP 7	1. **Established a second commitment period under the KP;** 2. Agreed on long-term cooperative action under the UNFCCC; 3. Operationalised the Technology Mechanism in 2012; 4. Launched the Green Climate Fund, the Adaptation Committee; and **a new subsidiary body under the Convention (ADP);** 5. **Amended the KP and its Annexes, added the 7th type of GHG—Nitrogen trifluoride (NF_3).**

TABLE 3.1 *Evolution of the International Climate Change Regime* (cont.)

Stage of Development	Major Outcomes	Meeting Date
	Doha Climate Gateway	26 Nov–7 Dec 2012
	Warsaw Outcomes	11–23 Nov 2013
	Lima Call for Climate Action	1–12 Dec 2014
	Paris Agreement	30 Nov–11 Dec 2015

3.2.2.1 The *UNFCCC* and its *Kyoto Protocol*

In contrast to the international regulation of transboundary air pollution and ozone depletion, climate change regulation is a much broader and more complex issue which involves all aspects of people's daily lives,[117] and has greater differentiated economic and political implications for both developed and developing countries. The success of the framework approach adopted by the *1979 CLRTAP* and *1985 Vienna Convention* led to a consensus in support of a similar approach to climate change. This consensus was formally confirmed by the 44th United Nations General Assembly (UNGA) in 1989.[118] As stated by

117 See Milke Hulme, 'The Idea of Climate Change' (2010) 19(3) *GAIA: Ecological Perspectives for Science & Society* 171, 171.

118 *Protection of Global Climate for Present and Future Generations of Mankind,* UNGA. A/RES/44/207 (22 December 1989).

COP/CMP or Organiser	Key Elements/Contributions
COP18 /CMP 8	1. Adopted the ***Doha Amendment to the KP***, which includes new commitments for annex I parties to the KP who agreed to take on commitments in a second commitment period (01/01/2013–31/12/2020), amended GHG list, and other amended articles of the KP. 2. **Terminated the AWG-LCA and AWG-KP**; 3. Transferred some issues to be considered by the SBSTA and SBI, e.g., MRV, REDD+, market and non-market mechanisms.
COP 19 /CMP 9	1. Established the Warsaw international mechanism for loss and damage associated with climate change impacts; 2. Established the Warsaw REDD+ framework.
COP 20 /CMP 10	1. Elaborated the '**Elements for a draft negotiating text**' for the 2015 agreement; 2. Adopted a decision on Intended Nationally Determined Contributions (INDCs).
COP 21/CMP 11	Adopted the *Paris Agreement* on 12 December 2015; Deferred the discussion of GHG emissions from international shipping to subsequent conferences.

the UNGA Resolution 44/207, 'a framework convention on climate was urgently required' and 'specific protocols with commitments could develop within this framework'.[119]

As discussed earlier in this chapter, the establishment of the IPCC served as a bridge between the scientific community and the political agenda. To incorporate the international consensus achieved into the drafting of a climate convention,[120] the Response Strategies Working Group of the IPCC drew up a 'Legal Measures' paper in 1989, which listed the possible elements for

119 Ibid.

120 The 'consensus' here includes both the fact that scientific evidence proves the existence of global warming and climate change, and the broad agreement that a Convention should follow the format of the *1979 CLRTAP* and the *1985 Vienna Convention*, namely the framework treaty model.

the drafting of a framework convention on climate change.[121] The document was endorsed by the Ministerial Declaration of the Second World Climate Conference in 1990.[122] In the same year, the Intergovernmental Negotiating Committee (INC) was established by the UN to host the negotiation and drafting of a framework convention on climate change, and was also required to take the work of the IPCC into consideration during this process.[123]

The *UNFCCC* was adopted in 1992 through the joint efforts of the international community. As the 'Constitution' for the international climate change regime,[124] to date the *UNFCCC* has 197 parties including 196 States and the European Union.[125] However, the bodies under the *UNFCCC* have been changing due to the termination of some temporary bodies mandated by the convention for certain periods. Figure 3.3 provides the current structure of the *UNFCCC* bodies. Except for the SBSTA and the Subsidiary Body for Implementation (SBI) which are permanent subsidiary bodies under the *UNFCCC*, most of the other subsidiary bodies exist temporarily.

The main contributions from the *UNFCCC* are shown in Table 3.1. First, the ultimate objective of the Convention and any related legal instruments is to stabilise GHG concentrations 'at a level that would prevent dangerous anthropogenic interference with the climate system' rather than reverse GHG emissions.[126] Stabilisation should be achieved within a time frame that allows ecosystems to adapt naturally, ensures that food production is not threatened and enables sustainable economic development.[127] It is difficult to infer from this Article what concentration levels and rates of change are 'safe'.[128] However, in assessing whether it is necessary to reduce GHG emissions from international shipping, for instance, the contributions of shipping GHG emissions to

121 Barrett, above n. 16, 184.

122 Ibid 184.

123 *Protection of Global Climate for Present and Future Generations of Mankind*, UNGA. A/RES/45/212 (21 December 1990) art. 1.

124 D.M. Bodansky, 'The Emerging Climate Change Regime' (1995) 20(1) *ANNUAL REVIEW OF ENERGY AND THE ENVIRONMENT* 425, 426.

125 United Nations Framework Convention on Climate Change (UNFCCC), *Status of Ratification of the Convention* <http://UNFCCC.int/essential_background/convention/status_of_ratification/items/2631.php> accessed 30 April 2016.

126 *UNFCCC* art. 2.

127 *UNFCCC* art. 2.

128 Bodansky, above n. 124, 433.

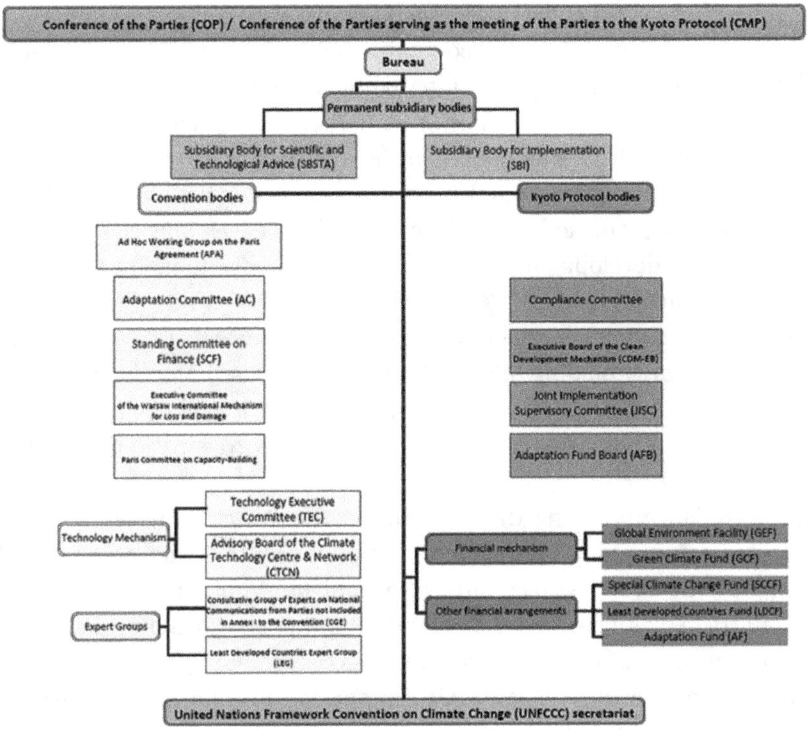

FIGURE 3.3 *Structure of the UNFCCC bodies.*[129]

global anthropogenic GHG emissions could be an important criterion. The discussions in Chapter 1 indicate that this was 2.7 per cent in 2007 and is projected to be 5.7 per cent by 2050 if such growth remains unchecked.[130] This percentage is high enough to be regulated. Second, the Convention raises several guiding principles for the parties to achieve the above objective. These principles include the principle of equity,[131] the precautionary principle,[132] the CBDR

129 UNFCCC, *Bodies* (2016) <http://UNFCCC.int/bodies/items/6241.php> accessed 22 April 2016.

130 ø. Buhaug et al., 'Second IMO GHG Study 2009' (International Maritime Organization (IMO), 2009) 1; see also ch. 1, 1.1.1.2.

131 Or called 'intra- and intergenerational equity'. See *UNFCCC* art. 3(1).

132 *UNFCCC* art. 3(3).

principle,[133] the sustainable development principle[134] and promoting a supportive and open international economic system.[135] Most of these principles reflect more general principles of international environmental law and apply to the issue of GHG emissions from international shipping.[136] Among them, the CBDR was explicitly mentioned for the first time although it was actually applied in the *Montreal Protocol*. This principle, together with the right to sustainable development, addressed the concern from developing countries that their economic development would not be impeded due to their engagement in fighting climate change.[137] Nevertheless, the use of words 'guided' at the beginning and 'should' throughout Article 3 also indicate that these principles are 'not necessarily binding rules', although they are useful for the interpretation and implementation of the Convention.[138] Thus, it might be inferred that the application of the CBDR principle to the issue of GHG emissions from ships could be flexible.[139]

In accordance with the CBDR principle, the *UNFCCC* divides countries into three categories, namely, Annex I (OECD countries and economies in transition), Annex II (OECD countries only) and non-Annex I (mostly developing countries).[140] All parties have general commitments, including developing national inventories of anthropogenic emissions, promoting sustainable management, and reporting obligations.[141] The specific aim of returning emissions to 1990 levels by 2000 is imposed on Annex I Parties,[142] while Annex II Parties must provide financial assistance and promote technology transfer to developing countries.[143] These commitments, however, have been criticised as 'neither

---

133 *UNFCCC* art. 3(1)(2).

134 *UNFCCC* art. 3(4).

135 *UNFCCC* art. 3(5).

136 See ch. 2.

137 Bodansky, above n. 124, 435.

138 Birnie, Boyle and Redgwell, above n. 1, 359. Based on article 31(1) of the *1969 Vienna Convention on the Law of Treaties*, article 3 of the *UNFCCC* could be regarded as 'the context' for interpreting the *UNFCCC* or its related legal instruments including its *Kyoto Protocol*.

139 See ch. 2, 2.5.

140 However, some countries which became OECD members after the adoption of the *UNFCCC* in 1992 are non-Annex I countries (such as the Republic of Korea and Israel), and the Annex I list has not been updated. See also ch. 5, 5.4.2.

141 *UNFCCC* art. 4(1).

142 *UNFCCC* art. 4(2).

143 *UNFCCC* art. 4(3)(4)(5).

strong nor clear'.[144] This was mainly because under the Convention Annex I Parties are free to choose their different starting points, resources, economies and other individual circumstances without a uniform requirement,[145] and a certain degree of flexibility in implementing their commitments is allowed for economies in transition.[146] Furthermore, the obligations by all parties to promote and cooperate in the reduction of GHG emissions in the transport sector were included in the *UNFCCC*,[147] although they only served as general commitments. In this sense, the *UNFCCC* could also be deemed as the 'Constitution' of the GHG reduction regime from international shipping.

These commitments were identified by the First Conference of the Parties of the *UNFCCC* (COP 1) in 1995 as 'not adequate' in its strong mandate (commonly known as the *Berlin Mandate*) and negotiations on a protocol or another legal instrument were launched.[148] More importantly, it was in this conference that the SBSTA and the SBI were requested to address the allocation and control of GHG emissions from international shipping for the first time. Decision 4/CP.1 (methodological issues) of the *UNFCCC COP 1 Report* in 1995 decided:

> 1. (f) That the Subsidiary Body for Scientific and Technological Advice and the Subsidiary Body for Implementation, *taking fully into account ongoing work* in Governments and international organizations, including the *International Maritime Organization* and the International Civil Aviation Organization, address the issue of *the allocation and control of emissions from international bunker fuels*, and report this work to the Conference of the Parties at its second session.[149]

---

144 Birnie, Boyle and Redgwell, above n. 1, 360.

145 *UNFCCC* art. 4(2).

146 *UNFCCC* art. 4(6).

147 *UNFCCC* art. 4(1)(c). The article provides:

> All Parties, taking into account their common but differentiated responsibilities and their specific national and regional development priorities, objectives and circumstances, shall:
>
> (c) *Promote and cooperate* in the development, application and diffusion, including transfer, of technologies, practices and processes that control, reduce or prevent *anthropogenic emissions of greenhouse gases not controlled by the Montreal Protocol in all relevant sectors*, including the energy, ***transport***, industry, agriculture, forestry and waste management sectors.

148 *The Berlin Mandate*, Report of the Conference of the Parties on its First Session, FCCC/CP/1995/7/Add.1 (28 March–7 April 1995) preamble.

149 *Methodological Issues*, Decision 4/CP.1, Report of the Conference of the Parties on its First Session, FCCC/CP/1995/7/Add.1 (28 March–7 April 1995) art. 1(f).

Bunker fuel is degraded residue heavy fuel oil, which has been widely used by international shipping due to its low cost.[150] In accordance with the above decision, the SBSTA and SBI were requested to work out how to allocate GHG emissions from international shipping to individual States so as to regulate this GHG emissions issue through the scheduled *Kyoto Protocol*. This is because the *UNFCCC* and its scheduled *Kyoto Protocol* are State-based agreements whereas ships engaged in international shipping are emitting during the whole voyage. However, it is technically difficult and politically sensitive to allocate shipping GHG emissions to individual States due to the transboundary nature of GHG emissions and the close linkage of international shipping with international trade. Due to the significance of this work on the allocation of emissions, it is arguable that the *UNFCCC* started its work on addressing GHG emissions from international shipping at COP 1 in 1995. To address this difficult problem, at the SBSTA 4th meeting (SBSTA 4) in 1996 the *UNFCCC* Secretariat prepared a paper that included eight allocation options for consideration by the SBSTA.[151] Then, SBSTA identified five options from these eight choices as the basis for future work on the allocation of GHG emissions from international shipping. These five options are:

---

150 Md. Saiful Karim and Shawkat Alam, 'Climate Change and Reduction of Emissions of Greenhouse Gases from Ships: An Appraisal' (2011) 1(1) *Asian Journal of International Law* 131, 131.

151 These eight options are:
'(1) No allocation, as in the current situation.
(2) Allocation of global bunker sales and associated emissions to Parties in proportion to their national emissions.
(3) Allocation to Parties according to the country where the bunker fuel is sold.
(4) Allocation to Parties according to the nationality of the transporting company, or to the country where a ship or aircraft is registered, or to the country of the operator.
(5) Allocation to Parties according to the country of departure or destination of an aircraft or vessel. Alternatively, the emissions related to the journey of an aircraft or vessel could be shared by the country of departure and the country of arrival.
(6) Allocation to Parties according to the country of departure or destination of passenger or cargo. Alternatively, the emissions related to the journey of passengers or cargo could be shared by the country of departure and the country of arrival.
(7) Allocation to Parties according to the country of origin of passengers or owner of cargo.
(8) Allocation to the Party of all emissions generated in its national space.'
Communications from Parties Included in Annex I to the Convention: Guidelines, Schedule and Process for Consideration, SBSTA Fourth Session, Agenda Item 5(a), Doc FCCC/SBSTA/1996/9/Add.1 (24 October 1996) p. 11.

> Option 1: no allocation;
> Option 2: allocation to the country where the bunker fuel is sold;
> Option 3: allocation to the country of the transporting company, the country of registration of the aircraft/vessel, or the country of the operator;
> Option 4: allocation to the country of departure or destination of the aircraft/vessel (including some kind of sharing of emissions between them); and
> Option 5: allocation to the country of departure or destination of the passenger/cargo (including some kind of sharing of emissions between them).[152]

However, States failed to reach a consensus on selecting the most-favoured option for the allocation and control of GHG emissions from international shipping. This is probably because countries that would have been allocated substantial amounts of emissions from bunker fuels would be in a disadvantageous situation in international trade, and these options are not feasible for domestic implementation.[153] Due to the deadlock on the allocation issue, the *UNFCCC* delegated responsibility to the IMO to regulate the issue for shipping under Article 2(2) of its *Kyoto Protocol*. The mandate that the IMO has from Article 2(2) of the protocol not only gives the IMO such authority or responsibility, it also underpins the application of principles from both the *UNFCCC* and the *Kyoto Protocol* to this GHG emissions issue.[154] Meanwhile, GHG emissions from international shipping are neither part of national emissions nor the subject of the emission targets agreed in the *Kyoto Protocol*.[155] Since then, the SBSTA has not discussed substantial issues with regard to the regulation of GHG emissions from international shipping. However, cooperation between the Secretariats of the *UNFCCC* and the IMO has been ongoing since 1998, and these two organisations regularly exchange information regarding the regulation of GHG emissions from ships.[156]

The Geneva Ministerial Declaration adopted by the *UNFCCC* COP 2 in 1996 endorsed the newly published IPCC Second Assessment Report of 1995, and called for accelerating negotiations on a 'legally-binding protocol or another

152 Ibid. 11–13; Oberthür, above n. 34, 193.

153 Oberthür, above n. 34, 193.

154 See ch. 2, 2.5.

155 Oberthür, above n. 34, 193.

156 *Report of the Marine Environmental Protection Committee on Its Fifty-Fifth Session*, MEPC 55th Session, Agenda Item 23, IMO Doc MEPC 55/23 (16 October 2006) para. 4.28.

legal instrument'.[157] These efforts eventually led to the adoption of the *Kyoto Protocol* in 1997. The *Kyoto Protocol* is the only protocol of the UNFCCC; however, its entry into force experienced a lengthy and painful process until 2005. Despite this, through placing quantitative restrictions on emissions from industrialised economies, the *Kyoto Protocol* has been regarded as the culmination of international efforts to date to address the climate change problem.[158] Due to the contributions from the *Montreal Protocol*, only six types of unregulated GHGs were listed in the Protocol at that time.[159] Comparable with the UNFCCC, the *Kyoto Protocol* divides its parties into two groups: Annex I and non-Annex I, or generally developed and developing States.[160] The Protocol sets legally binding targets on the reduction of anthropogenic GHG emissions from Annex I States for the first commitment period from 2008 to 2012.[161] To reflect differentiated circumstances between the main industrial actors,[162] a system of differentiated targets within the rolling time scale was also agreed as Annex B to the Protocol.[163] Given the different historical and current contributions to global GHG emissions from both developed and developing States, these targets incorporate the CBDR principle and are thus acceptable for most States.

The most innovative aspect of the *Kyoto Protocol* is its so-called 'flexibility mechanisms', which were created for Parties to achieve their targets. These market-based mechanisms include Joint Implementation (JI),[164] the Clean

---

157 *The Geneva Ministerial Declaration*, Report of the Conference of the Parties on its Second Session, FCCC/CP/1996/15/Add.1 (8–19 July 1996), item 8.

158 D.I. Hodgkinson and R. Garner, *Global Climate Change: Australian Law and Policy* (LexisNexis Butterworths, 2008) 34–64.

159 But a seventh GHG was added to the list by the *Doha Amendment to the Kyoto Protocol* in 2012. See *Doha Amendment to the Kyoto Protocol*, adopted 8 December 2012, Decision 1/CMP.8, C.N.718.2012.TREATIES-XXVII.7.c (not yet in force).

160 To date 192 parties, including Australia, China (but excluding the USA), and the European Union have either ratified, acceded to, approved or accepted the Protocol. See <http://unfccc.int/kyoto_protocol/status_of_ratification/items/2613.php> accessed 20 June 2014.

161 *Kyoto Protocol* art. 3(1).

162 UNFCCC art. 4(2)(a). 'The differentiated circumstances' here refer to the 'differences in these Parties' starting points and approaches, economic structures and resource bases, the need to maintain strong and sustainable economic growth, available technologies and other individual circumstances, as well as the need for equitable and appropriate contributions by each of these Parties to the global effort regarding that objective.'

163 See Ong, above n. 112, 456.

164 *Kyoto Protocol* art. 6.

Development Mechanism (CDM)[165] and Emissions Trading (ET).[166] JI allows Annex I States to trade emission reduction units (ERUs) among themselves. ERUs can be obtained either by implementing cooperative projects to reduce GHG emissions or through establishing GHG sinks.[167] As the only flexibility mechanism available to developing States, CDM enables Annex I Parties to provide for actual GHG emission reduction projects in non-Annex I Parties, and thus receive the generated Certified Emission Reductions (CERs), either through financial sponsoring, or by technology transfer. In this way Annex I Parties can meet their emissions targets while the non-Annex I Parties will benefit from such projects.[168] ET generally allows Annex I Parties to purchase emissions credits from other Annex I Parties so as to fulfil their commitments provided that such trading is supplemental to their domestic actions.[169] In contrast to traditional 'command and control' type regulations, the above three mechanisms are more market-based. This change was interpreted as a response to the shift of the US position in dealing with the *Kyoto Protocol*.[170] However, from the perspective of international environmental law, this shift may reflect the trend of international environmental regulation. For example, to cope with GHG emissions from ships more effectively, extensive discussions on MBMs have been under way within the IMO.[171]

3.2.2.2 Post-Kyoto Efforts and Outcomes

3.2.2.2.1 *International Bunker Fuels Negotiation under the AWG-LCA*

As noted earlier, the *UNFCCC*'s SBSTA was responsible for the issue of allocation and control of GHG emissions from international shipping before the adoption of the *Kyoto Protocol* in 1997. However, after the *Kyoto Protocol* authorised the IMO to regulate shipping GHG emissions, the *UNFCCC* did not completely rely on the IMO's work in this regard. Rather, the AWG-LCA of the *UNFCCC* also discussed the issue of international bunker fuels in the context of paragraph 1b(iv) of the *Bali Action Plan*,[172] cooperative sectoral approaches

165 *Kyoto Protocol* art. 12.

166 *Kyoto Protocol* art. 7.

167 Ong, above n. 112, 456.

168 Ibid 457.

169 Birnie, Boyle and Redgwell, above n. 1, 367.

170 Ong, above n. 112, 456.

171 See, e.g., *Ensuring No Net Incidence on Developing Countries from a Global Maritime Market-Based Mechanism*, submitted by World Wide Fund for Nature (WWF), IMO Doc MEPC 63/5/6 (22 December 2011).

172 *Bali Action Plan*, Decision 1/CP.13, Report of the Conference of the Parties on its Thirteenth Session, Doc FCCC/CP/2007/6/Add.1 (14 March 2008) para. 1b(iv) ('*Bali Action Plan*').

and sector-specific actions. As seen from Table 3.1, the AWG-LCA was established as a subsidiary body under the Convention at COP 13 and CMP 3 of the *UNFCCC* process in 2007, working for long-term cooperative action under the *UNFCCC*. As such its work in relation to international bunker fuels, or GHG emissions from international shipping, did not conflict with the IMO's work. The IMO's work primarily focuses on specific technical and operational measures, whereas the AWG-LCA's work essentially involves regulatory principles, the setting of reduction targets, climate financing, preventing competitive distortions and carbon leakage, and the regulatory competence of the IMO. Theoretically, the work of the AWG-LCA could complement the IMO's work and possibly address some controversial issues that the IMO is facing, such as the regulatory principles and the IMO's regulatory competence. Nevertheless, no substantial outcomes relating to GHG emissions from international shipping had been achieved before the AWG-LCA terminated its five-year work at the Doha Climate Change Conference in 2012.

The AWG-LCA organised 15 sessions from March 2008 to December 2012.[173] Although no consensus was achieved, some of the proposals and options discussed under the AWG-LCA may contribute to the current discussions and negotiations within the IMO. Firstly, it was proposed that both the CBDR and NMFT principles could be applied to this GHG emissions issue but may not be treated equally. For instance, one option suggests that '[the regulation of GHG emissions from international shipping should be] in accordance with the principles and customary practices of the IMO, taking into account [the CBDR principle]'.[174] This means that the NMFT principle should apply, but the CBDR principle could be applied in different forms since it is only '[taken] into account' in this context. Similarly, some options suggest global levies on maritime bunker fuels, and propose that traffic on routes to and from Small Island Developing States (SIDS) and the Least Developed Countries (LDCs) should be exempt.[175] Although it is arguable that ships may change their routes

---

173 UNFCCC, *International Bunker Fuels under the AWG-LCA* <http://UNFCCC.int/methods/emissions_from_intl_transport/items/6141.php> accessed 20 June 2014.

174 *Cooperative Sectoral Approaches and Sector—Specific Actions in order to Enhance the Implementation of Article 4, Paragraph 1(c), of the Convention*, AWG-LCA 14th Session (Third Part), Panama City (1–7 October 2011) Option 2, p. 5.

175 *Report of the Ad Hoc Working Group on Long-Term Cooperative Action under the Convention on Its Seventh Session, Held in Bangkok from 28 September to 9 October 2009, and Barcelona from 2 to 6 November 2009*, Doc FCCC/AWGLCA/2009/14 (20 November 2009) Option 7 (para. 26), p. 102.

to avoid more stringent rules, this option provides a base for similar discussions on MBM proposals within the IMO.[176]

Secondly, it was suggested that the global emissions targets relating to international shipping be set by the *UNFCCC*, and this target could be 20 per cent below 2005 levels in 2020 on a scale consistent with the agreed two degrees objective, and this target may be achieved by means of MBMs.[177] This proposal involves both the reduction target and MBMs. Since MBMs often involve emissions reduction from different sectors,[178] it is reasonable for the *UNFCCC* to set this target due to the IMO's limited capacity in other sectors.

Thirdly, it was suggested that the IMO's regulatory competence, in particular relating to MBMs, should be clarified by the *UNFCCC*. One option provides that, 'all Parties in Annex I to the Convention shall pursue limitation or reduction of [GHG emissions from international shipping], working exclusively through the [IMO]'.[179] In this case, the proponents attempt to exclude other competent international bodies from regulating the shipping GHG emissions issue by any means, including MBMs. Theoretically, it is controversial whether the IMO has competence in regulating MBMs, so it might be necessary for the *UNFCCC* to clarify its view on this debate.[180] It may be inferred that these options discussed under the AWG-LCA were not adopted due to their lack of support from the main stakeholders. The responses from the main stakeholders of this issue are thus provided in the following chapters.

3.2.2.2.2 *Other Post-Kyoto Achievements and Their Implications on Shipping GHG Emissions*

Although the negotiation of GHG emissions from international shipping under the AWG-LCA was unsuccessful, the COPs and CMPs of the *UNFCCC* process have achieved outstanding outcomes and some of them may have significant implications for the regulation of shipping GHG emissions. After the adoption of the *Kyoto Protocol* in 1997, as of April 2016, 21 COPs and 11 CMPs have been

176 See ch. 4, 4.3.4.2.

177 *Cooperative Sectoral Approaches and Sector—Specific Actions in order to Enhance the Implementation of Article 4, Paragraph 1(c), of the Convention*, AWG-LCA 14th Session (Third Part), Panama City (1–7 October 2011) Options 2, 3, p. 5.

178 See ch. 4, 4.3.4.2.

179 *Cooperative Sectoral Approaches and Sector—Specific Actions in order to Enhance the Implementation of Article 4, Paragraph 1(c), of the Convention*, AWG-LCA 14th Session (Third Part), Panama City (1–7 October 2011) Option 8, p. 6.

180 The discussion of this issue is provided in Chapter 4, 4.2.

held. A broad range of matters have been discussed and various decisions made during these conferences. Table 3.1 lists the major outcomes and contributions achieved in these conferences. Based on distinct missions and achievements, the development of the climate change regime under the post-Kyoto era can be divided into three stages. The first stage includes the periods from COP 4 to COP 10, during which various unresolved issues within the *UNFCCC* and its *Kyoto Protocol* were first raised by the *Buenos Aires Plan of Action* (*BAPA*) and then discussed and supplemented in the subsequent COPs. The second stage commenced in 2005 when the *Kyoto Protocol* entered into force and the 'twin track' Convention and Protocol negotiations were launched. One of the focuses of the work in this stage had been to establish a second commitment period by means of a new Protocol, an amendment to the *Kyoto Protocol*, or a new climate change agreement after the first commitment period indicated in the *Kyoto Protocol* expired on 31 December 2012. The third stage commenced in 2015 when the *Paris Agreement* was adopted. This agreement employs a new reduction approach by means of 'Nationally Determined Contributions (NDCs)'. This section examines the first and second stages, but defers the discussion of the third stage to the next section.

In the first stage, seven important issues were put forward in the *Buenos Aires Plan of Action* (*BAPA*) and most of them were required to be finished before COP 6 in 2000.[181] To address these matters, the *Bonn Agreements* adopted in COP 6 enacted the 'core elements for the implementation of the BAPA', providing specific approaches and requirements for such implementation.[182] Nevertheless, due to the diverse interests from developed and developing States, the tasks set in the BAPA were not finished until COP 10 in 2004. During this process, the *Marrakesh Accords* adopted in COP 7 made vital contributions in successfully drafting detailed rules, procedures, technical guidelines and work programs.[183] Thus, the post-Kyoto cycle of policy-making launched by the BAPA was basically fulfilled with only minor matters

181 *The Buenos Aires Plan of Action*, Decision 1/CP.4, Report of the Conference of the Parties on its Fourth Session, FCCC/CP/1998/16/Add.1 (25 January 1999) ('*BAPA*'). These seven issues include financial mechanisms, technology transfer, adverse effects of climate change and implementation of response measures, activities implemented jointly, flexibility mechanisms, and the preparation for future COPs/CMPs.

182 *The Bonn Agreements on the Implementation of the Buenos Aires Plan of Action*, Decision 5/CP.6, Report of the Conference of the Parties on the Second Part of its Sixth Session, FCCC/CP/2001/5 (25 September 2001).

183 See *The Marrakesh Accords*, Decisions 2–14/CP.7, Report of the Conference of the Parties on its Seventh Session, FCCC/CP/2001/13/Add.1 (21 January 2002).

supplemented by the subsequent three COPs.[184] During this stage, GHG emissions from ships were not specifically discussed and no outcomes were achieved on this issue.

In the second stage, most of the COPs and CMPs have been working along with four subsidiary bodies: the AWG-LCA which was launched in Bali (COP 13) in 2007 and terminated in Doha (COP 18) in 2012, the Ad Hoc Working Group on Further Commitments for Annex I Parties under the *Kyoto Protocol* (AWG-KP) which was established in Montreal (CMP 1) in 2005 and terminated in Doha (CMP 8) in 2012, and the SBSTA and SBI (see Table 3.1 and Figure 3.3). As the two mechanisms worked in parallel, the AWG-KP worked for a second commitment for Annex B Parties of the *Kyoto Protocol* beyond the end of the first commitment period in 2012,[185] while the AWG-LCA primarily worked for long-term cooperative action under the *UNFCCC*.[186] As discussed above, the SBSTA and SBI (mainly SBSTA) organised the negotiation on the allocation and control of GHG emissions from international shipping from 1995 to 1996, and since 1998 the SBSTA has mainly exchanged information with the IMO on the regulation of shipping GHG emissions.

It was not until the Bali Climate Change Conference in 2007 that the development of a post-2012 climate change legal framework began, although the establishment of the AWG-KP at the CMP 1 in 2005 launched the negotiations for the next phase of the *Kyoto Protocol*. This was not only because of the establishment of the twin-track negotiation process,[187] but also due to the substantial contributions from the *Bali Road Map*. The *Bali Road Map* constitutes a set of decisions that represent the work to be done under various negotiating 'tracks'. In particular, the AWG-LCA's work on international bunker fuels, as discussed above, was guided by paragraph 1b(iv) of the *Bali Action Plan*, which is a part of the *Bali Road Map*.

As seen from Table 3.1, a number of outcomes have been achieved in the climate change conferences following the Bali conference in 2007. Examples are

184 Yamin and Depledge, above n. 28, 28.

185 *Consideration of Commitments for Subsequent Periods for Parties Included in Annex I to the Convention under Article 3, Para. 9 of the Kyoto Protocol*, Decision 1/CMP.1, Doc FCCC/KP/CMP/2005/8/Add.1 (2006) art. 1.

186 *Bali Action Plan* art. 1.

187 The twin-track negotiation process refers to the simultaneous negotiations under the COPs and CMPs of the *UNFCCC*.

the *Copenhagen Accord*,[188] the *Cancun Agreements*,[189] the *Durban Package*,[190] the *Doha Climate Gateway*,[191] and the *Warsaw Outcomes*.[192] The outcomes listed in Table 3.1 reflect decreased political support for the CBDR principle during global climate change negotiations. The *Copenhagen Accord* explicitly provides that combating climate change should be conducted in accordance with the CBDR principle,[193] and asserts that mitigation actions will be 'voluntary and on the basis of support' for least developed countries and small island developing States.[194] The *Cancun Agreements* require both developed and developing countries to exercise the communications obligation on measurement, reporting and verification (MRV). In particular, it even imposes greater burdens on developing countries than developed countries.[195] The *Durban Package* has been regarded as an advance to the climate regime on the

188 *Copenhagen Accord*, Decision 2/CP.15, Report of the Conference of the Parties on its Fifteenth Session, FCCC/CP/2009/11/Add.1 (30 March 2010) ('*Copenhagen Accord*').

189 *The Cancun Agreements*, Decisions 1–2/CMP.6, Report of the Conference of the Parties serving as the Meeting of the Parties to the Kyoto Protocol on its Sixth Session, FCCC/KP/CMP/2010/12/Add.1 (15 March 2011); Decision 1/CP.16, Report of the Conference of the Parties on its Sixteenth Session, FCCC/CP/2010/7/Add.1 (15 March 2011) ('*Cancun Agreements*').

190 See *Report of the Conference of the Parties on Its Seventeenth Session, Held in Durban from 28 November to 11 December 2011*, Doc FCCC/CP/2011/9/Add.1 (15 March 2012); *Report of the Conference of the Parties Serving as the Meeting of the Parties to the Kyoto Protocol on Its Seventh Session, Held in Durban from 28 November to 11 December 2011*, Doc FCCC/KP/CMP/2011/10/Add. 1 (15 March 2012).

191 See *Report of the Conference of the Parties on Its Eighteenth Session, Held in Doha from 26 November to 8 December 2012*, Doc FCCC/CP/2012/8 (28 February 2013); *Report of the Conference of the Parties Serving as the Meeting of the Parties to the Kyoto Protocol on Its Eighth Session, Held in Doha from 26 November to 8 December 2012*, Doc FCCC/KP/CMP/2012/13 (28 February 2013).

192 See *Report of the Conference of the Parties on Its Nineteenth Session, Held in Warsaw from 11 to 23 November 2013*, Doc FCCC/CP/2013/10 (31 January 2014); *Report of the Conference of the Parties Serving as the Meeting of the Parties to the Kyoto Protocol on Its Ninth Session, Held in Warsaw from 11 to 23 November 2013*, Doc FCCC/KP/CMP/2013/9 (31 January 2014).

193 *Copenhagen Accord* art. 1.

194 *Copenhagen Accord* art. 5.

195 Rajamani, above n. 113, 509, 513. In accordance with the *Cancun Agreements*, UNFCCC non-Annex I States are required to submit their national communications every four to five years, whereas Annex I States only need to do this every four years. Stathis Palassis, 'Climate Change and Shipping' in Robin Warner and Clive Schofield (eds), *Climate Change and the Oceans: Gauging the Legal and Policy Currents in the Asia Pacific and Beyond* (Edward Elgar Publishing Limited, 2012) 200, 206. However, Palassis argues that

grounds that: it ended the uncertainty of the future of the *Kyoto Protocol* by extending it for a second commitment period from 2013 to 2020, established a roadmap for adopting a post-2020 climate regime applicable to all and fulfilled the promise of the *Cancun Agreements*.[196] However, the *Durban Package* decisions do not contain a reference to the CBDR principle or even 'equity'.[197] It is thus argued that the weakened role of the CBDR principle, in particular the interpretation of 'differentiation', in the above decisions or statements represents 'a shift towards greater parallelism between developed and developing countries'.[198] Or in other words, 'differentiated responsibility' might be replaced by 'symmetry' as a guide for a future climate regime.[199]

From an international law perspective, most of the above decisions and statements are soft law in nature and the CBDR principle is currently applicable to the issue of climate change based on the *UNFCCC* and its *Kyoto Protocol*. Furthermore, most developed countries support a broader interpretation of 'differentiation' rather than to simply abandon the CBDR principle. For example, at the Durban climate change conference in 2011, developed countries insisted that any reference to the CBDR principle must be qualified with a statement that 'this principle must be interpreted in the light of contemporary economic realities'.[200] This might be interpreted as meaning that the current Annex I countries list should be updated to suit changed economic situations. Based on the current Annex I countries list, some OECD countries (such as the Republic of Korea and Israel) and well-developed countries (Singapore as an example) are treated as developing countries.[201] Additionally, non-Annex I States may also need to be sub-categorised into SIDS, LDCs, large developing countries and other developing countries to reflect their differing economic situations and regulatory interests.[202]

---

the *Cancun Agreements* affirmed the role of the IMO as the appropriate international organisation regulating GHG emissions from international shipping.

196 Lavanya Rajamani, 'The Durban Platform for Enhanced Action and the Future of the Climate Regime' (2012) 61(2) *International and Comparative Law Quarterly* 501, 515.

197 Ibid. 507.

198 Lavanya Rajamani, 'The Climate Regime in Evolution: The Disagreements that Survive the Cancun Agreements' (2011) 5(2) *Carbon & Climate Law Review* 136, 144.

199 Rajamani, above n. 196, 502.

200 Ibid. 508.

201 See ch. 7, 7.6.3.

202 See Rajamani, above n. 196, 517–518. Due to differing situations of developing countries, Rajamani asserts that the differentiation between developing countries should be based on 'self-perception' by individual developing countries.

3.2.2.3 The *Paris Agreement* and its Implications for GHG Emissions from International Shipping

At the Durban Climate Change Conference, held in November and December 2011, parties to the *UNFCCC* agreed to launch the Ad Hoc Working Group on the Durban Platform for Enhanced Action (ADP) with a mandate to develop a 'protocol, another legal instrument or an agreed outcome with legal force under the Convention applicable to all Parties'.[203] This universal climate change agreement was scheduled to be adopted in the Paris climate change conference in December 2015, and to be implemented from 2020.[204] After years of doubt and indecision, the *Paris Agreement* was eventually adopted on 12 December 2015. Meanwhile the Ad Hoc Working Group on the Paris Agreement (APA) was established to prepare for the entry into force of the Agreement and for the convening of the first section of the Conference of the Parties serving as the meeting of the Parties to the Paris Agreement (CMA) as well as other scheduled work.

The *Paris Agreement* has been regarded as successful for a number of reasons. The Agreement is the first multilateral environmental agreement to recognise human rights, climate justice and the right to health, and it explictly acknowledges climate change as 'a common concern of humankind'.[205] Besides, it aims to hold the increase of global average temperature to well below 2 degree Celsius and pursue efforts to limit the temperature increase to 1.5 degree Celsius above pre-industrial levels.[206] In addition to the committed financing of US$100 billion per year starting in 2020 to be provided to developing countries, the Agreement creates the concepts of the Intended Nationally Determined Contributions (INDCs) and Nationally Determined Contributions (NDCs) to enable countries to submit their self-determined national climate commitments. According to the Agreement, each Party's NDCs shall be reviewed every five years and should 'represent a progression' from previous

203 *Establishment of An Ad Hoc Working Group on the Durban Platform for Enhanced Action*, UNFCCC Decision 1/CP.17, Doc FCCC/CP/2011/9/Add.1 (2011) para. 2. There are different views and interpretations on these forms of agreement. See, e.g., 'another legal instrument' could be an amendment under Article 15, a new or an amended annex under Article 16, an amendment to the Kyoto Protocol, or an implementation agreement similar to a protocol; 'an agreed outcome with legal force' could be unilateral declarations by Parties, or COP decisions. Xolisa Ngwadla, Achala C. Abeysinghe and Adéyêmi Freitas, *The 2015 Climate Agreement: Lessons from the Bali Road Map* (2012) <http://www.eurocapacity.org/downloads/2015ClimateAgreement.pdf> accessed 21 June 2014, p. 7–8.

204 UNFCCC, *The Doha Climate Gateway* (2012) <http://UNFCCC.int/key_steps/doha_climate_gateway/items/7389.php> accessed 21 June 2014.

205 *Paris Agreement* preamble.

206 *Paris Agreement* art. 2(1)(a).

commitments.[207] Parties to the Agreement are also bound to transparency framework,[208] which together with the obligation to undertake and communicate the NDCs represents the legally-binding portion of the Agreement. The above features of the Agreement send a clear signal for decarbonisation to policy makers, investors and the business community.

The participation of 195 countries assured the adoption of the *Paris Agreement* on 12 December 2015. Before that, 161 countries and country blocs, representing some 95 per cent of global GHG emissions and 98 per cent of global population, had submitted their INDCs.[209] As of 22 April 2016, 188 countries that had submitted their INDCs were collectively responsible for over 99 per cent of global GHG emissions in 2010 of all the Parties to the *UNFCCC*.[210] It appears that the *Paris Agreement* has received wide support from international community in jointly tackling climate change. On 22 April 2016, 175 Parties (174 countries and the European Union) signed the *Paris Agreement* and 15 countries deposited their instruments of ratification during the signing ceremony, which reveals that largest number of countries to ever sign a multilateral agreement on the day of its opening for signature.[211]

The *Paris Agreement* is not perfect and there are some challenges attached to its implementation. First, the non-legally binding nature of the NDCs regulated under the *Paris Agreement*, together with the insufficient sum of all existing NDCs, makes it difficult for the international community to reach its goal of GHG emissions reduction. Theoretically the *Paris Agreement* is a treaty within the meaning of the *Vienna Convention on the Law of Treaties*,[212] and it can contain both binding and non-binding provisions based on whether legal obligations are created.[213] According to the *Paris Agreement*, the communication

207 *Paris Agreement* art. 4(3)(9).

208 *Paris Agreement* art. 13.

209 Bianka Kretschmer, Felix Fallasch, *From Paris to Songdo: How the Green Climate Fund's New Strategic Vision Supports the Paris Agreement* (7 April 2016) < http://climateanalytics.org/blog/2016/from-paris-to-songdo> accessed 2 May 2016.

210 Australian-German Climate and Energy College, *INDC FactSheets* (22 April 2016) <http://www.climate-energy-college.net/indc-factsheets> accessed 2 May 2016.

211 UNFCCC, *Record Support for Advancing Paris Climate Agreement Entry into Force* (22 April 2016) <http://newsroom.UNFCCC.int/paris-agreement/closing-paris-agreement-signing-press-release/> accessed 2 May 2016.

212 *Vienna Convention on the Law of Treaties*, opened for signature 23 May 1969, 8 ILM 679 (entered into force 27 January 1980) art. 2(1)(a).

213 Daniel Bodansky and Lavanya Rajamani, *Key Legal Issues in the 2015 Climate Negotiations* (June 2015) <http://www.c2es.org/docUploads/legal-issues-brief-06-2015.pdf> accessed 2 May 2016.

of NDCs is legally binding, but their contents and targets are not and there is no legal mechanism to ensure the implementation of these commitments. Meanwhile, research indicates that the sum of all existing NDCs, if fully implemented, leads to an increase in global temperatures of approvimately between 2.7 and 3 degrees Celsius above pre-industrial levels by the end of this century.[214] These will be serious limiting factors to achieving the goal of limiting temperature rise to 2 or 1.5 degrees Celsius.

Second, uncertainties remain as to the grant of financial assistence and technological transfer provided under the *Paris Agreement*. The US$100 billion per annum funds, which was set to be reached by developed countries in 2020, has been extended in the Decision 1/CP.21 through 2025.[215] This indicates less financial support from developed countries compared to their previous commitments, and there is uncertainty as to whether the fiancial assistence will be continued or increased beyond 2025. Furthermore, it has not been made clear where the funds will come from, how much contributions are to be provided by developed countries or industrialising countries, and how to allocate them between poor countries.[216] Regarding the transfer of technologies, the absence of the provisions on protecting the intellectual property rights under the Agreement might make it difficult for the private sector in developed countries to invest in or transfer the technologies that they own to developing countries.[217] Given that many developing countries, India as an example, have set their INDCs conditional on finance and technologies being made available, it will be difficult for the Agreement to be implemented.

The lack of substantive review and strengthening INDC ambition levels for the 2020–2025 period is another deficiency existed in the *Paris Agreement*. It is arguable that this period could become 'lost period' for increasing mitigation

---

214 Maxim Shrestha, 'COP 21 and the Paris Agreement: Achievement or Half measure?' (5 February 2016) (No. 028) *RSIS Commentary* 2; Marie Kurdziel, Thomas day, et al., *Challenges and Lessons Learned in the Preparation of Intended Nationally Determined Contributions (INDCs)* (1 March 2016) <http://mitigationpartnership.net/sites/default/files/challenges_lessons_indcs.pdf> accessed 2 May 2016, p. 3.

215 *Decision 1/CP.21 Aoption of the Paris Agreement*, Doc FCCC/CP/2015/10/Add.1 (29 January 2016) para. 53.

216 Shrestha, above n. 214.

217 Munjurul Hannan Khan, *Opinion: Paris Agreement—Opportunities and Challenges for Developing Countries* (23 February 2016) <http://cdkn.org/2016/02/opinion-paris-agreement-opportunities-and-challenges/?loclang=en_gb> accessed 2 May 2016.

ambition.[218] Similarly, Article 8 of the Agreement addresses the Loss and Damage associated with climate change impacts. However, this Article does not 'provide a basis for any liability or compensation' due to the strong requirement from the US.[219]

One crucial issue that the *Paris Agreement* does not address is GHG emissions from international shipping. Nevertheless, this does not mean that GHG emissions from international shipping will be completely subject to the IMO rather than the *Paris Agreement*. Instead, the *Paris Agreement* has some important implications for this shipping emissions issue.

Firstly, the 'Negotiation Text' for the *Paris Agreement* provides provisions relating to GHG emissions from international shipping, and it is projected that the subsequent conferences of the *UNFCCC* and its *Paris Agreement* will continue to discuss this matter.

At the Geneva Climate Change Conference in February 2015, States agreed on the 'Negotiating Text' for the Paris Climate Agreement (Nov/Dec 2015). The text provides:

> 23bis. [In meeting the 2 degree objective, Parties agree on the need for global sectoral emission reduction targets for international aviation and maritime transport and on the need for all Parties to work through the International Aviation Organization (ICAO) and the International Maritime Organization (IMO) to develop global policy frameworks to achieve these targets].[220]

This is an unequivocal endorsement of the setting of reduction targets for the international shipping sector and the central role of the IMO in achieving reductions. The February 2015 meeting also gave support for the establishment of a levy scheme:

---

218 PPMC, *Transport @ COP 21 Paris* (6 December 2015) <http://www.kas.de/wf/doc/kas_43927-1522-2-30.pdf?160119103248> accessed 2 May 2016, p. 20.

219 *Decision 1/CP.21 Aoption of the Paris Agreement*, Doc FCCC/CP/2015/10/Add.1 (29 January 2016) para. 51; International Center for Climate Governance, *International Climate Policy* (No. 39, January 2016) <http://www.iccgov.org/wp-content/uploads/2016/01/ICCG-International-Climate-Policy-Magazine-39.pdf> accessed 2 May 2016, p. 5.

220 *Outcomes of the United Nations Climate Change Conferences held in Lima in December 2014 and in Geneva in February 2015*, Note by the Secretariat, MEPC 68th Session, Agenda Item 5, IMO Doc MEPC 68/5 (18 February 2015) para. 17.

> 47.5 Option (a):
> c. In establishing the levy scheme, ICAO and IMO are encouraged to take into consideration the needs of developing countries, particularly the LDCs, SIDS and countries in Africa heavily reliant on tourism and international transport of traded goods.[221]

It seems increasingly likely that a MBM/Levy scheme will be adopted. If the subsequent *Paris Agreement* had adopted a levy scheme to reduce GHG emissions from ships, the IMO would have utilised this mandate to develop implementing regulations. This would also terminate the current debates on whether the IMO has the competence to regulate MBMs.[222] Therefore, the 'Option' above is attractive because it can address the current legal debate but also contribute to the further reduction of GHG emissions from international shipping. However, after being included in earlier drafts of the Agreement, all references to these rapidly growing GHG emissions from international shipping had been removed from the final *Paris Agreement*. As things currently stand, it can be inferred that the UNFCCC defers the dicussion of this shipping GHG issue to the subsequent COPs, CMPs or CMAs. The efforts of climate change mitigation in the land transport sector might be compromised if equally abitious actions on tackling GHG emissions from international shipping are not taken.[223]

Secondly, during the period of post-2020 and beyond the CBDR principle will still be applicable to the IMO's regulation on GHG emissions from international shipping although the *Kyoto Protocol* that provided the IMO with a GHG mandate will expire then. As discussed in Chapter 2, the CBDR principle is running through the *UNFCCC* and its *Kyoto Protocol*, which enables the application of the CBDR principle to the shipping GHG emissions issue due to the IMO's GHG mandate delegated by Article 2(2) of the *Kyoto Protocol*. With the expiry of the second commitment period in 2020, the *Kyoto Protocol* will be replaced by the *Paris Agreement*. However, the IMO's GHG mandate will not be terminated before it is accomplished even though the *Kyoto Protocol* that gave it this mandate will expire. Furthermore, the *Paris Agreement* is still an agreement under the *UNFCCC* and the CBDR principle has been corporated into this Agreement in a consistent manner. For instance, the CBDR principle

---

221 Ibid. para. 18.

222 Article 2(2) of the *Kyoto Protocol* has been regarded as a bit vague in that it does not recognise the explicit competence of the IMO, or define the precise measures that the IMO might adopt to address the GHG emissions issue. Thus some shipping associations have requested the *UNFCCC* to address this problem. See ch. 4, 4.2.

223 PPMC, above n. 218, 21.

is explicitly stipulated in the preamble, Article 2(2) and Article 4(3)(19) and implied in Article 4(4) of the *Paris Agreement*. Accordingly, the CBDR principle that applies in the *UNFCCC*, its *Kyoto Protocol* and *Paris Agreement* should continue to be applicable to the IMO's further regulation on GHG emissions from international shipping.

Thirdly, the *Paris Agreement* represents the evolution of the CBDR principle from the differentiation based on Annexes to the *UNFCCC* to a regime of flexible self-differentiation,[224] which will have an impact on the reduction of GHG emissions from international shipping. Under the *UNFCCC* regime, Annex I States to the *UNFCCC* (developed States) have compulsory reduction obligations whereas non-Annex I States to the *UNFCCC* (developing States) are exempt from compulsory reduction commitments. However, some developed States asserted that this type of differentiation should not be applied to the adoption of the Paris climate agreement. For example, On 12 February 2014 the US released its positions on a 2015 climate agreement suggesting that the CBDR principle should be interpreted differently during the period of post-2020 and beyond. It asserted that national efforts will be differentiated based on a range of factors, including 'circumstances, level of development, mitigation opportunities, capabilities', and so on, but it would not support 'a bifurcated approach to the new agreement, particularly one based on groupings that may have made sense in 1992 but that are clearly not rational or workable in the post-2020 era'.[225] Meanwhile, developing States insisted that the CBDR principle should be incorporated into the forthcoming Paris climate agreement. For instance, as the largest GHG emitter in the world, China declared in 2011 that it would not participate in a legally binding climate agreement before 2020, but would agree to participate in such an agreement after 2020 under certain conditions including the grant of financial assistance and technological transfer provided by developed States.[226]

224 Clément Bultheel, Romain Morel, et al., *COP21: Success at 'the End of the Beginning'* (18 December 2015) <http://www.i4ce.org/wp-core/wp-content/uploads/2015/12/15-12-18-I4CE-Climate-Brief-38-COP211.pdf> accessed 3 May 2016, p. 2.

225 The United States of America, *U.S. Submission on Elements of the 2015 Agreement* (12 February 2014) <https://UNFCCC.int/files/documentation/submissions_from_parties/adp/application/pdf/u.s._submission_on_elements_of_the_2105_agreement.pdf> accessed 3 May 2016.

226 China's Xinhua News Agency, *China Sets Conditions on Binding Climate Change Commitment after 2020* (6 December 2011) <http://english.peopledaily.com.cn/90883/7667257.html> accessed 3 May 2016. During the Durban climate change conference in 2011, China asserted that these conditions include 'new carbon-cutting pledges by rich nations in the second commitment period under the Kyoto Protocol, a fast launch

Eventually a compromise concerning the proper interpretation of the CBDR principle was reached in 2015. The *Paris Agreement* adds the element of 'in light of different national circumstances',[227] which drew from a 2014 US-China joint announcement on climate change,[228] to the end of the CBDR and respective capablities. While the Agreement requires developed States to 'take the lead' in tackling climate change,[229] the concept of NDC has been created to reflect a more flexible regime that is based on specific conditions of each State rather than on groups of States. Namely, each State determines its 'fair contribution' based on its respective capabilities and in light of its 'different national circumstances'.[230] Accordingly, the *Paris Agreement* only mentions developed and developing States rather than referring to the Annexes to the *UNFCCC*. Furthermore, the Agreement encourages developing State Parties to provide financial support to other developing State Parties voluntarily.[231] These are the new features of the CBDR principle embodied in the *Paris Agreement*.

The international shipping industry has responded to the adoption of the *Paris Agreement* in a timely manner. In February 2016, the International Chamber of Shipping (ICS) submitted a proposal to the 68th MEPC meeting of the IMO. It suggested that the IMO should adopt an Intended IMO Determined Contribution on behalf of the international shipping industry and report it to the *UNFCCC* meetings in the future. It took the view that this would mirror the commitments or the INDCs that the Parties to the *UNFCCC* have made for their national economies and this language is also consistent with the language used in the *Paris Agreement*.[232] It is understandable for the ICS to propose this non-legally binding commitment rather than the sector's binding reduction target which the IMO had been discussed about previously. It becomes less likely for

of the Green Climate Fund agreed on in Cancun under a supervisory regime, implementing the consensus of adaptation, technology transfer, transparency, capability building and other points agreed upon in the former conferences as well as appraising developed countries' commitment during the first period of the Kyoto Protocol'.

227 See, e.g., *Paris Agreement* preamble, arts. 2(2), 4(3)(19).

228 The White House of the USA, *U.S.-China Joint Announcement on Climate Change* (12 November 2014) <https://www.whitehouse.gov/the-press-office/2014/11/11/us-china-joint-announcement-climate-change> accessed 3 May 2016, para. 2.

229 *Paris Agreement* art. 4(4).

230 See, e.g., *Paris Agreement* arts. 2,3.

231 *Paris Agreement* art. 9(2).

232 *Proposal to Develop an 'Intended IMO Determined Contribution' on CO2 Reduction for International Shipping*, submitted by International Chamer of Shipping (ICS), MEPC 69th Session, Agenda Item 7, IMO Doc MEPC 69/7/1 (12 February 2016) paras 7,8.

the international shipping industry, a comparatively cost effecitve and clean method of transportation, to adopt more strigent measures than those in other sectors to reduce its GHG emissions. To some extent this can be seen as a negative impact of the *Paris Agreement* on the reduction of GHG emissions from international shipping within the IMO. Currently it is unknown which step that the IMO is going to take as to the setting of the reduction target for the international shipping sector. However, it is clear that the evolved CBDR principle under the *Paris Agreement* will play a role in the IMO's further regulation on the reduction of shipping GHG emissions.

Finally, the objective of reduction set by the *Paris Agreement* reveals the significance of reducing GHG emissions from interational shipping, and the IMO has responded to the enhanced transparency requirement stipulated under the Agreement. As discussed above, current exisiting NDCs will lead to the temperature rise of 2.7 or 3 degrees Celsius above pre-industrial levels by 2100. In order to reach the 2 degrees Celsius threshold, ambitious reductions efforts made by the international shipping and aviation industries, which are currently excluded from the reduction commitment under the *Paris Agreement*, would possibly fill in the gaps existed in current NDCs-based regime. Meanwhile, Article 13 of the *Paris Agreement* specifies the establishment of 'an enhanced transparency framework for action and support, with built-in flexibility which takes into account Parties' different capacities and builds upon collective experience'.[233] This provision aims to address the widespread concerns on the absence of a detailed MRV mechanism as required under the *Cancun Agreements*.[234] It is projected that this transparency framework will be elaborated in the subsequent *UNFCCC* and its *Paris Agreement* process. Furthermore, this framework also creates the basis for a similar transparent MRV system in the international shipping industry.

Indeed, as a response to the unilateral EU Regulation on Monitoring, Reporting and Verification (MRV),[235] the IMO had been discussing a similar data collection system for global application before the adoption of the EU MRV Regulation in July 2015. The IMO had obtained general support from its member States and agreed that it was imperative to develop a global data

233 *Paris Agreement* art. 13(1).

234 Bultheel, Morel, et al., above n. 224.

235 In June 2013, the European Commission developed a proposal for an EU Regulation on Monitoring, Reporting and Verification (MRV) and submitted it to the European Parliament and the Council. Consequently, the MRV Regulation 2015/757 was adopted and came into force on 1 July 2015.

collection system for international shipping.[236] However, it was ageed at the IMO in May 2015 that the development of a data collection system for ships should employ a three-step approach (data collection, data analysis and decision making), and it is premature to decide whether this system should be voluntary or mandatory.[237] It seems that the regulatory process on the data collection system within the IMO will be lengthy. Nevertheless, the adoption of the *Paris Agreement* has accelerated this process. At the 69th MEPC meeting in April 2016, the IMO approved mandatory data collection requirements for ships engaged in international shipping. Under this system, ships of 5,000 gross tonnage (GT) or above will be required to collect consumption data for each type of fuel they use, and the data will then be reported to flag State after the end of each year. Flag State will issue a Statement of Compliance to the ship, and tranfer the data to the IMO. Based on the data analysis, the IMO will determine whether any further measures are needed to enhance energy efficiency and address GHG emissions from international shipping.[238] This move has been regarded by the IMO as sending 'a clear and positive signal about the Organization's continuing commitment to climate change mitigation'.[239] This probably also indicates that the interaction between the *UNFCCC* process and the IMO GHG regime will continue. The absence of GHG emissions from international shipping under the *Paris Agreement* will not prevent it from discussing and addressing this shipping GHG issue in its subsequent conferences.

3.4 Conclusion

This chapter has examined the responses from the UN to the issue of GHG emissions from international shipping. Faced with the aggravating situations of climate change around the world, the UN made timely institutional and legal responses and these responses have significant implications for shipping emissions. To cope with climate change, the UN established a number of institutions. Among them, the IPCC underpins the combating of GHG emissions from shipping by means of its Assessment Reports; the UNEP and WMO raise

236 *Report of the Marine Environment Protection Committee on Its Sixty-Seventh Session*, MEPC 67th Session, Agenda Item 20, IMO Doc MEPC 67/20 (31 October 2014) para. 5.7.

237 *Report of the Marine Environment Protection Committee on Its Sixty-Eighth Session*, MEPC 68th Session, Agenda Item 21, IMO Doc MEPC 68/21 (29 May 2015) paras 4.8, 4.10.

238 IMO, *IMO Takes Further Action on Climate Change* (22 April 2016) <http://www.IMO.org/en/MediaCentre/PressBriefings/Pages/11-data-collection-.aspx> accessed 4 May 2016.

239 Ibid.

awareness of the need to tackle the issue and implement outcomes within the international climate change regime; the *UNFCCC* and its COPs and CMPs, especially its subsidiary SBSTA and AWG-LCA, provide crucial platforms for different countries to negotiate on the issue of international bunker fuels. Furthermore, unilateral actions by individual States or the EU to reduce shipping emissions should be considered.

The UN came to establish an international legal framework on climate change with the *UNFCCC* and its *Kyoto Protocol* as its core element. During this process, the *1979 CLRTAP* and *1985 Vienna Convention* provided a 'framework treaty' model for future international environmental agreements. The precautionary principle and the spirit of cooperation revealed from the *Vienna Convention* paved the way for future conventions. Additionally, three innovative approaches adopted by the *1987 Montreal Protocol* might be particularly valuable to the issue of shipping emissions, namely, a more flexible arrangement, a well-designed application of the CBDR principle and gaining participation and investment from industry.

An international climate change regime has been established and continues its development. In particular, the SBSTA worked on the allocation and control of GHG emissions from international shipping from 1995 to 1996, but failed to reach consensus in adopting an option on the allocation of GHG emissions. Since 1998 the SBSTA has exchanged information with the IMO as to the regulation of GHG emissions from ships. The subsequent AWG-LCA started to work on international bunker fuels under the *Bali Action Plan* in 2008. It worked on regulatory principles, the setting of reduction targets and the IMO's competence. However, no substantial outcome had been achieved before the AWG-LCA terminated its work at the Doha Climate Change Conference in 2012. Currently, regulatory measures to reduce shipping GHG emissions mainly rely on the work of the IMO.

The adoption of the *Paris Agreement* in December 2015 has been treated as a success in that it is an important step in the evolution of climate governance and a reaffirmation of environmental multilateralism. However, there are some challenges attached to its implementation. Particularly the non-legally binding nature of the NDCs has made it difficult for the international community to realise the purpose of the Agreement. The *Paris Agreement* does not address GHG emissions from international shipping. Nevertheless, it has some significant implications for this shipping emissions issue. Among them, the CBDR principle will still be applicable to the IMO's regulation on GHG emissions from international shipping during the post-2020 era; the NDC-based responsibilities represent the evolution of the CBDR principle from the Annexes-based differentiation to the flexible self-differentiation. As responses to the *Paris Agreement*, the international shipping industry is considering to adopt an Intended IMO Determined Contribution while the IMO has approved mandatory data collection system for the ships of 5,000 GT and above. It is expected that the interaction between the *UNFCCC* process and the IMO GHG regime will continue.

CHAPTER 4

The IMO Response to the Issue of Greenhouse Gas Emissions from International Shipping

4.1 Introduction

The previous chapters revealed that the problem of greenhouse gas (GHG) emissions from international shipping is an issue with international dimensions. While a number of emerging principles on international environmental liability might apply to the issue, the UN has also made active institutional and legal responses. The *United Nations Framework Convention on Climate Change* (*UNFCCC*)[1] and its *Kyoto Protocol* and *Paris Agreement*,[2] together with its Subsidiary Body on Scientific and Technological Advice (SBSTA), the Ad Hoc Working Group on Long-term Cooperative Action (AWG-LCA), Conferences of Parties (COPs), the COPs serving as the Meeting of the Parties to the Kyoto Protocol (CMPs), and the COPs serving as the Meeting of the Parties to the Paris Agreement (CMAs), have contributed to the international tackling of the issue, although the effectiveness of their efforts has been questioned.[3]

---

1 *United Nations Framework Convention on Climate Change*, opened for signature 9 May 1992, 31 ILM 848 (entered into force 21 March 1994) ('*UNFCCC*').

2 *Kyoto Protocol to the United Nations Framework Convention on Climate Change*, opened for signature 16 March 1998, 37 ILM 22 (entered into force 16 February 2005) ('*Kyoto Protocol*'); *Paris Agreement to the United Nations Framework Convention on Climate Change* (2015), opened for signature 22 April 2016, FCCC/CP/2015/L.9 (not yet in force) ('*Paris Agreement*').

3 See, e.g., Sebastian Oberthür, 'Institutional Interaction to Address Greenhouse Gas Emissions from International Transport: ICAO, IMO and the Kyoto Protocol' (2003) 3(3) *Climate Policy* 191, 193. Oberthür asserts that Parties to the *UNFCCC* had wide discussions on the GHG emissions from international transport, and its SBSTA selected five options as the basis for further work on the allocation of emissions from aviation and marine bunker fuels in 1996, which failed later in reaching consensus among different States. Due to the political deadlock on the allocation issue, the *UNFCCC* had to turn to the IMO for the future regulation of the issue. But see Bernd Hackmann, 'Analysis of the Governance Architecture to Regulate GHG Emissions from International Shipping' (2012) 12(1) *International Environmental Agreements: Politics, Law and Economics* 85, 90. Hackmann purports that the work by the *UNFCCC* on the issue is still proceeding, and regulating shipping GHG emissions should fall under the scope of the Bali Action Plan.

 | DOI 10.1163/9789004329317_005

As a specialised agency of the UN, the International Maritime Organization (IMO) has recognised the problem and has acted on it based on Article 2(2) of the *Kyoto Protocol* as well as the *IMO Convention* and the *LOSC*.[4] In contrast to the efforts made within the international climate change regime, the IMO's work is worthy of higher expectation due to the IMO's mandate and strength in regulating GHG emission-related technical matters. In particular, the newly-adopted amendments of Annex VI to the *International Convention for the Prevention of Pollution from Ships* (*MARPOL 73/78*)[5] and guidelines by the IMO have assured the international community of progress regarding the adoption of energy efficiency measures.

The first part of this chapter looks briefly at the establishment of the IMO, and its mandate and competence to regulate GHG emissions from international shipping. Having established the central role of the IMO in providing a solution to the problem, the chapter then examines and assesses the IMO GHG regime from four perspectives, namely, the evolution of the regime, *MARPOL* Annex VI and its amendments, the adopted technical and operational measures within the IMO and market-based measures (MBMs) currently under discussion.

4.2 The IMO's Mandate and Competence to Regulate Greenhouse Gas Emissions from International Shipping

To cope with the increasingly serious issues of safety at sea and marine pollution,[6] the United Nations (UN) held a Maritime Conference in Geneva on 6 March 1948. This conference adopted a convention that formally established the Inter-Governmental Maritime Consultative Organization (IMCO), which

4 See ch. 2, 2.5.3.

5 *International Convention for the Prevention of Pollution from Ships* (*MARPOL 73/78*), signed 2 November 1973, 12 ILM 1319, as amended by the 1978 Protocol to the 1973 Convention, 1341 UNTS 3, 17 ILM 546 (entered into force 2 October 1983). To date, *MARPOL 73/78* has adopted six annexes and their revisions, namely, Annex I on Oil (entered into force 2 October 1983); Annex II on Noxious Liquid Substances Carried in Bulk (entered into force 6 April 1987); Annex III on Harmful Substances Carried in Packaged Form (entered into force 1 July 1992); Annex IV on Sewage (entered into force 27 September 2003); Annex V on Garbage (entered into force 31 December 1988); and Annex VI on Air Pollution from Ships (entered into force 19 May 2005). Of these 5 annexes, only Annexes 1 and 2 are compulsory, whereas the other 3 annexes were adopted as voluntary annexes to *MARPOL 73/78* which means they are only binding on those States which ratify them.

6 G.P. Pamborides, *International Shipping Law: Legislation and Enforcement* (Kluwer Law International, 1999) 79–80.

subsequently changed its name to the IMO in May 1982.[7] Article 1 of the *IMO Convention* outlines five purposes of the organisation,[8] which can be broadly summarised into its jurisdiction on technical and commercial matters relating to shipping. It appears that paragraphs (b) and (c) of Article 1 give the IMO competence to regulate commercial aspects of shipping aiming at removing discriminatory and 'unfair restrictive practices'. However, due to the potential threat to the practice of free enterprise through the IMO's regulation of the commercial aspects of shipping, many States have united to limit the purposes of the IMO to technical aspects.[9] To date the IMO has never been allowed to exercise its full economic mandate.[10] Therefore, the main purpose of the IMO is 'to encourage the general adoption of the highest practicable standards in matters concerning maritime safety, efficiency of navigation and the prevention and control of marine pollution from ships' primarily through technical means.[11]

---

7 *Convention on the Inter-Governmental Maritime Consultative Organization*, opened for signature 6 March 1948, 289 UNTS 3 (entered into force 17 March 1958), amended and renamed as *Convention on the International Maritime Organization*, opened for signature 14 November 1975, 9 UTS 61 (entered into force 22 May 1982) ('*IMO Convention*').

8 Article 1 of the *IMO Convention* provides that,
'The purposes of the Organization are:
(a) To provide machinery for co-operation among Governments in the field of governmental regulation and practices relating to technical matters of all kinds affecting shipping engaged in international trade; to encourage and facilitate the general adoption of the highest practicable standards in matters concerning the maritime safety, efficiency of navigation and prevention and control of marine pollution from ships; and to deal with administrative and legal matters related to the purposes set out in this Article;
(b) To encourage the removal of discriminatory action and unnecessary restrictions by Governments affecting shipping engaged in international trade so as to …
(c) To provide for the consideration by the Organization of matters concerning unfair restrictive practices by shipping concerns in accordance with Part II;
(d) To provide for the consideration by the Organization of any matters concerning shipping and the effect of shipping on the marine environment that may be referred to it by any organ or specialized agency of the United Nations;
(e) To provide for the exchange of information among Governments on matters under consideration by the Organization.'

9 Pamborides, above n. 6, 83.

10 Alan Khee-Jin Tan, *Vessel-Source Marine Pollution: the Law and Politics of International Regulation* (Cambridge University Press, 2006) 75.

11 *IMO Convention* art. 1(a). Although the Facilitation Committee of the IMO has regulated some matters involving elements of trade, the purposes of these regulations are generally to be achieved by technical means.

The IMO currently has 171 member States, three associate members, and 77 international non-governmental organisations (NGOs) under consultative status.[12] The IMO's structure has developed, reflecting its evolving mandate.[13] Initially, the IMO only comprised the Assembly, the Council and the Maritime Safety Committee (MSC).[14] Currently, the IMO consists of an Assembly, a Council and five main Committees: the MSC, the Marine Environment Protection Committee (MEPC), the Legal Committee, the Technical Co-operation Committee and the Facilitation Committee. Among these, the MEPC, which comprises all member States, is responsible for the reduction of GHG emissions from international shipping. In addition, the inter-sessional meeting of the IMO's Working Group on GHG Emissions from Ships (GHG-WG) was established between two MEPC sessions. These changes indicate the growing significance of marine environmental protection and co-operation among various institutions.[15] To accomplish its mission, the IMO makes full use of different instruments, namely resolutions, codes, guidelines and conventions.

It was argued in Chapter 2 that the IMO's mandate to regulate GHG emissions from international shipping comes from the *IMO Convention*, the *LOSC* and the *Kyoto Protocol*. While the *IMO Convention* and *LOSC* provide the IMO with general competence to regulate GHG emissions from ships, the *Kyoto Protocol* gives the IMO a specific mandate to regulate this matter. It was thus inferred that this interpretation justifies the application of both the Common but Differentiated Responsibility (CBDR) and No More Favourable Treatment (NMFT) principles to the regulation of GHG emissions from international shipping.[16] Indeed another implication that can be drawn from this interpretation is that the IMO may not be the sole competent international

12 International Maritime Organization (IMO), *Member States, IGOs and NGOs* <http://www.imo.org/en/About/Membership/Pages/Default.aspx> accessed 10 May 2016.

13 Tan, above n. 10, 76.

14 Lawrence Juda, 'IMCO and the Regulation of Ocean Pollution from Ships' (1977) 26(3) *The International and Comparative Law Quarterly* 558, 559.

15 For instance, the establishment of the Facilitation Committee was to harmonise shipping procedures and eliminate unnecessary formalities and 'red tape' in international shipping. Tan, above n. 10, 76.

16 See ch. 2, 2.5.3. The CBDR principle requires both developed and developing States to contribute to addressing environmental problems, but imposes the primary responsibility on developed States due to their different historical contribution to the problems and the differentiated capability of developed and developing States. The NMFT principle refers to 'port States enforcing applicable standards in a uniform manner to all ships in their ports, regardless of flag'. ø. Buhaug et al., 'Second IMO GHG Study 2009' (International Maritime Organization (IMO), 2009) 20; see also *MARPOL 73/78* art. 5(4).

organisation in regulating GHG emissions from ships. Currently there are three routes in regulating GHG emissions from international shipping within the IMO: technical measures, operational measures and MBMs. As discussed above, regulating technical and operational measures falls under the competence of the IMO based on the *IMO Convention*, and currently no other international body possesses the IMO's technical expertise. In practice, the global shipping industry, including national shipping industries from the *UNFCCC*'s non-Annex I States, takes the view that the IMO is the only competent organisation to regulate the issue from a technical and operational perspective.[17] Against this backdrop, technical and operational measures were regulated by the IMO in the form of energy efficiency measures under the revised Annex VI of *MARPOL 73/78* in July 2011. Therefore, given that the *Kyoto Protocol* has delegated to the IMO the competence to regulate this GHG issue, there is little doubt that the IMO is the sole competent international organisation to regulate technical and operational measures to reduce shipping GHG emissions.

In accordance with the origin of the IMO's mandate, it is arguable that the MBMs currently being discussed within the IMO could be considered as being beyond the competence that the IMO has received from the *IMO Convention* and the *LOSC*. This is because some of these measures involve global emissions reduction from different sectors,[18] and also go beyond the scope of

17 For instance, the four Round Table members, namely the International Chamber of Shipping (ICS), the Baltic and International Maritime Council (BIMCO), the International Association of Independent Tanker Owners (INTERTANKO) and the International Association of Dry Cargo Shipowners (INTERCARGO), asserted that both technical and operational measures and MBMs should be governed by the IMO. *Future IMO Regulation regarding Greenhouse Gas Emissions from International Shipping*, submitted by Denmark, Marshall Islands, BIMCO, ICS, INTERCARGO, INTERTANKO and OCIMF, MEPC 57th Session, Agenda Item 4, IMO Doc MEPC 57/4/2 (21 December 2008) para. 10; Round Table of International Shipping Associations, *Round Table Associations Position Paper on GHG+MBMs* (22 February 2012) <https://www.bimco.org/About/Press/Press_Releases/2012/2012_02_22_Round_Table_MBM.aspx> accessed 1 June 2014. China's shipping industry generally supports the IMO's role in regulating technical and operational measures to tackle shipping GHG emissions. 王尔德 [Wang Erde], '《减排谈判首次共识，中国减排将始于 2019 年》 [Agreement Achieved through Reduction Negotiations and China Is to Start Its Reduction in 2019]', *21st Century Business Herald* (Beijing), 26 July 2011 <http://stock.sohu.com/20110726/n314586469.shtml> accessed 22 June 2014. See ch. 5, 5.2, 5.4.1.

18 The third Intersessional Meeting of the Working Group on GHG Emissions from Ships grouped the MBMs proposals into two categories, namely 'focus on in-sector' and 'in-sector and out-of-sector'. Based on this grouping, current MBM proposals involving out-of-sector emission reductions are the International GHG Fund, the Emissions Trading

technical matters relating to shipping. Theoretically, the mandate that the IMO has from the *Kyoto Protocol* also gives it the competence for such work. However, Article 2(2) of the *Kyoto Protocol* is vague in that it does not recognise the explicit competence of the IMO, or define the precise measures that the IMO might adopt to address the GHG issue. Therefore, in 2011 the International Chamber of Shipping (ICS) called on the participants in the Durban Climate Change Conference to give the IMO a clear mandate to reduce emissions through MBMs.[19] Nevertheless, this request was not addressed by the Durban Conference. Furthermore, the 2015 *Paris Agreement* also keeps silence on the GHG mandate of the IMO. Consequently, it is possible that in the future in order to regulate MBMs involving out-of-sector emissions reduction and international trade, the IMO will collaborate with other international organisations, such as the *UNFCCC* or the World Trade Organization (WTO), due to their broader competence or expertise in international trade.[20]

4.3 The IMO Greenhouse Gas Emissions Regime

The IMO has partially regulated GHG emissions from international shipping by means of negotiations and discussions within its MEPC. The conventions, codes, resolutions, guidelines and discussions achieved or conducted during this process constitute the IMO regime on the reduction of shipping GHG emissions or, in other words, the IMO GHG Emissions regime. This part firstly reviews the development of this regime and then examines Annex VI to *MARPOL 73/78* and its amendments—the major achievement of this regime to date. Then this part continues to summarise and assess the adopted technical and operational measures within the MEPC. Finally the theoretical foundation, the necessity, and the feasibility and impact assessment of MBMs are analysed.

Scheme, the Port State Levy and the Rebate Mechanism. *Report of the Third Intersessional Meeting of the Working Group on Greenhouse Gas Emissions from Ships*, MEPC 62nd Session, Agenda Item 5, IMO Doc MEPC 62/5/1 (8 April 2011) annex 3.

19 Kevin Cooper, *Reducing Shipping Emissions: An Overview of Recent International Initiatives* (25 September 2012) <http://www.safety4sea.com/analysis/89/134/reducing-shipping-emissions-> accessed 24 June 2014.

20 In practice, some *UNFCCC* non-Annex I States, such as China, India, and Malaysia, doubt the competence of the IMO in regulating MBMs and assert that MBMs should be determined by the *UNFCCC*. *Report of the Marine Environment Protection Committee on Its Sixtieth Session*, MEPC 60th Session, Agenda Item 22, IMO Doc MEPC 60/22 (12 April 2010) annex 4, pp. 2, 8, 10. See also ch. 6, 6.3.2.2.

4.3.1 *Evolution of the IMO Greenhouse Gas Emissions Regime*

Although discussions on GHG emissions from ships within the IMO started in the late 1980s, it is generally accepted that the IMO's work on this issue formally commenced in 1997.[21] During that year, the MARPOL Conference not only adopted a protocol on Annex VI to *MARPOL 73/78* but also adopted Resolution 8 on 'CO_2 emissions from ships', which requested the IMO to undertake a study on GHG emissions from ships and consider feasible CO_2 reduction strategies.[22] Following joint efforts by several internationally renowned research institutes,[23] a Study of GHG Emissions from Ships was published in 2000.[24] This study not only answered the question of why GHG emissions from shipping should be reduced, but it also explored how to deal with the issue. It canvassed the reduction potential of different technical, operational and market-based approaches, which to some extent provide a 'road map' for future policies within the IMO. In 2003, a resolution was adopted by the IMO Assembly on 'IMO policies and practices related to the reduction of greenhouse gas emissions from ships', urging the MEPC to 'identify and develop the mechanism or mechanisms needed to achieve the limitation or reduction of GHG emissions from international shipping'.[25] Since then, the MEPC has been working on this issue by means of negotiations and discussions within its series of session meetings, as well as in its GHG-WGs. The main events during this process are listed in Table 4.1.

The evolution of the IMO GHG Emissions regime has been lengthy and intermittent. During a 14-year period from September 1997 to July 2011, no binding agreements regarding GHG emissions from international shipping were reached within the IMO, and producers of emissions were exempt from

21 International Maritime Organisation (IMO), 'Main Events in IMO's Work on Limitation and Reduction of Greenhouse Gas Emissions from International Shipping' (2011) <http://www.imo.org/MediaCentre/resources/Pages/Greenhouse%20gas%20emissions.aspx> accessed 16 September 2012, p. 3.

22 Ibid.

23 These institutes were from Norway and the USA, namely MARINTEK (Norway), Det Norske Veritas (Norway), ECON, Center for Economic Analysis (Norway), and Carnegie Mellon University (United States). Kjell Skjølsvik et al., 'Study of Greenhouse Gas Emissions from Ships' (International Maritime Organization (IMO), 2000) 7.

24 These institutes were from Norway and the United States, namely MARINTEK, Norway, Det Norske Veritas, Norway, ECON, Centre for Economic Analysis, Norway, and Carnegie Mellon University, United States.

25 *IMO Policies and Practices Related to the Reduction of Greenhouse Gas Emissions from Ships*, IMO Assembly 23rd Session, Agenda Item 19, IMO Doc Res A.963(23) (5 December 2003).

liabilities under this regime. Intensive discussions on this issue have only occurred since 2008. GHG emissions from international shipping were partially regulated by technical and operational measures in July 2011 for the first time, and the regulatory scope was further extended in April 2014. Table 4.1 indicates that the IMO reiterated at least three times its role as the most competent international body in regulating GHG emissions from ships. This claim might have resulted from competitive institutional pressure from other international organisations such as the *UNFCCC* and the possibility of unilateral measures being adopted by individual States and the European Union (EU).[26]

The debate on the incorporation of either the CBDR principle or the NMFT principle into the reduction of GHG emissions from ships has run through all of the negotiations and discussions within the IMO. The conflict of the two principles has delayed the advancement of the negotiations within the MEPC.[27] To expedite the negotiation process within the MEPC, the 52nd MEPC meeting adopted a two-step strategy, according to which the MEPC was to deal with all technical matters related to GHG limitations or reductions first and then resolve the politically related issues including the application of the CBDR or NMFT principles.[28] However, it has been difficult to separate the two steps. In 2008, the 57th MEPC meeting adopted nine fundamental principles as a basis for future regulations, although they were opposed by many developing countries. Principle 2 provides that the future IMO framework should be 'binding and equally applicable to all flag States in order to avoid evasion',[29] which incorporates the NMFT principle. The 58th MEPC meeting in 2008 discussed a proposed change to Principle 2, which was amended to read that it was, 'binding and equally applicable to all ships, without this requiring States to accept similar regulations/standards in other fora.'[30] In this case, the responsible entity shifts from flag States to all ships and the application of this principle is limited to either the MEPC or the IMO. However, the NMFT principle remained in this version and, consequently, no consensus was achieved. As to the proposed MBMs, the application of the CBDR or the NMFT principle has been the

26 See Oberthür, above n. 3, 202–203.

27 See Hackmann, above n. 3, 96.

28 *Report of the Marine Environment Protection Committee on Its Fifty-Second Session*, MEPC 52nd Session, Agenda Item 24, IMO Doc MEPC 52/24 (18 October 2004) para. 4.35.

29 *Report of the Marine Environment Protection Committee on Its Fifty-Seventh Session*, MEPC 57th Session, Agenda Item 21, IMO Doc MEPC 57/21 (7 April 2008) para. 4.73.

30 *Identifying Consensus on IMO Principles on Addressing Greenhouse Gas Emissions from International Shipping*, submitted by Australia, Canada, Denmark, Germany, Japan, Marshall Islands, Norway, Panama and the United States, MEPC 58th Session, Agenda Item 4, IMO Doc MEPC 58/4/16 (1 August 2008) para. 5.

main focus of debate, which can be seen from the succeeding rounds of negotiations. It is concluded that the evolution of the IMO GHG Emissions regime is a process where various technical and operational measures and MBMs have been discussed and negotiated in an attempt to reach agreement between developed and developing countries. To date this regime is still under development. However, some outcomes, including the amendments of Annex VI to *MARPOL 73/78* and some guidelines, have been achieved within the MEPC.

TABLE 4.1 *Main events of IMO's work on GHG emissions from international shipping*[31]

Meetings	Meeting Date	Major Outcomes	Base Documents
MARPOL Conference	**Sep 1997**	Resolution 8 on 'CO_2 emissions from ships' requested the IMO to undertake a study on GHG emissions from ships and consider feasible CO_2 reduction strategies.	MARPOL Annex VI; **Resolution 8**
MEPC 45	Jun 2000	First IMO Study on GHG Emissions from Ships.	MEPC 45/8
IMO Assembly Meeting	**Dec 2003**	**Urged the MEPC to develop mechanisms to tackle the GHG emissions issue;** Gave priorities to establish a GHG baseline, develop a methodology and guidelines, and evaluate technical, operational and market-based solutions; called for the MEPC's draft of a GHG work plan.	Assembly **Res. A.963(23)**
MEPC 52	Oct 2004	Agreed on a two-step strategy: firstly deal with technical matters, then politically related issues.	MEPC 52.24
MEPC 53	Jul 2005	Approved IMO's 'Interim Guidelines for Voluntary Ship CO_2 Emission Indexing for Use in Trials', aiming to provide a voluntary system for the ship operators during a trial period.	MEPC/Circ.471 MEPC 53/24

31 This table is compiled by the author based on the following sources: International Maritime Organisation (IMO), above n. 21; reports of MEPC meetings from the 45th MEPC meeting in 2000 to the 66th MEPC meeting in 2014.

TABLE 4.1 *Main events of IMO's work on GHG emissions from international shipping* (cont.)

Meetings	Meeting Date	Major Outcomes	Base Documents
MEPC 55	Oct 2006	Confirmed the leading role of the IMO in the work of the issue; Adopted a work plan; Agreed to update the *2000 IMO GHG Study*.	Assembly Res. A.963(23) MEPC 55/23
MEPC 57	Apr 2008	**Adopted nine fundamental principles as a basis for future regulations, incorporating the NMFT principle.**	MEPC 57/21
GHG-WG 1	Jun 2008	Developed a mandatory regime to control shipping GHG emissions.	MEPC 58/4
MEPC 58	Oct 2008	Adopted revised MARPOL Annex VI (designated Emission Control Area for SO_X and NO_X); Made progress in developing technical and operational measures, including the formula, mandatory or voluntary, either the CBDR or NMFT principle, etc; Discussed MBMs; Agreed on the role of the IMO as the most competent international body in regulating shipping GHG emissions.	MEPC.176(58); MEPC 58/23/ Add.1, Annex 13 MEPC 58/23 MEPC 58/4/21
GHG-WG 2	Mar 2009	Further refinement of the EEDI; Considered how to improve the EEOI; Debated over a draft SEEMP.	MEPC 59/4/2
MEPC 59	Jul 2009	Agreed to disseminate a package of interim and voluntary technical and operational measures; Agreed a work plan of proposed MBMs for further consideration—in-depth debate held; Presented *Second IMO GHG Study 2009*; Reaffirmed the role of the IMO in regulating the GHG issue.	MEPC 59/ INF.10 MEPC 59/24 MEPC 59/24/ Add.1

Meetings	Meeting Date	Major Outcomes	Base Documents
MEPC 60	Mar 2010	Discussed the relations between attained EEDI and required EEDI; Agreed to establish an Expert Group (EG) to undertake a feasibility study and impact assessment of various proposed MBMs.	MEPC 60/22 MEPC 60/ INF.21
MEPC 61	Oct 2010	Debated on proposed amendments to MARPOL Annex VI; Debated on MBMs, GHG-EG submitted a study report; Agreed to hold a GHG-WG on MBMs in March 2011.	MEPC 61/24 MEPC 61/INF.2
GHG-WG 3	Mar 2011	Concluded that regarding the MBMs for international shipping, no incompatibility existed between IMO and other customary international law; Called for a further impact study by the MBM-EG.	MEPC 62/5/1
MEPC 62	Jul 2011	**Adopted amendments of Annex VI to *MARPOL 73/78* for inclusion of energy efficiency measures**; Adopted a work plan to include further guidelines to EEDI and SEEMP, remaining EEDI and SEEMP related guidelines, an EEDI framework for ship types, sizes and propulsion systems not covered by current EEDI requirements.	MEPC.203(62) MEPC 62/5/29
MEPC 63	Mar 2012	Adopted four guidelines to assist the implementation of the mandatory regulations on Energy Efficiency for Ships in Annex VI to *MARPOL 73/78*.	MEPC.212(63); MEPC.213(63); MEPC.214(63); MEPC.215(63)
MEPC 64	Oct 2012	Implementation of energy efficiency; MBMs: updated emission estimate, impact on developing States are scheduled to be conducted.	MEPC.224(64) MEPC 64/23

TABLE 4.1 *Main events of IMO's work on GHG emissions from international shipping* (cont.)

Meetings	Meeting Date	Major Outcomes	Base Documents
MEPC 65	May 2013	Further discussions on EEDI, particularly on US's proposal on enhancing energy efficiency of ships; discussion on MBMS suspended; **adopted Resolution on Promotion of Technical Cooperation and Transfer of Technology;** adopted the *2013 Interim Guidelines for Determining Minimum Propolsion Power to Maintain the Manoeuvrability of Ships in Adverse Conditions.*	MEPC.229(65) MEPC 65/22 MEPC.232(65)
MEPC 66	Apr 2014	**(1) adopted amendments of Annex VI to *MARPOL 73/78* extending the application scope of the EEDI to include an extra five types of ships, adding a Chapter 5 to make the IMO Audit Scheme mandatory;** (2) adopted the *2014 Guidelines on the Method of Calculation of the Attained EEDI for New Ships*; (3) established the Working Group on Further Technical and Operational Measures for Enhancing Energy Efficiency of International Shipping to develop a data collection system for fuel consumption of ships; established the Ad Hoc Expert Working Group on Facilitation of Transfer of Technology for Ships (AHEWG-TT); agreed to establish an EEDI database.	MEPC.242(66); MEPC.245(66); MEPC.247(66); MEPC.251(66); MEPC 66/21

Meetings	Meeting Date	Major Outcomes	Base Documents
MEPC 67	Oct 2014	(1) approved the **Third IMO GHG Study 2014**; (2) adopted the *2014 Guidelines on Survey and Certification of the EEDI*; adopted Amendments to the *2013 Interim Guidelines for Determining Minimum Propulsion Power to Maintain the Manoeuvrability of Ships in Adverse Conditions*; (3) agreed to establish a Correspondence Group on EEDI review to review the status of technological developments as required by Regulation 21.6 of MARPOL Annex VI; (4) noted that in principle there was a clear agreement to develop a data collection system.	MEPC 67/INF.3 Annex MEPC.254(67) MEPC.255(67)
MEPC 68	May 2015	(1) adopted Amendments to the *2014 Guidelines on Survey and Certification of the EEDI*; (2) adopted Amendments to the *2013 Interim Guidelines for Determining Minimum Propulsion Power to Maintain the Manoeuvrability of Ships in Adverse Conditions*; (3) adopted *Amendments to the 2014 Guidelines on the Method of Calculation of the Attained EEDI for New Ships*; (4) agreed that development of a data collection system for ships should progress and follow a three-step approach, and further developed the text for the data collection system for fuel consumption.	MEPC.261(68) MEPC.262(68) MEPC.263(68)
MEPC 69	Apr 2016	**Approved the mandatory data collection system which serves as the first step of a three-step process** (data collection, data analysis, and decision-making on what further measures, if any, are required).	MEPC 69/6

4.3.2 *Annex VI to MARPOL 73/78 and its Amendments*

As a means of reducing shipping GHG emissions, technical and operational measures were examined in the report entitled *Study of Greenhouse Gas Emissions from Ships* in 2000 (the *First IMO GHG Study*). Based on both short-term and long-term perspectives, the report classified CO_2 reduction potential by technical measures into two categories: measures for new ships and measures for existing ships.[32] It identified various technical and operational measures and asserted that these measures have limited potential for reducing shipping emissions. It concluded that it might be 'more feasible' for the shipping industry to implement these measures primarily through new ship construction.[33] The *Second IMO GHG Study* in 2009 emphasised the role of new ship construction in increasing efficiency and reducing emissions. It proposed a mandatory energy efficiency design index (EEDI) for new ships as an incentive to improve the design efficiency of these ships.[34] Based on this work and on intensive discussions and negotiations on various technical, operational and MBMs within the IMO, the 62nd MEPC meeting adopted the revised Annex VI to *MARPOL 73/78* on 15 July 2011. This amendment represents 'the first ever mandatory global [and legally binding] GHG reduction regime for an international industry sector.'[35] Since that time, GHG emissions from shipping have been partially regulated. However, this amendment to Annex VI only regulates a package of mandatory technical and operational measures to reduce GHG emissions from international shipping. By adding a new Chapter 4 to Annex VI on the regulation of energy efficiency for ships, the amendment makes mandatory the EEDI for new ships and the Ship Energy Efficiency Management Plan (SEEMP) for all ships.

A breakthrough on other measures including MBMs has not yet been achieved due to the deadlock in the negotiation of the CBDR and NMFT principles and the divergent views on the necessity of MBMs within the IMO. The successful outcome on the technical and operational regulation can be attributed to the following two factors. First, the energy efficiency measures were included in the revised Annex VI rather than creating a new Annex VII to

32 Skjølsvik et al., above n. 23, 14.

33 Ibid. 8–9.

34 Buhaug et al., above n. 16, 1. The report concludes that if technical and operational measures are implemented together, 'these measures could increase efficiency and reduce the emissions rate by 25% to 75% below the current levels.'

35 International Maritime Organization (IMO), *Mandatory Energy Efficiency Measures for International Shipping Adopted at IMO Environment Meeting* (15 July 2011) <http://www.imo.org/MediaCentre/PressBriefings/Pages/42-mepc-ghg.aspx> accessed 1 June 2012.

MARPOL 73/78. For this amendment to Annex VI, the 'tacit acceptance' procedure applied. According to this procedure, amendments of the *MARPOL 73/78* annexes or appendices to such annexes enter into force on a specified date unless a specific number of State parties object by an agreed date.[36] Due to the technical nature of these annexes and appendices, it might be inferred that the 'tacit acceptance' procedure basically applies to technical amendments. In this case, the 'silence' of a member State represents its approval and makes a formal acceptance unnecessary.[37] This procedure, however, does not apply to either the articles of the convention or to the introduction of a new annex.[38] The main benefit of the 'tacit acceptance' procedure lies in the expedited entry into force of the amendments. This procedure partially explains why these revisions entered into force on 1 January 2013 shortly after their adoption, despite the opposition of many developing States.[39] Second, the voting mechanism within the MEPC accelerates the adoption of these measures. Rule 27 of the *Rules of Procedures of the MEPC* provides that decisions of the committee and of its subsidiary bodies are made by a majority of the members present and voting rather than by a consensus.[40] This policy ensures that a resolution can be adopted by the MEPC even if some States with large owned fleets oppose it. To change this situation, at the 64th MEPC meeting many developing States proposed that all decisions of the MEPC on GHG emissions from ships should be adopted by consensus. However, the debate on this issue has been postponed.[41]

The energy efficiency measures apply to all ships of 400 gross tonnage and above.[42] Due to the global financial crisis since 2009, this new regulation has

---

36 *MARPOL 73/78* art. 16.

37 Pamborides, above n. 6, 101.

38 *MARPOL 73/78* arts. 16(2)(f), 16(5).

39 See, e.g., *Comments on the Proposed Mandatory Energy Efficiency Regulations*, submitted by China, Saudi Arabia and South Africa, MEPC 62nd Session, Agenda Item 5, IMO Doc MEPC 62/5/10 (5 May 2011). The co-sponsors of this proposal, namely China, Saudi Arabia, and South Africa, opposed the adoption of the amendment because of its exclusion of the CBDR principle, technical uncertainty and other factors. They also opposed the inclusion of energy efficiency measures in *MARPOL Annex VI* due to the different nature of GHGs and other air pollutants.

40 International Maritime Organization (IMO), *Basic Documents Volume I* (International Maritime Organization, 2010) 113.

41 *Further Work on GHG Emissions from Ships*, submitted by Brazil, China, India, Peru, Saudi Arabia and South Africa, MEPC 64th Session, Agenda Item 5, IMO Doc MEPC 64/5/9 (27 July 2012) para. 8.1.

42 *MARPOL 73/78 Annex VI* (2011 amendments) reg 19.1.

imposed significant pressure on global shipping industries, in particular, those from developing States. Nevertheless, under Regulation 19, there is flexibility in the application of the EEDI:

> Regulation 19
> 1. This chapter shall apply to all ships of 400 gross tonnage and above …
> 4. Notwithstanding the provisions of paragraph 1 of this regulation, *the Administration may waive the requirement for a ship of 400 gross tonnage and above from complying* with regulation 20 and regulation 21.
> 5. The provision of *paragraph 4* of this regulation *shall not apply to ships of 400 gross tonnage and above*:
> (1) for which the building contract is placed on or after 1 January 2017; or
> (2) in the absence of a building contract, the keel of which is laid or which is at a similar stage of construction on or after 1 July 2017; or
> (3) *the delivery of which is on or after 1 July 2019*; or
> (4) in cases of a major conversion of a new or existing ship, as defined in regulation 2.24, on or after 1 January 2017, and in which regulation 5.4.2 and regulation 5.4.3 of chapter 2 apply. [emphasis added]

Regulation 19.4 and 19.5 indicate that for some States the actual commencement date of the EEDI might be postponed for six and a half years from 1 January 2013.[43] Since the administration is generally the flag State of a ship,[44] this regulation gives the ships from developing States a long lead time for their preparation and adjustment. This treatment is still non-differentiated between developed and developing States and thus does not apply the CBDR principle.[45] In practice, this waiver might be used primarily by ships flying the flags of developing States due to the much more stringent requirements of developed

43 Or such ships will be exempt from complying with EEDI until 1 January 2017 based on contract date. *MARPOL 73/78 Annex VI* (2011 amendments) reg 19.5.2.

44 *MARPOL 73/78* art. 2(5).

45 This regulation was misinterpreted by some media and scholars as solely applying to the developing countries. See, e.g., John Vidal, *Maritime Countries Agree First Ever Shipping Emissions Regulation* <http://www.guardian.co.uk/environment/2011/jul/18/maritime-countries-shipping-emissions-regulation> accessed 1 January 2012; Laura Boone, 'Reducing Air Pollution from Marine Vessels to Mitigate Arctic Warming: Is it Time to Target Black Carbon?' (2012) (1) *Carbon & Climate Law Review* 13, 18. Vidal commented that 'China, Brazil, Saudi Arabia and South Africa have secured a six and a half year delay for new ships registered in developing countries.'

States. Nevertheless, prior to the adoption of this amendment, Vanuatu submitted a proposal on possible exemptions from the EEDI requirements for ships trading to the least developed States and small island developing States, but it was not accepted at the 61st MEPC meeting.[46] In this sense, this waiver clause could be deemed to be a compromise between developed States and developing States.

According to an assessment by Lloyd's Register and Det Norske Veritas, the impact of the waiver clause (Regulation 19.5) is estimated to be low on the total emission reduction potential.[47] This is because low compliance costs and the commercial disadvantages associated with non-compliance make it unattractive for flag States or shipowners to opt for an EEDI waiver.[48] Given the situations of the States supporting this waiver clause, notably Brazil, China, and Saudi Arabia, the most likely level of waiver is only 5 per cent.[49] To assist with the implementation of the mandatory regulations on energy efficiency for ships in the 2011 Annex VI, the 63rd MEPC meeting in March 2012 adopted four important guidelines,[50] and the 65th MEPC meeting in May 2013

46 *Report of the Marine Environment Protection Committee on Its Sixty-First Session*, MEPC 61st Session, Agenda Item 24, IMO Doc MEPC 61/24 (6 October 2010) ss 5.32–5.33. Before the meeting, Vanuatu proposed to include a provision in the draft regulation on an exemption for these vessels trading to least developed countries and small island developing states (SIDS). However, the committee did not agree with the proposal on the grounds that the adoption of this provision could mean that 'the least efficient ships would serve these trades/routes indefinitely' and would prejudice the benefits of developing countries due to higher transportation costs resulted as such. See also ch. 6, 6.3.2.3.

47 Zabi Bazari and Tore Longva, '*Assessment of IMO Mandated Energy Efficiency Measures for International Shipping*' (IMO Doc MEPC 63/INF.2, Annex, 31 October 2011) 6–7, Appendix 1, 3.

48 Ibid. Appendix 1, 1–3. Appendix 1 of the report analyses that technology cost of compliance to EEDI will be low due to such factors as EEDI reference lines, ship hydrodynamic optimization, and preparation for future more stringent Phases 2 and 3; and an EEDI non-compliance ship is projected to suffer from certain commercial costs including higher ship fuel cost, cost of re-verification, second hand value, opportunity costs, and charter-ability.

49 Ibid. According to the report, these three countries supported the waiver clause at the 62nd MEPC meeting. If the waiver will be taken up by these countries, as of October 2011, the current tonnage and number of ships for these three flags totally cover 4.6 percent of the global fleet.

50 These four guidelines are: *2012 Guidelines on the Method of Calculation of the Attained Energy Efficiency Design Index (EEDI) for New Ships*, Resolution MEPC.212(63), IMO Doc MEPC 63/23 Annex 8 (2 March 2012) annex 8 ('*EEDI Calculation Guidelines*'); *2012 Guidelines for the Development of a Ship Energy Efficiency Management Plan (SEEMP)*, Resolution

adopted a Resolution on Promotion of Technical Co-operation and Transfer of Technology relating to the Improvement of Energy Efficiency of Ships.[51] In April 2014, Annex VI to *MARPOL 73/78* was amended to extend the application scope of the EEDI to include an extra five types of ships and add a Chapter 5 to make the IMO Audit Scheme mandatory.[52] Consequently the four guidelines adopted in 2012 have been further amended in the 67th and 68th MEPC meetings in 2014 and 2015 respectively. These amendments, together with various technical, operational measures and MBMs, are discussed in the following sections.

4.3.3 *Outcomes within the Marine Environment Protection Committee*

Technical, operational measures and MBMs have been widely discussed and negotiated within the MEPC since the adoption of Resolution 8 on 'CO_2 emissions from ships' in 1997. Currently, technical and operational measures are included in Annex VI to *MARPOL 73/78* in the forms of the EEDI and the SEEMP requirements respectively, whereas MBMs are still unregulated. This section introduces the adopted technical and operational measures, followed by an assessment of their benefits and deficiencies.

4.3.3.1 Technical Measures

The EEDI is the main technical measure regulated by the revised Annex VI to *MARPOL 73/78* in 2011. The EEDI provides a specific figure representing a minimum energy efficiency level for certain ship types and size segments, expressed in grams of CO_2 per ship's capacity-mile (for example, gross tonne nautical mile). The lower EEDI indicates better energy efficiency of ship design. Regulations 20 and 21 divide it into attained EEDI and required EEDI,[53] and

---

MEPC.213(63), IMO Doc MEPC 63/23 Annex 9 (2 March 2012) annex 9 ('*SEEMP Guidelines*'); *2012 Guidelines on Survey and Certification of the Energy Efficiency Design Index (EEDI)*, Resolution MEPC.214(63), IMO Doc MEPC 63/23/Add.1 Annex 10 (2 March 2012) annex 10 ('*EEDI Survey and Certification Guidelines*'); *Guidelines for Calculation of Reference Lines for Use with the Energy Efficiency Design Index (EEDI)*, Resolution MEPC.215(63), IMO Doc MEPC 63/23/Add.1 Annex 11 (2 March 2012) annex 11 ('*EEDI Reference Lines Guidelines*').

51 *Promotion of Technical Co-operation and Transfer of Technology relating to the Improvement of Energy Efficiency of Ships*, IMO Doc Res MEPC.229(65) (17 May 2013).

52 *Amendments to MARPOL Annex VI and the NO_x Technical Code 2008*, IMO Doc Res MEPC.251(66) (4 April 2014) ('*MARPOL 73/78 Annex VI (2014 amendments)*').

53 Attained EEDI refers to the EEDI value achieved by an individual ship in accordance with Regulation 20 of Chapter 4, *MARPOL 73/78* Annex VI; while required EEDI is the maximum value of attained EEDI that is allowed by Regulation 21 of Chapter 4 for the specific ship type and size. *MARPOL 73/78 Annex VI (2011 amendments)* regs 2.36–37.

both of them are calculated by a formula based on the technical design parameters for a given ship.[54] Based on the formula, the attained EEDI should be less than or equal to the required EEDI.[55] As a 'non-prescriptive' and 'performance-based' mechanism, the EEDI only requires a minimum energy efficiency level.[56] Provided the EEDI requirement is achieved, ship designers and shipbuilders are free to choose the most cost-efficient solutions for the ship to meet the regulations. The EEDI could provide a strong incentive for the shipping industry to improve ship fuel consumption with updated technical development. This approach, similar to those adopted in the 1987 *Montreal Protocol* and its amendments,[57] would encourage the adoption of the most cost-efficient technologies by industry. Consequently, non-compliant ships would suffer from significant opportunity costs and become less competitive in the international shipping market.[58] Meanwhile, the EEDI is essentially a 'hard rule' rather than a commercial incentive scheme. Based on the mandatory EEDI requirements, substandard ships might be detained, fined by port States, or even forbidden to trade, although the way to achieve the emissions reduction is left to the shipping industry.[59]

Aside from the EEDI waiver clause under Regulation 19 of Annex VI, the EEDI does not apply to all ship types or to all types of propulsion systems. Under the 2011 amendments of Annex VI to *MARPOL 73/78*, Regulation 21 (Required EEDI) only lists seven types of ships, namely bulk carriers, gas carriers, tankers, container ships, general cargo ships, refrigerated cargo ships and combination carriers.[60] Ships with diesel-electric propulsion, turbine propulsion and hybrid propulsion are excluded from the EEDI requirement.[61] The exemptions for these types of ships can be mainly attributed to the technical

---

54 The formula of required EEDI is indicated by Regulation 21 of *MARPOL 73/78* Annex VI, while the formula of Attained EEDI is provided by its guidelines. *2012 Guidelines on the Method of Calculation of the Attained Energy Efficiency Design Index (EEDI) for New Ships*, Resolution MEPC.212(63), IMO Doc MEPC 63/23 Annex 8 (2 March 2012) art. 2.

55 *MARPOL 73/78 Annex VI (2011 amendments)* reg 21.1.3.

56 IMO, above n. 21, 12.

57 See ch. 3, 3.2.1.2.

58 Opportunity cost is an economics term, which in this context might include the loss of future EEDI-based incentives. For instance, EEDI might be used for chartering, port discounts, flag registration discounts by ports, flag States, charters and Port States, where these non-EEDI ships are excluded from application. Bazari and Longva, above n. 47, Appendix 1, p. 3.

59 王尔德 [Wang Erde], above n. 17.

60 *MARPOL 73/78 Annex VI (2011 amendments)* reg 21, table 1, 2.

61 *MARPOL 73/78 Annex VI (2011 amendments)* reg 19.3.

difficulty of incorporating them into the EEDI formulae due to the complexity of their shipping emissions. Nevertheless, as the first step in reducing the majority of shipping emissions sources, the EEDI coverage as regulated in the 2011 amendments to Annex VI accounts for 70 per cent of emissions from new ships.[62] The regulated seven types of ships are essentially those designed to transport cargos, representing 'the largest and most energy intensive segments of the world merchant fleet'.[63] Under the 2014 amendments of Annex VI to *MARPOL 73/78*, the application scope of the EEDI has been extended to cover an extra five types of ships. They are Liquefied Natural Gas (LNG) carrier, roll-on/roll-off (ro-ro) cargo ship (vehicle carrier), ro-ro cargo ship, ro-ro passenger ship, and cruise passenger ship having non-conventional propulsion.[64] However, the amended Regulation 19 of Annex VI still exempts ships not propelled by mechanical means, platforms including Floating Production Storage and Offloading Facilities (FPSOs) and Floating Storage Units (FSUs) and drilling rigs, regardless of their forms of propulsion.[65] Cargo ships having ice-breaking capability and most ships which have non-conventional propulsion are also exempted from the EEDI.[66] Passenger ships other than cruise passenger ships will also remain unregulated by the EEDI.

This step-by-step approach was utilised by the IMO to relieve strong opposition from developing countries and expedite the regulation progress within the MEPC. However, the 2011 amendments were not reached by consensus within the MEPC, which indicates a challenge for their future implementation.[67] As a global mandatory instrument, the amended Annex VI requests port States to

62 IMO, above n. 21, Annex 1, p. 32.

63 Ibid. 12.

64 Res MEPC.251(66) reg 21 ('*MARPOL 73/78 Annex VI (2014 amendments)*').

65 *MARPOL 73/78 Annex VI (2014 amendments)* reg 19(2).

66 *MARPOL 73/78 Annex VI (2014 amendments)* reg 19(3). Regulation 19(3) of Annex VI provides that '[the EEDI] shall not apply to ships which have non-conventional propulsion, except that regulations 20 and 21 shall apply to cruise passenger ships having non-conventional propulsion and LNG carriers having conventional or non-conventional propulsion, delivered on or after 1 September 2019, as defined in paragraph 43 of regulation 2. [The EEDI] shall not apply to cargo ships having ice-breaking capability'.

67 Md. Saiful Karim, 'IMO Mandatory Energy Efficiency Measures for International Shipping: The First Mandatory Global Greenhouse Gas Reduction Instrument for an International Industry' (2011) 7(1) *Macquarie Journal of International and Comparative Environmental Law* 111, 113. See also James Harrison, 'Recent Developments and Continuing Challenges in the Regulation of Greenhouse Gas Emissions from International Shipping' (2012) *University of Edinburgh Research Paper Series* <http://ssrn.com/abstract=2037038> accessed 1 June 2013, p. 19.

verify whether there is a valid International Energy Efficiency Certificate (IEE Certificate) on ships calling at their ports so as to monitor the compliance of ships.[68] Nevertheless, some flag States may not join the instrument, and the ships flying their flags may seek suitable routes to avoid the regulation. To facilitate the enforcement of EEDI requirements, the amendments and subsequent guidelines have provided a phased approach and an IMO commissioned report offers technological options.

First, Regulation 21 provides four phases for the implementation of the EEDI. Phase 0 (1 January 2013–31 December 2014) provides a two-year grace period for all ships regardless of their flags to be exempt from EEDI requirements. This regulation gives the shipping industry lead time to make necessary preparations such as technology research and development and staff training. This measure was initially proposed by China as a five-year Phase 0 and was supported by other developing States.[69] Thus, it is actually a compromise achieved between developed States and developing States. In Phase 1 (1 January 2015–31 December 2019), a CO_2 reduction level of 10 per cent is mandated, and this percentage will become higher every five years to be consistent with updated technological developments in efficiency and reduction measures. In Phase 3 (1 January 2025 onwards), a 30 per cent reduction is set for most ship types calculated from a reference line for ships built between 2000 and 2010.[70] This schedule for implementation follows a step-by-step approach and provides differentiated requirements for different ship sizes. Generally, the EEDI requirements on ship size below certain capacities are lower. This arrangement meets the special demand by various States for trade considerations, physical port limitations and cargo logistic issues since not all States need large-size ships. Although, according to the economics theory of 'economies of scale,' at a given speed, the larger the ship the lower the fuel consumption per unit of cargo.[71]

---

68 *MARPOL 73/78 Annex VI (2011 amendments)* reg 10.5.

69 *Report of the Outcome of the Intersessional Meeting of the Working Group on Energy Efficiency Measures for Ships*, MEPC 61st Session, Agenda Item 5, IMO Doc MEPC 61/5/3 (7 July 2010) para. 2.19.2. In Doc EE-WG 1/2, China argued that the reduction rate X should not be implemented immediately once the mandatory EEDI takes effect and that is X=0 for the first phase for five years.

70 A reference line refers to 'a curve representing an average index value fitted on a set of individual index values for a defined group of ships,' so it represents the average efficiency for ships. The reference line was also called baseline but was later abandoned by the MEPC as the reference line can better reflect its purpose and function. *EEDI Survey and Certification Guidelines* art. 4; IMO, above n. 21, 12.

71 IMO, above n. 21, 34.

Second, the selection of technologies is vital for ship designers and shipbuilders to meet the EEDI requirements for new ships. An assessment report commissioned by the IMO provides 15 types of technologies for reducing the EEDI of ships in the future as shown in Table 4.2.[72] These technologies can be classified into five groups, namely:

- ship capacity enhancement;
- hull and propeller;
- engines, waste heat recovery and propulsion system;
- alternative fuels; and
- alternative sources of energy.[73]

Compared with the seven types of technologies available for new ships as indicated in the *First IMO GHG Study*, there are currently more choices available for the shipping industry.[74] It is projected that during Phases 0 and 1 (1 January 2013–31 December 2019), hull, propeller and main engine optimisation will contribute more to EEDI compliance, while during Phases 2 and 3 (1 January 2020 onwards), new technologies and design speed reduction will be utilised more to meet the EEDI requirements.[75] The order of these technologies does not imply any prioritisation. However, it is of 'critical importance' to ensure safe navigation under adverse conditions, while energy efficiency of international shipping is promoted.[76] Based on this understanding, the need for a minimum speed is incorporated into the EEDI formula and into Regulation 21.5 of Annex VI, although reducing speed is generally regarded as the easiest way to improve a ship's fuel efficiency.[77]

72 Bazari and Longva, above n. 47, 14–15.

73 Ibid.

74 Skjølsvik et al., above n. 23, 14.

75 Bazari and Longva, above n. 47, 15.

76 IMO, above n. 21, 34.

77 *MARPOL 73/78 Annex VI (2011 amendments)* regs 21.1, 21.5. Regulation 21.5 provides: 'For each ship to which this regulation applies, the installed propulsion power shall not be less than the propulsion power needed to maintain the manoeuvrability of the ship under adverse conditions as defined in the guidelines to be developed by the Organization.'

TABLE 4.2 *Technologies for EEDI reduction*[78]

No.	EEDI Reduction Measures	Remark
1	Optimised hull dimensions and form	Ship design for efficiency via choice of main dimensions (port and canal restrictions) and hull forms.
2	Lightweight construction	New lightweight ship construction material.
3	Hull coating	Use of advanced hull coatings/paints.
4	Hull air lubrication system	Air cavity via injection of air under/around the hull to reduce wet surface and thereby ship resistance.
5	Optimisation of propeller-hull interface and flow devices	Propeller-hull-rudder design optimisation plus relevant changes to ship's aft body.
6	Contra-rotating propeller	Two propellers in series; rotating at different direction.
7	Engine efficiency improvement	De-rating, long-stroke, electronic injection, variable geometry turbocharging, etc.
8	Waster heat recovery	Main and auxiliary engines' exhaust gas waste heat recovery and conversion to electric power.
9	Gas fuelled (LNG)	Natural gas fuel and dual fuel engines.
10	Hybrid electric power and propulsion concepts	For some ships, the use of electric or hybrid would be more efficient.
11	Reducing on-board power demand (auxiliary system and hotel loads)	Maximum heat recovery and minimising required electrical loads flexible power solutions and power management.
12	Variable speed drive for pumps, fans, etc.	Use of variable speed electric motors for control of rotating flow machinery leads to significant reduction in their energy use.
13	Wind power (sail, wind engine, etc.)	Sails, flettner rotor, kites, etc. These are considered as emerging technologies.
14	Solar power	Solar photovoltaic cells.
15	Design speed reduction (new builds)	Reducing design speed via choice of lower power or de-rated engines.

78 Bazari and Longva, above n. 47, 14–15.

Thirdly, Annex VI and *EEDI Survey and Certification Guidelines* regulate a two-stage survey and verification process to ensure the smooth and uniform implementation of the EEDI. Based on regulations 5, 6, 7, 8 and 9 of *MARPOL* Annex VI and *EEDI Calculation Guidelines, EEDI Survey and Certification Guidelines* divide the process into two stages: preliminary verification at the design stage, and final verification at the sea trial. Their working mechanism is illustrated in Figure 4.1. At the first stage, a report of pre-verification will be provided by the verifier to the submitter once the verification is complete. At the second stage, a certificate will be issued if a ship has passed the certification. Through this process, verifiers of the EEDI of ships ensure that the ships under survey and certification comply with the EEDI requirements.[79]

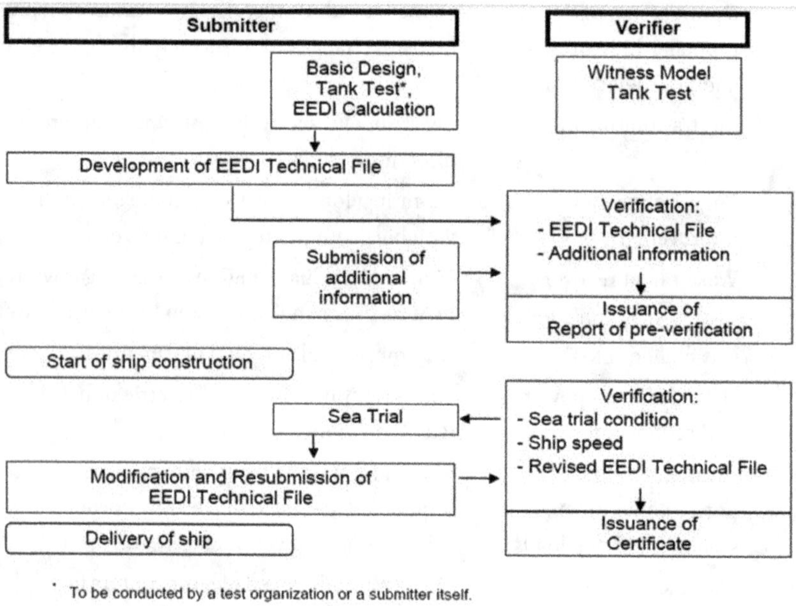

FIGURE 4.1 *Survey and certification process.*[80]

79 Verifier means 'an Administration or organization duly authorized by it', or in other words, flag State or organizations duly authorized by it. *EEDI Survey and Certification Guidelines* art. 2.1.

80 *EEDI Survey and Certification Guidelines* art. 4.1.1, Figure 1.

4.3.3.2 Operational Measures

The SEEMP is the operational measure regulated by Annex VI to *MARPOL 73/78*. It constitutes the other component of the energy efficiency measures besides the EEDI. This plan provides a flexible mechanism for shipowners and ship operators to monitor ship and fleet efficiency performance over time in a cost-effective manner. The main objective of the plan is to minimise shipping GHG emissions by means of reducing fuel consumption,[81] while the Energy Efficiency Operational Indicator (EEOI) is often utilised as a monitoring tool and to establish benchmarks related to energy efficiency of ships.[82] Regulation 22 of Annex VI briefly regulates the SEEMP, which provides that,

> Each ship shall keep on board a ship specific Ship Energy Efficiency Management Plan (SEEMP). This may form part of the ship's Safety Management System (SMS).[83]

Based on Regulations 19.1 and 22.1 of Annex VI, the SEEMP applies to all existing and new ships of 400 gross tonnage and above on a mandatory basis. As a 'ship specific' plan, the SEEMP adopts a four-step approach to improve a ship's energy efficiency, namely planning, implementation, monitoring, and self-evaluation and improvement. The *SEEMP Guidelines* introduce procedures and measures at each stage. In the planning stage, the Guidelines recognise that the specific measures for a ship to improve energy efficiency should be identified first, and then company-specific measures, human resource development and goal setting issues are to be addressed.[84] In the implementation stage, it is essential to establish an implementation system of the above identified and selected measures by drafting procedures for energy efficiency and defining and allocating tasks, and record-keeping should be done simultaneously.[85] In the monitoring stage, a monitoring tool, as well as the establishment of a monitoring system, is to be selected.[86] In particular, it recommends that the EEOI be utilised in this context as an internationally recognised primary

81 Bazari and Longva, above n. 47, Appendix 4, p. 12.

82 The energy efficiency operational indicator (EEOI) can be applied to almost all new and existing ships and is generally used to measure ships energy efficiency at each voyage or over a certain period of time. It enables ship operators to measure the fuel efficiency of a ship in operation and to gauge the effect of any changes in operation. Currently, the EEOI is circulated to encourage shipowners and ship operators to use it on a voluntary basis.

83 *MARPOL 73/78 Annex VI (2011 amendments)* reg 22.1.

84 *SEEMP Guidelines* art. 4.1.

85 *SEEMP Guidelines* art. 4.2.

86 *SEEMP Guidelines* art. 4.3.

monitoring tool.[87] The final stage deals with self-evaluation and improvement, which aims to produce meaningful feedback for the next improvement cycle.[88] Furthermore, the SEEMP incorporates best practices for the fuel-efficient operation of ships.[89] Similar to the *EEDI Guidelines*, the *SEEMP Guidelines* also highlight the importance of safe navigation.[90] By introducing specific procedures, measures and best practices along with the four stages of the plan, the SEEMP urges shipowners and ship operators at each stage to consider new technologies and practices when seeking to optimise the energy efficiency performance of a ship. Additionally, the guidelines provide a reference for the classification societies and shipping companies to make their own SEEMPs.

It is important for a ship to adopt specific operational measures for each voyage to meet the SEEMP requirements. The main SEEMP-related measures are listed in Table 4.3. These operational measures aim at reducing fuel consumption and CO_2 emissions, and they can be classified into three categories.[91] The first category is enhanced technical and operational management. This category includes measures related to enhanced weather routing, hull and propeller cleaning, better main and auxiliary engine maintenance and turning, and efficient operation of larger electrical consumers. The second category is enhanced logistics and fleet planning. For instance, combining cargoes to achieve a higher utilisation rate, optimisation of logistic chains, larger cargo batches, adjustments for optimised arrival times and slower steaming and changed contract formats between charter and shipowner. The third category is port-related measures. Examples include larger port capacity, quicker loading and discharging, flexible design of cargo handling equipment, more efficient port clearance and slot time allocation and fewer restrictions on ship draft, beam or length.

A well-implemented SEEMP might lead to enhanced technical and operational management as illustrated earlier in the first category.[92] The second and third categories, however, are less influenced by the SEEMP since they involve many stakeholders, which makes their implementation rely heavily on the co-operation of many people and groups.

87 *SEEMP Guidelines* art. 4.3.1.

88 *SEEMP Guidelines* art. 4.4.

89 *SEEMP Guidelines* art. 5.

90 *SEEMP Guidelines* art. 3.7.

91 The information on this classification comes from the IMO *Assessment Report on Energy Efficiency Measures for International Shipping*, reprinted in Bazari and Longva, above n. 47, Annex, p. 15.

92 Ibid.

TABLE 4.3 *SEEMP related measures*[93]

No.	Energy Efficiency Measure	Remark
1	Engine tuning and monitoring	Engine operational performance and condition optimisation.
2	Hull condition	Hull operational fouling and damage avoidance.
3	Propeller condition	Propeller operational fouling and damage avoidance.
4	Reduced auxiliary power	Reducing the electrical load via machinery operation and power management.
5	Speed reduction (operation)	Operational slow steaming.
6	Trim/draft	Trim and draft monitoring and optimisation.
7	Voyage execution	Reducing port times, waiting times, etc. and increasing the passage time, just in time arrival.
8	Weather routing	Use of weather routing services to avoid rough seas and head currents, to optimize voyage efficiency.
9	Advanced hull coating	Re-paint using advanced paints.
10	Propeller upgrade and aft body flow devices	Propeller and after-body retrofit for optimisation. Also addition of flow improving devices (e.g. duct and fins.)

4.3.3.3 Assessment of Current Technical and Operational Measures

The EEDI and the SEEMP are the main technical and operational measures adopted by amendments of Annex VI to *MARPOL 73/78* in July 2011 as the first mandatory and legally binding energy efficiency standards.[94] The adoption of these measures was a breakthrough in the lengthy deadlock of the political negotiations on shipping GHG emissions within the IMO and also confirmed the leading role of the IMO in regulating the issue.[95] According to an IMO

93 Ibid. 16.

94 IMO, above n. 35.

95 For example, Ban Ki-moon, UN Secretary-General said in acknowledgment of the decision of the parties to Annex VI of *MARPOL 73/78* to adopt mandatory energy efficiency

assessment report, the combined EEDI and SEEMP will lead to significant emission reductions.[96] This reduction, if valued in terms of annual fuel cost savings, will reach about US $50 billion in 2020 and increase to US $200 billion by 2030.[97] Meanwhile, the cost of EEDI compliance for an 'average ship' will not be significant, although this cost will be higher in Phases 2 and 3 than in Phases 0 and 1 due to possible investment in design-speed reduction.[98] Therefore, the overall CO_2 reduction resulting from the implementation of current technical and operational measures will be not only 'positive' but also economically sound for the shipping industry.

Both the EEDI and the SEEMP highlight the importance of safe navigation of ships while also improving the energy efficiency of shipping.[99] The EEDI and SEEMP requirements are linked to other IMO treaties on maritime safety and security, such as the *1972 Convention on the International Regulations for Preventing Collisions at Sea* (*COLREG*).[100] To meet the safety requirements, a technological threshold is to be achieved by shipowners and ship operators. Additionally, both the EEDI and the SEEMP provide a strong incentive for the shipping industry to choose and update cost-efficient technologies to meet the criteria set under the EEDI and the SEEMP. The shipping industry can freely choose the technologies provided that they meet the requirements. This 'freedom from prescription' approach is vital for the success of this mechanism on the ground that it was strongly supported by the global shipping industry before it was adopted by the IMO.[101] Since it is almost impossible to implement these IMO instruments, including the technical and operational measures, without compliance by the shipping industry, their active participation is essential.

---

measures during the 62nd MEPC meeting, that: 'I would like to congratulate you on this significant outcome reached at IMO's MEPC 62. This underscores the fact that the IMO is best positioned to play a leadership role in addressing GHG emissions from international shipping. This is indeed very welcome progress.' IMO, above n. 21, para. 99.

96 Bazari and Longva, above n. 47, 8, executive summary.

97 Ibid.

98 Ibid.

99 See, e.g., *MARPOL 73/78 Annex VI* (*2011 amendments*) regs 21.5, 22.1; *SEEMP Guidelines* art 3.7.

100 *Convention on the International Regulations for Preventing Collisions at Sea*, opened for signature 20 October 1972, UKTS 77 (entered into force 15 July 1977) ('*COLREG*').

101 See, e.g., at the 57th MEPC meeting, the International Chamber of Shipping (ICS) proposed five principles for guiding the amendment of Annex VI to *MARPOL 73/78*, and one of them is that the ship operators should have the freedom to choose their compliance mechanism so as to protect the shipping industry from monopolistic situations. It treated the 'freedom from prescription' as the most effective means for stimulating future innovation. *The Revision of MARPOL Annex VI*, submitted by the International Chamber of Shipping (ICS),

Another example of these efforts is the negotiation on the possible approval of the SEEMP by flag State administrations. During the 60th and 61st MEPC meetings, many State delegations supported the proposal that the contents of the SEEMP should as a rule be examined by the administration or organisation recognised by the administration,[102] while other State delegations and many industry representatives stressed that the SEEMP should not be approved by the administration but may be audited as a part of the ship's safety management systems.[103] Eventually, it was agreed that approval of the SEEMP by flag State administrations would not be required. This result was achieved by many international shipping associations and could be deemed a victory for the shipping industry over the flag States. Where the SEEMP of a ship needs to be approved by its flag State, it will be often less efficient and more costly for the shipowners, whereas it may be beneficial for flag States to better manage their ships. Moreover, in view of the current 'Flag of Convenience' (FOC) problem,[104] to rely on the examination of a ship's SEEMP in the audit of a ship's safety management system rather than on the approval of a flag State's administration is also beneficial for the reduction of GHG emissions from international shipping. This is because many open registry States often do not have enough motive and expertise to organise this examination of the SEEMP.

Despite the benefits of these technical and operational measures, some deficiencies remain and impose challenges on the future implementation of these measures. Regarding the EEDI, it only applies to certain types of new ships; existing ships are not covered by the EEDI. This situation, if combined with the lenient timetable as introduced in the Regulation and the projected growth in international trade,[105] may significantly reduce the effectiveness of the EEDI. Additionally, future regulation for the remaining types of new ships may adopt different energy efficiency standards, which would increase the difficulty of

---

MEPC 57th Session, Agenda Item 4, IMO Doc MEPC 57/4/28 (13 February 2008) para. 5.3. The shipping industry is the main stakeholder of regulating GHG emissions from international shipping, and its response to this GHG issue is examined in Chapter 5.

102 Generally, the 'Administration' refers to a flag State and the 'organisation recognised by the Administration' refers to the classification society in that flag State.

103 *Report of the Outcome of the Intersessional Meeting of the Working Group on Energy Efficiency Measures for Ships*, MEPC 61st Session, Agenda Item 5, IMO Doc MEPC 61/5/3 (7 July 2010) para. 2.24.

104 A 'flag of convenience' refers to 'the flag of any country allowing the registration of foreign-owned and foreign-controlled vessels under conditions which for whatever the reasons, are convenient and opportune for the persons who are registering the vessels.' B.A. Boczek, *Flags of Convenience: An International Legal Study* (Oxford University Press, 1962) 117.

105 *MARPOL 73/78 Annex VI (2011 amendments)* reg 21.1–2, Table 1.

enforcement. With respect to *MARPOL 73/78*, implementation mainly relies on flag States and port States. The IMO sets energy efficiency standards itself through Annex VI to *MARPOL 73/78*. However, the authority it gives to port States is limited. The added paragraph 5 of regulation 10 of Annex VI provides that,

> In relation to chapter 4, any port State inspection shall be limited to verifying, when appropriate, that there is a valid International Energy Efficiency Certificate on board, in accordance with article 5 of the Convention.[106]

This regulation is a standard phrase for port State control, but it excludes unilateral actions by port States in dealing with shipping GHG emissions. Nevertheless, it will be beneficial for the global reduction of GHG emissions from ships if some States take further steps in this regard. It is also believed that potential regulatory competition between different institutions will provide a significant motivation for the IMO to facilitate its work.[107] An example of unilateral action is the inclusion of the emissions from the international aviation industry into the emission trading scheme of the EU. Although this initiative has been suspended due to opposition from various developed and developing countries,[108] it has motivated the efforts of the International Civil Aviation Organization (ICAO) in accelerating its work under the *Kyoto Protocol*.[109]

The SEEMP is introduced as representing a reduction measure for existing and new ships. Essentially, it is a management scheme that entails no reduction requirements. The lack of reduction target setting and monitoring reduces the effectiveness of the SEEMP.[110] This deficiency needs to be rectified by means of other incentives.[111] An IMO assessment report also recommends that the EEOI

106 *MARPOL 73/78 Annex VI (2011 amendments)* reg 10.5.

107 Sebastian Oberthür, 'The Climate Change Regime: Interactions with ICAO, IMO, and the EU Burden-Sharing Agreement' in Sebastian Oberthür and Thomas Gehring (eds), *Institutional Interaction in Global Environmental Governance* (The MIT Press, 2006) 53, 202.

108 The European Union (EU) has included the emissions from the international aviation sector into its emissions trading scheme since 1 January 2012, which charges carbon tax to all airlines that fly in and out of the EU. This policy was suspended in December 2012 due to strong opposition from many countries including the United States, Russia, China, and India. See Elena Ares, *EU ETS and Aviation* (23 May 2012) <www.parliament.uk/briefing-papers/SN05533.pdf> accessed 1 June 2013.

109 *Kyoto Protocol* art. 2(2).

110 Bazari and Longva, above n. 47, 7, executive summary.

111 Ibid. Annex 15. The report lists some of the drivers for more effective use of the SEEMP, including high fuel and carbon prices, more vigorous awareness building, and cultural change on board ships, more collaboration between industry stakeholders and a solution

should be encouraged or mandated as a performance indicator for the SEEMP rather than remain as a voluntary provision.[112]

Another challenge comes from the future enforcement of these measures by developing countries that opposed the adoption of the measures. Regulation 23 of Annex VI to *MARPOL 73/78* underscores the promotion of technical co-operation and transfer of technology, aiming to strengthen the capacity-building of developing countries. This mechanism, if well designed, could be regarded as a type of differentiated treatment. Since common responsibility and differentiated responsibility are two core elements of the CBDR principle, and common responsibility has been incorporated in this context via the NMFT principle,[113] the design and implementation of the technical co-operation and transfer of technology elements of this mechanism might constitute the application of the CBDR principle. However, Regulation 23 lacks 'concrete obligations' on any State[114] and stipulates that this technical co-operation is subject to national laws, regulations, and policies.[115] It is likely that the transfer of technology from developed countries to developing countries will not be straightforward due to various domestic regulations on intellectual property protection in developed countries.[116] In developed countries, most energy-efficient technologies are owned by private shipping companies, so how to achieve the successful transfer of technologies in a cost-effective manner remains a difficult question.[117]

Triggered by a proposal submitted by South Africa,[118] the 65th MEPC meeting in May 2013 adopted a MEPC Resolution on Promotion of Technical

---

to the issue of split incentives, and effective monitoring of SEEMP implementation via rigorous audits and reviews.

112 Ibid. 7.

113 *MARPOL 73/78* art. 5(4).

114 Harrison, above n. 67, 16.

115 *MARPOL 73/78 Annex VI (2011 amendments)* reg 23.2.

116 Harrison, above n. 67, 17.

117 This question does not occur in the *International Convention for the Control and Management of Ships' Ballast Water and Sediments* (*BWM Convention*) adopted by the IMO in 2004, where there are similar technological standards. This is because the IMO was not mandated by the *Kyoto Protocol* to adopt the *BWM Convention* and accordingly only the NMFT principle applies to this matter instead of the CBDR principle. Therefore, Article 13 of the *BWM Convention* only stipulates technical assistance and co-operation generally rather than facilitating the transfer of technology from developed countries to developing countries with the recognition of the CBDR principle. *International Convention for the Control and Management of Ships' Ballast Water and Sediments*, opened for signature 13 February 2004, IMO Doc. BWM/CONF/36 (not yet in force).

118 See, e.g., *Report of the Marine Environment Protection Committee on Its Sixty-First Session*, MEPC 61st Session, Agenda Item 24, IMO Doc MEPC 61/24 (6 October 2010) Annex 5, p. 1; *Amendment to Draft Compromise MEPC Resolution on Promotion of Technical Co-operation*

Co-operation and Transfer of Technology relating to the Improvement of Energy Efficiency of Ships to address the criticism from developing countries. This Resolution explicitly recognises both the NMFT principle and the CBDR principle,[119] and requests the IMO to provide technical assistance and funding for developing countries.[120] According to this Resolution, an expert working group will be established to facilitate the transfer of technology for ships.[121] However, this Resolution still does not impose concrete obligations on any State to transfer such technology, but rather underscores respect for intellectual property rights.[122] The protection of intellectual property rights has often been regarded as a formidable obstacle to the transfer of technologies,[123] which, if combined with the non-binding nature of a Resolution, would make the implementation of this Resolution by developed countries difficult. Therefore it appears that a market-based approach to technology acquisition might be a better option for developing countries.[124] In 2014 the IMO established the Ad Hoc Expert Working Group on Facilitation of Transfer of Technology for Ships (AHEWG-TT) and agreed to establish an EEDI database, aiming to expedite the tranfer of technology in the international shipping industry. Nevertheless, to date these initiatives have progressed very slowly.

At the 66th MEPC meeting in April 2014, the IMO adopted amendments to Annex VI of *MARPOL 73/78* which makes the IMO Audit Scheme mandatory through adding a Chapter 5 entitled 'verification of compliance with the provisions of this annex'. Under these amendments, the IMO shall conduct periodic audits in accordance with the audit standard as specified in *IMO Instruments Implementation Code (III Code)*[125] to verify compliance with and implementation of this Annex by flag States, coastal States and port States which have ratified the amendments.[126] The amendments would impose pressure on States, in particular FOC States, to exercise their obligations and responsibilities

and Transfer of Technology relating to the Improvement of Energy Efficiency of Ships, submitted by South Africa, MEPC 65th Session, Agenda Item 4, IMO Doc MEPC 65/4/33/Corr.1 (16 May 2013).

119 *Resolution MEPC.229(65)* preamble paras. 3–4.

120 *Resolution MEPC.229(65)* art. 1.

121 *Resolution MEPC.229(65)* art. 3.

122 *Resolution MEPC.229(65)* art. 4.

123 Nitya Nanda and Nidhi Srivastava, 'Clean Technology Transfer and Intellectual Property Rights' (2009) 9(3) *Sustainable Development Law & Policy* 42–69, at 46.

124 This issue is further examined and a proposal is provided in Chapter 7, 7.4.2.

125 *IMO Instruments Implementation Code (III Code)*, IMO Doc Res A.1070(28) (4 December 2013).

126 *MARPOL 73/78 Annex VI (2014 amendments)* reg 25(1).

contained in this Annex. However, as only States which have ratified these amendments would be legally bound, it appears vital to encourage more States to ratify the amended Annex VI to *MARPOL* 73/78.[127]

Figure 4.2 is one scenario devised by a recent IMO Assessment Study on Energy Efficiency Measures.[128] The figure shows that based on the 2010 CO_2 emissions level, it is almost impossible to achieve absolute emission reduction from 2010 to 2050 using the EEDI and SEEMP alone. This is because new emissions produced by increased world trade outweigh the emissions reductions achieved by these two measures. For all scenarios, this conclusion is the same. Therefore, in addition to technical and operational measures already adopted in Annex VI to *MARPOL* 73/78, the international community has turned to MBMs to explore their utility in the possible reduction of GHG emissions from international shipping.

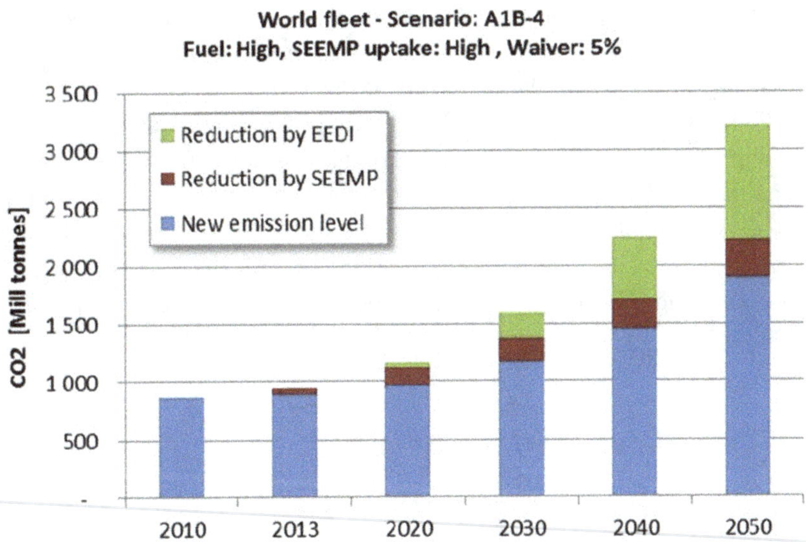

FIGURE 4.2 *Annual emission reduction by 2050 and new emissions levels (scenario A1B-4).*[129]

127 IMO, *Status of Conventions* (25 June 2014) <http://www.imo.org/About/Conventions/StatusOfConventions/Pages/Default.aspx> accessed 28 June 2014. As of 25 June 2014, 75 countries which account for 94.77 percent of world tonnage had ratified Annex VI to *MARPOL* 73/78.

128 Bazari and Longva, above n. 47, 8, executive summary.

129 Ibid. 5.

4.3.4 *Market-Based Measures*

MBMs, which are also referred to as market-based instruments or market-based mechanisms, are generally regarded as an important supplement to the technical and operational measures already in place in reducing GHG emissions from international shipping. In accordance with the polluter-pays principle, MBMs aim to provide the polluters (shipowners and ship operators) with an economic incentive to reduce their GHG emissions.[130] As a comparatively new concept in the shipping context, MBMs have been controversial since they were formally put forward in the *First IMO GHG Study* in 2000. The IMO has endeavoured to promote the awareness of stakeholders in applying MBMs. The *Scientific Study on International Shipping and Market-Based Instruments* (*Scientific Study*) was published in December 2009 as a follow up to the 2000 GHG study commissioned by the IMO.[131] In August 2010, another IMO-commissioned report undertaken by the Expert Group on Feasibility Study and Impact Assessment of Possible Market-based Measures (Expert Group) was released, namely the *Full Report of the Work Undertaken by the Expert Group on Feasibility Study and Impact Assessment of Possible Market-based Measures* (*Feasibility Study and Impact Assessment Report*).[132] Currently a report commissioned by the IMO on possible impacts on consumers and industries in developing countries is under way. Through this work, most countries have come to accept MBMs. Seven main types of proposals have been submitted to the IMO for discussion, although some countries still oppose the adoption of any MBM.[133] The following sections explore the necessity of adopting MBMs and then provide a feasibility and impact assessment of current MBM options.

---

130 Harilaos N. Psaraftis, 'Market-Based Measures for Greenhouse Gas Emissions from Ships: A Review' (2012) 11(2) *WMU Journal of Maritime Affairs* 211, 213.

131 *Scientific Study on International Shipping and Market-based Instruments*, MEPC 60th Session, Agenda Item 4, IMO Doc MEPC 60/INF.21 (15 January 2010).

132 *Full Report of the Work Undertaken by the Expert Group on Feasibility Study and Impact Assessment of Possible Market-based Measures*, MEPC 61st Session, Agenda Item 5, IMO Doc MEPC 61/INF.2 (13 August 2010).

133 See, e.g., *Uncertainties and Problems in Market-based Measures*, submitted by China and India, MEPC 61st Session, Agenda Item 5, IMO Doc MEPC 61/5/24 (5 August 2010); *Market-based Measures—Inequitable Burden on Developing Countries*, submitted by India, MEPC 61st Session, Agenda Item 5, IMO Doc MEPC 61/5/19 (2 August 2010).

4.3.4.1 The Market-Based Measure as a Policy Option

The emergence of MBMs has been interpreted in economics as an approach to overcome the problem of environmental externalities.[134] MBMs are one of the main types of environmental policies,[135] and they have been employed by many countries to regulate adverse environmental impacts resulting from anthropogenic activities. As defined by the Organization for Economic Cooperation and Development (OECD),

> [MBMs] seek to address the market failure of 'environmental externalities' either by incorporating the external cost of production or consumption activities through taxes or charges on processes or products, or by creating property rights and facilitating the establishment of a proxy market for the use of environmental services.[136]

MBMs can be classified into three groups, namely, environmental fees (contribution), tradable permit (allowance) schemes and liability rules.[137] There is no 'one-size-fits-all' MBM. In practice, different MBMs provide solutions for different problems, and some issues might need a mix of two or three types of MBMs.

MBMs can be designed to internalise the external cost of GHG emissions from international shipping by means of a GHG fund or different emission trading schemes. However, the first step is to decide whether MBMs are needed for the reduction of GHG emissions from international shipping. It is a difficult question. Many developing countries, in particular China, India, and Brazil, oppose the adoption of any MBMs. Their argument has mainly been underpinned by three reasons. Firstly, they assert that there are uncertainties associated with MBMs. Examples are the uncertainties of the carbon market, the calculation of the emissions from international shipping, the impact of a

134 Environmental externalities 'refer to the economic concept of uncompensated environmental effects of production and consumption that affect consumer utility and enterprise cost outside the market mechanism.' OECD, *Environmental Externalities* (4 March 2003) <http://stats.oecd.org/glossary/detail.asp?ID=824> accessed 1 June 2013.

135 Environmental policies are often classified as command and control, market-based, education, provision of information, and voluntary measures. Thomas Dietz and Paul C. Stern, 'Exploring New Tools for Environmental Protection' in Thomas Dietz and Paul C. Stern (eds), *New Tools for Environmental Protection: Education, Information, and Voluntary Measures* (National Academies Press, 2002) 4.

136 OECD, *Market-based Instruments* (23 July 2007) <http://stats.oecd.org/glossary/detail.asp?ID=7214> accessed 1 July 2014.

137 *Scientific Study on International Shipping and Market-based Instruments*, MEPC 60th Session, Agenda Item 4, IMO Doc MEPC 60/INF.21 (15 January 2010) annex, 14.

carbon tax on ships on the export industry and the future development of the shipping industry and world trade.[138] Secondly, there are fundamental inadequacies both in theory and in principle relating to MBMs. Developing States argue that the implementation of current MBM proposals requires several prerequisites so as to avert the distortion of competition, such as the same or similar level of economic and technological development realised among all participating countries, some convergence of political power and the deployment of a common central institution.[139]

Thirdly, they also assert that the NMFT principle incorporated in the majority of current MBM proposals ignores historical responsibility and is a disadvantage for developing countries.[140] Moreover, some of the proposed MBMs are regarded by some developing countries as being likely to violate WTO rules.[141] For instance, the MBM on Port State Levy proposed by Jamaica envisages levying a globally uniform emissions charge on all vessels calling at their respective ports, based on the amount of pollution produced by the vessel during the voyage (see Table 4.4). This proposal measures the amount of pollution by the amount of fuel consumed, which may not be accurate due to different ship types and operational methods. In this case, it actually leads to differentiated treatment of different ships, which might contravene the general most-favoured-nation (MFN) treatment as incorporated in Article I of the *General Agreement on Tariffs and Trade*.[142]

In contrast to developing countries, most developed countries and NGOs are in favour of certain types of MBMs, although they disagree on what type of

---

138 *Uncertainties and Problems in Market-based Measures*, submitted by China and India, MEPC 61st Session, Agenda Item 5, IMO Doc MEPC 61/5/24 (5 August 2010) p. 2.

139 Ibid. 3.

140 Ibid; *Market-based Measures—Inequitable Burden on Developing Countries*, submitted by India, MEPC 61st Session, Agenda Item 5, IMO Doc MEPC 61/5/19 (2 August 2010) p. 3.

141 See, e.g., *Possible Incompatibility between the WTO Rules and Market-based Measures for International Shipping*, submitted by India and Saudi Arabia, MEPC 64th Session, Agenda Item 5, IMO Doc MEPC 64/5/3 (29 June 2012).

142 Ibid. para. 25. *General Agreement on Tariff and Trade*, opened for signature 30 October 1947, 55 UNTS 194 (in force provisionally since 1 January 1948 under the 1947 Protocol of Application, 55 UNTS 308) ('*GATT*'). According to Article I of the *GATT*, the general most-favoured-nation (MFN) treatment requires WTO Members to accord the most favourable tariff and regulatory treatment that is granted to the product of any Member at the time of import or export of 'like' products to all other WTO Members immediately and unconditionally. The MFN treatment is a part of the non-discrimination principle. See *World Trade Organization's Views on Document MEPC 64/5/4 Submitted by India and Saudi Arabia*, note by the Secretary-General, MEPC 65th Session, Agenda Item 5, IMO Doc MEPC 65/INF.18 (21 February 2013) annex, p. 1.

MBM is best. The reason is simple: the current EEDI and SEEMP are not sufficient for effective reduction of GHG emissions from international shipping due to the projected growth of international seaborne trade. Thus, specific types of MBMs are needed to supplement the energy-efficiency measures.

As intensive discussions on MBMs have been held within the IMO, it is anticipated that a form of MBM will be adopted in the near future by the IMO or other international institutions to reduce emissions from ships.[143] First, as discussed earlier, in practice it is not possible to achieve absolute emissions reduction using EEDI and SEEMP alone, which has been proven by a number of scenario modellings, revealed in many assessment reports.[144] Moreover, the EEDI and SEEMP regulations only entered into force on 1 January 2013, so in practice compliance with these regulations by various States and their emissions reduction potential cannot be identified in the short term. In April 2016, the IMO approved the mandatory data collection system which serves as the first step of a three-step process. This system would help the IMO to decide whether any further measures, including MBMs, are needed to enhance energy efficiency. However, it would be lengthy to see this three-step process accomplished. The shipping industry has recognised the deficiencies of these measures and work on their improvement has been conducted within the IMO. However, given the intricacies of ship types and shipping features, a technical breakthrough is hardly likely to be achieved soon. Currently, global emissions are 'considerably higher' than the level consistent with the two degree Celsius target in 2020, and this trend continues.[145] In these circumstances, it is necessary for the international shipping industry to explore the possibility of adopting MBMs for more GHG reduction rather than waiting for the effects of applying energy-efficiency measures to be identified.

Second, it is technically possible to incorporate the CBDR principle into a future MBM, and proposals applying the principle have been submitted to the IMO by different countries and NGOs.[146] As shown in the comments by

143 Other international institutions may include the *UNFCCC*, the EU or the WTO. See ch. 4, 4.2; ch. 7, 7.6.

144 Bazari and Longva, above n. 47, 8, executive summary.

145 United Nations Environment Programme (UNEP), 'The Emissions Gap Report 2012: A UNEP Synthesis Report' (November 2012) <http://www.unep.org/pdf/2012gapreport.pdf> accessed 1 June 2013, p. 1, executive summary.

146 For example, the WWF suggested that a MBM that is both global and differentiated was possible to develop incorporating both the CBDR and the NMFT principles simultaneously, and it also put forward a specific revenue allocation scheme for different countries. *Benefits and Possible Adverse Impacts of Market-based Instruments*, submitted by World Wide Fund for Nature (WWF), MEPC 58th Session, Agenda Item 4, IMO Doc MEPC 58/4/39 (15 August 2008). This approach was also adopted by the Scientific Study on International

some developing countries, the core debate within the MEPC lies in the ignorance of the CBDR principle reflected in many MBM proposals. Once this problem is resolved, it may be possible to adopt MBMS that are accepted by most countries. It seems that any MBM proposal that ignores the CBDR principle would not be feasible on the ground that the CBDR principle in the shipping context has been supported by 'the majority of delegations' within the MEPC.[147] In recent years, some international shipping organisations, as well as the shipowners' associations in States listed in Annex I to the *UNFCCC*, have also come to accept the incorporation of the CBDR principle into a proposed MBM.[148]

The possible adoption of MBMS could reduce shipping GHG emissions in two respects: in-sector reduction and out-of-sector reduction.[149] In the first case, a MBM may provide an economic incentive (for example, a charge on fuel, a refund to 'good performance ships'[150]) for the shipping industry to reduce

---

Shipping and Market-based Instruments, a study led by the University of Cambridge in partnership with Cambridge Econometrics, MARINTEK, Manchester Metropolitan University, and Deutsches Zentrum für Luft-und Raumfahrt e.V. *Scientific Study on International Shipping and Market-based Instruments*, MEPC 60th Session, Agenda Item 4, IMO Doc MEPC 60/INF.21 (15 January 2010).

147 *Report of the Marine Environment Protection Committee on its 58th Session*, IMO Doc MEPC 58/23 (16 October 2008) para. 4.45.

148 See, e.g., at the 59th MEPC meeting, the Cruise Lines International Association (CLIA) proposed three principles for tackling GHG emissions from ships, namely the NMFT principle, the principle of high quality, multiple benefit carbon mitigation investment, and the CBDR principle. It further explained that a framework established by the IMO on combating climate change should respect both the NMFT principle and the CBDR principle. *Consideration of Adoption of Three Principles for Market-based Instruments*, submitted by Cruise Lines International Association (CLIA), MEPC 59th Session, Agenda Item 4, IMO Doc MEPC 59/4/32 (8 May 2009) para. 1. Another example is from Australian Shipowners Association (ASA). In a proposal drafted by the ASA and some of other shipowners association, the ASA supported a 'cap-and-trade' emission-trading scheme for shipping and recognized that the CBDR principle 'may also need to be reflected' in this scheme. Australian Shipowners Association et al., *A Global Cap-and-Trade System to Reduce Carbon Emissions from International Shipping* (2009) <http://www.asa.com.au/wp-content/uploads/Joint-Industry-ETS-Discussion-PapervFINAL1.pdf> accessed 1 June 2013.

149 International Maritime Organisation (IMO), *Market-based Measures* <http://www.imo.org/OurWork/Environment/PollutionPrevention/AirPollution/Pages/Market-Based-Measures.aspx> accessed 1 June 2013.

150 'Good performance ships' refer to ships that operate in a fuel-efficient manner. Good performance ships are rewarded in different ways under different MBM proposals. For example, the leveraged incentive scheme proposed by Japan in 2010 required all ships to pay contributions based on the marine bunker that they purchase, but a part of the GHG contributions is refunded to ships labelled as 'good performance ships' in order to produce

its fuel consumption. The industry might invest in more fuel efficient ships or technologies or operate ships in a more energy-efficient manner. In the second case, the money collected from a MBM could be utilised to reduce GHG emissions outside the marine sector. In this way, growing shipping emissions could be offset by emission reduction in other sectors.

4.3.4.2 Assessment of Current Market-Based Measure Proposals

In order to adopt MBMs for the reduction of GHG emissions from international shipping, it is important to know what choices exist and the impacts that they may have on the shipping industry and different countries, in particular, developing countries. Based on these analyses, the selection and adoption of a suitable MBM is possible. Currently, there are seven types of MBM proposals being discussed and debated within the IMO. A brief introduction of these proposals is provided in Table 4.4. At the third GHG-WG meeting in 2011, intensive debate on the grouping of these proposals was held to simplify future assessments and facilitate the decision-making process of the MEPC. It was concluded that MBM proposals should be grouped into two categories, the first focused on in-sector reduction and the second focused on in-sector and out-of-sector reduction, as indicated in Table 4.4.[151] This grouping is based on the areas in which the reduction of GHG emissions from ships will mainly take place and has received many comments on their strengths and weaknesses from different delegations.[152] This section divides these MBM options into three groups. They are environmental fee-related MBM proposals, tradable permit scheme-related MBM proposals, and hybrid MBM proposals.

---

incentives for efficiency improvement. The *2012 SEEMP Guidelines* provides guidance on best practices for fuel-efficient operation of ships. See *Consideration of A Market-Based Mechanism: Leveraged Incentive Scheme to Improve the Energy Efficiency of Ships Based on the International GHG Fund*, submitted by Japan, MEPC 60th Session, Agenda Item 4, IMO Doc MEPC 60/4/37 (15 January 2010) para. 3; *2012 SEEMP Guidelines* reg 5.

151 *Report of the Third Intersessional Meeting of the Working Group on Greenhouse Gas Emissions from Ships*, MEPC 62nd Session, Agenda Item 5, IMO Doc MEPC 62/5/1 (8 April 2011) para. 3.39.

152 Ibid. para. 3.40–3.46; see also *Grouping and Evaluation of Proposed MBMs*, submitted by Greece, Intersessional Meeting of the Working Group on GHG Emissions from Ships 3rd Session, Agenda Item 3, IMO Doc GHG-WG 3/3 (24 February 2011); *The Evaluation on the Relative Strengths and Weaknesses of the Reduction Mechanisms Employed by the MBM Proposals*, submitted by the Republic of Korea, Intersessional Meeting of the Working Group on GHG Emissions from Ships 3rd Session, Agenda Item 3, IMO Doc GHG-WG 3/3/1 (25 February 2011).

TABLE 4.4 *Seven types of MBM troposals submitted to the IMO (as of May 2013)*[153]

MBM proposals	Proponents	Working mechanisms / Grouping of emission reduction	Base documents
GHG Fund	Cyprus, Denmark, the Marshall Islands, Nigeria, and the International Parcel Tankers Association (IPTA)	Establishes a global reduction target for international shipping, set by either the *UNFCCC* or the IMO. Emissions above the target line would be offset largely by purchasing approved emission reduction credits. The offsetting activities would be financed by a contribution paid by ships on every tonne of bunker fuel purchased. (*Grouping: In-Sector and Out-of-Sector*)	MEPC 59/4/5, MEPC 60/4/8, GHG-WG 3/2/1, GHG-WG 3/3/4
	Clean Shipping Coalition (CSC)	Establishes a speed-related GHG or compensation fund to include regulated slow steaming in the design and impact assessment of any MBM proposals. It set average target speeds for different types and sizes of ships in order to meet the agreed emissions reduction target set by the IMO for an MBM. Additional speed levy or contribution would be payable for ships having higher average speeds. Revenues could be used to purchase offsets. (*Grouping: Focus on In-Sector*)	MEPC 64/5/8, MEPC 64/INF.14

153 This table was compiled by the author based on the following sources: IMO, above n. 21; *Report of the Third Intersessional Meeting of the Working Group on Greenhouse Gas Emissions from Ships*, MEPC 62nd Session, Agenda Item 5, IMO Doc MEPC 62/5/1 (8 April 2011) annex 2. The IMO suspended its discussion on MBM proposals in May 2013. Therefore, the proposals in this table have not been updated yet.

MBM proposals	Proponents	Working mechanisms / Grouping of emission reduction	Base documents
Port State Levy (PSL)	Jamaica	Levies a uniform emissions charge on all vessels calling at their respective ports based on the amount of fuel consumed by the respective vessel on its voyage to that port. The CBDR principle could be achieved through a self-administered fund and/or some international mechanism. (*Grouping: 'Focus on In-Sector' and 'In-Sector & Out-of-Sector'*)	MEPC 60/4/40, MEPC 64/5/4
Efficiency Incentive Scheme (EIS)	Japan and World Shipping Council (WSC)	All new ships, except for those that meet pre-set EEDI thresholds and existing ships, are required to make payment contributions based on the amount of the bunker fuel consumed/purchased and the degree to which the ship's efficiency falls short of a specific standard. Funds collected go to an International GHG Fund for further allocation. (*Grouping: Focus on In-Sector*)	MEPC 60/4/37, MEPC 60/4/39, GHG-WG 3/3/2, MEPC 63/5/3, MEPC 64/5/2, MEPC 64/INF.15
Ship Efficiency and Credit Trading (SECT)	United Sates	Subjects all ships to mandatory energy-efficiency standards. As one means of complying with the standard, an efficiency credit trading program would be established. These standards would become more stringent over time. (*Grouping: Focus on In-Sector*)	MEPC 60/4/12, MEPC 61/5/16, MEPC 61/ IMF.24

TABLE 4.4 *Seven types of MBM troposals submitted to the IMO (as of May 2013)* (cont.)

MBM proposals	Proponents	Working mechanisms / Grouping of emission reduction	Base documents
Global Emissions Trading System (ETS) for international shipping	Norway (later added as co-sponsor, Germany)	Sets a sector-wide cap on net emissions from international shipping. A number of allowances (ship emission units) corresponding to the cap would be released into the market each year via a global auctioning process. The units could then be traded.	MEPC 60/4/22; MEPC 60/4/26; MEPC 60/4/41; MEPC 60/4/54; GHG-WG 3/3/5; GHG-WG 3/3/6; GHG-WG 3/3/8
	United Kingdom	Differs from the Norwegian ETS proposal in two respects: the method of allocating emissions allowances (national instead of global auctioning) and the approach for setting the emissions cap (set with a long-term declining trajectory).	
	France	Sets out additional details on auction design under a shipping ETS. In all other respects, the proposal is similar to the Norwegian ETS proposal.	
	(*Grouping: In-Sector & Out-of-Sector*)		
Penalty on Trade and Development	Bahamas	The imposition of any costs should be proportionate to the contribution by international shipping to global carbon dioxide emissions. The reduction will apply to individual ships and not member states, and developing states will not be faced with a penalty upon trade and development. (*Grouping: Focus on In-Sector*)	MEPC 60/4/10, GHG-WG 3/2

MBM proposals	Proponents	Working mechanisms / Grouping of emission reduction	Base documents
Rebate Mechanism for a market-based instrument for international shipping	IUCN (WWF provides add-on options)	Compensate developing countries for the financial impact of an MBM. It could be either applied to any maritime MBM that generates revenue (add-on option) or integrated with the International Maritime Emission Reduction Scheme. (integrated option) *(Grouping: 'Focus on In-Sector' and 'In-Sector & Out-of-Sector' (add-on); In-Sector and Out-of-Sector (integrated))*	MEPC 60/4/55, MEPC 61/5/33; MEPC 64/5/10, MEPC 64/5/12

4.3.4.2.1 *Environmental Fee-Related Market-Based Measure Proposals*

The GHG Fund, Port State Levy and Penalty on Trade and Development are types of environmental fee-related MBM proposals. They provide the polluter with an incentive to reduce GHG emissions in order to pay lower fees. Among the three proposals, the GHG Fund has received the most attention. The *Scientific Study on International Shipping and Market-Based Instruments* asserts that all emissions covered by the GHG Fund will raise revenue for a central governing body, and the amount depends on the carbon price per tonne of CO_2 and on the amount of emissions.[154] The higher carbon price generally indicates more reduction of CO_2 emissions.[155] In this case, the carbon price, or the 'contribution,' is actually a levy on fuels since it needs to be imposed on ships if these MBMs apply.[156] In this way, the shipping GHG emissions could be reduced, and the revenues raised could be utilised to either compensate developing countries or reduce out-of-sector emissions through purchasing 'offsets.' Nevertheless, the utilisation of revenues for reducing out-of-sector GHG emissions does not indicate that in-sector emission reduction is less significant. Rather, the in-sector reduction can be achieved through the collection of a

154 *Scientific Study on International Shipping and Market-based Instruments*, MEPC 60th Session, Agenda Item 4, IMO Doc MEPC 60/INF.21 (15 January 2010) p. 3.

155 Ibid.

156 Psaraftis, above n. 130, 223.

contribution or levy.[157] This proposal seems feasible and easy to implement since shipowners generally respond to prices quickly.[158] The main concern about this proposal lies in its dealing with revenue and how the special conditions of developing countries are taken into account. There might be another concern about the increased cost, including the extra administrative burden, associated with the GHG Fund proposal. The *Feasibility Study and Impact Assessment Report*, undertaken by the Expert Group and commissioned by the IMO, provides a comprehensive assessment of proposed MBMs. This report reveals that the increased cost for the GHG Fund is the second lowest among the current MBM proposals.[159]

The Port State Levy proposed by Jamaica levies a uniform emissions charge on all vessels calling at ports, based on the amount of fuel consumed by the vessels on their voyage to that port. This option can be easily implemented and is consistent with the polluter-pays principle due to its inclusion of all emissions produced by the ship during that journey. However, as mentioned earlier, this option might neither be accurate nor fair for all ships since it measures the ship's actual emissions solely by the fuels that have been consumed. This measurement ignores other pertinent parameters such as different types of ships and the location of ships (at ports or in other maritime zones) and, thus, is not conducted in a 'cost-effective' manner.[160] Meanwhile, since port States play a crucial role in the enforcement of this MBM, it is important to ensure that all port States collaborate in implementing it, including those that choose not to participate in the system and those that lack proper monitoring and enforcement mechanisms.[161] Otherwise, some ships may opt for routes through ports that lack monitoring and enforcement mechanisms to avoid the levy. This may lead to competitive distortion, distortion in trade flows and a 'non-level playing field' among shipping companies and ports. Additionally, under this scheme, the increased cost option is estimated to be the highest among the seven types

---

157 Ibid.

158 Ibid. 225.

159 *Full Report of the Work Undertaken by the Expert Group on Feasibility Study and Impact Assessment of Possible Market-based Measures*, MEPC 61st Session, Agenda Item 5, IMO Doc MEPC 61/INF.2 (13 August 2010) pp. 14–16. Based on this report, the MBM on Penalty on Trade and Development proposed by the Bahamas has the lowest increased cost.

160 *Grouping and Evaluation of Proposed MBMs*, submitted by Greece, Intersessional Meeting of the Working Group on GHG Emissions from Ships 3rd Session, Agenda Item 3, IMO Doc GHG-WG 3/3 (24 February 2011) para. 44.

161 Psaraftis, above n. 130, 222.

of MBM proposals.[162] At the 64th MEPC meeting in 2012, the International Chamber of Shipping (ICS) announced that its preferred MBM is a levy or compensation fund-based scheme which relates to the actual fuel consumption of individual ships in service.[163] This preference has also been followed by the shipping industries in some countries such as Greece and Korea.[164]

The Penalty on Trade and Development proposed by the Bahamas aims to reduce shipping GHG emissions through the imposition of a penalty (cost) and insists that such costs should be proportionate to the GHG emissions from international shipping. It seeks to collect emission statistics from either the EEOI or ship funnels using a suitable sensor. According to the proposal, the ship is required to submit data to its flag State or recognised organisation for annual verification. No extra cost would be generated under this scheme. However, the EEOI is not available for all types of ships, and, currently, EEOI baselines are also impossible to establish.[165] The application of this proposal to the GHG issue will not be feasible if this problem cannot be resolved.

4.3.4.2.2 *Tradable Permit Scheme-Related Market-Based Measure Proposals*

The three types of global emissions trading systems (ETS) for international shipping submitted by Norway, the United Kingdom and France are tradable permit scheme-related MBM proposals. The ETS mechanism was first regulated by the *Kyoto Protocol* and is utilised in the EU. The EU ETS scheme has applied since 1 January 2005 as the world's largest company-level 'cap-and-trade' system. As of 14 November 2012, all 27 EU member states and three

162 *Full Report of the Work Undertaken by the Expert Group on Feasibility Study and Impact Assessment of Possible Market-based Measures*, MEPC 61st Session, Agenda Item 5, IMO Doc MEPC 61/INF.2 (13 August 2010) p. 14.

163 *Operational Energy Efficiency of New and Existing Ships*, submitted by the International Chamber of Shipping (ICS), MEPC 64th Session, Agenda Item 5, IMO Doc MEPC 64/5/11 (27 July 2012) para. 11.

164 Union of Greek Shipowners, *Prevention of Environmental Pollution by Ships: Regulation and Compensation Regimes and Industry Standards* (2011) <http://www.nee.gr/default.asp?t=anakoinoseisDetails&id=13> accessed 1 June 2013, p. 29; George A. Gratsos, *Green and More Profitable Shipping* (13 November 2012) <http://www.nee.gr/downloads/183NEWSFRONT%20NAFTILIAKI%2013-11-12.pdf> accessed 1 June 2013; Sang-Yoon Lee and Young-Tae Chang, 'Shipping Companies' Awareness and Preparedness for Greenhouse Gas Regulations: A Korean Case' in Theo Notteboom (ed), *Current Issues in Shipping, Ports and Logistics* (2011) 25, 44.

165 Psaraftis, above n. 130, 221.

other European countries have participated in the scheme.[166] There are only minor differences between the three ETS proposals for international shipping. Compared with the Norwegian ETS, the proposal by the United Kingdom has a different method of allocating emissions allowances and a different approach to setting the emissions cap, while the French proposal provides details on auction design. The main strength in relation to these ETSs lies in their higher certainty of CO_2 reduction. Although no international ETS has been implemented, a regional EU ETS might provide a 'prototype' from which international shipping can learn.[167]

There are significant challenges in implementing these ETS proposals. First, significant carbon leakage and distortion of competition risks exist under the proposals. Carbon leakage generally refers to differentiated carbon policies and their subsequent impacts on GHG emissions.[168] Since carbon leakage might hinder the success of a global GHG emissions reduction and thus distort global competition, it is important for the ETS to be applied to the international transportation sector, including international aviation, rather than solely to the shipping industry or even part of the shipping industry.[169] The Norwegian ETS provides two exemptions from applying the scheme, namely ships below certain sizes and ships on international voyages to Small Island Developing States (SIDS).[170] The design of the voyage exemption was to meet the needs of developing countries, even though it is a common practice within the IMO regime to set a threshold for ship size. However, this proposal may also make it possible for some shipowners or ship operators to opt for certain ship sizes or certain shipping routes through the SIDS in order to obtain emission exemptions. In this case, competition will be distorted, and the reduction goal may also be difficult to achieve. Second, compared with the GHG Fund proposal, an ETS incurs much higher administrative costs to track, monitor and enforce as well as to avoid evasion and fraud. Third, the current development

166 These three states are Iceland, Liechtenstein and Norway. European Commission, *Emissions Trading System (EU ETS)* <http://ec.europa.eu/clima/policies/ets/index_en.htm> accessed 1 June 2013.

167 Psaraftis, above n. 130, 223.

168 Larry Parker and John Blodgett, '"Carbon Leakage" and Trade: Issues and Approaches' (19 December 2008) <http://www.fas.org/sgp/crs/misc/R40100.pdf> accessed 1 June 2013.

169 Jodie Moffat, 'Arranging Deckchairs on the Titanic: Climate Change, Greenhouse Gas Emissions and International Shipping' (2010) 24(2) *Australian and New Zealand Maritime Law Journal* 104, 121.

170 *A Further Outline of A Global Emission Trading System (ETS) for International Shipping*, submitted by Norway, MEPC 60th Session, Agenda Item 4, IMO Doc MEPC 60/4/22 (15 January 2010) p. 12, Annex 2.

of the EU ETS provides more uncertainty for the future development of an ETS for international shipping. As discussed earlier, the inclusion of the emissions from the international aviation industry into the EU ETS was suspended in December 2012, which, to some extent, makes the ETS less attractive for the shipping industry. International shipping organisations are generally opposed to an ETS, whereas the shipping associations in some of the Annex I States to the *UNFCCC* support it.[171]

4.3.4.2.3 *Hybrid Market-Based Measure Proposals*

The Efficiency Incentive Scheme, Ship Efficiency and Credit Trading and Rebate Mechanism belong to the category of hybrid MBM proposals. The Efficiency Incentive Scheme and Ship Efficiency and Credit Trading can be regarded as hybrid MBMs with the EEDI as a benchmark, whereas the Rebate Mechanism is a hybrid MBM that can be built into any other MBM. One common feature between the Efficiency Incentive Scheme and Ship Efficiency and Credit Trading is that they both reward good performance ships, and the EEDI is used for measurement. However, it is the EEDI that makes the two hybrid MBM proposals less attractive. Two factors contribute to this argument. First, low EEDI indicates high energy efficiency, whereas a ship with a low EEDI does not necessarily mean that it has the lowest GHG emissions.[172] Its emissions might be more than those from a ship with a larger engine (high EEDI), which the smaller ship needs to maintain certain speed to ensure safety in bad weather.[173] In this case, the EEDI measurement does not work well. Second, the two hybrid MBM proposals, if adopted, will apply to both new ships and existing ships, whereas the EEDI adopted by Annex VI to *MARPOL 73/78* applies only to new ships.[174] There has been no research indicating the possible application of the EEDI to existing ships. After testing and verification, the International Association of Dry Cargo Shipowners (INTERCARGO) asserts that the EEDI 'does not apply to, and hence it cannot and should not be used

171 See, e.g., the Round Table of International Shipping Associations, which opposed any emission trading system (ETS) in that it would be 'unworkable' for the shipping industry. Round Table of International Shipping Associations, above n. 17. The national shipowners associations in Australia, Belgium, Norway, Sweden, and the United Kingdom released a discussion paper in 2009 supporting a global cap-and-trade ETS to reduce GHG emissions from international shipping. Australian Shipowners Association et al., above n. 148. See ch. 5, 5.2–5.3.

172 Psaraftis, above n. 130, 222.

173 Ibid.

174 *MARPOL 73/78 Annex VI (2011 amendments)* regs 20–21.

for, existing ships.'[175] Therefore, the adoption of these hybrid MBM proposals is not straightforward.

The Rebate Mechanism consists of two options: an add-on option by integrating with any revenue-raising MBM and an integrated option incorporated with the International Maritime Emission Reduction Scheme, which is a levy-on-fuel scheme. The main feature of this hybrid MBM is its compatibility with the CBDR principle. Under the add-on option, all ships pay for their emissions. However, a developing country obtains an annual rebate based on its share of global seaborne imports first, and then the remaining revenue from developed countries will be disbursed through the *UNFCCC*. In this way, the 'no net incidence' on developing countries can be ensured.[176] In other words, developing countries will not suffer any loss, but they will benefit from participating in the Rebate Mechanism. The first draft of the legal text for the Rebate Mechanism was submitted to the 64th MEPC meeting by the World Wide Fund for Nature in October 2012. It stipulates that '[e]ach Party not included in annex II of the *UNFCCC*, or any successor annex, shall be eligible to an apportioned rebate [from a potential MBM Convention].'[177] It further provides that this rebate could be foregone as its contribution to international co-operation.[178] This proposed regulation would expand the scope of the beneficiaries of this scheme from SIDS and least developed countries as proposed by some countries to all non-Annex II States to the *UNFCCC*. It is expected that this mechanism will be attractive for developing countries due to its incorporation of the CBDR principle. Compared with other proposals, this proposal better reflects the interests of both developing countries and developed countries. Nevertheless, if the add-on option is built into any other MBM proposal, such as a GHG Fund or ETS, the administrative costs will probably be higher due to the possible increased number of administrative bodies. Therefore, it will be challenging to control these costs. Furthermore, there is another concern about whether the available data are accurate and reliable in terms of calculating a developing country's share of global imports by value.

---

175 *Application of the EEDI to Existing Ships*, submitted by INTERCARGO, MEPC 63rd Session, Agenda Item 5, IMO Doc MEPC 63/5/12 (6 January 2012) summary.

176 *Ensuring No Net Incidence on Developing Countries from A Global Maritime Market-Based Mechanism*, submitted by World Wide Fund for Nature (WWF), MEPC 63rd Session, Agenda Item 5, IMO Doc MEPC 63/5/6 (22 December 2011).

177 *Draft Legal Text on Uses of Financing Generated from A Maritime MBM*, submitted by the World Wide Fund for Nature (WWF), MEPC 64th Session, Agenda Item 5, IMO Doc MEPC 64/5/10 (27 July 2012) Annex, art. 4(1).

178 Ibid.

4.4 Conclusion

The purpose of this chapter is to identify the IMO's response to the issue of GHG emissions from international shipping. As a specialised agency of the UN, the IMO received its mandate for regulating shipping GHG emissions from the *Kyoto Protocol*. Meanwhile, the *IMO Convention* and the *LOSC* also provided the IMO with competence in this area. These competences make it possible for the IMO to apply both the CBDR and NMFT principles in addressing GHG emissions from ships. It is also arguable that the IMO has an exclusive role in regulating technical and operational measures and a non-exclusive role in regulating MBMs.

To accomplish its mandate for regulating GHG emissions from international shipping, the IMO has been working on this issue since the adoption of Resolution 8 and Annex VI to *MARPOL 73/78* in 1997. The main achievements by the IMO in this regard include amendments to Annex VI of *MARPOL 73/78* adopted in 2011 and 2014, as well as various guidelines, codes and resolutions. The amendments to Annex VI make the EEDI mandatory for new ships, and the SEEMP for all ships. The EEDI is the most important technical measure that aims at promoting the use of energy-efficient equipment and engines, whereas the SEEMP is an operational measure with the purpose of improving the energy efficiency of ship operations. The energy efficiency measures, including the EEDI and SEEMP, serve as the first legally binding standards tackling shipping GHG emissions. They maintain a balance between safe navigation and energy efficiency. Furthermore, these regulations ensure significant emissions reduction and provide a strong incentive for the shipping industry to update cost-efficient technologies. However, the limited EEDI coverage and the lack of an SEEMP reduction target need to be addressed. In particular, the lack of full incorporation of the CBDR principle makes the future enforcement of these regulations questionable particularly for developing country fleets.

In furtherance of reducing GHG emissions from ships, the IMO has organised various discussions and negotiations on potential MBMs. Of the current seven types of MBMs proposed to the IMO, each of them has its advantages and disadvantages. Generally, the GHG Fund has low administrative costs and has been welcomed by most of the shipping industry, whereas the Rebate Mechanism serves as the only MBM that effectively incorporates the CBDR principle. An ETS, as a widely discussed option, is facing significant challenges in its future implementation.

CHAPTER 5

Response from the Shipping Industry to the Issue of Greenhouse Gas Emissions from International Shipping

5.1 Introduction

The regulation of GHG emissions from international shipping is a complex process in which different State and non-State actors advance their own interests either politically or commercially. This can be seen from the debate within the IMO on the issue. The response from the shipping industry is critical in incorporating its sectoral interest into the IMO's regulation. The representatives from the shipping industry, in particular the shipping industry associations that have consultative status with the IMO as Non-Government Organisations (NGOs), have influenced the direction of policy-making by contributing to the debate in the IMO. Voluntary or even mandatory instruments by the IMO can never be implemented effectively without compliance by the shipping industry.[1] Thus, in order to find an effective solution for the reduction of GHG emissions from ships, an analysis of the views from the shipping industry is indispensable.

The shipping industry may be grouped into different subcategories based on differing criteria. According to the purpose of shipping, it consists of a cruise sector and a cargo sector.[2] The cruise sector provides passenger and ferry services while the cargo sector transports cargo through designated sea routes. The shipping industry can also be categorised into bulk shipping and liner shipping based on the goods that are transported.[3] The bulk sector, which includes wet/liquid bulk and dry bulk, transports raw materials such as crude oil and other petroleum products or iron ore and coal. Whereas the liner sector mainly engages in small shipments of general commercial freight and it transits

---

1 Costas Giziakis and Anastasia Christodoulou, 'Environmental Awareness and Practice concerning Maritime Air Emissions: the Case of the Greek Shipping Industry' (2012) 39(3) *Maritime Policy & Management* 353, 354.

2 Costas T. Grammenos and Chong Ju Choi, 'The Greek Shipping Industry: Regulatory Change and Evolving Organizational Forms' (1999) 29(1) *International Studies of Management & Organization* 34, 38.

3 H.E. Haralambides, 'Structure and Operations in the Liner Shipping Industry' in David A. Hensher and Kenneth J. Button (eds.), *Handbook of Transport Modelling* (Elsevier Ltd, 2007) 607, 607.

 | DOI 10.1163/9789004329317_006

regular routes on fixed schedules. Regulation 21 of Annex VI to *International Convention for the Prevention of Pollution from Ships* (*MARPOL* 73/78) lists 12 types of ships to be regulated by the EEDI.[4] These categories cover the majority of the international shipping industry, although the implementation of these ships is phased depending on specific ship types.[5] In accordance with different stakeholders of shipping, the shipping industry constitutes six sectors: ship design, shipbuilding, ship insurance, cargo owner, bunker supplier and classification societies. The response from these sectors is analysed in this chapter.

The origin of the modern shipping industry can be traced back to the year 1787 when steamships emerged. With the development of information and communications technologies (ICT) and the opening of the Suez Canal in 1869, the shipping industry grew exponentially to facilitate faster trade between Europe and Asia. To address the 'unsustainable predatory competition' associated with this process,[6] cartels or 'liner conferences' and various shipping associations were established. As discussed in Chapter 4, the reduction of GHG emissions from international shipping is primarily regulated by the IMO. The main approach adopted by the global shipping industry to participating in the regulatory process is to submit their written proposals to the IMO and attend its discussions. This work can mainly be achieved by shipping NGOs.

This chapter has three major parts. The first part examines the response from international and regional shipping organisations by summarising and analysing the documents submitted by them to the IMO as well as their reports obtained from other sources. The organisations examined include those representing ship designers, shipbuilders, shipowners and ship operators, cargo owners, ship insurers, the classification societies and the bunker suppliers. The second part assesses the response from the shipping industry in *UNFCCC* Annex I States with Australia, Greece and the United Kingdom as examples. The third part explores the response from the shipping industry in *UNFCCC* non-Annex I States with China, Republic of Korea and India as examples. It is generally less feasible for the shipping industry in individual States, in particular those from non-Annex I States, to submit proposals directly to the IMO. Therefore, the methodology adopted in the second and third parts combines the report assessment and the surveys of shipping companies that have been prepared by other institutions or individuals.

4 *International Convention for the Prevention of Pollution from Ships* (*MARPOL* 73/78), signed 2 November 1973, 12 ILM 1319, as amended by the 1978 Protocol to the 1973 Convention, 1341 UNTS 3, 17 ILM 546 (entered into force 2 October 1983) Annex VI (2014 amendments) reg. 21.

5 Ibid. Table 1.

6 Grammenos and Choi, above n. 2, 42.

5.2 Response from International and Regional Shipping Organisations

This part examines the views and actions of the global shipping NGOs on the issue of GHG emissions from international shipping. It analyses six types of stakeholders: ship designers and shipbuilders, shipowners and ship operators, cargo owners, ship insurers, classification societies, and bunker suppliers. The issues that these international or regional shipping organisations have responded to include:

- whether the reduction of GHG emissions from ships is necessary;
- views and practice on proposed technical and operational measures;
- views and practice on proposed market-based measures (MBMs); and
- views and practice on other issues in relation to the reduction of GHG emissions from ships.

5.2.1 *Ship Designers and Shipbuilders*

Ship design and shipbuilding play an important role in the maritime sector because they influence the supply of various types of ships. The price of shipbuilding is an important factor based on which the shipowner makes investment decisions and shipyards win orders.[7] Meanwhile, the cost of shipbuilding is determined by many factors, including ship design options, market demand and shipyard capacity.[8] The reduction of GHG emissions from ships will generally increase the cost of shipbuilding in that the Energy Efficiency Design Index (EEDI) requires the adoption of new technologies.[9] Theoretically speaking, ship designers and shipbuilders do not need to be concerned about the cost of ship designing and shipbuilding since the shipowners bear the cost. However, the increased cost might challenge the market competitiveness of their products and thus influence the number of orders that they receive. Another concern will be whether their technological capability can meet the GHG emissions reduction requirements.

The International Paint and Printing Ink Council (IPPIC) is the only NGO focusing on ship design. It has achieved consultative status within the IMO and has submitted a proposal to the IMO. The Community of European Shipyards'

7 Liping Jiang and Jørgen T. Lauridsen, 'Price Formation of Dry Bulk Carriers in the Chinese Shipbuilding Industry' (2012) 39(3) *Maritime Policy & Management* 339, 339. Generally a high freight rate spurs shipowners to order new ships. A positive market expectation boosts the price of shipbuilding, and vice versa.

8 Ibid. 340.

9 See ch. 4, 4.3.3.1.

Associations (CESA) is the main regional shipping NGO representing the interests of shipbuilders within the IMO. The Tripartite Working Group of shipyard operators, shipowners and classification societies (Tripartite Working Group) was formed in 2007 to pool resources, share knowledge and make joint proposals for achieving reductions for new ship building.[10] Since CESA is also a member of the Tripartite Working Group, the views from this Group also reflect those from the shipbuilding sector. As an influential shipbuilders' association, the Japan, Europe, China, Korea and USA Shipbuilders' Association (JECKU), in particular its Committee for Expertise of Shipbuilding Specifics (CESS), also participates in discussion on GHG emissions from ships. However, it has not obtained consultative status within the IMO and has not submitted any proposal to the organisation.

Ship designers and shipbuilders generally welcome the reduction of GHG emissions from ships. At the 57th Marine Environment Protection Committee (MEPC) meeting in 2008, the Tripartite Working Group was formed and a consensus was reached that the shipping industry should contribute to the reduction of shipping GHG emissions. To achieve this goal the Group noted that, 'a broad, inclusive and goal-based approach' should be adopted to facilitate this process.[11] At the 58th MEPC meeting, the CESA, on behalf of European shipbuilders and ship repairers, welcomed the efforts made by the IMO and asserted that 'a convincing and effective approach towards reduction of the specific CO_2 emissions from international maritime transport is urgently needed'.[12] In 2010 the IPPIC commented that it is important to control GHG emissions from ships since the shipping industry must ensure that 'its operations are as efficient as possible'.[13] At the 18th JECKU Top Executive Meeting

---

10 *A Cross-industry Goal-based Approach to Reduction of GHG Emissions from New Ships*, submitted by the ICS, BIMCO, CESA, INTERCARGO, INTERTANKO and OCIMF, MEPC 57th Session, Agenda Item 4, IMO Doc MEPC 57/4/8 (23 January 2008) para. 3. The members of the Tripartite Working Group include the CESA, International Chamber of Shipping (ICS), the Baltic and International Maritime Council (BIMCO), International Association of Dry Cargo Shipowners (INTERCARGO), International Association of Independent Tanker Owners (INTERTANKO) and Oil Companies International Marine Forum (OCIMF).

11 Ibid. para. 4.

12 *Development of a CO_2 Design Index for New Ships*, submitted by the Community of European Shipyards' Association (CESA), MEPC 58th Session, Agenda Item 4, IMO Doc MEPC 58/4/12 (1 August 2008) para. 1.

13 *The Importance of Using Effective Anti-fouling Coatings in relation to Greenhouse Gas Emissions from Shipping*, submitted by the International Paint and Printing Ink Council (IPPIC), MEPC 60th Session, Agenda Item 4, IMO Doc MEPC 60/4/21 (15 January 2010) para. 7.

in 2009, the chairman welcomed the efforts of the IMO in reducing GHG emissions from ships and asserted that this 'can only be secured' through the shipbuilding industry around the world.[14] It is clear that ship designers and shipbuilders recognise the importance of reducing GHG emissions from ships and are ready to contribute to this work.

The CESA actively participated in the debate and trial application of the draft EEDI regarding the adoption of proposed technical and operational measures by the IMO. The development of its views can be divided into three stages. The period from the 57th MEPC meeting to the 58th MEPC meeting (March–October 2008) belongs to the first stage. During this period, the CESA acknowledged that technical innovation would be useful for the reduction of GHG emissions from ships, but stressed that 'operational measures have an even higher reduction potential compared to the available options at new building stage'.[15] It insisted that 'effective measures should focus on existing ships firstly and the experience gained should be used for the development of measures for new ships that will come into operation in the future'.[16] As for the proposed new ship design CO_2 index, the CESA considered it 'inappropriate and premature' to be utilised 'in a prescriptive way'.[17] It can be deduced that in this stage the CESA was more interested in operational measures (Ship Energy Efficiency Management Plan (SEEMP)) than technical measures (EEDI), and preferred voluntary measures to mandatory measures. This preference is consistent with its own interests. The SEEMP relates to the operational measures which are mainly utilised by ship operators, while the EEDI relates to the technical measures on new ships which require shipbuilders to invest more on research and development (R&D), in particular, the upgrading of technology so as to meet the EEDI requirements. The cost of shipbuilding might not increase at least in the short term if the IMO relies more on operational measures rather than technical measures or considers proposed EEDI requirements to be voluntary.

The period from the second GHG-WG meeting to the 59th MEPC meeting (March–July 2009) constitutes the second stage. During this stage the CESA came to accept the concept of EEDI but asserted that this concept was not

---

14 JECKU, *Chairman's Note (18th JECKU Top Executive Meeting October 29, 2009, Berlin)* <http://www.sajn.or.jp/e/press/Press_Berlin_TEM.pdf> accessed 8 April 2013.

15 *Development of a CO_2 Design Index for New Ships*, submitted by the Community of European Shipyards' Association (CESA), MEPC 58th Session, Agenda Item 4, IMO Doc MEPC 58/4/12 (1 August 2008) para. 1.

16 Ibid. para. 2.

17 Ibid. para. 5.

technically mature. It undertook a trial application of the draft EEDI and found shortcomings with the proposed baseline. New vessels, especially in short sea shipping, may have to reduce their speed under a mandatory EEDI regime incorporated with the proposed baseline. Nevertheless, older vessels may be in service much longer than they should be since they are not subject to the speed limitation.[18] To address this problem, the CESA put forward a three-stage phase-in approach to implementing the EEDI at the 59th MEPC meeting.[19] That is to start with standard ship types, such as large carriers, tankers and container vessels of more than 20,000 mt dwt. Then develop the indexes for vessels smaller than 20,000 mt dwt. Finally it will develop indexes for the more complex ship types. These proposals by CESA pointed out the deficiencies that existed in the proposed draft EEDI. Its proposals were later adopted by the IMO. The period after MEPC 59 (2009–2014) belongs to the third stage. In this stage the CESA has worked on a new reference line to include other types of ships, such as the ro-ro cargo and ro-ro passenger ship types into the EEDI framework, and facilitated the implementation of the adopted Annex VI to *MARPOL 73/78*. In particular, at the 62nd MEPC meeting in 2011, the CESA lodged a statement to the MEPC to highlight the significance of protecting intellectual property rights during the course of the transfer of technology. It asserted that advanced emissions reduction technology is the key to environmental protection and competitiveness of the shipbuilding industry, which demands a high level of intellectual property right protection.[20] In other words, the EEDI technology in shipbuilding has to be legally protected from any possible free utilisation. This view reveals the complexity of the transfer of technology as regulated under Regulation 23(2) of Annex VI to *MARPOL 73/78*. It appears that the regulation on technology transfer needs to be more specific and improved to build sufficient capacity for developing countries to comply with the EEDI.[21]

The IPPIC asserted that it is economically and technologically impractical to apply these technologies to the existing fleet. It proposed the use of anti-fouling

18 *CO_2 Reduction Requires Efficient Instruments Based on Sound Technical Solutions*, submitted by the Community of European Shipyards' Association (CESA), Intersessional Meeting of the Greenhouse Gas Working Group 2nd Session, Agenda Item 2, IMO Doc GHG-WG 2/2/22 (6 February 2009) Annex 1, para. 25–26.

19 *Phase-in Implementation of the Energy Efficiency Design Index for Standard and Complex Ship Types*, submitted by the Community of European Shipyards' Associations (CESA), MEPC 59th Session, Agenda Item 4, IMO Doc MEPC 59/4/38 (20 May 2009) para. 10.

20 *Report of the Marine Environment Protection Committee on Its Sixty-Second Session*, MEPC 62nd Session, Agenda Item 24, IMO Doc MEPC 62/24 (26 July 2011) annex 11, p. 1.

21 A proposed solution to address this transfer of technology problem is provided in Chapter 7. See ch. 7, 7.4.2.

coatings/paints to prevent additional GHG emissions from shipping on the ground that this is an economically and environmentally better method.[22] The application of anti-fouling paints to immersed areas of ships can prevent the colonisation and growth of marine organisms such as algae, tubeworms and barnacles. It has been proved that a ship not applying an anti-fouling system to its hull may require up to 70 per cent extra propulsion power when compared to those which apply this paint.[23] This approach, however, has not aroused much attention within the IMO since anti-fouling paints themselves may not be an independent measure. It can be regarded as a type of EEDI technology.[24]

The Committee for Expertise of Shipbuilding Specifics (CESS) is a JECKU committee. It supports the adoption of the EEDI and promises that the shipbuilding industry will work closely with shipowners and classification societies to ensure the smooth implementation of the scheme.[25] Meanwhile, it also underscores the importance of intellectual property protection as an essential element of technology development in this regard.[26] This view is consistent with that of the CESA as discussed above.

The CESA believes that the EEDI cannot achieve any short-term reduction of GHG emissions because it only applies to new ships. Instead, the CESA regards MBMs as a 'more effective solution' to address the GHG emissions issue,[27] and 'strongly proposes' the implementation of MBMs for ships engaging in voyages in the short sea shipping sector.[28] Due to its specialisation in shipbuilding, however, the CESA has not provided its comments on specific MBMs.

---

22 *The Importance of Using Effective Anti-fouling Coatings in relation to Greenhouse Gas Emissions from Shipping*, submitted by the International Paint and Printing Ink Council (IPPIC), MEPC 60th Session, Agenda Item 4, IMO Doc MEPC 60/4/21 (15 January 2010) paras 7, 13.

23 Ibid. para. 11.

24 See ch. 4, 4.3.3.1, Table 4.2.

25 JECKU, *CESS 2012 Press Release* (23 August 2012) <http://www.sajn.or.jp/e/press/CESS_AM2012_en,pdf> accessed 8 April 2013.

26 JECKU, *CESS Annual Meeting 2009 Press Release* (2 September 2009) <http://www.sajn.or.jp/e/press/CESSPressRelease2009.pdf> accessed 8 April 2013.

27 *Phase-in Implementation of the Energy Efficiency Design Index for Standard and Complex Ship Types*, submitted by the Community of European Shipyards' Associations (CESA), MEPC 59th Session, Agenda Item 4, IMO Doc MEPC 59/4/38 (20 May 2009) para. 3.

28 *Comments related to Trial Calculations of the EEDI for Subgroups of Ro-ro Cargo Ships*, submitted by INTERFERRY and CESA, MEPC 60th Session, Agenda Item 4, IMO Doc MEPC 60/4/48 (29 January 2010) para. 22.

It is concluded that at the international and regional level ship designers and shipbuilders support the reduction of GHG emissions from international shipping. They argue that the SEEMP will be more effective than the EEDI. However, they have come to accept the concept of EEDI. They assert that the deficiencies in the current EEDI should be addressed and they have conducted research and trial application to improve the EEDI. Although the European shipbuilders' association has highlighted the importance of MBMs in tackling the GHG issue, a wider Japan, Europe, China, Korea and USA Shipbuilders' Association, in which some *UNFCCC* non-Annex I States are members, has stressed the function of technical and operational measures and has ignored the MBMs in this regard. The above response from ship designers and shipbuilders generally supports the IMO's regulatory initiatives. However, the fact that these organisations are dominated by *UNFCCC* Annex I States makes the above conclusions less reliable. Figure 5.1 demonstrates the change in the shipbuilding sector's global share of shipbuilding between the Far East and the rest of the world from 1964 to 2008. Since the Far East surpassed Western Europe in 1966, the percentage of the Far East in the world's shipbuilding deliveries has been growing. Currently the Far East accounts for more than 90 per cent of the global market in shipbuilding.[29] However, no international or regional shipping NGOs from the Far East have achieved consultative or observer status at the IMO.[30] Consequently, their views have not been expressed in international fora. This situation does not match the contribution made by this region to the global shipping industry. To compensate for the lack of the voice from the region on the issue, the response from the shipbuilding industry in China and the Republic of Korea, two main countries of the Far East, is examined in the third part of this chapter.

---

29 Lloyd's Register, *World Shipbuilding Deliveries* (11 October 2010) <http://shipbuildinghistory.com/today/statistics/world.htm> accessed 21 January 2013. However, European countries and the US still largely dominate other important sectors of the shipping industry, such as ship brokering, ship financing, maritime arbitration, insurance and claims, as well as global shipping pricing. See, e.g., Suranjana Roy Bhattacharya, 'Chinese Shipping Industry Is Big but Not Powful', *Gulf News* 19 July 2010 <http://gulfnews.com/business/opinion/chinese-shipping-industry-is-big-but-not-powerful-1.656076> accessed 1 June 2013.

30 Article A(IV) of the IMO *Guidelines for the Granting of Consultative Status* provides that consultative status shall only be granted to NGOs which are truly international. In practice the CESA, as a regional shipping organisation, has been granted a consultative status at the IMO. International Maritime Organization (IMO), *Basic Documents Volume I* (International Maritime Organization, 2010) 165.

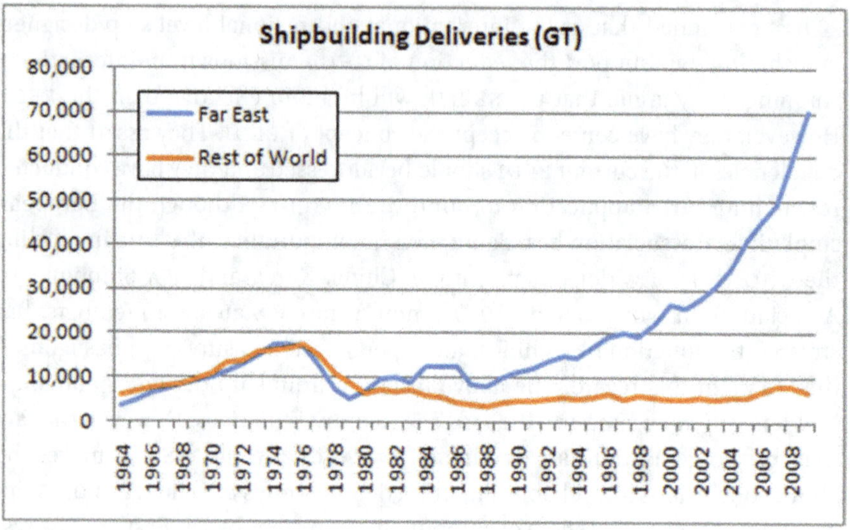

FIGURE 5.1 *World shipbuilding deliveries.*
SOURCE: LLOYD'S REGISTER OF SHIPPING'S "WORLD FLEET STATISTICS".[31]
NOTE: FAR EAST COUNTRIES AND REGIONS MAINLY INCLUDE JAPAN, REPUBLIC OF KOREA, CHINA AND CHINESE TAIPEI.

5.2.2 *Shipowners and Ship Operators*

Shipowners refer to 'individuals, companies and state-owned enterprises which own, manage and operate the commercial shipping fleets of the world'.[32] This means that a shipowner may also be a ship operator, which constitutes one of the reasons why they are both discussed in this section. The shipowner is the ship's registered legal owner which appears in the registry of one country.[33] However, the real shipowner is not always the same as the one that appears in the registry. This is because some of the true beneficial owners want to conceal their identity to avoid their exposure to liability. The secrecy surrounding ship ownership should not influence the burden of responsibility for the possible transboundary harm resulting from the GHG emissions from ships. Ship operators usually include ship managers and charterers, and they are often regarded

31 Lloyd's Register, above n. 29.

32 Alan Khee-Jin Tan, *Vessel-Source Marine Pollution: the Law and Politics of International Regulation* (Cambridge University Press, 2006) 34.

33 Ibid.

as polluters who should be held responsible for employing the services of substandard vessels.[34]

As discussed in Chapter 2, the polluter-pays principle applies to the reduction of GHG emissions from international shipping. The polluter may include the flag State, ship owners and operators, or in some cases, individuals. Among them, shipowners and ship operators, in particular the shipowners, are treated as the primary polluters in the shipping sector.[35] The flag States, however, tend to transfer the costs of environmental compliance to the shipping industry.[36] Against this backdrop, shipowners and ship operators are key stakeholders in the reduction of GHG emissions from ships, so their responses warrant examination.

Currently the interests of shipowners are mainly represented by the Round Table of International Shipping Associations (Round Table) within the MEPC. The four member organisations of the Round Table are the International Chamber of Shipping (ICS)/the International Shipping Federation (ISF),[37] the Baltic and International Maritime Council (BIMCO), the International Association of Independent Tanker Owners (INTERTANKO), and the International Association of Dry Cargo Shipowners (INTERCARGO). The ICS is the principal international trade association for the shipping industry, representing all sectors and trades. Its membership constitutes 47 national shipowners' associations from more than 30 countries, whose member shipping companies operate two thirds of the world's merchant tonnage. Although the ICS is dominated by members from developed countries,[38] developing countries, such as China, India and Cyprus, also have influence within the organisation. As the largest private shipping organisation in the world, BIMCO controls around 65 per cent of the world's shipping tonnage, and has members in more than 120 countries, including shipowners, operators, managers, brokers, agents, the Protection and Indemnity (P&I) Clubs and other commercial actors.[39] INTERTANKO represents the interests of independent tanker owners and operators of oil and chemical tankers, namely non-oil companies and non-State

34 G.P. Pamborides, *International Shipping Law: Legislation and Enforcement* (Kluwer Law International, 1999) 145.

35 Tan, above n. 32, 36.

36 Ibid.

37 The ISF is the international employers' association for shipping companies, and its secretariat is provided by the ICS.

38 Tan, above n. 32, 37.

39 BIMCO, *About BIMCO* <https://www.bimco.org/About/About_BIMCO.aspx> accessed 22 January 2013.

controlled tanker owners. It controls 80 per cent of the world's tanker fleet and the vast majority of tanker owners are members.[40] INTERCARGO mainly advocates the interests of the bulk carrier owners in the international dry bulk trades, such as coal, grain, iron ore and other commodities.[41]

In addition to BIMCO and INTERTANKO that include ship operators as their members, there are a number of global associations of ship operators which have consultative status at IMO. Examples are Cruise Lines International Association (CLIA), INTERFERRY, the World Shipping Council (WSC), the Society of International Gas Tanker and Terminal Operators (SIGTTO) and the International Parcel Tankers Association (IPTA). These ship operator associations, together with global shipowner associations, have taken an active part in the discussions and debate within the IMO on the reduction of GHG emissions from ships. This section also canvasses the views of other regional shipowners associations which do not have consultative status at the IMO, such as the Asian Shipowners Forum (ASF).[42]

At the 57th MEPC meeting in 2008, the four Round Table members (ICS, BIMCO, INTERCARGO and INTERTANKO), co-sponsored by other countries and organisations, submitted a proposal to the IMO to support the reduction of GHG emissions from international shipping and encourage the IMO to take earlier action. Additionally, they put forward nine fundamental principles as a base for future discussions on reducing GHG emissions.[43] Meanwhile, ICS

---

40 INTERTANKO, *About Us* <http://www.intertanko.com/About-Us/> accessed 22 January 2013.

41 INTERCARGO, *About INTERCARGO* <http://www.intercargo.org/about/78-about-intercargo.html> accessed 22 January 2013.

42 The ASF was founded in April 1992 and consists of eight members from the shipowners' associations of Asia Pacific nations, namely Australia, China, Hong Kong, India, Japan, Korea, Chinese Taipei and the Federation of ASEAN Shipowners' Associations (FASA), consisting of Indonesia, Malaysia, Myanmar, the Philippines, Singapore, Thailand and Vietnam. Half of the world merchant fleet is believed to be controlled by the ASF membership.

43 *Future IMO Regulation regarding Greenhouse Gas Emissions from International Shipping*, submitted by Denmark, Marshall Islands, BIMCO, ICS, INTERCARGO, INTERTANKO and OCIMF, MEPC 57th Session, Agenda Item 4, IMO Doc MEPC 57/4/2 (21 December 2008) para. 11. The nine principles are:

'(1) effective in contributing to the reduction of total global greenhouse gas emissions;
(2) binding and equally applicable to all flag States in order to avoid evasion;
(3) cost-effective;
(4) able to limit—or at least—effectively minimize competitive distortion;
(5) based on sustainable environmental development without penalizing global trade and growth;

proclaimed that it would be 'fully committed to the adopting of stringent and effective regulations' for the reduction of shipping emissions.[44] In the next year, INTERTANKO asserted that it would 'strongly support' the IMO's efforts in regulating GHG emissions from ships.[45] The IPTA also expressed its support for the EEDI concept.[46] A ship operator or shipowner generally can save money by using substandard ships.[47] So, at first sight the natural inclination of shipowners and ship operators would be to oppose the reduction of GHG emissions from ships. However, this cutting of costs gained from the utilisation of substandard ships can only be maintained on a short-term basis since the introduction of more stringent regulations will be enforced by either flag States or port States, which eventually will narrow their profit margin.[48] In this sense, the reduction of shipping GHG emissions will be to the benefit of the shipping industry as a whole from a long-term perspective, and thus 'provide a common basis for fair competition in a free market'.[49] Therefore, it is not surprising that shipowners and ship operators welcome the IMO's regulatory efforts in reducing GHG emissions from international shipping.

Various global shipowners and ship operators associations made prompt responses to the technical and operational measures before and after the adoption of the EEDI and SEEMP by the IMO in July 2011. The response from these organisations mainly focused on regulatory principles, methodologies, application scope and other technical aspects of the proposed measures. They are examined in the following paragraphs.

---

(6) based on a global-based approach and not prescribe specific methods;
(7) supportive of promoting and facilitating technical innovation and R&D in the entire shipping sector;
(8) accommodating to leading technologies in the field of energy efficiency; and
(9) practical, transparent, fraud free and easy to administer.'

44 *The Revision of MARPOL Annex VI*, submitted by the International Chamber of Shipping (ICS), MEPC 57th Session, Agenda Item 4, IMO Doc MEPC 57/4/28 (13 February 2008) para. 3.

45 *Comments on MEPC 59/4/8 and MEPC 59/4/9 relating to the Energy Efficiency Design Index, the Ship Energy Management Plan and Possible Market-based Instruments*, submitted by INTERTANKO, MEPC 59th Session, Agenda Item 4, IMO Doc MEPC 59/4/43 (22 May 2009) para. 2.

46 *Introduction of A Cubic Capacity Correction Factor into the EEDI Formula*, submitted by the International Parcel Tankers Association (IPTA), MEPC 62nd Session, Agenda Item 6, IMO Doc MEPC 62/6/13 (5 May 2011) para. 12.

47 Pamborides, above n. 34, 139–140.

48 Ibid. 140.

49 Ibid.

At the 57th MEPC meeting, the ICS proposed five principles to guide the amendment of Annex VI to *MARPOL 73/78*. One of them is that ship operators should have freedom to choose their compliance mechanism, namely they should be allowed to adopt appropriate technologies they select, so as to protect the shipping industry from monopolistic situations.[50] It regarded this 'freedom from prescription' as the most effective means for stimulating future innovation but also asserted that there is a need to specify performance criteria for exhaust gas monitoring and measuring equipment to be carried by ships.[51] This principle is generally positive for the development of the shipping industry since it provides significant encouragement for technological innovation by the shipowners and ship operators. However, performance criteria should be in place not only for exhaust gas monitoring and measuring equipment, but also for all other equipment involving the efficient operation of ships. As noted in Chapter 1, exhaust gas is the main but not sole source of GHG emissions from ships. Other substances and equipment, such as refrigerants and fire-fighting equipment, may also emit GHGs.[52] The 'freedom of prescription' principle, however, does not mean that shipping companies may adopt any technology they choose. Instead, the technological criteria should be regulated in a prescriptive way so as to underpin the enforcement of these measures. The amended Annex VI in 2011 prescribed that the criteria should also be reviewed and updated regularly with the availability of better technologies.[53] In this way, marine reduction of GHG emissions could be achieved as technology advances.

The ICS suggested that stringent regulations shall be adopted 'without delay'.[54] This view probably reflects the desire of shipowners and ship operators from developed countries to expand their global market share through technological upgrading in the international shipping market. Furthermore, the ICS opposed the proposal which permits port States to set different limits

---

50 *The Revision of MARPOL Annex VI*, submitted by the International Chamber of Shipping (ICS), MEPC 57th Session, Agenda Item 4, IMO Doc MEPC 57/4/28 (13 February 2008) para 5.3.

51 Ibid.

52 See ch. 1, 1.2.2.1.

53 *MARPOL 73/78 Annex VI (2011 amendments)* reg 21.6.

54 *The Revision of MARPOL Annex VI*, submitted by the International Chamber of Shipping (ICS), MEPC 57th Session, Agenda Item 4, IMO Doc MEPC 57/4/28 (13 February 2008) para. 5.

by informing the IMO in advance.[55] This opinion is consistent with the 2011 amendments to Annex VI of *MARPOL 73/78*.[56]

Shipowners' and ship operators' associations also extensively debated the methodology relating to GHG emissions from ships. At the 57th MEPC meeting, BIMCO, ICS, INTERCARGO and INTERTANKO, together with other co-sponsors, proposed that any CO_2 indexing method for new ships should be based on a 'generally accepted methodology'.[57] But at the 58th MEPC meeting, BIMCO and INTERTANKO asserted that 'different methodologies will be required by different owners to match the very different efficiency assessment needs of different ships engaged in different trades'.[58] This difference of opinion reflects the complex situations of various ship types. Sometimes even for sister ships, or two fleets of the same type, 'vastly differing and varying results' may appear.[59] This discussion on methodology also led to another controversy on the application scope of the proposed EEDI and SEEMP.

It is generally accepted by the shipping industry that the SEEMP could be applied to all ships. At the 58th MEPC meeting, Round Table members submitted to the IMO their draft Ship Efficiency Management Plan (SEMP), an earlier vision of the SEEMP. The draft provides guidance on best practices for the efficient operation of ships. Many options are listed, such as fuel-efficient operations, optimised ship handling, more efficient propulsion systems, improved fleet management, improved cargo handling, energy management and fuel type. However, it also mentions that these measures are often 'area and trade dependent and are likely to require the agreement and support of a number of different stakeholders if they are to be utilized most effectively'.[60] This expression, however, reveals that these measures may not be compatible. Or in other words, these shipping organisations believe that operational measures may apply to all ship types, but should not be applied to all areas [or countries] due to their unique trading routes or geographical characteristics. This

55 Ibid. 9.

56 See *MARPOL 73/78 Annex VI (2011 amendments)* art. 10.21(5); see also ch. 4, 4.3.3.3, ch. 6, 6.6.

57 *A Mandatory CO_2 Design Index for New Ships*, submitted by Denmark, Marshall Islands, BIMCO, ICS, INTERCARGO, INTERTANKO and OCIMF, MEPC 57th Session, Agenda Item 4, IMO Doc MEPC 57/4/3 (21 December 2007) para. 12.

58 *Guidelines for the Implementation of the Ship Operational Index—Ship Efficiency Management Tool*, submitted by INTERTANKO, OCIMF and BIMCO, MEPC 58th Session, Agenda Item 4, IMO Doc MEPC 58/4/13 (1 August 2008) para. 3.

59 Ibid.

60 *Ship Efficiency Management Plan*, submitted by ICS, BIMCO, Intercargo, Intertanko and OCIMF, MEPC 58th Session, Agenda Item 4, IMO Doc MEPC 58/INF.7 (28 July 2008) annex, para. 3.35.

is consistent with the view by BIMCO and INTERTANKO that the Operational Index, a former expression of the SEEMP, will prove to be a useful tool for ship operators, but is 'not suited for mandatory application'.[61] This opinion was reflected in the revised Annex VI to *MARPOL 73/78* by explicitly regulating the mandatory SEEMP but not mentioning the operational index as the mandatory benchmark tool. In a joint statement made after the adoption of the SEEMP, the Round Table reiterated its full support for the SEEMP in that it 'provide[s] the framework allowing shipowners a degree of flexibility in how best to adapt it to individual ships'.[62]

In contrast to the discussion on the application of the proposed SEEMP, more international shipowners' and ship operators' associations have been engaged in research and testing of the application of the proposed EEDI. Since 2009, the application scope of the proposed EEDI has been discussed at all MEPC and GHG-WG meetings of the IMO. The ICS, the CLIA and INTERFERRY asserted that the proposed EEDI is 'not applicable to all ships, especially to those with complex and sophisticated machinery installations that are not "conventional" in nature'.[63] The CLIA commented that the draft EEDI should not be applied directly to passenger ships and electrically propelled ships. It proposed a revision to the EEDI calculation.[64] The ICS recognised the difficulty in applying EEDI to ro-ro vessels, high speed craft, smaller vessels (less than 20,000 dwt), passenger vessels, general cargo ships and steam turbine ships.[65] It proposed the phased approach, and suggested that each phase of

61 *Guidelines for the Implementation of the Ship Operational Index—Ship Efficiency Management Tool*, submitted by INTERTANKO, OCIMF and BIMCO, MEPC 58th Session, Agenda Item 4, IMO Doc MEPC 58/4/13 (1 August 2008) para. 2.

62 Round Table of International Shipping Associations, *Round Table Associations Position Paper on GHG+MBMs* (22 February 2012) <https://www.bimco.org/About/Press/Press_Releases/2012/2012_02_22_Round_Table_MBM.aspx> accessed 28 January 2013.

63 *Application of EEDI to Ships other than Those Operating with Conventional Machinery and Power Distribution Arrangements*, submitted by ICS, CLIA, INTERFERRY and the Marshall Islands, Intersessional Meeting of the Greenhouse Gas Working Group 2nd Session, Agenda Item 2, IMO Doc GHG-WG 2/2/19 (6 February 2009) para. 14.

64 *Refinements to the "Draft Guidelines on the Method of Calculation of the Energy Efficiency Design Index for New Ships" for Conventional Passenger Ships*, submitted by Cruise Lines International Association (CLIA), MEPC 59th Session, Agenda Item 4, IMO Doc MEPC 59/4/12 (9 April 2009).

65 *The Need for Refinement of the Energy Efficiency Design Index (EEDI)*, submitted by the International Chamber of Shipping (ICS), MEPC 59th Session, Agenda Item 4, IMO Doc MEPC 59/4/13 (9 April 2009) para. 4.

application be reviewed before it is completed.[66] At the 59th MEPC meeting, INTERTANKO considered that the EEDI formula had 'matured enough' for ship operators to test after it was tested by its members,[67] and at the 60th MEPC meeting, it asserted that 'the EEDI formula would not be appropriate for a certain category of tankers, such as the diesel-electrically powered tankers and dual-engine shuttle tanker designs'.[68] The SIGTTO proposed that ships carrying liquefied gases in bulk should be divided into liquefied natural gases (LNG) and liquefied petroleum gas (LPG) for the purposes of implementing the EEDI,[69] and found after assessment that the draft EEDI works well for LPG ships but not for the LNG ships.[70] Due to the technical contributions by these shipping organisations, the current application scope of the EEDI has covered 12 types of new ships and will be implemented in four phases. In terms of safety aspects of the EEDI, INTERFERRY proposed a modification to the circulated EEDI with regard to redundant propulsion systems.[71] This view was later incorporated in the 2011 amendments of Annex VI to *MARPOL 73/78*.

---

66 *Amendments to MARPOL Annex VI—Inclusion of Regulations on Energy Efficiency for Ships*, submitted by the International Chamber of Shipping (ICS), MEPC 62nd Session, Agenda Item 6, IMO Doc MEPC 62/6/24 (20 May 2011) para. 4.

67 *Comments on MEPC 59/4/8 and MEPC 59/4/9 relating to the Energy Efficiency Design Index, the Ship Energy Management Plan and Possible Market-based Instruments*, submitted by INTERTANKO, MEPC 59th Session, Agenda Item 4, IMO Doc MEPC 59/4/43 (22 May 2009) para. 4.

68 *Energy Efficiency Design Index for Tankers*, submitted by INTERTANKO, MEPC 60th Session, Agenda Item 4, IMO Doc MEPC 60/4/3 (18 December 2009) para. 6; *Energy Efficiency Design Index for Propulsion Redundancy*, submitted by INTERTANKO, MEPC 60th Session, Agenda Item 4, IMO Doc MEPC 60/4/4 (18 December 2009) para. 3.

69 *Considerations of the Application of the EEDI Reference Lines to LNG Vessels*, submitted by Society of International Gas Tanker and Terminal Operators (SIGTTO), MEPC 62nd Session, Agenda Item 6, IMO Doc MEPC 62/6/20 (20 May 2011) para. 21.

70 *Results of Data Gathering Exercise for the Assessment of the Energy Efficiency Design Index (EEDI) for Ships Carrying Liquefied Gases in Bulk*, submitted by the Society of International Gas Tanker and Terminal Operators (SIGTTO), MEPC 60th Session, Agenda Item 4, IMO Doc MEPC 60/4/44 (22 January 2010) para. 8,9.

71 *Application of Power Correction Factor F_j for Enhanced Safety*, submitted by INTERFERRY, MEPC 60th Session, Agenda Item 4, IMO Doc MEPC 60/4/20 (15 January 2010) para. 8. This view of INTERFERRY can also be seen from an interview with Johan Roos, INTERFERRY's executive director of EU and IMO affairs. He says that,

'Our approach to a workable EEDI solution for ro-ro vessels was two-fold. Obviously it had to bring about absolute efficiency gains and a consequent reduction in greenhouse gas emissions. But we also had to ensure that new ships can be built for all ro-ro markets

The response from the shipowners and ship operators associations also includes their joint project/research and individual trials aiming to address the GHG issue. For example, INTERTANKO established a Virtual Arrival Project aiming to reduce actual GHG emissions and fuel usage from ships by optimising the vessel speed,[72] together with representatives from other sectors such as the Oil Companies International Marine Forum (OCIMF), a cargo owner. After the adoption of the technical and operational measures, these organisations have continued to contribute to the improvement of the EEDI, in particular, they have proposed various methods to include the remaining ship types into the EEDI regime. Among these efforts, it is worth noting that after testing and verification, INTERCARGO asserted that the application of the EEDI should be limited to new ships, and insisted that the EEDI 'does not apply to, and hence it cannot and should not be used for, existing ships'.[73] If this proves to be the case, the hybrid MBMs will not work due to their setting of the EEDI as their benchmark.[74] At the 66th MEPC meeting in 2014, the ICS proposed to establish a mandatory global system for monitoring and reporting fuel consumption in response to the US's proposal of enhancing technical and operational measures raised at the 65th MEPC meeting in 2013.[75] Based on this proposal, the ICS agreed to further reduce shipping GHG emissions by enhancing the energy efficiency of ships while temporarily suspending the discussion on MBMs.

The views of shipowners' and ship operators' associations on MBMs mainly focus on their necessity, principles, advantages and disadvantages of different

---

taking into account any external factors such as limitations on draught or length, or the need for having enhanced power to operate in tidal areas or across very busy straits.'

INTERFERRY, *Interferry Welcomes EEDI Decision* (12 October 2012) <http://www.marinelink.com/news/interferry-welcomes348466.aspx> accessed 25 January 2013.

72 *Comments on the Outcomes of the United Nations Climate Change Conference Held in Copenhagen*, Denmark, submitted by OCIMF and INTERTANKO, MEPC 60th Session, Agenda Item 4, IMO Doc MEPC 60/4/46 (28 January 2010) para. 4. 'Virtual Arrival' refers to a process whereby 'inefficiencies within the maritime supply chain are identified and, within mutual agreement, removed by agreeing a revised arrival time at a port and then optimizing the vessel speed to achieve that time'.

73 *Application of the EEDI to Existing Ships*, submitted by INTERCARGO, MEPC 63rd Session, Agenda Item 5, IMO Doc MEPC 63/5/12 (6 January 2012) para. 19.

74 See ch. 4, 4.3.4.2.

75 *Proposal of the United States to Enhance Energy Efficiency in International Shipping*, submitted by the United States, MEPC 65th Session, Agenda Item 4, IMO Doc MEPC 65/4/19 (8 March 2013). In this proposal the US suggested to establish attained energy efficiency standards for new and existing ships through a phased approach. This proposal aroused the interest of many countries and the 65th MEPC meeting in May 2013 thus agreed to suspend the scheduled discussion on MBMs.

schemes. As mentioned earlier in this section, shipowners' and ship operators' associations generally support the efforts of the IMO in reducing GHG emissions from ships.[76] However, in terms of the timing of regulating MBMs, the ICS asserted that an MBM for shipping is currently not appropriate due to 'the state of the global economy and the impact on shipping markets'.[77] BIMCO is of the view that MBMs 'do not appear warranted at this particular time'.[78] In February 2012, the Round Table asserted that MBMs 'are not justified at this particular time', but it also admitted that MBMs might eventually be introduced for shipping.[79] These views reveal the concerns of the shipping industry on the possible negative influence of proposed MBMs on the international shipping market. It is therefore necessary to conduct research on the possible impacts of these measures on the industry, the global supply chain and developing countries.

CLIA, INTERTANKO and the ICS submitted their proposals to the IMO on the principles for choosing and implementing MBMs to reduce GHG emissions from international shipping. At the 59th MEPC meeting, CLIA proposed three principles on this issue, namely the IMO principle of no more favourable treatment of ships (NMFT), the principle of high quality, multiple benefit carbon mitigation investments, and the principle of common but differentiated responsibilities and respective capabilities (CBDR). It explained that a framework established by the IMO for tackling climate change should respect both the NMFT and the CBDR principles. This could be accomplished by 'not unduly penaliz[ing] vessels based upon their trading routes or flag' and 'ensuring a portion of the redistributed funds are applied to those areas where a net benefit is achieved by non-Annex I parties through a market-based instrument' (MBI).[80]

---

76 This view is also agreed by the Asian Shipowners' Association (ASF). At its 24th Interim Meeting in 2011, its delegates 'fully agreed that realistic and effective measures to reduce GHG emissions should be dealt with in the IMO'. Asian Shipowners' Forum (ASF), *Note of Understanding* (2 November 2011) <http://www.jsanet.or.jp/e/pressrelease_e/2011/pdf/a20111108e.pdf> accessed 7 April 2013, para. 5.1.

77 *Operational Energy Efficiency of New and Existing Ships*, submitted by the International Chamber of Shipping (ICS), MEPC 64th Session, Agenda Item 5, IMO Doc MEPC 64/5/11 (27 July 2012) para. 11.

78 BIMCO, *GHG and Market-based Measures: BIMCO's Position* (October 2011) <https://www.bimco.org/en/About/Viewpoint/07_Greenhouse_Gases.aspx> accessed 23 January 2013.

79 Round Table of International Shipping Associations, above n. 62.

80 *Consideration of Adoption of Three Principles for Market-based Instruments*, submitted by Cruise Lines International Association (CLIA), MEPC 59th Session, Agenda Item 4, IMO Doc MEPC 59/4/32 (8 May 2009) para. 1.

INTERTANKO treated the nine IMO principles agreed by the 57th MEPC meeting as the criteria for selecting the MBMs, and asserted that a MBI should be:

1. governed by the IMO and be specific for the shipping industry;
2. effective in contributing to the reduction of total GHG emissions;
3. environmentally sustainable without negative impact on global trade and growth; and
4. efficient and credible enforcement & monitoring.[81]

Under point 4, it reiterated that a MBI 'should be binding and equally applicable to all ships'. The ICS also embraced the nine IMO principles and highlighted the status of the IMO as the only competent body to regulate MBMs, although other international administration bodies may operate on behalf of the IMO.[82] However, while insisting on the NMFT principle, it also made a compromise. It stated that,

> If there is a need to find a means to accommodate the UNFCCC principle of "common but differentiated responsibility" then this must be achieved at a level above that of the individual shipping company.[83]

It also set two priorities for the disbursement of funds generated from a MBI: a mitigation and adaption scheme, and research and development.[84]

Although the CLIA, INTERTANKO and the ICS all represent the interests of shipowners or/and ship operators, the principles put forward by them have different emphases. They all respect the role of the IMO as the regulator of any MBM, and respect the NMFT principle. However, INTERTANKO did not explicitly raise the CBDR principle because it was not included in the nine IMO principles at the 57th MEPC meeting. The CLIA underscored both the NMFT principle and the CBDR principle, and identified the means for achieving these principles. Whereas the ICS still stressed the NMFT principle, and seemed reluctant to accept the CBDR principle which can be seen from the

81 *Comments on MEPC 59/4/8 and MEPC 59/4/9 relating to the Energy Efficiency Design Index, the Ship Energy Management Plan and Possible Market-based Instruments*, submitted by INTERTANKO, MEPC 59th Session, Agenda Item 4, IMO Doc MEPC 59/4/43 (22 May 2009) para. 9.

82 *Control of Greenhouse Gas Emissions from International Maritime Transport*, submitted by the International Chamber of Shipping (ICS), MEPC 60th Session, Agenda Item 4, IMO Doc MEPC 60/4/13 (15 January 2010) para. 9.1.

83 Ibid. para. 5.

84 Ibid. para. 10.

expression 'if there is a need' as cited above. Also the two priorities identified by the ICS did not specifically mention the interests of developing countries. This view, however, is slightly different from that of the Round Table, although the ICS is a member of the latter. In a joint official statement, the Round Table asserted that the CBDR principle 'cannot be practically applied to ships in light of the very nature of international shipping operations'.[85] Instead, it preferred the NMFT principle to maintain a level playing field for international shipping. It can be deduced that within the shipowners' and ship operators' sector there has been no consensus reached on whether the CBDR principle should be applied to the reduction of GHG emissions from ships, although currently it seems opponents of this principle dominate in this sector. As discussed in Chapter 2, the principles raised by CLIA are more feasible if a compromise is to be achieved between developed countries and developing countries.

Some shipowners and ship operators associations have analysed MBM proposals. At the first GHG-WG meeting in 2008, INTERFERRY proposed a maritime carbon reduction scheme. This scheme was a type of MBM based on the cap and trade principle, and it was actually the application of the European Union Emissions Trading Scheme (EU ETS) to the context of international shipping. Since the proposed scheme was essentially an idea without specific designs, the proposal did not attract much attention from the international community. At the 60th MEPC meeting in 2010, the WSC proposed a Vessel Efficiency System (VES), which was later merged with the Leveraged Incentive Scheme proposed by Japan into a new Efficiency Incentive Scheme (EIS) in 2011. As discussed in Chapter 4, the EIS works through an International GHG Fund. Substandard ships pay contributions based on the amount of the bunker fuel consumed or purchased. However, as a hybrid MBM, the EIS treats the EEDI as the benchmark for all ships, which is arguably to be infeasible.[86] At the 62nd MEPC meeting in 2011, IPTA commented on the proposed MBM on the International GHG Fund. It asserted that, as a scheme based on a set target line, the International GHG Fund does not cap shipping activities.[87] The Fund can embrace both the NMFT and the CBDR principles by 'reserving most of the

85 Round Table of International Shipping Associations, above n. 62.

86 Harilaos N. Psaraftis, 'Market-Based Measures for Greenhouse Gas Emissions from Ships: A Review' (2012) 11(2) *WMU Journal of Maritime Affairs* 211, 222. See ch. 4, 4.3.4.2.

87 *The International Greenhouse Gas Fund—Strengths and Weaknesses*, submitted by Cyprus, Denmark, the Marshall Islands, Liberia, Nigeria, the Republic of Korea and the International Parcel Tankers Association (IPTA), MEPC 62nd Session, Agenda Item 5, IMO Doc MEPC 62/5/33 (20 May 2011) para. 4. The IPTA asserted that under this scheme, ships must pay the GHG contribution for the fuel that they consumed and those revenues will be used to buy offsets through the *UNFCCC* framework. Meanwhile ships have the incentive to reduce fuel consumption. So, it does not cap shipping activities.

revenues to climate change purposes in developing countries'.[88] Meanwhile, under this scheme, it is the ship rather than the flag State that needs to pay the GHG contribution. It also commented that this MBM is compatible with WTO's non-discrimination principle. According to the proposed GHG Fund scheme, an international convention will be established to provide requirements for the payment of a uniform GHG contribution where no discrimination is foreseen. Nevertheless, a conflict between Parties that include an element of discrimination is an indispensable element for any incompatibility with WTO rules.[89] At the 64th MEPC meeting in 2012, the ICS announced that its preferred MBM would either be based on a levy or a compensation fund and should relate to the actual fuel consumption of individual ships in service. ICS asserted that such a scheme would ensure that:

1. a level playing field is maintained;
2. serious market distortion is avoided;
3. management of the system will be easier; and
4. the desired transparency will be provided.[90]

Compared with the views from ICS, the Round Table opposed any ETS because it would be 'unworkable' for the shipping industry.[91] Similarly, the Asian Shipowners Forum (ASF) has also opposed an ETS on the grounds that it is practically less applicable, unreasonably costly[92] and might become a major source for funding non-shipping sectors.[93] However, it overestimates the functions of technical and operational measures.[94] It asserts that MBMs, such as the proposed global bunker levy and ETS, are premature so further study is

88 Ibid. para. 6.

89 Ibid. para. 19. See *General Agreement on Tariff and Trade*, opened for signature 30 October 1947, 55 UNTS 194 (in force provisionally since 1 January 1948 under the 1947 Protocol of Application, 55 UNTS 308) arts. I, III ('*GATT*').

90 *Operational Energy Efficiency of New and Existing Ships*, submitted by the International Chamber of Shipping (ICS), MEPC 64th Session, Agenda Item 5, IMO Doc MEPC 64/5/11 (27 July 2012) para. 11.

91 Round Table of International Shipping Associations, above n. 62.

92 Asian Shipowners' Forum (ASF), *The 18th ASF Joint Statement* (27 May 2009) <http://www.jsanet.or.jp/e/pressrelease_e/2009/pdf/20090529e_2.pdf> accessed 7 April 2013, p. 9.

93 Asian Shipowners' Forum (ASF), above n. 76, para. 5.1.

94 Asian Shipowners' Forum (ASF), above n. 92. In this statement, SAF asserts that substantial reduction of total GHG emissions can 'only be' achieved by technical and operational measures.

needed.[95] The ASF has not expressed any of its preferred MBMs and it seems that MBMs have still not been accepted by the Asian shipping industry.

It can be deduced that the international shipowners' and ship operators' associations are more interested in levy or compensation fund-based MBMs than an Emission Trading Scheme (ETS) and other MBM proposals. Two factors might contribute to this preference. First, being profit-oriented, the shipping industry is more concerned about the sound development of the global shipping market than the reduction of GHG emissions from ships. A levy or a compensation fund—based MBM can achieve this goal more easily than a global ETS. This is because the ETS sets a sector-wide cap on net emissions from international shipping, which may limit the development of international shipping and lead to carbon leakage.[96] Second, the regulation of a levy or a compensation fund—based MBM may result in the reduction of the shipping sector (in-sector reduction). This is consistent with the competence of the IMO. On the other hand, a global, cross-sectoral emissions trading market is needed in order to regulate an ETS and avoid carbon leakage. This is beyond the competence of the IMO.[97] For this reason and to maintain the status of the IMO as the only regulator of a MBM, the global shipowners' and ship operators' associations prefer a levy or a compensation fund—based MBM. Surprisingly, while international shipowners' and ship operators' associations oppose an ETS, some national ship owners' associations in Europe, such as Germany and Norway, support an ETS being applied to the shipping industry.[98] This disparity of views reveals the complexity of this issue, and indicates that further study is needed to enhance understanding of the issues so as to find a solution.

In summary, the international shipowners and ship operators generally welcome the IMO's efforts in reducing GHG emissions from international shipping. Regarding the proposed technical and operational measures, they contributed a great deal to the improvement of the EEDI and SEEMP through their proposals to the IMO. They take the view that they should have the

---

95 Asian Shipowners' Forum (ASF), *Press Release* (15 September 2008) <http://www.jsanet.or.jp/e/pressrelease_e/2008/pdf/20080930.pdf> accessed 7 April 2013, p. 2.

96 Carbon leakage generally refers to differentiated carbon policies and their subsequent impacts on GHG emissions. In this context, reduction of GHG emissions from international shipping may contribute to the faster development of other transportation means (e.g., road transportation or aviation) if similar policies do not apply in these sectors. See also ch. 4, 4.3.4.2.

97 *Emission "Caps" and Reduction Targets*, submitted by the World Shipping Council (WSC), MEPC 60th Session, Agenda Item 4, IMO Doc MEPC 60/4/28 (15 January 2010) para. 9.

98 Psaraftis, above n. 86, 231. But, the Greek shipping industry has been against an ETS. See ch. 5, 5.3.2.

freedom to choose their compliance mechanisms, and that specific technical requirements, including the Operational Index, should not be regulated in a prescriptive way. They suggested that the EEDI only cover new ships of certain types and be implemented in four phases. Currently they are working to include the unregulated types of ships into the EEDI regime and provide options for enhancing energy efficiency measures. As to the proposed MBMs, international shipowners and ship operators insist on the NMFT principle and assert that the MBM should be governed by the IMO. However, some of them also accept the application of the CBDR principle to the regulation of GHG emissions from international shipping, although the Round Table opposes its application. Of the current MBM options, most of the shipowners' and ship operators' organisations prefer a levy or a compensation fund—based MBM and oppose an ETS.

5.2.3 *Cargo Owners*

Cargo owners are exempt from any compensation liability to pollution victims that results from incidents involving their cargos. This responsibility is generally attributed to the shipowners or ship operators due to their direct operational role.[99] It is believed that this liability arrangement originates from the strong influence of the cargo owners, particularly oil companies. In the 1960s and 1970s, pollution-related regulations in various States, including preventive and remedial measures and damage compensation, were generally imposed upon tanker owners.[100] At present, although some of the cargo owners have their own ships, most of them use 'spot' charters and tend to employ substandard operators so as to avoid the fluctuating freight rates and minimise transportation costs.[101] In the context of international shipping, as discussed in Chapter 1, cargo emissions and emissions from cargo ships constitute important sources of GHG emissions from ships.[102] For this reason, cargo owners are an important stakeholder in the reduction of GHG emissions from ships. It is debated whether cargo owners should share the regulatory costs, including preventive and remedial measures and damage compensation for pollution victims, with shipowners and ship operators, due to their choice of employing substandard ships.[103] However, because of the contribution of cargo ships to

99 Tan, above n. 32, 38; Pamborides, above n. 34, 144–145.

100 Tan, above n. 32, 39.

101 Ibid. 40.

102 See ch. 1, 1.2.2.1.

103 Pamborides, above n. 34, 145.

GHG emissions, it is necessary to examine the response of the cargo owners to the reduction of GHG emissions from international shipping.

The Oil Companies International Marine Forum (OCIMF) is the main NGO which represents the interests of cargo owners within the IMO, particularly those from 93 oil companies worldwide.[104] Similar to international shipowners and ship operators associations, OCIMF welcomes the efforts of the IMO in reducing GHG emissions from ships. It insists that the IMO is the sole competent international organisation to regulate GHG emissions from ships. To justify the exclusion of the CBDR's application to this issue, it supports the legal advice from the IMO's Sub-Division for Legal Affairs in document MEPC 58/4/20 that IMO's mandate in regulating shipping GHG emissions comes from the *IMO Convention* itself to regulate all aspects of international shipping rather than the *UNFCCC*.[105]

The OCIMF prepared most of its proposals on EEDI and SEEMP in collaboration with international shipowners and ship operators associations due to its lack of expertise in the technical and operational measures. For instance, it asserted that different methodologies for shipping efficiency assessment should be required by different owners to suit the needs of different trades, thus the operational index should not be mandatory.[106] It believed that SEEMP measures should be area and trade dependent.[107] At the 59th MEPC meeting in 2009, the OCIMF asserted that 'operational and technical improvements will be gradual rather than revolutionary' due to the long-lived shipping fleet.[108] This indicates that it prefers a step-by-step approach to proposed technical and operational measures. Moreover, the OCIMF has supported the trial of the proposed EEDI and SEEMP measures. It applied a SEEMP to improve the

104 Oil Companies International Marine Forum (OCIMF), *Introduction* (2010) <http://www.ocimf.com/Organisation/Introduction> accessed 26 January 2013.

105 *Comments related to the Outcome of Informal Consultations Conducted by the Chairman and Proposals for Further Progress*, submitted by Oil Companies International Marine Forum (OCIMF), MEPC 59th Session, Agenda Item 4, IMO Doc MEPC 59/4/46 (22 May 2009) para. 8. See also ch. 2, 2.5.3.1.

106 *Guidelines for the Implementation of the Ship Operational Index—Ship Efficiency Management Tool*, submitted by INTERTANKO, OCIMF and BIMCO, MEPC 58th Session, Agenda Item 4, IMO Doc MEPC 58/4/13 (1 August 2008) para. 2,3.

107 *Ship Efficiency Management Plan*, submitted by ICS, BIMCO, Intercargo, Intertanko and OCIMF, MEPC 58th Session, Agenda Item 4, IMO Doc MEPC 58/INF.7 (28 July 2008) para. 3.35.

108 *Technical Evaluation of Market-based Instruments*, submitted by Oil Companies International Marine Forum (OCIMF), MEPC 59th Session, Agenda Item 4, IMO Doc MEPC 59/4/17 (7 May 2009) para. 4.

energy efficiency of a fleet of Very Large Crude Carriers (VLCC) and claimed that compared with 2010, emissions of CO_2 from that fleet reduced by 10 per cent.[109]

The OCIMF has expressed its views on the assessment principles of the proposed MBMs and commented on the options. It believes that the nine principles for shipping GHG legislation agreed by the 57th MEPC meeting should form the basis for any GHG reduction measures.[110] Any selected MBMs should have the direct effect of reducing GHG emissions from the shipping industry without restricting world trade and leading to carbon leakage.[111] In particular, it asserts that 'a substantial proportion of the revenue generated [50%], from any [MBM]' should be utilised to 'promote and facilitate marine R&D aimed at reducing shipping GHG emissions'.[112] These views emphasise the potential reduction of GHG emissions from ships but ignore the CBDR principle insisted on by the shipping industries from developing countries. The shipping industries in many developing countries are more concerned about their market share than the reduction of GHG emissions and they generally lack energy efficient technologies. Therefore, it is difficult for developing countries to accept these views.

The OCIMF has not explicitly expressed its preference among current MBM proposals. Instead, it examines the advantages and disadvantages of MBMs such as the ETS, GHG Compensation Fund, Leveraged Incentive Scheme and Efficiency Standards Index. At the 59th MEPC meeting in 2009, it commented that both the ETS and the GHG Fund have challenges in their design and will inevitably lead to an increase in cost to consumers. It also commented

---

109 *Project to Develop a SEEMP Using a Structured Methodology and the Resulting Improvement in Energy Efficiency*, submitted by Oil Companies International Marine Forum (OCIMF), MEPC 62nd Session, Agenda Item 5, IMO Doc MEPC 62/INF.12 (6 May 2011) para. 10.

110 *Technical Evaluation of Market-based Instruments*, submitted by Oil Companies International Marine Forum (OCIMF), MEPC 59th Session, Agenda Item 4, IMO Doc MEPC 59/4/17 (7 May 2009) para. 18; *Updated Technical Evaluation of Proposed GHG Reduction Measures*, submitted by Oil Companies International Marine Forum (OCIMF), MEPC 60th Session, Agenda Item 4, IMO Doc MEPC 60/4/50 (29 January 2010) para. 3.

111 *Technical Evaluation of Market-based Instruments*, submitted by Oil Companies International Marine Forum (OCIMF), MEPC 59th Session, Agenda Item 4, IMO Doc MEPC 59/4/17 (7 May 2009) para. 6.4, 17; *Updated Technical Evaluation of Proposed GHG Reduction Measures*, submitted by Oil Companies International Marine Forum (OCIMF), MEPC 60th Session, Agenda Item 4, IMO Doc MEPC 60/4/50 (29 January 2010) para. 14.

112 *Technical Evaluation of Market-based Instruments*, submitted by Oil Companies International Marine Forum (OCIMF), MEPC 59th Session, Agenda Item 4, IMO Doc MEPC 59/4/17 (7 May 2009) para. 7.4.

that the Compensation Fund proposal lacks a linkage to the market price of Carbon, while the Marine ETS will impose a heavy administrative burden on shipowners.[113] At the 60th MEPC meeting in 2010, it compared the four MBM proposals and concluded that these proposals require further development and it would be premature to judge whether they meet the nine IMO fundamental principles.[114]

It is concluded that international cargo owners, the OCIMF being the principal representative body, generally support the work of the IMO in reducing GHG emissions from international shipping. They believe that IMO is the only competent international organisation to regulate this GHG emissions issue, but favour exclusion of the CBDR principle. This obviation of the interests of developing countries probably results from the dominance of developed countries within the organisation.[115] In their opinion, the nine IMO fundamental principles are the only criteria for assessing any GHG reduction measures. They welcome the proposed technical and operational measures and have contributed to their improvement together with international shipowners and ship operators associations. To date they have not expressed their preference for any MBM, but they have asserted that further development of the MBM proposals is required.

5.2.4 *Ship Insurers*

The ship insurers are probably the only parties among stakeholders of the shipping industry who do not welcome any upgrading of ship standards on the ground that it might increase the number of claims.[116] Generally there are two types of risks against which shipowners or ship operators insure their operations, namely hull and machinery (H&M) and third party liability risks.[117] H&M coverage deals with the damage to the shipowner's ship through possible collisions, groundings and other accidents, whereas third party liability coverage is to compensate the shipowner against claims by third parties for damage incurred to their interests due to the operation of the ship.[118]

113 Ibid. para. 15.

114 *Updated Technical Evaluation of Proposed GHG Reduction Measures*, submitted by Oil Companies International Marine Forum (OCIMF), MEPC 60th Session, Agenda Item 4, IMO Doc MEPC 60/4/50 (29 January 2010) para. 17.

115 Tan, above n. 32, 40.

116 Pamborides, above n. 34, 141.

117 Tan, above n. 32, 40.

118 Ibid.

The shipping insurance industry is represented at the IMO mainly by the International Union of Marine Insurance (IUMI) and the International Group of P&I Associations (P&I Clubs).[119] As a member of BIMCO, P&I Clubs might have expressed its views on the reduction of shipping GHG emissions through BIMCO. However, as independent NGOs, the IUMI and P&I Clubs have not submitted any proposal to the IMO on the issues under discussion. This is probably because it would not be viable for the ship insurers to be involved in the reduction of GHG emissions from international shipping, including the EEDI, SEEMP and proposed MBMs. As discussed in Chapter 2, the transboundary harm resulting from GHG emissions from ships is a cumulative process, which in any case would not lead to actual damage either to the shipowner's ship or to any third party in the short term. In this case, it would only be possible for ship insurers to be involved in the reduction of GHG emissions from ships when an innovation or significant change in marine insurance policy occurs. This is currently a remote possibility.

5.2.5 *Classification Societies*

The classification society is a significant sector of the shipping industry. It plays an important role by providing various classification and statutory services and assistance, which cover almost all technical aspects of shipping.[120] The role of classification and classification societies has been recognised by a number of international conventions. Article 94 of the *1982 United Nations Convention on the Law of the Sea (LOSC)* requires a flag State to 'effectively exercise its jurisdiction and control in administrative, technical and social matters over ships flying its flag' and take 'such measures for ships flying its flag as are necessary to ensure safety at sea'.[121] These requirements are generally referred to as 'statutory requirements', which range from the ship's design and its structural integrity to pollution control, accident prevention and emergency handling.[122]

119 Currently IUMI has 55 national associations of hull and cargo insurers from all over the world, while P&I Clubs mainly deal with third party liability insurance and other types of insurance, and members of the Group cover about 90 per cent of world shipping tonnage. International Union of Marine Insurance (IUMI), *Membership* <http://www.iumi.com/about-iumi/membership> accessed 27 January 2013; Tan, above n. 32, 41.

120 Pamborides, above n. 34, 143.

121 *United Nations Convention on the Law of the Sea*, opened for signature 10 December 1982, 1833 UNTS 3 (entered into force 16 November 1994) arts. 94.1, 94.3 ('*LOSC*').

122 IACS, *Classification Societies: What, Why and How?* (March 2011) <http://www.iacs.org.uk/document/public/explained/class_whatwhy&how.pdf> accessed 18 January 2013.

The *International Convention for the Safety of Life at Sea (SOLAS)*,[123] the *1988 Protocol to the International Convention on Load Lines*,[124] and some other international conventions permit the flag State to delegate the verification of ships to a Recognised Organisation (RO) to verify whether the ship has met these requirements. These conventions also recognise the classification society as a competent RO which can be nominated by flag States.

The International Association of Classification Societies (IACS) was established by seven major societies in 1968. Currently the IACS, and two of its 13 members, Lloyd's Register and the Royal Institution of Naval Architects (RINA), are the main classification societies involved in the IMO's work of reducing GHG emissions from ships. However, due to their expertise in technical aspects of shipping, the work of classification societies mainly contributes to the verification, safety and interpretation of the proposed EEDI. The International Organization for Standardization (ISO) and International Towing Tank Conference (ITTC), a type of RO in this context,[125] also participated in some of this work.

The IACS treats the IMO's work in addressing GHG emissions from ships as its highest priority and has been contributing to the advancement of this work within the IMO.[126] Regarding the verification of the proposed EEDI, the IACS provided a methodology for a CO_2 design index for new ships at the 58th MEPC meeting, and suggested that the verification process of the EEDI be performed in two phases: data examination and sea trials verification at the 59th MEPC meeting. The ITTC strongly supported the EEDI, but asserted that 'it is

123 *International Convention for the Safety of Life at Sea*, opened for signature 1 November 1974, 1184 UNTS 2 (entered into force 25 May 1980) (SOLAS) Ch. II-1, reg. 3–1. Regulation 3-1 reads that,

> 'In addition to the requirements contained elsewhere in the present regulations, ships shall be designed, constructed and maintained in compliance with the structural, mechanical and electrical requirements of a classification society which is recognized by the Administration in accordance with the provisions of regulation XI-1/1, or with applicable national standards of the Administration which provide an equivalent level of safety.'

124 *International Convention on Load Lines*, opened for signature 5 April 1966, 640 UNTS 133 (entered into force 21 July 1968) (Protocol of 11 November 1988, entered into force 3 February 2000) art. III.

125 The report of MEPC 59 ascertained that the verifiers of the EEDI could be administrations, classification societies, and any other RO, including the ITTC.

126 *Report of the Marine Environment Protection Committee on its 58th Session*, IMO Doc MEPC 58/23 (16 October 2008) annex 9, p. 25.

not a substitute for taking action in reducing emissions from the existing fleet'.[127] The ITTC also proposed a verification and approval process for the proposed EEDI. Meanwhile, RINA, a member of IACS, also proposed a modification to the EEDI to improve its application scope and monitoring methodology at the second GHG-WG meeting in 2009. To ensure the safety of navigation in adverse conditions, at the 61st MEPC meeting in 2010 the IACS proposed adding a paragraph to Regulation 4 of the draft legal text for the required EEDI, stressing the need to maintain a minimum speed requirement. This proposal was later adopted by IMO without any further modification. Furthermore, after the adoption of EEDI by Annex VI to *MARPOL 73/78*, the IACS provided a unified interpretation of the terms 'new ships' and 'major conversion' and clarified the relationship between the International Energy Efficiency Certificate (IEE Certificate) and the International Air Pollution Prevention Certificate (IAPP) as requested by the 63rd MEPC meeting in 2011. The next year the IACS agreed to strengthen the implementation of the EEDI by using the industry developed guidelines as a Procedural Requirement for its members. At the 65th and 66th MEPC meetings, the IACS proposed to establish an 'EEDI database' so as to facilitate reviews of the EEDI as regulated under Regulation 21.6 of the 2011 amendments to Annex VI.[128] In this way, the IACS has been significantly contributing to the development, interpretation and implementation of the EEDI.

---

127 *Proposal for an Energy Efficiency Design Index Verification Process*, submitted by International Towing Tank Conference (ITTC), MEPC 60th Session, Agenda Item 4, IMO Doc MEPC 60/4/45 (25 January 2010) para. 2.

128 *MARPOL 73/78 Annex VI* (*2011 amendments*) reg. 21.6. This regulation provides,

'At the beginning of Phase 1 and at the midpoint of Phase 2, the Organization shall review the status of technological developments and, if proven necessary, amend the time periods, the EEDI reference line parameters for relevant ship types and reduction rates set out in this regulation.'

Establishment of An 'EEDI Database', submitted by the International Association of Classification Societies (IACS), MEPC 65th Session, Agenda Item 4, IMO Doc MEPC 65/4/31 (8 March 2013); *Establishment of An 'EEDI Database'*, submitted by Liberia, IACS, and INTERCARGO, MEPC 66th Session, Agenda Item 4, IMO Doc MEPC 66/4/13 (24 January 2014).

However, the proposal of IACS on developing an EEDI database was not agreed by the 66th MEPC meeting in 2014. Some countries were concerned about the protection of intellectual property rights and commercially sensitive information, whereas other countries asserted that the database may be partially open due to the confidentiality of some information. *Report of the Marine Environment Protection Committee on Its Sixty-sixth Session*, MEPC 66th Session, Agenda Item 21, IMO Doc MEPC 66/21 (25 April 2014) para. 4.35.

The Lloyd's Register is a member of the IACS. It drafted an assessment framework for the Institute of Marine Engineering Science and Technology (IMarEst) aiming to provide a mechanism to analyse various MBM proposals. However, classification societies have not been involved in MBM-related work.

5.2.6 *Bunker Suppliers*

Bunker suppliers generally are not a main member of the shipping industry. Nevertheless, through providing lower emitting fuels to ships and participating in the discussion of MBMs, bunker suppliers have become one of the stakeholders in the work on reducing GHG emissions from ships. At the IMO the main representatives for this sector are the International Bunker Industry Association (IBIA) and the International Petroleum Industry Environmental Conservation Association (IPITCA). Among the proposals that they have submitted to the IMO, the one that IBIA submitted in 2008 is relevant to the reduction of GHG emissions from ships.

At the 58th MEPC meeting in 2008, the IBIA predicted that with the projected growth in the shipping sector there would be an absolute increase in CO_2 emitted by the industry between 2008 and 2020 and thus supported the reduction effort by the international community.[129] It believed that proposed design indexes could guide shipbuilders and ship buyers to reduce GHG emissions.[130] The IBIA also explicitly expressed its preference for the CO_2 Cap and Trade scheme to the shipping industry and put forward principles for the proposed mechanisms, including setting a shipping sector CO_2 cap (a cap on total CO_2 emissions), allocating sector emissions entitlement under the cap, and establishing a global registry. Indeed, this scheme is a type of ETS, and the IBTA suggested that the global registry should be operated by the bunker industry and shipping sector so as to 'retain control of [their] investments and of the scheme integrity on a global basis'.[131]

5.2.7 *Conclusion*

In addition to the above discussions from six organisational perspectives, other international and regional shipping organisations have also expressed their views on the regulation of GHG emissions from international shipping.

129 *Response to the Outcome of the First Intersessional Meeting of the Working Group on Greenhouse Gas Emissions from Ships*, submitted by the International Bunker Industry Association (IBIA), MEPC 58th Session, Agenda Item 4, IMO Doc MEPC 58/4/19 (1 August 2008) para. 3.

130 Ibid. 2.

131 Ibid. 17.

For instance, the Clean Shipping Coalition (CSC) argues that 'speed reduction should be pursued by the IMO as a regulatory option in its own right and not just as a possible consequence of MBIs or the EEDI.'[132] This proposal has not attracted much attention at the IMO because speed reduction is often treated as one of the operational measures or a part of the SEEMP. Some industry groups have also been established to cope with the challenges that they are facing in reducing shipping GHG emissions. For example, the Tripartite Working Group was established in 2008 and the Joint Industry Working Group (JWG) on the EEDI was organised in 2010. The JWG drafted industry guidelines on calculation and verification of the EEDI based on IMO resolutions and submitted it to the IMO for further discussion. The Tripartite Working Group also conducted joint research and expressed the views of its members as a whole.

It is concluded that generally international and regional shipping organisations welcome the IMO's efforts in reducing GHG emissions from international shipping, and assert that the IMO is the only competent organisation to regulate the issue. However, the various stakeholders in the global shipping industry have not achieved consensus on how to reduce GHG emissions from ships. Based on current discussions within the IMO, their views mainly reflect the following four perspectives. Firstly, stakeholders of the global shipping industry have different preferences for the proposed technical and operational measures and MBMs. Generally ship designers and shipbuilders, the CESA as an example, prefer operational measures (SEEMP) to technical measures (EEDI) because they believe that the SEEMP will be more effective than the EEDI. They also highlight the importance of MBMs. Other stakeholders, such as shipowners, ship operators and cargo owners, welcome proposed technical and operational measures and MBMs and have contributed to the improvement of these measures through their proposals to the IMO. Secondly, most stakeholders, in particular shipowners and ship operators, support the freedom to choose their compliance mechanisms in relation to the EEDI and SEEMP, and suggest that the issues not be regulated in a prescriptive way. Thirdly, most stakeholders insist on the application of the NMFT principle to the issues and exclude the application of the CBDR principle. Nevertheless, as an international shipowners' association, the ICS accepts the application of the CBDR principle to the GHG emissions issue provided that this principle does not apply directly to individual shipping companies. This compromise, however, has much to do with the presence of shipowners' associations from developing countries in

---

132 *Speed Reduction—the Key to the Fast and Efficient Reduction of Greenhouse Gas Emissions from Ships*, submitted by the Clean Shipping Coalition (CSC), MEPC 61st Session, Agenda Item 5, IMO Doc MEPC 61/5/10 (23 July 2010) executive summary.

the ICS. Fourthly, stakeholders of the global shipping industry have different preferences for the proposed MBMs. Shipowners and ship operators prefer a levy or compensation fund-based MBM and oppose any ETS, whereas the bunker suppliers support the application of an ETS to the shipping sector.

5.3 Response from the Shipping Industry in the *UNFCCC* Annex I States

The development of the shipping industries in different countries, particularly between developed and developing countries, has been imbalanced due to historical, economic and geographic factors. Many of the shipping organisations from developing countries are not members of international and regional shipping organisations. Accordingly, the response from global shipping organisations analysed in the previous sections of this chapter does not completely represent the views from the shipping sectors in all countries. Therefore, it is necessary to examine the response from the individual shipping sectors in some of the developed and developing countries to the issue of GHG emissions from international shipping.

There are no uniform definitions or lists of what constitutes developed countries and developing countries. A number of lists are used by various organisations for different purposes. The World Bank, the United Nations Development Programme (UNDP), the World Factbook produced by the US Central Intelligence Agency (CIA), the American Mathematical Society (AMS), and the Organisation for Economic Co-operation and Development (OECD) all have their own definitions or lists of developed countries and developing countries. Thus in the context of climate change, based on different criteria, *UNFCCC* Annex I States are not all developed States and *UNFCCC* non-Annex I States may not be widely-recognised developing States. For example, Turkey is a *UNFCCC* Annex I State, but it is identified as a developing State by the World Bank, AMS and OECD.[133] Singapore, the Republic of Korea and Cyprus are all non-Annex I States, but they are identified by the World Bank as developed States.[134] However, since the classification of Annex I States and non-Annex I

133 The World Bank, *Country and Leading Groups* (2012) <http://data.worldbank.org/about/country-classifications/country-and-lending-groups> accessed 24 March 2013; American Mathematical Society, *Developing Countries List* <http://www.ams.org/membership/individual/types/mem-develop> accessed 24 March 2013; OECD, *List of Developing Countries in Alphabetical Order* <http://www.icml9.org/public/documents/pdf/es/OECD.pdf> accessed 24 March 2014.

134 The World Bank, ibid.

States has been adopted by the UNFCCC and its *Kyoto Protocol* (including its eight-year second commitment period which commenced on 1 January 2013), it can be deduced that the international community has generally recognised these categories of country groups for the purpose of bearing different responsibilities in tackling global climate change. For the purpose of this book, this classification of different countries is utilised to examine the response from the shipping industries in individual States. The States examined as examples of Annex I States are Australia, Greece and the United Kingdom, while China, the Republic of Korea and India are analysed as examples of non-Annex I States.

5.3.1 *Australia*

Australia is a UNFCCC Annex I State surrounded by sea and all of its large cities are coastal. International shipping plays a significant role in Australia's economy because most of Australia's imports and exports are moved by ships.[135] Australia is the world's fifth largest shipping nation in terms of tonnes of cargo shipped and kilometres travelled.[136] In 2013–2014, 1221.8 million tonnes of goods valued at $243.6 billion were exported by international shipping to other countries, while 99.1 million tonnes of goods valued at $198.1 billion were imported to Australia by sea.[137] During this period the value of Australia's exports and imports by sea increased by 10.1 per cent and 7.4 per cent respectively on 2012–2013, whereas the weight of its exports and imports by sea increased by 14.2 per cent and 0.1 per cent respectively.[138]

The Australian trading fleet is relatively small in contrast to the vital role of international shipping in Australia's economy. In 2013–2014, the Australian trading fleet was comprised of 109 ships with a total 4.4 million deadweight tonnage (dwt).[139] Meanwhile, the average age of ships in the trading fleet was 14.8 years during this period.[140] By comparison, in January 2014, globally

135 Craig Forrest, 'Shipping and the Marine Environment in Australia' in Warwick Gullett, Clive Schofield and Joanna Vince (eds.), *Marine Resources Management* (LexisNexis Butterworths, 2011) 123, 123.

136 Australian Maritime Safety Authority, *Vessels* (2013) <https://www.amsa.gov.au/vessels/> accessed 1 June 2014.

137 Australian Government: Department of Infrastructure and Transport, 'Australian Sea Freight 2013–14' (2015) <http://bitre.gov.au/publications/2015/files/asf_2013_14.pdf> accessed 12 May 2016, p. vii.

138 Ibid.

139 Ibid. viii.

140 Ibid. However, the average age of ships in Australian trading fleet was 14.7 years in 2012–2013 and 16.7 years five years ago.

47,601 seagoing vessels were in service with 1,676 million dwt, and the global average age per ship was 20.18 years.[141] The size of Australia's shipping industry is small but the industry has been growing. However, more than 99 per cent of Australia's international trade is carried by foreign-flagged vessels,[142] which some analysts have assessed as being harmful to the development of Australia's shipping industry.[143] For these reasons, in September 2011, the Australian Department of Infrastructure and Transport announced a 'Stronger Shipping for a Stronger Economy' shipping reform package which commenced on 1 July 2012. Consequently, an Australian International Shipping Register (AISR) was established to increase Australia's involvement in the international shipping trade by offering Australian registration which is believed to be globally competitive and highly regarded.[144] In addition to the Australian government's efforts in promoting its shipping industry, in recent years Australia has also made rapid progress in tackling climate change. Examples include its ratification of the *Kyoto Protocol* in 2007 and the adoption of a carbon pricing scheme (commonly referred to as a 'Carbon Tax') in July 2012.[145] The shipping industry in Australia contributed to these achievements by means of participation in the discussion and debate.

---

141 United Nations Conference on Trade and Development (UNCTAD), 'Review of Maritime Transport 2014' (2014) <http://unctad.org/en/PublicationsLibrary/rmt2014_en.pdf> accessed 12 May 2016, pp. 31, 37.

142 The Parliament of the Commonwealth of Australia, *Rebuilding Australia's Coastal Shipping Industry: Inquiry into Coastal Shipping Policy and Regulation* (October 2008) <http://www.aph.gov.au/Parliamentary_Business/Committees/House_of_Representatives_Committees?url=itrdlg/coastalshipping/report.htm> accessed 12 May 2016, p. 8.

143 Ibid. Foreword.

144 Australian Shipowners Association, *Reforming Australia's Shipping: Stakeholder Discussion Paper* (31 January 2011) <http://www.asa.com.au/wp-content/uploads/ASA-Submssion-Shipping-Reform-Discussion-Paper.pdf> accessed 28 March 2013, p. 5.

145 The carbon pricing scheme was introduced by the Gillard Government on 1 July 2012. Under the mechanism, around 500 of the country's largest polluters will be required to pay for each tonne of pollution they emit. The price starts at $23 per tonne and will rise at 2.5 per cent per annum in real terms. On 1 July 2015, the carbon price will transition to a fully flexible price under an emissions trading scheme where price will be determined by the market. So, the carbon tax will be an ETS applying to all sectors in the long term. Nevertheless, the Abbott Government abolished this carbon pricing scheme on 17 July 2014. See Lenore Tayor, *Australia Kills Off Carbon Tax* (17 July 2014) <http://www.theguardian.com/environment/2014/jul/17/australia-kills-off-carbon-tax> accessed 17 July 2014. See also ch. 2, 2.1.2.2.

The Australian Federal Government is responsible for regulating ships engaged in international trade.[146] However, it has not responded quickly to the issue of regulating GHG emissions from international shipping. The Australian shipping industry, represented by various shipping organisations, has expressed its views on this GHG emissions issue. The main shipping associations in Australia are the Australian Shipowners Association (ASA), Shipping Australia Limited (SAL),[147] Maritime Union of Australia (MUA) and the Australian Peak Shippers Association Inc. (APSAI). These organisations generally support the efforts of the IMO in regulating GHG emissions from ships. In a document submitted by the ASA to the Australian Federal Government, the ASA has agreed to the nine principles for the development of regulating GHG emissions from ships agreed at the 57th MEPC meeting of the IMO. It agreed that these measures should be 'binding and include all flag States', and should be 'goal-based and not prescribe particular methods'.[148] Regarding the mandatory energy efficiency measures adopted by the IMO in 2011, many members of SAL have incorporated the EEDI into their new building designs and adopted the SEEMP prior to its implementation which commenced on 1 January 2013.[149] SAL believes that these measures will achieve a 20 per cent CO_2 emissions reduction from ships by 2020 and this number will increase to 50 per cent by 2030.[150]

The ASA has actively responded to the proposed MBMs. Firstly, it agrees that a MBM will be beneficial for the reduction of GHG emissions from international shipping, but it also asserts that it will be difficult to adopt appropriate MBMs.[151] Secondly, it purports that any MBM to be adopted must be 'flag

146 Forrest, above n. 135, 125–126. Regarding the division of powers to regulate ships and shipping between the Commonwealth and States, the Commonwealth is responsible for the regulation of all trading ships on an interstate or overseas voyage, whereas each State regulates the water off its coastline up to a limit of three nautical miles from the territorial sea baseline, and other jurisdictions authorised by relevant federal acts.

147 Shipping Australia Limited was formed in 2001 as a result of a merger between the Australian Chamber of Shipping (ACS) and Liner Shipping Services Ltd (LSSL).

148 Australian Shipowners Association, *Proposed Carbon Tax Considerations and Implications for the Australian Shipping Industry* (10 May 2011) <http://www.asa.com.au/wp-content/uploads/ASA-Submission-Carbon-Price-May-2011.pdf> accessed 29 March 2013, p. 7.

149 Shipping Australia Limited, *Shipping Australia Supports Studies into Ship Emissions in the Australian Region* (31 August 2012) <http://shippingaustralia.com.au/wp-content/uploads/2012/01/20123108ShippingAustraliasupportsstudiesintoshipemissionsintheAustralianregion.pdf> accessed 29 March 2013.

150 Ibid.

151 Australian Shipowners Association, above n. 148.

neutral' so as to avoid market distortions and maximise its effectiveness in reducing emissions.[152] Thirdly, it suggests that the engagement of the shipping industry in discussions on the proposed MBMs is vital to ensure that any MBM to be adopted is workable and effective.[153] Fourthly, it categorises the proposed MBMs into two types, namely, 'cap and trade' Emission Trading Schemes (ETS) and a carbon levy or tax system which is linkable to a Fund to help emissions reductions.[154] Furthermore, it expresses its preference for an ETS to be applied to the reduction of GHG emissions from ships. In a discussion paper submitted by the ASA, co-sponsored by the national associations of Belgium, Norway, Sweden and the United Kingdom, the ASA explains why a global ETS will be effective in reducing shipping GHG emissions, and how such a mechanism can work in practice at the shipowner/operator level.[155] It lists five key advantages of an ETS. Namely an ETS:

- provides for certainty of environmental outcome;
- allows the market to set the price of carbon;
- allows the shipping company to find the most cost-effective solutions;
- resonates with other legislative developments around the world; and
- fits well with other existing carbon reduction infrastructure, such as Clean Development Mechanisms and Joint Implementation processes under the Kyoto Protocol.[156]

In illustrating the working mechanism of the cap-and-trade scheme, the ASA put forward a two-step strategy for the implementation of an ETS. The first step is to agree on the baseline, which represents the total emissions from international shipping at a given point in time. Then, the IMO, possibly in conjunction with the *UNFCCC*, sets the cap by reference to the baseline.[157] The second step is the allocation of allowances, or in other words, the obtaining of carbon credits. During this process, monitoring, reporting and verification (MRV) are vital.[158] In this way, an absolute emission reduction could be achieved via the cap by

152 Ibid.

153 Ibid.

154 Ibid.

155 Australian Shipowners Association et al., *A Global Cap-and-Trade System to Reduce Carbon Emissions from International Shipping* (2009) <http://www.asa.com.au/wp-content/uploads/Joint-Industry-ETS-Discussion-PapervFINAL1.pdf> accessed 29 March 2013, p. 3.

156 Ibid. 5.

157 Ibid. 9.

158 Ibid. 9–19.

means of the trading of carbon credits. In this document, the ASA realises that the setting of a global cap requires the collaboration of the IMO and the *UNFCCC*, which would be a pragmatic way of dealing with this issue. This is probably because coordinating different global sectors might be beyond the mandate and competence of the IMO[159] that is generally responsible for regulating maritime issues within the shipping sector. The ASA does not exclude the application of the CBDR principle to the issue. Instead, it recognises that the CBDR principle may need to be reflected in an ETS for shipping, and in this case 'there are more sophisticated ways of meeting this principle than by simply excluding ships of a given flag'.[160]

Similarly to the ASA, the MUA also favours an ETS for the reduction of GHG emissions from ships. In particular, the MUA proposes the inclusion of transport fuels in the ETS scheme, which imposes reduction obligations on upstream fuel suppliers.[161] Additionally, MUA suggests that an ETS for the international shipping industry not be regulated under the *Kyoto Protocol* or its amendments.[162] This view is consistent with those of international and regional shipping associations, and reveals their support for the leading role of the IMO in regulating GHG emissions from international shipping.

It is concluded that the shipping industry in Australia welcomes the mandatory technical and operational measures adopted by Annex VI to *MARPOL 73/78*. Many shipping companies had incorporated them into their building designs and operational plans prior to the commencement of these measures on 1 January 2013. This response coheres with the recently-launched shipping revival reform by the Australian government, indicating that both the shipping industry and Australian government are paying attention to the development of their engagement in international shipping. As for the proposed MBMs to tackle GHG emissions from ships, the Australian shipping industry prefers a cap-and-trade ETS and has explored how such a system could work at the shipowner or ship operator level. In order to achieve absolute GHG emissions reduction from ships, the shipping industry in Australia insists on the application of the NMFT principle to the regulation of GHG emissions from international shipping. Nevertheless, it does not exclude the possible application of the CBDR principle. Compared with the rigid positions of some other shipping

159 An ETS involves the reduction of GHG emissions from both in and outside the shipping sector. See ch. 4, 4.3.4.2.

160 Australian Shipowners Association et al, above n. 155, 8.

161 Timothy Nast, 'The Reponse of the International Shipping Industry to Global Climate Change' (2013) 44(1) *Journal of Maritime Law and Commence* 29, 39.

162 Ibid. 40.

organisations, the flexibility of views reflected in Australia's shipping industry might lead to more fruitful negotiations with developing countries on furthering the emissions reduction of international shipping in international fora.

5.3.2 *Greece*

Greece is a country with a long and distinguished maritime tradition.[163] It has been prominent in the world maritime industry for decades. Greek shipowners have established the largest merchant fleet in the world. As of 1 January 2015, Greek interests controlled 4,017 vessels of various categories, comprising 279 million total deadweight tonnage (dwt).[164] This accounts for about 8.28 per cent of the world's total number of vessels in service and 16.11 per cent of the world fleet dwt, making Greece the country which owns the largest fleets in the world.[165] Meanwhile, Greece is also the seventh flag State with the largest registered fleets.[166] The data indicate that Greece plays an important role in international shipping. The response from its shipping industry is significant for the efforts of the international community in reducing GHG emissions from ships.

The main shipping associations in Greece are the Union of Greek Shipowners (UGS), the Hellenic Chamber of Shipping (HCS), the Greek Shipowners Association for Passenger Ships (GSAPS), and the Hellenic Shortsea Shipowners Association (HSSA). Among them, the UGS and the HCS have expressed their views on the reduction of GHG emissions from international shipping. In 2009, a survey was carried out by Costas Giziakis and Anastasia Christodoulou to examine the response of various shipping companies in Greece on the issue.[167] Part of the results of that survey were also utilised in this section.

The shipping industry in Greece generally welcomes the regulatory efforts of the international community in reducing GHG emissions from ships. The 2009 survey in Greece reveals that 73 per cent of the shipping companies surveyed know *very well* the efforts of the international community in regulating GHG emissions from ships, and 23 per cent of these companies know that there have

163 Grammenos and Choi, above n. 2, 38.

164 United Nations Conference on Trade and Development (UNCTAD), 'Review of Maritime Transport 2015' (2015) <http://unctad.org/en/PublicationsLibrary/rmt2015_en.pdf> accessed 12 May 2016, p. 36.

165 Ibid.

166 Ibid. 42.

167 Totally 35 Greek shipping companies of various sizes and industry segments, including the dry bulkers, the tankers, the containerships, the general cargo carriers and the ferries, participated in the survey. Giziakis and Christodoulou, above n. 1, 358.

been efforts made on this issue.[168] Regarding the institutional arrangements for regulating the GHG issue, the support for the IMO, EU and *UNFCCC* roles covers 93 per cent, 23 per cent and 13 per cent of all involved shipping companies respectively.[169] It can be deduced that the shipping companies in Greece have been actively engaged in the work of reducing GHG emissions from ships, and, compared with the EU and the *UNFCCC*, the IMO has been regarded by them as the most suitable institution to regulate GHG emissions from international shipping.

The Greek shipping industry has actively participated in the discussion of the technical and operational measures before and after their adoption in July 2011. From the perspective of Greek shipping companies, the EEDI for new ships is an environmentally effective measure. It can promote Research & Development in the maritime sector, and can be easily implemented.[170] However, the President of the HCS asserted that the current EEDI formulation indicates that the same ship will have a different EEDI at different speeds.[171] If this is the case, in order to make the EEDI an effective tool for reducing GHG emissions from ships, it might be necessary to improve or interpret the EEDI formula so that it can compare ship design energy efficiency at a fixed speed. Otherwise, under current Annex VI to *MARPOL 73/78*, a ship which complies with the EEDI requirement may actually reduce its GHG emissions more if it speeds up. This scenario is quite possible since generally a ship will automatically decelerate in times of poor markets or high bunker prices and accelerate in times of good markets or low bunker prices.[172] As to the SEEMP applicable for all ships, 52 per cent of the participant shipping companies in Greece believe that an Energy Efficiency Operational Indicator (EEOI) is 'probably necessary' for better implementation of the SEEMP.[173] Although currently the SEEMP is mandatory, it is only a management scheme which entails no reduction requirement (e.g., reduction target or monitoring requirement). A SEEMP requires that what is done to manage the operational efficiency of a ship should be documented. Nevertheless, an EEOI could be utilised as a tool to

---

168 Ibid.

169 Ibid. 359.

170 Giziakis and Christodoulou, above n. 1, 365.

171 George A. Gratsos, *Green and More Profitable Shipping* (13 November 2012) <http://www.nee.gr/downloads/183NEWSFRONT%20NAFTILIAKI%2013-11-12.pdf> accessed 2 April 2013, p. 9.

172 Ibid. 15.

173 Giziakis and Christodoulou, above n. 1, 361.

monitor a ship's performance, which combined with the SEEMP could achieve better efficiency.[174]

The MBMs have aroused more attention from the Greek shipping industry than the technical and operational measures. First, both national shipping organisations and individual shipping companies generally oppose an ETS to be adopted to reduce GHG emissions from ships. Whether an ETS is cost-effective and operationally feasible has been widely discussed. The UGS asserted that an ETS system ignores the 'structural, operational and contractual complexities of bulk shipping' and this will create a heavy and unwarranted administrative burden on the bulk shipping sector.[175] The adoption of an ETS needs to address many issues, such as the criteria of emissions allocation, thresholds, global cap setting, ship types and evasion possibilities via transhipment.[176] It will be challenging and time-consuming to address these issues due to the nature and pattern of the bulk/tramp shipping. It is also argued that the party paying for the fuel, such as charterers or cargo receivers, should also be responsible for GHG emissions from ships.[177] This argument is based on the fact that the charterer decides the voyage, the ship's speed and itinerary while the cargo receiver decides the cargo's origin and volume, the date of its shipping and delivery.[178] The opposition to an ETS by shipping organisations is consistent with the results of the survey of Greek shipping companies in 2009. According to the survey, 47 per cent of the participant companies believe that an ETS is not a good solution for addressing the GHG issue, while 27 per cent regard it as totally ineffective.[179]

The UGS favours the International Fund for GHG Emissions from Ships (GHG Fund) as its preferred MBM,[180] whereas the HCS prefers a Global Levy Scheme on Marine Bunkers (Bunker Levy).[181] Under the GHG Fund, contributions would be collected through bunker fuel suppliers or via direct payment from shipowners. In this way, the GHG Fund could be applicable to all ships

---

174 Ibid. 357.

175 Union of Greek Shipowners, *Prevention of Environmental Pollution by Ships: Regulation and Compensation Regimes and Industry Standards* (2011) <http://www.nee.gr/default.asp?t=anakoinoseisDetails&id=13> accessed 3 April 2013, p. 28.

176 Ibid. 29.

177 Union of Greek Shipowners, *MBIs for Reduction of CO_2 Emissions from Tramp Shipping* (2010) <http://www.shippingandco2.org/UGSMBIsForReductionOfCO2EmissionsFromTrampShipping%20.pdf> accessed 3 April 2013, p. 3.

178 Ibid.

179 Giziakis and Christodoulou, above n. 1, 360.

180 Union of Greek Shipowners, above n. 175, 29.

181 Gratsos, above n. 171, 13.

worldwide based on their fuel consumption, which makes the cost predictable to shipowners and is consistent with the money raised 'going directly for the benefit of the environment'.[182] Comparable with the UGS, the HCS strongly supports the Bunker Levy, which is believed to trigger 'an automatic profitable speed balancing mechanism' through directly applying to the cost of fuel for any trip.[183] That is, if a levy on bunker is collected, the ship operator may automatically adjust the speed based on the new bunker price so as to reduce GHG emissions from ships. This mechanism, however, attributes emissions reduction to the adjusting of ship's speed while ignoring the speed requirement from the cargo receiver, which in practice may make it difficult to achieve any emissions reduction from ships. Essentially, the above two proposals are the same type of MBM except for the different names utilised by different organisations. In the 2009 survey of Greek shipping companies, 53 per cent considered that a Bulk Levy is not a good solution for shipping, and only 10 per cent regarded it as an effective measure for tackling shipping GHG emissions.[184] This disparity reveals that a consensus on a suitable MBM for reducing global shipping emissions has not been achieved in the Greek shipping industry.

In summary, the Greek shipping industry has taken an active part in the regulation of GHG emissions from ships. It welcomes the efforts of the international community and insists that this process should be governed by the IMO. It has contributed to the development and improvement of the EEDI and SEEMP, and has suggested that EEOI should be made mandatory for better implementation of the SEEMP. Generally the Greek shipping industry is against an ETS, and prefers a bunker levy or GHG Fund related MBM, through which GHG emissions from ships could be reduced by adjusting driving speeds through increased bunker prices. Consensus on this mechanism, however, has not been achieved within the Greek shipping sector.

5.3.3 *The United Kingdom*

The United Kingdom (UK) is a traditional maritime power surrounded by the sea. It has been active in developing its shipping industry and participating in international and regional debate on the reduction of GHG emissions from ships. As at 1 January 2015, there were 1227 vessels in the UK trading fleet, with a total dwt of 48 million tonnes which accounted for 2.79 per cent of the world fleet.[185] Although the percentage of UK vessels in the world fleet is not high,

182 Union of Greek Shipowners, above n. 175, 29.

183 Gratsos, above n. 171, 13.

184 Giziakis and Christodoulou, above n. 1, 360.

185 UNCTAD, above n. 164, 36.

the UK controlled the ninth largest owned fleet in the world,[186] and the UK shipping industry has contributed to many aspects of the UK's economy. In 2013, the UK shipping industry directly employed 133,900 people, which created greater employment than the air transport and the support activities for land and water transport.[187] In the same year, the shipping sector made a £3.0 billion gross value-added contribution to the GDP of the UK, constituting 0.2 per cent of the UK economy.[188] In addition, 95 per cent of the UK's international trade is transported by ships.[189]

The shipping industry in the UK, including the Shipbuilders and Ship Repairers Association (SSA) and the UK Chamber of Shipping (UCS), has actively responded to global discussion on the reduction of GHG emissions from shipping.

Many UK shipping companies had adopted the voluntary energy efficiency measures before the adoption of the mandatory EEDI and SEEMP in July 2011. As reported by the SSA, customers (shipowners) have been increasingly valuing sustainable performance, and thus requiring more energy efficient ships.[190] For instance, the Maersk Line, a Danish shipping company, after adopting energy efficiency measures, met its 25 per cent CO_2 emissions reduction target in 2013 which it had set for itself eight years earlier.[191] In 2011, the UCS expressed its full support for the adoption of the EEDI and SEEMP by the IMO. However, it also commented that technical and operational measures alone will not be sufficient to achieve the absolute reduction of GHG emissions from shipping, and that the introduction of a MBM will be indispensable.[192]

186 Ibid.

187 Oxford Economics, 'The Economic Impact of the UK Maritime Services Sector: Shipping' (May 2015) <http://www.alinkfor.me/CoS/02/Shipping_The_economic_impact_of_the_UK_Maritime_Services_Sector.pdf> accessed 12 May 2016, p. 2.

188 Ibid. 6.

189 Maritime Journal, *Maritime Industry Supports UK* (21 Dec 2012) <http://www.maritime-journal.com/news101/industry-news/maritime-industry-supports-uk> accessed 11 April 2013.

190 The Shipbuilders and Shiprepairers Association (SSA), *Maersk Line's Focus on Energy Efficiency Pays Off* (8 March 2013) <http://www.ssa.org.uk/news/industry/maersk-lines-focus-on-energy-efficiency-pays-off> accessed 11 April 2013.

191 Ibid. Maersk Line promised to make a 40 per cent CO_2 reduction per container kilometre by 2020 with the year of 2007 as the baseline.

192 See, e.g., UK Chamber of Shipping, 'Shipping's Carbon Emissions—Design and Implementation of Market-based Measure: Part 1: A Cap-and-Trade Emissions Trading System' (2011) <http://www.ukchamberofshipping.com/media/filer/2012/10/10/shippings_carbon_emissions_-_cap-and-trade_2011.pdf> accessed 11 April 2013, p. 5.

The UCS released a discussion paper in 2009 entitled 'A global cap-and-trade system to reduce carbon emissions from international shipping', in collaboration with the national shipowners associations from Australia, Belgium, Norway and Sweden. In this paper, the UCS proposed a global ETS as its preferred MBM because it would create greater incentives for shipowners to cut emissions in the long term. This proposal did not obtain much support due to its lack of a long-term target and vague descriptions of its mechanisms. In 2011, the UCS published two papers to facilitate further debate by providing more information on the structure and establishment of a possible ETS, and an International GHG Contribution Fund.[193] Although both MBMs would apply the IMO NMFT principle to all ships over 400 GT, they also leave room for the incorporation of the CBDR principle. Specifically, under an ETS, ship operators need to buy and then surrender offset credits from a *UNFCCC* certified project in proportion to their bunker fuel purchases.[194] Since the *UNFCCC* certified project lies in an open market, the funds raised can also lead to out-of-sector reduction and satisfy the CBDR principle via their utilisation.[195] Similarly, under an international GHG Contribution Fund, *UNFCCC* non-Annex I States can receive the benefits of projects funded through carbon offset purchases, a 'net profit' from the system, although they also need to pay the contributions first.[196] In this way, the differentiated responsibility of developing countries can be reflected in both the ETS and the GHG Fund options. This discussion paper and its proposals indicate that the incorporation of the CBDR principle into a MBM is not only feasible but is also becoming more accepted by the shipping industry in *UNFCCC* Annex I States.

GHG emissions from international shipping currently are not regulated under the *UK Climate Change Act 2008*. The question of how to regulate this issue, and in particular whether an ETS should be adopted, has been debated in the UK. In 2011, the UCS expressed its view that an EU ETS is not suitable for shipping on the ground that a global solution is the only way to address

---

193 Ibid. 6. The UCS divides current MBM proposals into three categories, namely taxation (levy) arrangement, an ETS, and hybrid forms of the above two. The international GHG Contribution Fund is a type of levy arrangement.

194 Ibid. 5, 7.

195 Ibid.

196 UK Chamber of Shipping, 'Shipping's Carbon Emissions—Design and Implementation of Market-based Measures: Part 2: An International GHG Contribution Fund' (2011) <http://www.ukchamberofshipping.com/media/filer/2012/10/10/shippings_carbon_emissions_-_levy_2011.pdf> accessed 11 April 2013, p. 17.

this problem due to the nature of shipping.[197] The Energy and Climate Change Committee, a body appointed by the UK House of Commons, released a report in January 2012, asserting that UK shipping emissions should be included in an EU ETS. It commented that it is a 'delaying tactic' by the UK shipping industry to call for a global ETS.[198] Furthermore, this report underscores the environmental benefits of unilateral action at the EU level and encourages other countries to adopt their own measures to tackle GHG emissions.[199] The view of the shipping industry, however, has been endorsed by the Committee on Climate Change, an independent statutory body established under the UK *Climate Change Act 2008*.[200]

In summary, the UK shipping industry supports the work of the IMO in reducing GHG emissions from international shipping. Many shipping companies had already adopted the energy efficiency measures before they were regulated as amendments of Annex VI to *MARPOL 73/78*. The industry also argues that technical and operational measures alone will be insufficient to achieve the absolute reduction of GHG emissions from ships and advocates the introduction of MBMs. Of the current MBM proposals, they prefer a global ETS, which they believe will provide more incentives for shipowners to reduce GHG emissions.

5.3.4 *Conclusion*

The three States reviewed in this section are all *UNFCCC* Annex I States, and the shipping industries in these States all support the effort of the IMO in

197 Fiona Harvey, *UK Shipping Industry Rejects EU's Carbon Reduction Programme* (9 August 2011) <http://www.guardian.co.uk/environment/2011/aug/09/shipping-industry-rejects-carbon-trading> accessed 11 April 2013.

198 Energy and Climate Change Committee House of Commons, 'The EU Emissions Trading System: Tenth Report of Session 2010–12' (2012) <http://www.publications.parliament.uk/pa/cm201012/cmselect/cmenergy/1476/1476.pdf#page=50&zoom=auto,0,186> accessed 11 April 2013, p. 38.

199 Ibid. Summary.

200 See, e.g., Committee on Climate Change, 'International Aviation & Shipping Review: Scope of Carbon Budgets—Statutory Advice on Inclusion of International Aviation and Shipping' (1 April 2012) <http://www.theccc.org.uk/publication/international-aviation-shipping-review/> accessed 11 April 2013. Also, see, UK Chamber of Shipping, *UK Chamber of Shipping Refutes Conclusion of Energy and Climate Change Committee That International Solution on Reducing Carbon Is A 'Delaying Tactic'* (27 Jan 2012) <http://www.ukchamberofshipping.com/news/2012/01/27/uk-chamber-shipping-refutes-conclusion-energy-and-climate-change-committee-international-solution-reducing-carbon-delaying-tactic/> accessed 11 April 2013.

reducing GHG emissions from international shipping. They value highly the EEDI and SEEMP measures adopted by the IMO, and many of their shipping companies had already incorporated them into the building and operation of their ships prior to the regulation of these measures. However, their attitudes vary towards the proposed MBMs. The shipping sectors in Australia and the UK prefer an ETS and accept the incorporation of the CBDR principle into a MBM, whereas the shipping sector in Greece opposes an ETS and supports a bunker levy or GHG Fund—related MBM. The view of the Greek shipping industry on the CBDR principle is unclear. Nevertheless, the fact that two EU member States have totally different preferences for MBMs indicates the complexity of adopting MBMs. It seems it will not be straightforward to reach consensus on adopting MBMs not only globally but also at the regional level.

5.4 Response from the Shipping Industry in the *UNFCCC* Non-Annex I States

The regulatory measures that have been adopted by the IMO, such as the EEDI and SEEMP, or MBMs possibly to be adopted in the future, will increase transportation costs for the shipping industry, and may also have an impact on international trade.[201] These impacts will be greater for *UNFCCC* non-Annex I States (developing countries) than *UNFCCC* Annex I States (developed countries).[202] Therefore, whether the shipping industry in a country can absorb these higher costs will influence the degree to which they comply with these measures. The development of the shipping industries in *UNFCCC* non-Annex I States generally commenced later and has lagged behind in many respects when compared with the development of these industries in *UNFCCC* Annex I States. The shipping industries in these non-Annex I States are facing barriers resulting from historical, financial and technological gaps. This section takes China, the Republic of Korea and India as examples to examine the responses from the shipping industries in *UNFCCC* non-Annex I States.

201 See Kevin P. Gallagher, 'International Trade and Air Pollution: Estimating the Economic Costs of Air Emissions from Waterborne Commerce Vessels in the United States' (2005) 77(2) *Journal of Environmental Management* 99, 103.

202 Haifeng Wang, 'Economic Costs of CO_2 Emissions Reduction for Non-Annex I Countries in International Shipping' (2010) 14(4) *Energy for Sustainable Development* 280, 285.

5.4.1 *China*

China has a lengthy continental coastline of approximately 18,000 kilometres.[203] A number of excellent natural ports are located around the coast; in 2003 the coastal areas contributed 50 per cent of China's GDP and constituted 80 per cent of the country's international trade value.[204] Until 2010, over 90 per cent of China's imports and exports were moved by international shipping.[205] These advantageous natural resources, together with preferential policies by central and local governments, have made possible the rapid development of China's shipping industry. As at 1 January 2015, China controlled the third-largest owned fleet (in dwt) in the world with 2970 vessels registered in China and 1996 registered in other flag States.[206] The dwt controlled by China in that year covered 9.08 per cent of the world total.[207] Established in 1961, the China Ocean Shipping (Group) Company (COSCO) is the first Chinese international shipping company. However, China's international shipping sector started its rapid development only after 1978 when China adopted its reforms and opening-up policies.[208] In 2010 China's shipbuilding sector ranked first in the world in three categories, namely its accomplished shipbuilding output, volume of new ship orders and holding orders, which accounted for 43 per cent, 54 per cent and 41 per cent of the world market, respectively.[209] According to research jointly undertaken by Lloyd's Register, QinetiQ and the University of Strathclyde, by 2030, the Chinese-owned fleet will probably reach 19–24 per cent of the world fleet, rivalling Greece and other European countries.[210] As

---

203 This section only discusses the mainland China, excluding Chinese Hong Kong, Macau, and Chinese Taipei.

204 Weijie Gao, *Development Strategy of Chinese Shipping Company under the Multilateral Framework of WTO* (2003) <http://www.cosco.com/en/pic/forum/654923323232.pdf> accessed 1 June 2013.

205 彤新春 [Rong Xinchun],《试论新中国海运事业的发展和变迁 (*1949–2010*)》[*Development and Evolution of China's Shipping Industry* (*1949–2010*)] (31 October 2012) <http://economy.guoxue.com/?p=7575> accessed 1 June 2013.

206 UNCTAD, above n. 164, 36.

207 Ibid.

208 Gao, above n. 204, 2.

209 解玉真等 [Xie Yuzhen et al.], '《EEDI 对中国造船及航运业的影响》[The Impacts of the EEDI on the Chinese Shipbuilding and Shipping Industries]' (2011) 11 中国海事 *China Maritime* 23, 24.

210 Lloyd's Register QinetiQ, and University of Strathclyde, 'Global Marine Trends 2030' (2013) <https://www.dropbox.com/sh/ysc3kkspzsxs6de/n9hnjB3CQf/GMT2030%20LowRes.pdf> accessed 1 July 2014, p. 83.

a UNFCCC non-Annex I State, China has promoted its shipping industry to a high level in terms of its shipbuilding capability and shipping fleet.

The shipping associations in China mainly include the China Classification Society (CCS), the China Association of the National Shipbuilding Industry (CANSI), and the China Shipowners Association (CSA). Due to China's unique political structure, work on the reduction of GHG emissions from ships is dominated or guided by the government, mainly implemented by shipping companies, supported by the shipping industry and participated in by the public.[211] Most shipping companies, in particular large-scale companies, are state-owned. This means that the response of the shipping industry in China is often consistent with the positions of the Chinese Government.

In 2009, the Chinese Government announced its GHG emissions control target, before the 2009 Copenhagen climate change conference. It stated that it would cut its CO_2 emissions per unit of GDP by 40 per cent to 45 per cent by 2020 from the 2005 level. Against this backdrop, the Ministry of Industry and Information Technology of China released a 'Development Plan for the Ship Industry during the 12th Five-Year Plan (2011–2015)' in 2012. This plan seeks a greater role for China's shipping industry around the world by making it 'powerful' rather than simply 'big'. It also puts forward some requirements to reach this goal. For example, the current structure of the shipping sector needs to be optimised and upgraded, its technological innovation and overall quality should be improved, and the energy efficiency requirement in ship design and ship building should be strengthened.[212]

As a follow-up to this Development Plan, China's shipping industry was allocated its sector reduction target by the Ministry of Transport. The target requires the shipping companies to reduce their energy consumption and CO_2 emissions per unit turnover by 15 per cent and 16 per cent respectively from the 2005 level by the end of 2015.[213] China's shipping industry still considers the IMO's regulatory initiatives in reducing shipping emissions, although GHG

---

211 朱建华 [Zhu Jianhua], '《我国船舶运输温室气体减排对策探讨》 Approaches for the Reduction of Greenhouse Gas Emissions from Ships in China' (2010) 1 水运科学研究 *Research on Waterborne Transportation* 1, 1.

212 中国工业和信息化部 [Ministry of Industry and Information Technology of the People's Republic of China], '《船舶工业"十二五"发展规划》 [Development Plan for the Ship Industry during the 12th Five-Year Plan (2011–2015)]' (2012) <http://www.china.com.cn/policy/txt/2012-03/12/content_24876042.htm> accessed 15 April 2013, pp. 5–6.

213 谢宗惠[Xie Zonghui], '《合力打造绿色航运产业链》 Jointly Establish A Green Industry Chain for the Shipping Industry', 中国水运报 *China Waterborne Transportation Newspaper* 22 September 2011 <http://www.zgsyzz.com/Article/ShowInfo.asp?ID=5318> accessed 15 April 2013.

emissions from international shipping are excluded from this target. This is because the shipbuilding sector, as a key sector of China's shipping industry participating in international business, needs to comply with the IMO rules to meet the requirements of its current and potential customers. Another benefit is that while complying with international rules the industry can also meet the domestic reduction target. Under the joint efforts of different sectors, China had successfully cut its CO_2 emissions per unit of GDP by 33.8 per cent by 2014 from the 2005 level.[214] There was only a step away from the accomplishment of its Copenhagen reduction target.

In June 2015, China submitted its Intended Nationally Determined Contributions (INDCs) to the Secretariat of the *UNFCCC*. Based on the INDCs, China pledges to achieve the peaking of CO_2 emissions around 2030 and make best efforts to peak early; meanwhile China vows to lower its CO_2 emissions per unit of GDP by 60 per cent to 65 per cent by 2030 from the 2005 level.[215] Compared with the Copenhagen commitment, China's INDCs are more ambitious. Accordingly China has treated the reduction of GHG emissions from the shipping industry as one of the priorities of its development.[216]

China's shipping industry has different views towards the proposed technical, operational and MBMs under discussion within the IMO. As far as the technical and operational measures are concerned, the shipping industry welcomes the efforts of the IMO and agrees that the IMO is the most competent institution to regulate this issue. Representatives from shipping companies actively participated in the discussions on the proposed EEDI and SEEMP, and submitted their proposed modification of the EEDI formula to the IMO through the Chinese government. After the adoption of the energy efficiency rules by the IMO in July 2011, CCS released its *Rules for Green Ships* on 10 July 2012, the first rules of this kind in the world,[217] as well as the *Attained EEDI Calculation Guide* and *EEDI Verification Guide*. These rules serve as the industry's compliance with IMO regulations through the incorporation of the EEDI and SEEMP requirements into China's domestic ship classification. They

214 National Development and Reform Commission of China, *Enhanced Actions on Climate Change: China's Intended Nationally Determined Contributions* (30 June 2015) <http://www4.unfccc.int/submissions/indc/Submission%20Pages/submissions.aspx> accessed 12 May 2016, p. 3.

215 Ibid. 5.

216 Ibid. 9.

217 China Classification Society (CCS), 《绿色船型认可 *EEDI*》 [*Green Ships Recognition EEDI*] (2012) <http://www.ccs.org.cn/ccswz/font/fontAction!moudleIndex.do?moudleId=61> accessed 15 April 2013.

provide information that China's shipbuilding industry will abide by international shipbuilding standards in building its ships for international buyers.

Regulation 19 of Annex VI to *MARPOL 73/78* allows all flag States, including the Chinese Government, to postpone its implementation of the EEDI regulation until 2019.[218] However, it appears that China's shipbuilding industry will not utilise this right due to the fierce competition in the international shipbuilding market. Even before the adoption of *Rules for Green Ships* by the CCS, the French classification society had issued an EEDI Certificate, the first such certificate in Asia, to a bulk carrier with 63,500 dwt built by the SinoPacific Shipbuilding Group.[219] In practice, however, Chinese shipping companies often reduce their GHG emissions by employing large vessels and slowing their speed, a strategy which is said to be commonly utilised by other large shipping companies, such as the Maersk Line.[220]

Although China's shipbuilding industry has responded positively to the newly-adopted EEDI and SEEMP regulation by the IMO, it also recognises that these new rules, in particular the EEDI, have imposed great challenges on it. First, as stated by the Chinese delegation at the IMO, the CBDR principle that it asserted during the IMO negotiations was not reflected 'in a full and objective manner' within the amended Annex VI to *MARPOL 73/78* in 2011.[221] The consequence of this is that the CBDR principle, confirmed as a basic principle in the global climate change regime by the *UNFCCC* and its *Kyoto Protocol*, has been further weakened in the shipping arena from the technical perspective. Although technically it is more feasible to apply the NMFT principle in this regard, ignoring the historical contribution of GHG emissions by developed countries will lead to an unfair extra burden for the shipping industry in developing countries, including China.

Second, it will be more costly for China's shipping industry to achieve the regularly-upgraded EEDI standards. The new Chapter 4 of Annex VI to *MARPOL 73/78* provides a four-phased reduction schedule. During the period

---

218 See ch. 4, 4.3.2.

219 大公报 [Takungpao], 《太平洋造船集团新船首获能效设计指数证书》 [*New Ship Built by SinoPacific Shipbuilding Group First Obtained the EEDI Certificate*] (24 February 2012) <http://www.csoa.cn/xuliebd/jienengjp/201202/t20120224_1198135.html> accessed 16 April 2013.

220 王尔德 [Wang Erde], '《减排谈判首次共识，中国减排将始于2019年》 [Agreement Achieved through Reduction Negotiations and China Is to Start Its Reduction in 2019]', *21st Century Business Herald* (Beijing), 26 July 2011 <http://stock.sohu.com/20110726/n314586469.shtml> accessed 16 April 2013.

221 *Report of the Marine Environment Protection Committee on Its Sixty-Second Session*, MEPC 62nd Session, Agenda Item 24, IMO Doc MEPC 62/24 (26 July 2011) annex 20, p. 1.

2013 to 2025, the EEDI reference line parameters for relevant ship types and reduction rates, as well as the time periods, will be reviewed and amended regularly to reflect the latest status of technological development.[222] However, it is often recognised that China's shipping industry is 'big but not powerful' and China 'does not have sufficient say in global shipping pricing'.[223] A report by CANSI reveals that, in comparison with European countries, the US, Japan and the Republic of Korea, the Chinese shipbuilding sector aims at building middle and low-level vessels and lacks core technologies and capability in its Research and Development (R&D). This objectively makes its profit rate quite low.[224]

A three-party conference, participated in by Chinese ship owners/operators, shipbuilders, and classification societies, was held in Beijing in November 2011. This conference aimed at meeting the EEDI requirements and reducing shipping cost. It also discussed how to research, design and build green ships jointly in order to maintain and promote the competitiveness of China's shipping industry. Aside from the increased cost in relation to R&D, China's shipbuilding companies may need to purchase energy-efficient technologies from other countries. In practice, it is difficult for them to obtain technologies through the transfer of technology arrangement as indicated in Regulation 23 of Annex VI to *MARPOL 73/78*, due to the so-called title 'newly industrialised [developing] country' imposed on China by some economists.[225]

Third, the Chinese shipping industry is concerned about whether the EEDI benchmark, a technological standard, will become a type of trade barrier for developing countries.[226] This is a valid concern. On the one hand, currently the core energy efficient technologies on shipbuilding are primarily controlled by a few developed countries or regional blocs like the EU, and China does not have sufficient say in the drafting of the EEDI formula and reference line.[227] On the other hand, the setting of this floatable and upgrading EEDI standard actually

222 *MARPOL 73/78 Annex VI* (2011 amendments) reg 21(6).

223 Bhattacharya, above n. 29. Bhattacharya asserts that mainly two factors contribute to the weak status of the Chinese shipping industry in global shipping pricing. One is that, the Chinese shipping industry is not familiar with maritime arbitration, insurance and claims, and the other lies in its lack of core competitiveness as regards shipping information, ship brokering, financing and leasing.

224 肜新春 [Rong Xinchun], above n. 205.

225 International Monetary Fund (IMF), 'World Economic Outlook, April 2011' (2011) <http://www.ioha2012.net/?page_id=1945> accessed 18 April 2013. See also Government of Canada, *Explanatory Notes—Asian Newly Industrialized Countries (NICs)* (14 June 2012) <http://www.ic.gc.ca/eic/site/tdo-dcd.nsf/eng/00042.html> accessed 18 April 2013.

226 肜新春 [Rong Xinchun], above n. 205.

227 Ibid.

raises the trading threshold for ships from most developing countries. Based on the mandatory EEDI requirements, substandard ships might be detained, fined by port States or even not allowed to trade.[228]

In contrast to the EEDI and SEEMP, MBMs are regarded as an even 'bigger challenge' by the Chinese shipping industry.[229] First, the Chinese shipping industry believes that it is premature to adopt any MBMs because many uncertainties and problems still relate to them,[230] and China's shipping sector also needs more time to conduct research to enhance its understanding of this issue. Meanwhile, China's shipping industry holds that a policy arrangement on financial, technological and capacity-building support from developed countries for the implementation of the EEDI and SEEMP by developing countries should be in place before a MBM is adopted.[231] It is expected that the benefits of the shipping industry in developing countries can be secured under this arrangement. In view of these factors, to date the Chinese shipping industry has not expressed its preference for any of the currently proposed MBMs.

Second, consistent with the position of the Chinese government, China's shipping industry supports the leading role of the IMO in regulating technically related issues, but doubts the competency of the IMO to regulate MBMs. From its point of view, the *IMO Convention* gives the IMO the competence to regulate technical issues but not trade-related issues. MBMs are trade-related measures and thus should be decided by the *UNFCCC*.[232] Additionally, China took the view that the CBDR principle needs to be incorporated if an MBM needs to be adopted in the future.[233]

Third, China's shipping industry opposes unilateral actions, in particular the proposed inclusion of the shipping GHG emissions into a EU ETS. If this happens and the EU ETS which includes GHG emissions from international shipping comes into force before July 2019, the lead period that China's shipping

---

228 王尔德 [Wang Erde], above n. 220.

229 Ibid.

230 *Uncertainties and Problems in Market-based Measures*, submitted by China and India, MEPC 61st Session, Agenda Item 5, IMO Doc MEPC 61/5/24 (5 August 2010).

231 *Further Work on GHG Emissions from Ships*, submitted by Brazil, China, India, Peru, Saudi Arabia and South Africa, MEPC 64th Session, Agenda Item 5, IMO Doc MEPC 64/5/9 (27 July 2012) para. 8.6.

232 *Report of the Marine Environment Protection Committee on its 59th Session*, Statement by the Delegation of China on GHG Issues, IMO Doc MEPC 59/24/Add.1 Annex 13 (2009) para. 5.

233 See, *Application of the Principle of "Common but Differentiated Responsibilities" to the Reduction of Greenhouse Gas Emissions from International Shipping*, submitted by China and India, MEPC 58th Session, Agenda Item 4, IMO Doc MEPC 58/4/32 (15 August 2008).

industry may get to phase in changes from Regulation 19 of Annex VI to *MARPOL 73/78* will become meaningless. In this case, an EU ETS may charge all ships calling at the ports of their member States, regardless of the flag that these ships are flying. In other words, the waiver that a flag State gives the ships flying its flag based on Regulation 19 of Annex VI may not be recognised by an EU ETS. Also, due to waning demand and higher costs resulting from the global financial crisis since 2009 and China's over-capacity, China's shipping industry, in particular its shipbuilding sector, is currently experiencing a recession. In 2012 China's completed shipbuilding output and holding orders were 60,210,000 dwt and 106,950,000 dwt each, which, compared with 2011, had decreased by 21.4 per cent and 28.7 per cent, respectively.[234]

Under the circumstances, any unilateral reduction actions will increase the shipping cost and weaken the development momentum of China's shipping industry, while at the same time the authority of the IMO's current work will also be diminished.[235] To date the EU has attributed its unilateral actions to the slow and unsatisfied regulatory process of emissions reduction under the relevant international authorities. On 1 January 2012 the EU included the emissions from the international aviation industry into the EU ETS due to slow progress within the International Civil Aviation Organization (ICAO). In December 2012 the EU suspended this policy either due to the improved performance from the ICAO or because of strong opposition from many countries including the US, Russia, China and India. In the same year, the EU published a consultation document asking for views on how best to reduce GHG emissions from ships so as to finally include GHG emissions from international shipping in a EU ETS.[236] Once shipping GHG emissions are included in the EU ETS, the co-existence of two regulatory mechanisms, namely the EU ETS and potential IMO MBMs, will make implementation and compliance by developing States shipping industries more challenging.

234 中国工业和信息化部 [Ministry of Industry and Information Technology of the People's Republic of China], '《2012 年全国造船完工量超过6000万载重吨新接订单同比下降四成》 [In 2002 National Accomplished Shipbuilding Output Exceeds 60000000 DWT and New Orders Decreased by 40%]' (14 January 2013) <http://www.miit.gov.cn/n11293472/n11293832/n11294132/n12858417/n12858628/15121853.html> accessed 16 April 2013.

235 刘能治 [Liu Nengye], '《航运减排国际法新动向》 [New Development of International Law on the Reduction of Greenhouse Gas Emissions from Shipping]' (2011) 34(9) 世界海运 [*World Maritime Shipping*] 11, 12.

236 Will Nichols, *EU Launches Attempt to Deliver Shipping Emissions Trading Scheme* (24 January 2012) <http://www.businessgreen.com/bg/news/2140997/eu-launches-attempt-deliver-shipping-emissions-trading-scheme> accessed 3 June 2013.

In summary, China's shipping industry has taken various measures to reduce its GHG emissions from ships. It supports the efforts of the IMO in adopting the EEDI and SEEMP measures and has enacted its own rules to incorporate them into China's ship classification, although these measures have not been as successful as expected by China's shipping industry in several respects. Regarding the proposed MBMs, the shipping industry in China claims that they are premature at this stage, and if they are to be adopted, they should be decided by the *UNFCCC* rather than the IMO itself. To secure benefits to China's shipping industry, the CBDR principle should be incorporated into MBMs to be adopted. To date China's shipping industry has not expressed its preference among the current MBM proposals.

5.4.2 *The Republic of Korea*

The Republic of Korea, also called South Korea, is located in the southern half of the Korean Peninsula in East Asia and has a coast line of more than 1,500 miles. Situated in a strategic neighbourhood between Russia, China and Japan, the South Korean Government pays much attention to the development of its shipping industry. South Korea has become one of the main maritime powers of the world, although the South Korean shipbuilding sector only began its development in the early 1970s. As at 1 January 2015, South Korea controlled the sixth largest owned fleet (dwt) in the world with 775 vessels registered under Korean flags and 843 registered in other flag States.[237] The dwt it owned in that year accounted for 4.62 per cent of the world total.[238] The South Korean shipbuilding sector has ranked first among South Korean exports since 2008,[239] and is now home to seven of the world's ten largest shipbuilding companies. Of the seven top shipbuilders, Hyundai Heavy Industries (HHI), Samsung Heavy Industries (SHI) and Daewoo Shipbuilding (DSB), also called the 'Big 3', are believed to have dominated the global market in terms of output.[240]

With its booming shipbuilding capability, the Republic of Korea overtook Japan to be the world's largest shipbuilding nation in 2000, after Japan surpassed its European counterparts in 1956. This title was taken over by China

237 UNCTAD, above n. 164, 36. As at 1 January 2015, the countries ranking from the 1st to 5th in terms of the largest controlled/owned fleets in the world are Greece, Japan, China, Germany and Singapore.

238 Ibid.

239 Michael E. Porter et al., 'Shipbuilding Cluster in the Republic of Korea' (Harvard Business School, 2010) <http://www.isc.hbs.edu/pdf/Student_Projects/Korea_Shipbuilding_2010.pdf> accessed 18 April 2014, p. 18.

240 Ibid.

in 2010 due to China's better performance in exports of ships, but in 2011 the Republic of Korea regained the top spot as global shipowners ordered more complex high-technology vessels, in the production of which currently the Republic of Korea has an absolute advantage against China.[241] In 2012 Korean exports of ships decreased by 30 per cent from the 2011 level, whereas its Chinese and Japanese counterparts only dropped 10.3 per cent and 14.6 per cent, respectively.[242] Against this slump, Korean shipbuilders attributed this slump to the lack of financial support by the Korean Government and were thus more affected by the global economic downturn than China.[243] The changes, however, indicate that the fierce competition within the global shipbuilding sector has imposed a great challenge on the South Korean shipping industry. Although the Republic of Korea has been regarded as a developed country by some countries and international organisations,[244] it is a *UNFCCC* non-Annex I State and the views of its shipping industry on the reduction of GHG emissions from ships are significant due to its status in the global shipping industry.

The main shipping associations in the Republic of Korea are the Korea Shipowners' Association (KSA), the Korea Offshore & Shipbuilding Association (KOSHIPA), the Korean Chamber of Commerce and Industry (KCCI) and the Korean Register of Shipping (KRS). These associations, together with shipping companies, have expressed their views towards the reduction of GHG emissions from ships.

The Korean shipbuilding sector believes that innovative technologies and creative production processes constitute its core competitiveness.[245] Due to its confidence in its shipping technologies, the Korean shipbuilding sector has been supportive of the efforts of the IMO in reducing GHG emissions from ships. It participated in the IMO discussions on the proposed EEDI and provided

241 Alex Lee, *South Korean Shipbuilding Faces Hard Times, Hyundai Heavy Reflects* (2 February 2012) <http://gcaptain.com/south-korean-shipbuilding-faces/> accessed 18 April 2013.

242 Korea Offshore & Shipbuilding Association (KOSHIPA), *Korean Shipbuilders Fall Behind Chinese Rivals* (2013) <http://www.koshipa.or.kr/eng/koshipa/koshipa3/news_view.jsp?kind=eng_n&idx=150&s_section=&s_keyword=> accessed 19 April 2013.

243 Ibid.

244 See, e.g., Australian Government: Ministry for Foreign Affairs, *List of Developing Countries as Declared by the Minister for Foreign Affairs* (2013) <http://www.ausaid.gov.au/ngos/Documents/list-developing-countries.pdf> accessed 18 April 2013; American Mathematical Society, above n. 133; International Monetary Fund (IMF), above n. 225, 172.

245 Korea Offshore & Shipbuilding Association (KOSHIPA), 'Review of 2012: Innovative Tech & Creative Production Processes' (2013) <http://www.koshipa.or.kr/eng/koshipa/koshipa3/introduction05.htm> accessed 19 April 2013.

a new concept approach and formula for the EEDI covering the various types of propulsion systems and power generation systems through the Korean government.[246] Unlike its Chinese counterparts, the Korean shipbuilding sector, in particular its large shipbuilding companies, welcomes the planned unilateral actions by the EU on strengthening environmental regulations.[247] Korea's large shipbuilders believe that they can gain more orders for constructing high-efficiency, eco-friendly ships once various EU technical and operational measures and MBMs are in place.[248] In contrast to the positive attitudes to reducing GHG emissions from ships by Korean shipping associations and large shipbuilding companies, small and medium-sized shipping companies, however, are not so supportive of this kind of regulation. A survey in 2011 reveals that Korea's small and medium-sized shipping firms were concerned that stricter environmental regulations on ships might increase their manufacturing costs and weaken their price competitiveness, while their Chinese counterparts might not be influenced in this way.[249] Due to this gap between different shipping firms, many small and medium-sized companies have not started their preparation for incorporating the EEDI and SEEMP measures,[250] while large companies have responded quickly to meet the new requirements. For instance, the HHI has been keen to develop its environmentally friendly high-value vessels, including drillships, liquefied natural gas (LNG) carriers, mega containerships and those using LNG as a fuel.[251] With this strategy, the HHI has achieved remarkable success in getting new orders.

The Proposed MBMs at the IMO have been classified by the Korean shipping industry into three categories, namely a GHG Fund (or levy), an ETS and hybrid schemes combined with the EEDI.[252] The strengths and weaknesses

---

246 *Proposal for New Concept Approach to EEDI for New Ships*, submitted by the Republic of Korea, Intersessional Meeting of the Greenhouse Gas Working Group 2nd Session, Agenda Item 2, IMO Doc GHG-WG 2/2/12 (6 February 2009).

247 Asia Shipbuilding & Offshore Information Service (ASIASIS), *Korea Welcomes EU Environmental Regulations* (17 November 2010) <http://www.simic.net.cn/news_show.php?lan=en&id=80211> accessed 19 April 2013.

248 Ibid.

249 Sang-Yoon Lee and Young-Tae Chang, 'Shipping Companies' Awareness and Preparedness for Greenhouse Gas Regulations: A Korean Case' in Theo Notteboom (ed.), *Current Issues in Shipping, Ports and Logistics* (2011) 25, 47.

250 Ibid.

251 Lee, above n. 241.

252 *Results of the Sixth Seoul International Maritime Forum*, submitted by the Republic of Korea, MEPC 65th Session, Agenda Item 5, IMO Doc MEPC 65/INF.6 (4 February 2013) Agenda 10(51).

of these MBMs have been analysed by the Korean Government in collaboration with its shipping industry.[253] Although the Korean shipping industry has not yet expressed its preference among the proposed MBMs, a 2010 survey of Korean shipping firms revealed that most Korean shipping firms support a carbon taxation scheme (or a levy) rather than an ETS as a MBM for the reduction of GHG emissions from international shipping. In that survey, 81 per cent of respondents welcomed carbon taxation because it is simple and easy to implement and the tax burden can be transferred to shippers.[254] In other words, it is practical and directly reflects the polluter-pays principle.[255] Only four carriers (19 per cent) preferred an ETS, and the opponents listed a number of reasons, such as:

- 'inferiority in technology and investment source';
- 'lack of ETS infrastructure and knowhow';
- 'difficulties in fair allocation of CO_2 emission rights to carriers';
- 'time and cost burden for calculating CO_2 emission amounts from present ships';
- 'complexity in defining responsible bodies about emission rights and obligations'; and
- 'uncertainty in ETS market'.[256]

This result is consistent with the opinions from the Korea Maritime Institute (KMI).[257] At the sixth Seoul International Maritime Forum held on 31 October 2012, the Vice President of the KMI asserted that among current proposed

253 See, e.g., *The Evaluation on the Relative Strengths and Weaknesses of the Reduction Mechanisms Employed by the MBM Proposals*, submitted by the Republic of Korea, Intersessional Meeting of the Working Group on GHG Emissions from Ships 3rd Session, Agenda Item 3, IMO Doc GHG-WG 3/3/1 (25 February 2011); *The International Greenhouse Gas Fund—Strengths and Weaknesses*, submitted by Cyprus, Denmark, the Marshall Islands, Liberia, Nigeria, the Republic of Korea and the International Parcel Tankers Association (IPTA), MEPC 62nd Session, Agenda Item 5, IMO Doc MEPC 62/5/33 (20 May 2011).

254 Lee and Chang, above n. 249, 44.

255 See ch. 2, 2.6.

256 Lee and Chang, above n. 249, 44.

257 Korea Maritime Institute (KMI) is a research centre working on South Korean policies on marine affairs and fisheries. It is run by the Korean Government through the Office of Government Policy Coordination. Partnerships in Environmental Management for the Seas of East Asia, *Korea Maritime Institute (KMI)* <http://www.pemsea.org/organization/kmi> accessed 1 June 2014.

MBMs, only the GHG Fund scheme meets all the nine principles adopted by the 57th MEPC meeting at the IMO.[258] While supporting the GHG Fund as an appropriate MBM, he also suggested that an ETS could be employed simultaneously as a hybrid scheme.[259] In this case, the GHG Fund serves as the in-sector GHG emissions reduction whereas the ETS will be responsible for the out-of-sector emissions reduction.[260] Although it seems that this arrangement can achieve absolute emissions reduction, the significant administrative cost associated with the scheme makes this solution less attractive.

In 2008, the KRS, Korea's classification society, established an Energy & Environmental Business Centre. Its purpose is to incorporate the IMO's regulations into domestic regulations, as well as introducing and operating possible MBMs whenever applicable.[261] This step, compared with the practice in other *UNFCCC* non-Annex I States where MBMs are not so welcomed, suggests that Korea has attempted to take a lead in reducing GHG emissions from ships.[262]

The Korean shipping industry has responded positively to the GHG emissions issue. It supports the efforts of the IMO in reducing GHG emissions from ships, and contributes to the improvement of the EEDI formula. In contrast to the shipping industries in the other main shipping nations, the Korean shipping sector even welcomes the unilateral actions to be possibly adopted by the EU. When incorporating and implementing IMO regulations, the Korean shipping industry outperforms some of its counterparts and tries to take the lead in reducing GHG emissions from ships. Of the current proposed MBMs, the Korean shipping industry prefers a GHG Fund or levy scheme and is against the ETS.

---

258 *Results of the Sixth Seoul International Maritime Forum*, submitted by the Republic of Korea, MEPC 65th Session, Agenda Item 5, IMO Doc MEPC 65/INF.6 (4 February 2013) Agenda 10(53).

259 Ibid. Agenda 10(54).

260 Ibid.

261 Korean Register of Shipping (KRS), *KR Expands into Green Growth Business* (September 2009) <http://www.krs.co.kr/kor/html/webzine/sub/SubRead_V2_e.asp?chasu=20&idx=7&gubun=4&lan=E> accessed 19 April 2013.

262 This can also be seen from the Co-operation Agreement that Korea signed with the IMO in April 2011 for the implementation of a technical cooperation project on building capacities in East Asian developing countries to reduce GHG emissions from ships. See IMO, *Republic of Korea to Assist IMO in Building Capacity in Developing Countries to Address Greenhouse Gas Emissions from Ships* (21 April 2011) <http://www.imo.org/MediaCentre/PressBriefings/Pages/Home.aspx> accessed 28 October 2013.

5.4.3 *India*

India is peninsular in nature having a long coastline of about 7517 km and 1197 islands. Its natural resources, together with 13 major ports and 176 notified non-major ports,[263] support the development of its seaborne trade and the shipping industry. As at 1 January 2015, India owned 844 vessels with 697 registered under Indian flags and 147 registered under other flags, ranking it as a State which has the 17th largest fleet (in dwt) in the world.[264] The dwt it owned in that year covered 1.26 per cent of the world total.[265] Meanwhile, with 1,174 vessels flying its flag, India also ranked as the 19th largest registered dwt flag State worldwide.[266] With this shipping fleet, around 95 per cent of India's trade in terms of volume and 68 per cent by value are transported by sea.[267] It is projected that India's share in global seaborne trade will rise to 9.3 per cent by 2020 from 3.66 per cent in January 2011.[268] Additionally, India has been keen to expand its global market share in shipbuilding and ship repair. In 2011 India had around 27 shipyards and 18 commercial dry docks for ship repairs. However, its shipbuilding or ship repair only covered about 1 per cent of global share in terms of value.[269] In order to expand its global market share in the shipping industry, the Indian Government has taken various measures, including introducing a shipbuilding subsidy scheme and transfer of technology,[270] to promote its shipping industry.

The Indian Government has responded to the mounting pressure to reduce GHG emissions from international shipping from the international community. It submitted a host of proposals to the IMO. Meanwhile, the shipping industry in India has also expressed its views towards this GHG issue through its

---

263 In India ports are divided into 'Major Ports' and 'Non-Major Ports'. Major Ports are under the jurisdiction of Central Government by means of policy and directives of Ministry of Shipping of Indian government, while Non-Major Ports are under the jurisdiction of State Government through policy and directives of State Government's nodal departments or agencies. Government of India Ministry of Finance, 'Position Paper on the Ports Sector in India' (December 2009) <http://pppinindia.com/pdf/ppp_position_paper_ports_122k9.pdf> accessed 20 April 2013, p. 8.

264 UNCTAD, above n. 164, 36.

265 Ibid.

266 Ibid. 42.

267 Ministry of Shipping Government of India, 'Annual Report 2012–2013' (2013) <http://www.shipping.nic.in/showfile.php?lid=1195> accessed 20 April 2013, p. 9.

268 Ministry of Shipping Government of India, 'Maritime Agenda: 2010–2020' (January 2011) <http://www.shipping.nic.in/showfile.php?lid=261> accessed 20 April 2013, p. 304.

269 Ibid. 383–384.

270 Ibid. 386–396.

shipping associations and shipping companies. They are the Indian National Shipowners' Association (INSA), the Shipyards Association of India (SAI), the Indian Coastal Conference Shipping Association (ICCSA) and some large shipping companies.

India is perhaps one of the most active developing countries which strongly support the incorporation of the CBDR principle into the global reduction of GHG emissions from ships. As the 'torchbearer' of the Indian shipping industry, the INSA mainly represents the interests of Indian shipowners. It believes that the CBDR principle should apply to this GHG issue. Accordingly, it asserts that the *UNFCCC* which backs the CBDR principle should be the central body regulating this GHG emissions issue, while the IMO which applies the NMFT principle should be responsible to the *UNFCCC*.[271] This argument, regarding the regulation of MBMs, is logical on the basis that MBMs are arguably beyond the competence of the IMO, which according to the *IMO Convention* is primarily responsible for technical affairs.[272] As to technical and operational measures, the EEDI and SEEMP, which represent the technical and operational measures respectively, have been solely regulated by the IMO.[273]

The shipping industry in India takes the view that the newly-adopted EEDI and SEEMP are 'not so benign' due to its disadvantages in relation to shipping technologies when compared with other countries.[274] This might be true with regard to the decreased number of orders that India's shipping industry has obtained in recent years. Consequently, the SAI suggested that domestic shipowners support Indian shipyards by placing their orders at home so as to change this situation.[275] Aside from its shipbuilding sector, the shipping companies in India have also encountered decreased business. As at 1 April 2011, Indian flagged vessels only carry 8.4 per cent of Indian trade cargo while the majority of seaborne trade was moved by overseas shipping companies.[276] Facing the stringent EEDI requirements, the Indian shipping associations, such as the ICCSA, encourage their member companies to employ qualified vessels. For example, the Shipping Corporation of India Ltd. (SCI) has made it

---

271 Anil Devli, *Overview of the Shipping Sector in India* (1 April 2011) <http://www.ahlers.com/images/news/2011/overview%20of%20the%20shipping%20sector%20in%20india%20%28mr%20anil%20devli%29.pdf> accessed 21 April 2013, p. 8.

272 *IMO Convention* art. 1.

273 See ch. 4, 4.2.

274 Devli, above n. 271, 9.

275 Perter Clarkson, *Indian Yards Ask Government Action* (6 April 2009) <http://www.globmaritime.com/news/shipbuilding-a-repair/4401-indian-yards-ask-government-action> accessed 21 April 2013.

276 Devli, above n. 271, 1.

a rule that the EEDI should be implemented at design stage for its ships so as to reduce GHG emissions from ships.[277]

Similar to its Chinese counterparts, the shipping industry in India does not welcome the adoption of any MBMs by the IMO.[278] The INSA asserts that these proposed MBMs will bring about 'adverse outcomes' for developing countries.[279] The Indian shipping industry and its government have provided many reasons for their opposition to the proposed MBMs. These reasons include the possible incompatibility between the WTO rules and MBMs,[280] lack of the CBDR principle,[281] uncertainties and problems with current MBMs,[282] an inequitable burden on developing countries[283] and the lack of specific and feasible financial, technological and capacity-building support from developed countries.[284]

In summary, the shipping industry in India has a complex attitude towards the reduction of GHG emissions from international shipping. It doubts the role of the IMO in regulating this GHG emissions issue, but still encourages its shipbuilding and shipping companies to meet the EEDI and SEEMP requirements

---

277 The Shipping Corporation of Indian Ltd. (SCI), 'Global Compact Initiative' (2012) <http://www.shipindia.com/pdf/MediaSpeak/COPfor2012-5c80e7.pdf.> accessed 21 April 2013, p. 7.

278 See, e.g., Devli, above n. 271, 9.

279 Ibid.

280 *Possible Incompatibility between WTO Rules and A Market-based Measure for International Shipping*, submitted by India, MEPC 62nd Session, Agenda Item 5, IMO Doc MEPC 62/5/27 (20 May 2011).

281 See, e.g., *Application of the Principle of "Common but Differentiated Responsibilities" to the Reduction of Greenhouse Gas Emissions from International Shipping*, submitted by China and India, MEPC 58th Session, Agenda Item 4, IMO Doc MEPC 58/4/32 (15 August 2008).

282 *Uncertainties and Problems in Market-based Measures*, submitted by China and India, MEPC 61st Session, Agenda Item 5, IMO Doc MEPC 61/5/24 (5 August 2010).

283 *Market-based Measures—Inequitable Burden on Developing Countries*, submitted by India, MEPC 61st Session, Agenda Item 5, IMO Doc MEPC 61/5/19 (2 August 2010). In 2011, the Indian government conducted a study on the possible impact of MBMs on India's shipping sector and trade. This study analyses iron ore exports from India to China, imports of coal to India from Australia and imports of crude oil to India from Saudi Arabia. It is concluded that the adoption of MBMs would definitely impact trade adversely, impose inequitable burden on Indian consumers, and lead to deleterious impacts on the environment as consumers of coal in India may choose to use cheaper poor quality Indian coal. *Market Based Measures—Impact on India's Shipping Trade*, submitted by India, IMO Doc MEPC 63/5/8 (23 December 2011) paras 1, 7, 13.

284 *Further Work on GHG Emissions from Ships*, submitted by Brazil, China, India, Peru, Saudi Arabia and South Africa, MEPC 64th Session, Agenda Item 5, IMO Doc MEPC 64/5/9 (27 July 2012) para. 8(6).

so as to expand its global market share. Meanwhile, it opposes the possible adoption of any MBMs by the IMO because it is feared that they would jeopardise the interests of the Indian shipping industry, as well as the benefits of other developing countries.

5.4.4 *Conclusion*

The three States examined in this section are all *UNFCCC* non-Annex I States. The responses from the shipping industries in these States on the GHG issue, however, are different. The shipping industries in China and the Republic of Korea support the efforts of the IMO in regulating the EEDI and SEEMP and have contributed much to the improvement of these requirements. The slight difference between them lies in the fact that China regrets that the CBDR principle was not fully incorporated into the amended Annex VI to *MARPOL 73/78* in July 2011. Nevertheless, the shipping industries in both China and the Republic of Korea are ready for the implementation of these rules. The shipping industry in India is generally against the EEDI and SEEMP; however, some large shipping companies have adjusted themselves to meet the new requirements. With regard to the proposed MBMs, the shipping industries in both China and India are opposed to any type of MBMs, whereas the shipping industry in the Republic of Korea welcomes the adoption of a MBM to reduce GHG emissions from ships. Of the proposed MBMs, the Korean shipping industry prefers a GHG Fund or levy-related MBM rather than an ETS. These findings indicate the complexity of the regulatory interests of the shipping industries from *UNFCCC* non-Annex I States as to the reduction of GHG emissions from international shipping. The shipping industries in China and India are in a similar development stage, whereas the shipping industry of the Republic of Korea, which is an advanced OECD member State, already owns better energy-efficient shipbuilding technologies than its Chinese and Indian counterparts. It is thus arguable that imbalanced regulatory interests lead to their differing responses to the regulation of GHG emissions from ships.[285]

5.5 Conclusion

The response from the shipping industry is important for any issues to be regulated by the IMO. The industry puts forward suggestions or provides feedback for the introduction of a new instrument, and comes up with possible

285 See ch. 6, 6.3.2.

initiatives to implement the instrument after it is adopted.[286] This also applies to the reduction of GHG emissions from international shipping.

The international and regional shipping organisations support the leading role of the IMO in reducing GHG emissions from ships. Although some of them prefer the adoption of operational measures rather than technical measures, these organisations have come to share a recognition that both measures would help to reduce GHG emissions. Currently they are still contributing to the further reduction of GHG emissions from international shipping, including extending the coverage of the EEDI to include most types of new ships, strengthening the implementation of adopted EEDI and SEEMP and enhancing the energy efficiency of international shipping by other means. As to the proposed MBMs, most international shipowners and ship operators associations prefer a GHG Fund or levy-related MBM be applied to the GHG issue, while the bunker sector supports an ETS. While most of these organisations claim that the NMFT principle should be solely applied to the issue, the international shipowners association also regards the incorporation of the CBDR principle into such measures as acceptable. It appears that the adoption of an MBM for the further reduction of GHG emissions from ships is acceptable to international and regional shipping organisations. Generally the global shipping industry prefers a GHG Fund or levy-related MBM, although it also asserts that a MBM should not be adopted for the time being.

At the national level, the shipping industries from various countries generally support the reduction of shipping GHG emissions by regulating the EEDI and SEEMP measures. The main disagreement among them occurs with regard to the proposed MBMs. Generally, the *UNFCCC* Annex I States support the adoption of an MBM but disagree on their preferred MBMs. Australia and the UK support an ETS and accept the application of the CBDR principle in this regard while Greece prefers a GHG Fund or levy relevant MBM. Within the *UNFCCC* non-Annex I States, the Korean shipping industry prefers a GHG Fund or levy-related MBM, whereas China and India believe that it is still premature to adopt any MBM at this stage. Therefore, it seems that the willingness of a country's shipping industry to accept an MBM depends on the development stage of the country. More developed countries possessing better technologies tend to accept an MBM more easily. Furthermore, as more developed countries and shipping associations accept the application of both the CBDR and the NMFT principles to the global regulation of GHG emissions from international shipping, it seems that finding ways to incorporate both principles into the issue under discussion will be the next step.

286 Pamborides, above n. 34.

CHAPTER 6

Response from Flag States and Port States to the Issue of Greenhouse Gas Emissions from International Shipping

6.1 Introduction

The previous chapters have identified the responses of the UN, the IMO and the shipping industry to the issue of regulating Greenhouse Gas (GHG) emissions from international shipping. As discussed in Chapter 2, flag States, coastal States and port States are also stakeholders in the GHG issue due to their prescriptive and enforcement jurisdiction on this matter. Theoretically, a State can be a flag State, a coastal State and a port State at the same time. However, in terms of GHG emissions from international shipping, coastal State jurisdictional authority over foreign flagged vessels in its maritime zones is limited, as compared with port State powers.[1] The 2011 amendment of Annex VI to the *International Convention for the Prevention of Pollution from Ships* (*MARPOL 73/78*)[2] secured important roles for flag States and port States in reducing GHG emissions from international shipping. The responses from the main flag States and port States to this issue are thus worthy of comprehensive assessment.

This chapter first discusses flag State control and the issue of 'flags of convenience'. Having established the central role of the flag State in ensuring the compliance of ships under its register with IMO regulations, this chapter then examines the response from selected main flag States to the issue of GHG emissions from international shipping. The States examined are Greece and Japan under Annex I to the *United Nations Framework Convention on Climate Change*

1 A coastal State's jurisdiction in its territorial sea and archipelagic waters is restricted by the *United Nations Convention on the Law of the Sea* (*LOSC*) in order to maintain navigational rights and an equitable balance between coastal States and foreign flag States, whereas its enforcement jurisdiction in the EEZ and high seas is generally not applicable to GHG emissions from international shipping. In regulating GHG emissions from ships, the *LOSC* provides flag States and port States with more jurisdictional authority. See ch. 2, 2.2.

2 *International Convention for the Prevention of Pollution from Ships* (*MARPOL 73/78*), signed 2 November 1973, 12 ILM 1319, as amended by the 1978 Protocol to the 1973 Convention, 1341 UNTS 3, 17 ILM 546 (entered into force 2 October 1983).

 | DOI 10.1163/9789004329317_007

(*UNFCCC*),[3] and Panama, China and Vanuatu which are not listed in Annex I to the *UNFCCC*. This chapter then examines the response from port States from two perspectives. The views from some influential global and regional port State organisations, such as the International Association of Ports and Harbors (IAPH) and regional Memoranda of Understanding (MOUs) on port State control, are examined first. This chapter then briefly examines the response from selected port States.

6.2 Flag State Control and the Issue of 'Flags of Convenience'

MARPOL 73/78 and the *United Nations Convention on the Law of the Sea* (*LOSC*)[4] have enabled flag States to have primary prescriptive and enforcement jurisdiction over the prevention of vessel source pollution, including GHG emissions from international shipping. Coastal States and port States have limited jurisdiction over the ships flagged by other countries.[5] Theoretically, a flag State will ensure that ships flying its flag comply with the technical and operational measures regulated by the IMO, namely the Energy Efficiency Design Index (EEDI) and the Ship Energy Efficiency Management Plan (SEEMP). However, in practice, a number of flag States are 'either unable or unwilling' to fulfil their duties in regulating and enforcing standards, similar to other vessel-source pollution measures.[6] This concern is often attributed to the issue of a 'flag of convenience' (FOC) or the 'open registration' of ships.

The terms FOC and 'open registration/registry' emerged in the practice of international shipping during the mid-1940s, and have been widely used since 1950.[7] Ships registered in such States are called 'ships under a FOC', and

3 *United Nations Framework Convention on Climate Change*, opened for signature 9 May 1992, 31 ILM 848 (entered into force 21 March 1994) ('*UNFCCC*').

4 *United Nations Convention on the Law of the Sea*, opened for signature 10 December 1982, 1833 UNTS 3 (entered into force 16 November 1994) ('*LOSC*').

5 Andrew Griffin, 'MARPOL 73/78 and Vessel Pollution: A Glass Half Full or Half Empty?' (1994) 1(2) *Indiana Journal of Global Legal Studies* 489, 506.

6 Ho-Sam Bang, 'Is Port State Control an Effective Means to Combat Vessel-Source Pollution? An Empirical Survey of the Practical Exercise by Port States of Their Powers of Control' (2008) 23(4) *The International Journal of Marine and Coastal Law* 715, 715; See also Amborse Rajadurai, 'Regulation of Shipping: The Vital Role of Port State Control' (2004) 18 *Australian and New Zealand Maritime Law Journal* 83, 86.

7 Francisco J. Montero Llácer, 'Open Registers: Past, Present and Future' (2003) 27(6) *Marine Policy* 513, 514; G.S. Egiyan, "'Flag of Convenience' or 'Open Registration' of Ships'" (1990) 14(2) *Marine Policy* 106, 106.

countries trading in their flag are called 'FOC countries' or 'open-registry countries'. Accordingly, the countries where the registers are open only to ships of their own country are often called 'close-registry countries'. However, in practice almost all registers cater for both national and foreign shipowners, and in recent years open registries have been increasingly utilised by shipowners.[8] Although there is no uniform definition of an FOC,[9] some of the features normally associated with an FOC have become generally accepted. These features include the lack of a 'genuine link' between the ship (or shipowner) and the State of registration (or ship flag State),[10] the shipowners' incentive of avoiding obstacles or restrictions in their own countries so as to obtain maximum

8 United Nations Conference on Trade and Development (UNCTAD), 'Review of Maritime Transport 2013' (2013) <http://unctad.org/en/PublicationsLibrary/rmt2013_en.pdf> accessed 1 June 2014, pp. 54–56.

9 See, e.g., Egiyan defines the FOC as 'national flags of those States in which shipowners register their ships so as to avoid: (a) financial obligations; and (b) the nature and conditions of shipping were their vessels registered in their own countries'. Egiyan, above n. 7, 107; Griffin defines the FOC as 'flags of certain countries whose laws make it easy and attractive for ships owned by foreign nationals or companies to fly these flags'. Griffin, above n. 5, 506.

10 See, e.g., Llácer, above n. 7, 520. See also *LOSC* art. 91(1). This provision requires a genuine link between the flag State and the ship. However, what constitutes the 'genuine link' and the effect of its absence has not been addressed in the *LOSC*. Some scholars, such as J.H.W. Verzijl and R.R. Churchill, suggest that the genuine link refers to 'the conditions of attribution of nationality'. Some competent international organisations, such as the International Tribunal for the Law of the Sea (ITLOS) and the IMO, underscore that the purpose of the genuine link is to 'secure more effective implementation of the duties of the flag State' over ships flying its flag. See, e.g., J.H.W Verzijl, *International Law in Historical Perspective, Part V, Nationality and Other Matters Relating to Individuals* (A.W. Sijthoff-Leiden, 1972) 149; R.R. Churchill, 'The Meaning of the "Genuine Link" Requirement in relation to the Nationality of Ships. A Study Prepared for the International Transport Workers' Federation' (2000) <http://www.itfglobal.org/seafarers/icons-site/images/ITF-Oct2000.pdf> accessed 1 July 2014, pp. 39, 69; Vincent P. Cogliati-Bantz, 'Disentangling the "Genuine Link": Enquiries in Sea, Air and Space Law' (2010) 79(3) *Nordic Journal of International Law* 383, 406, 411.

financial and administrative benefits[11] and the FOC State's motivation of obtaining income from the ship's registry.[12]

In terms of the relationship between flag State control and the FOC issue in reducing GHG emissions from international shipping, most current academic research indicates that the FOC has undermined the effectiveness of flag State control under *MARPOL* 73/78 and *LOSC*.[13] This argument is based on three factors. First, *MARPOL 73/78* and *LOSC* leave room for flexible flag State enforcement jurisdiction. Article 194(1) of the *LOSC* requires States to take 'the best practicable means at their disposal and in accordance with their capabilities' to prevent, reduce and control marine pollution.[14] Indeed, this article authorises the flag State, and in practice usually also the ship operators, to utilise 'the best practicable means' at their discretion. This regulation makes it possible for substandard vessel operations to be regarded as legal and appropriate under the *LOSC*. For this reason, a 'double standard' among the practice of different flag States is established.[15] Furthermore, Article 207(4) of the *LOSC* allows developing States to take into account their 'economic capacity' and 'their need for economic development' when they exercise their jurisdiction in relation to marine pollution.[16] This article does not incorporate the Common but Differentiated Responsibility (CBDR) principle,[17] and it only applies to pollution from land-based sources. Nevertheless, the fact that most FOC States are developing countries makes it possible for FOC States to utilise this provision

11 Through registering their ships in FOC countries, the shipowners may get such benefits as the easy registration of vessels, freedom from income taxes and lower other taxes, uncontrolled use of cash, possibility of hiring low-paid crew and reducing their numbers, reduced operating expenses and greater freedom from control by the flag State. See, e.g., G.S. Egiyan, 'The Principle of Genuine Link and the 1986 UN Convention on the Registration of Ships' (1988) 12(3) *Marine Policy* 314, 315; Egiyan, above n. 7, 107; Griffin, above n. 5, 506.

12 Bill Shaw, 'The Global Environment: A Proposal to Eliminate Marine Oil Pollution' (1987) 27(1) *Natural Resources Journal* 157, 160–163.

13 See, e.g., Egiyan, above n. 11, 315; Griffin, above n. 5, 507; Llácer, above n. 7, 520.

14 *LOSC* art. 194(1).

15 Erik Jaap Molenaar, *Coastal State Jurisdiction over Vessel-Source Pollution* (Kluwer Law International, 1998) 53.

16 *LOSC* art. 207(4).

17 The CBDR principle requires both developed and developing countries to contribute to addressing environmental problems and imposes the primary responsibility on developed countries due to their different historical contributions to the problems and the differentiated capability of developed and developing countries. A detailed discussion on the CBDR principle is provided in chapter 2, 2.5.

to protect their substandard ships.[18] Meanwhile, *MARPOL 73/78* also provides flag States with certain flexibility with respect to their enforcement jurisdiction. For instance, flag States shall investigate discharge violations and punish ships provided there is 'sufficient evidence', while the judging of 'sufficient evidence' is at the discretion of the flag States.[19]

The second factor for FOC's undermining flag State control is that, the FOC States do not have sufficient incentive to rigorously enforce pollution prevention measures over ships under their flag. This is because the primary motivation of an FOC State is to obtain income by registration fees. It is for this reason that some FOC States do not take their jurisdiction seriously. The third factor is that, most of the developing FOC States do not have sufficient capacity and resources to prescribe relevant pollution prevention measures, or to enforce, investigate or prosecute violations by ships under an FOC.[20] Partly due to these concerns, the *UN Convention on Conditions for Registration of Ships* (*UNCCRS*)[21] was adopted in 1986 by a conference of plenipotentiaries under the auspices of the United Nations Conference on Trade and Development (UNCTAD). This convention aims to strengthen maritime administrations and ship operators' identification through a 'genuine link' so as to address the FOC issue. To achieve this goal, the *UNCCRS* defines operator as the owner or any other national or juridical person to whom the responsibilities of the owner have been formally assigned.[22] Since the owner and the operator of a ship are not always the same person, company or organisation, the required concepts

18 As at 1 January 2012, ships registered in developing countries covered 26.41 per cent of the world fleet by deadweight tonnage, and this number reached 56.62 per cent for ships registered in the 10 major open and international registries. These 10 FOC countries are Antigua and Barbuda, Bahamas, Bermuda, Cyprus, Isle of Man, Liberia, Malta, Marshall Islands, Panama, St. Vincent and the Grenadines. United Nations Conference on Trade and Development (UNCTAD), 'Review of Maritime Transport 2012' (2012) <http://unctad.org/en/PublicationsLibrary/rmt2012_en.pdf> accessed 12 May 2016, p. 46, Table 2.8.

19 *MARPOL 73/78* art. 6(4)(5); Griffin, above n. 5, 508.

20 Gini Mattson, 'MARPOL 73/78 and Annex I: An Assessment of Its Effectiveness' (2006) 9(2) *Journal of International Wildlife Law & Policy* 175, 191.

21 *United Nations Convention on Conditions for the Registration of Ships*, opened for signature 7 February 1986, 26 ILM 1229 (not yet in force) ('*UNCCRS*').

22 *UNCCRS* art. 2.

of genuine link and beneficial ownership remain unaddressed under the *UNCCRS*.[23] To date the *UNCCRS* has not entered into force.[24]

Although FOC registries undermine flag State control regulations under *MARPOL 73/78* and *LOSC*, ships under an FOC are not necessarily substandard ships.[25] Rather, the negative impacts of an FOC could be addressed if mechanisms are strengthened. First, the No More Favourable Treatment principle (NMFT),[26] as discussed in previous chapters, could be applied to the prescription and enforcement of GHG emissions from international shipping to minimise the negative impact of an FOC. In this case, it would be less likely for ships to avoid the application of IMO GHG regulations by means of registering in FOC countries. Nevertheless, the application of the NMFT principle to this issue does not necessarily exclude the simultaneous application of the CBDR principle.[27] Second, some FOC countries exercise their jurisdiction and enforce compliance by ships flying their flags with IMO regulations. Indeed, many of the world's leading shipowners with excellent reputations for concern about environmental issues flag their ships with FOC countries, and many charterers with good reputations regularly charter FOC ships.[28] It is thus asserted by

23 George Kasoulides, 'The 1986 United Nations Convention on the Conditions for Registration of Vessels and the Question of Open Registry' (1989) 20(6) *Ocean Development and International Law* 543, 566. Kasoulides asserts that the link between a State and a ship flying its flag remains one of administrative competence under the 1986 UNCCRS, and for this reason the convention failed to clarify the most critical issues. Rather, the UNCCRS simply reinforced the *status quo* of the FOC.

24 Article 19 of the UNCCRS provides that 'the Convention shall enter into force 12 months after the date on which not less than 40 States, the combined tonnage of which amounts to at least 25 per cent of the world tonnage, have become Contracting Parties to it'. Nevertheless, as at 9 October 2013, only 15 States have ratified the convention.

25 See Kasoulides, above n. 23.

26 The NMFT principle refers to 'port States enforcing applicable standards in a uniform manner to all ships in their ports, regardless of flag'. A detailed discussion on the NMFT principle is provided in chapter 2, 2.5.

27 See ch. 2, 2.5.3.

28 Rajadurai, above n. 6, 86; Paris MOU, '2012 Annual Report on Port State Control' (2012) <https://www.parismou.org/Publications/Annual_report/> accessed 28 September 2013, p. 31. Based on the 2012 Paris Memorandum of Understanding (MOU) Annual Report on Port State Control, seven out of top 10 FOC States were in the 'White List'. These seven countries are Bahamas, Liberia, the Marshall Islands, Cyprus, Malta, Panama and Antigua and Barbuda. Under the Paris MOU, the 'White, Gray and Black List' represents three different levels of performance by flags, ranging from quality flags ('White Listed flags') to flags with a poor performance that are considered high ('Gray Listed flags') or very high risk ('Black Listed flags'). This ranking is based on the total number of inspections and

some scholars that substandard shipping does not have much to do with the concept of FOC but rather with 'the manner in which individual ship registers are administered'.[29] Therefore, to better implement the EEDI and SEEMP measures in reducing GHG emissions from ships by flag States, the key lies in the commitment of flag States to deal with this issue. The response from the main flag States, including an FOC State, is provided in the following section to examine the positions of flag States on the GHG issue.

6.3 Response from Main Flag States

As at 1 January 2015, ships registered in developed countries and countries with economies in transition accounted for 23.41 per cent of the world fleet by dead weight tonnage (dwt), whereas ships flying the flags of developing countries covered 76.36 per cent of the world fleet by dwt.[30] Of these developing countries, the three major FOC countries alone, namely Panama, Liberia and the Marshall Islands, covered 41.8 per cent of the world fleet by dwt.[31] However, these figures do not reveal the role developing flag States play within the IMO GHG regime. Although flagged by most of the world fleet by dwt, developing flag States, in particular the Least Developed Countries (LDCs) and Small Island Developing States (SIDS), have rarely expressed their views to the IMO.[32] These flag States typically lack capacity and resources, but their positions on the GHG issue are worthy of examination if uniform enforcement of the IMO's regulatory initiatives is to be achieved.

Developing flag States can be roughly classified into major FOC States,[33] major developing flag States, and other developing States. The views from the

detentions over a three-year rolling period for flags with at least 30 inspections in this period. Paris MOU, *New Targeting Lists Paris MOU* (16 June 2014) <https://www.parismou.org/new-targeting-lists-paris-mou> accessed 1 July 2014.

29 Rajadurai, above n. 6, 86.

30 United Nations Conference on Trade and Development (UNCTAD), 'Review of Maritime Transport 2015' (2015) <http://unctad.org/en/PublicationsLibrary/rmt2015_en.pdf> accessed 12 May 2016, p. 43.

31 Ibid. 41.

32 To date these large developing countries have actively participated in the discussions and negotiations on the GHG issue within the IMO, whereas small developing countries and FOC States have not contributed much to the IMO's current discussion. This can be seen from the number of proposals that these countries have submitted to the IMO.

33 Based on the groupings established by the UNCTAD, major FOC States refer to those countries where more than 90 per cent of their flagged ships by tonnage are owned by foreign nationals. UNCTAD, above n. 8, 46.

shipping industries in these countries on the GHG issue differ. As discussed earlier, major FOC States are often more interested in obtaining income from registering foreign ships than effectively managing the ships flying their flags, and the number of ships owned by their nationals is negligible. Major developing flag States are usually importing countries, such as China, India, Brazil, South Africa and Saudi Arabia. Most of these countries have a large owned fleet,[34] and as major importing countries their international trade can be easily affected by the IMO's regulatory measures.[35] Therefore, these countries, including their shipping industries, have actively participated in the IMO's regulatory discussions and contributed to this regulatory process. Apart from the above two types of developing countries, the remainder of the developing countries, including the LDCs and SIDS, are in the third category of other developing States. Generally these countries do not have long coastlines or significant shipping industries. The interests of the shipping industries in these countries are thus different from those of the shipping industries in other developing flag States. To examine the responses from the shipping industries in these three types of developing flag States, Panama, China and Vanuatu have been selected as illustrative examples. As to the case studies from *UNFCCC* Annex I States on their responses to this GHG issue, Greece and Japan have been selected as examples due to their geographical representativeness and important roles in global maritime transportation.

6.3.1 *The UNFCCC Annex I Flag States*

Many flag States under Annex I of the *UNFCCC*, such as Norway, Denmark and Germany, actively participated in the discussions and negotiations that led to the adoption of the amendments of Annex VI to *MARPOL 73/78* in 2011. Meanwhile the positions of these flag States on the proposed market-based measures (MBMs) also shaped the direction of MBMs regulation by the IMO. This section examines the response of Greece and Japan, the two influential flag States with the largest owned fleets in the world, to the regulation of GHG emissions from international shipping.

6.3.1.1 Greece

Greece is one of the major flag States in the world. As at 1 January 2015, Greece ranked seventh in the world among the flags of registration with the largest

34 For example, as of 1 January 2012 China and South Korea owned the fourth and fifth largest fleets in the world respectively while India, Brazil and Saudi Arabia ranked the 16th, 21st and 22nd largest fleets in the world respectively. UNCTAD, above n. 8, 41.

35 Tao Wang and Jim Watson, 'China's Carbon Emissions and International Trade: Implications for Post-2012 Policy' (2008) 8(6) *Climate Policy* 577, 585.

registered deadweight tonnage, following Panama, Liberia, the Marshall Islands, Chinese Hong Kong, Singapore and Malta.[36] The dwt of the ships flying the flag of Greece was 78,728,000, which accounted for 4.50 per cent of the world share.[37] In the same year, Greece owned the largest fleet in the world with 279,429,790 dwt, which covered 16.11 per cent of the world fleet.[38] Of these Greek owned fleets, 80.18 per cent of them (in dwt) flew the flags of foreign States.[39] As a traditional maritime power, Greece has participated actively in drafting international maritime regulations and has a solid record of complying with them. In contrast to the response from the Greek shipping industry discussed in Chapter 5, the position of the Greek Government on the GHG emissions issue is slightly different and has experienced an evolution.

Similar to many other countries, Greece acknowledged the need for further reduction of GHG emissions from international shipping due to projected growth in world trade and seaborne transportation, although it asserted that shipping is the most energy efficient mode of transport.[40] Greece strongly supported the leading role of the IMO in addressing the GHG issue on shipping due to the IMO's extensive technical and scientific expertise and the global application of the rules developed by the IMO.[41] It asserted that a reduction target, if it is to be set, should be prescribed by the IMO.[42] It is still controversial as to whether a reduction target should be set within the global shipping industry or global transportation sector, and how and by which international organisation such target should be established.[43] While insisting that a reduction target

---

36 UNCTAD, above n. 30, 42.

37 Ibid.

38 Ibid. 36.

39 Ibid.

40 *Report of the Marine Environment Protection Committee on its 58th Session*, IMO Doc MEPC 58/23 (16 October 2008) annex 9, p. 8.

41 Ibid. 9.

42 Ibid.

43 See, e.g., China and India asserted that a reduction target could be set by the *UNFCCC*; Norway argued that a cap on the shipping industry could be set associated with a market-based measure; the World Shipping Council considered a reduction target only applicable to maritime shipping inappropriate in the absence of a broader approach to regulating transportation emissions at the national and global level. *Report of the Marine Environment Protection Committee on Its Sixtieth Session*, MEPC 60th Session, Agenda Item 22, IMO Doc MEPC 60/22 (12 April 2010) p. 42. A detailed discussion on this issue is provided in Chapter 7.

should be set by the IMO, Greece also underscored that this target should not lead to distortions of competition in international trade.[44]

At the 58th MEPC meeting in 2008, Greece expressed its view that a holistic approach should be employed to effectively reduce GHG emissions from ships and protect the environment. It asserted that factors such as the availability of technologies to reduce emissions, the need to encourage innovation and the economics of world trade should be taken into account during the development of regulatory frameworks.[45] Additionally, Greece advocated that three fundamental principles should be applied to any future climate mitigation based regulation of shipping. They were:

- 'Regulation must be flag neutral to ensure a level playing field for shipping and agreed internationally to ensure consistency.
- Regulation must focus on relative reduction with a view to continuously improving efficiency of the individual ship and realize that absolute reduction objectives are not within reach given the growth in world trade.
- Regulation must ensure the free choice of method via goal based standard to promote innovation and cost effective solutions.'[46]

The first principle indicates that Greece supported the application of the NMFT principle to this GHG issue, which has received strong support from many countries due to the existence of FOCs. The first half sentence of the second principle aims at improving the energy efficiency of ships by means of technical and operational measures, which is technically appropriate. However, the second half sentence of this principle denies the possibility of absolute emissions reduction by any future regulation, which probably ignores the possibility of future adoption of MBMs. Indeed, the international community has not reached consensus on whether the technical and operational measures can achieve absolute emissions reduction. In 2011, an IMO-commissioned report concluded that the technical and operational measures adopted by the IMO (EEDI and SEEMP) alone would not achieve absolute emissions reduction.[47] Nevertheless, the conclusions of this report were regarded as 'doubtful' by China due to the 'significant uncertainties' and 'lack of transparency' relating

44 *Report of the Marine Environment Protection Committee on its 58th Session*, IMO Doc MEPC 58/23 (16 October 2008) Annex 9, p. 9.

45 Ibid.

46 Ibid.

47 Zabi Bazari and Tore Longva, '*Assessment of IMO Mandated Energy Efficiency Measures for International Shipping*' (IMO Doc MEPC 63/INF.2, Annex, 31 October 2011) para. 12.14.

to the calculation process of that study.[48] At the 65th MEPC meeting in May 2013, the IMO suspended its regulatory discussion on MBMs and attempted to establish attained energy efficiency standards for new and existing ships through a phased approach, as proposed by the US.[49] Therefore, it is premature for Greece to assert the second principle, which may reflect its priority in promoting its international trade rather than maximising global environmental protection. The third principle underscores the freedom from prescription in regulating the GHG issue. This position is consistent with the response from the global shipping industry as discussed in Chapter 5, and was adopted by the 2011 amendments of Annex VI to *MARPOL 73/78*.

Greece has actively participated in the discussions on the improvement of the technical measures relating to the EEDI. Greece's contribution mainly lies in four interrelated aspects of the EEDI. These are verification and certification of the EEDI, the EEDI formula, the method of calculation of the EEDI and safety issues related to EEDI. First, Greece proposed an improved procedure for the verification and certification of the EEDI by the Administration (flag State). It asserted that for any verification or even certification of the EEDI, sea trials should be conducted first so as to obtain measurements data, and then an approval by the Administration would be appropriate.[50] This procedure was adopted by the *2012 Guidelines on Survey and Certification of the Energy Efficiency Design Index*.[51] Second, Greece identified several deficiencies in the proposed EEDI formula, such as the EEDI reference line, lack of life-cycle considerations and possible misapplication of the EEDI for underpowered ships.[52] Additionally, Greece proposed improvement to the EEDI formula so as to ensure optimisation of ship design for energy efficiency and meeting the needs of an MBM with the EEDI as its benchmark.[53] These proposals

48 *Report of the Marine Environment Protection Committee on Its Sixty-Third Session*, MEPC 63rd Session, Agenda Item 23, IMO Doc MEPC 63/23 (14 March 2012) annex 7, p. 1.

49 *Report of the Marine Environment Protection Committee on Its Sixty-Fifth Session*, MEPC 65th Session, Agenda Item 22, IMO Doc MEPC 65/22 (24 May 2013) pp. 42–43, 47.

50 *Report on the Outcome of the Second Intersessional Meeting of the Working Group on Greenhouse Gas Emissions from Ships*, MEPC 59th Sesssion, Agenda Item 4, IMO Doc MEPC 59/4/2 (8 April 2009) p. 19, para. 2.57.

51 *2012 Guidelines on Survey and Certification of the Energy Efficiency Design Index (EEDI)*, Resolution MEPC.214(63), IMO Doc MEPC 63/23/Add.1 Annex 10 (2 March 2012) ('*EEDI Survey and Certification Guidelines*') Figure 1.

52 *Comments on the EEDI Baseline Formula*, submitted by Greece, MEPC 60th Session, Agenda Item 4, IMO Doc MEPC 60/4/15 (15 January 2010) para. 5.

53 Greece hoped to improve the EEDI formula to meet the requirements of a hybrid MBM. A detailed discussion on hybrid MBMs is provided in chapter 3. *Further Prospects for EEDI*

prompted significant discussions and contributed to the improvement of the EEDI formula. Third, Greece proposed to improve the method of calculation of the attained EEDI for new ships by means of including a correction factor to account for shallow water restrictions and innovative energy efficiency technologies.[54] Fourth, Greece was concerned that the proposed EEDI formula would work against safety by penalising safer or more robust structural features of ships or ignoring their manoeuvrability in adverse conditions,[55] so it suggested an improvement of the EEDI formula and regulation of minimum propulsion power.[56] These proposals were discussed within the IMO and were partially adopted by the IMO in the amendments of Annex VI to *MARPOL 73/78*.

The positions of Greece on the MBMs in tackling GHG emissions from ships have experienced an evolution. At the 58th MEPC meeting in 2008, Greece lodged two statements as Annexes to two IMO reports,[57] but these statements expressed the same view on the proposed MBMs. Greece asserted that it opposed MBMs and reserved its position on MBM proposals until full

---

Improvement, submitted by Greece, MEPC 62nd Session, Agenda Item 5, IMO Doc MEPC 62/5/6 (5 May 2011) p. 2.

54 Regulations 20 and 21 of the revised Annex VI to *MARPOL 73/78* divide the EEDI into attained EEDI and required EEDI, and both of them are calculated by a formula based on the technical design parameters for a given ship. Attained EEDI refers to the EEDI value achieved by an individual ship while required EEDI is the maximum value of attained EEDI that is allowed by Regulation 21 of the Annex VI for the specific ship type and size. See ch. 4, 4.3.3.1.

Guidance on the Treatment of Innovative Energy Efficiency Technologies for Calculation and Verification of the Attained EEDI, MEPC 64th Session, Agenda Item 4, IMO Doc MEPC 64/4/39 (9 August 2012); *Proposal for Amendments to the 2012 Guidelines on the Method of Calculation of the Attained Energy Efficiency Design Index (EEDI) for New Ships for Inclusion of A Correction Factor to Account For Shallow Water Restrictions*, submitted by Greece, MEPC 65th Session, Agenda Item 4, IMO Doc MEPC 65/4/17 (8 March 2013).

55 The proposed EEDI formula might lead to larger ships' steel weight and smaller deadweight tonnage, which involves safety risks. *Consideration of Safety Issues Related To EEDI*, submitted by Greece, MEPC 61st Session, Agenda Item 5, IMO Doc MEPC 61/5/23 (5 August 2010) para. 2.

56 Ibid.; *Minimum Propulsion Power to Maintain the Manoeuvrability in Adverse Conditions*, submitted by Greece, MEPC 64th Session, Agenda Item 4, IMO Doc MEPC 64/4/37 (9 August 2012).

57 *Report of the Outcome of the First Intersessional Meeting of the Working Group on Greenhouse Gas Emissions from Ships*, MEPC 58th Session, Agenda Item 4, IMO Doc MEPC 58/4 (4 July 2008) Annex 9, pp. 1–2; *Report of the Marine Environment Protection Committee on its 58th Session*, IMO Doc MEPC 58/23 (16 October 2008) Annex 9, pp. 8–10.

impact assessments had been conducted and their results made available.[58] The assessments should at least address three concerns, namely the added value in energy efficiency to be achieved by the world fleet, their multifaceted impact on international shipping, and their indiscriminate and smooth implementation.[59] These views indicate that at that time Greece was not supportive of MBMs due to the lack of detailed studies of MBM proposals. However, Greece changed its position in less than two years. At the 60th MEPC meeting in 2010, when the three concerns had not been addressed, Greece expressed its 'qualified support' for the proposed MBM on an International GHG Fund as 'an early indication of its preference' compared with the Emissions Trading Scheme (ETS).[60] It considered a carbon levy as a more effective MBM than an ETS, and asserted that for the same environmental impact a levy would be less costly to world trade.[61] Specifically, Greece treated the International GHG Fund as a type of levy scheme. This preference, however, is consistent with the view from the Greek shipping industry as discussed in Chapter 5. It is possible that the Greek shipping industry lobbied the Greek Government to adopt this change of position.

The views of Greece on its preferred MBM were further strengthened by two subsequent proposals that Greece submitted to the IMO in 2011.[62] It grouped the proposed MBM proposals into four categories, namely a levy scheme (the GHG Fund), an ETS, hybrid MBM proposals that include EEDI (including the proposals submitted by Japan, World Shipping Council and the United States), and all other proposals.[63] As discussed in Chapter 4, as of May 2013 the current MBM proposals can be summarised into seven types as follows:

---

58 *Report of the Outcome of the First Intersessional Meeting of the Working Group on Greenhouse Gas Emissions from Ships*, MEPC 58th Session, Agenda Item 4, IMO Doc MEPC 58/4 (4 July 2008) Annex 9, p. 1.

59 *Report of the Marine Environment Protection Committee on its 58th Session*, IMO Doc MEPC 58/23 (16 October 2008) Annex 9, p. 10.

60 *An International Fund for Greenhouse Gas Emissions from Ships*, submitted by Greece, MEPC 60th Session, Agenda Item 4, IMO Doc MEPC 60/4/49 (29 January 2010) para. 3.

61 Ibid. para. 5.3.

62 The two documents are: *Grouping and Evaluation of Proposed MBMs*, submitted by Greece, Intersessional Meeting of the Working Group on GHG Emissions from Ships 3rd Session, Agenda Item 3, IMO Doc GHG-WG 3/3 (24 February 2011); *MBM Proposals: A Way Ahead*, submitted by Greece, MEPC 62nd Session, Agenda Item 5, IMO Doc MEPC 62/5/7 (6 May 2011).

63 *Grouping and Evaluation of Proposed MBMs*, submitted by Greece, Intersessional Meeting of the Working Group on GHG Emissions from Ships 3rd Session, Agenda Item 3, IMO Doc GHG-WG 3/3 (24 February 2011) p. 7, para. 32.

- GHG Fund, one option was proposed by Cyprus, Denmark, the Marshall Islands, Nigeria and the International Parcel Tankers Association (IPTA), and the other option was proposed by the Clean Shipping Coalition (CSC);[64]
- Port State Levy, proposed by Jamaica;[65]
- Efficiency Incentive Scheme (EIS), proposed by Japan and World Shipping Council (WSC);[66]
- Ship Efficiency and Credit Trading (SECT), proposed by the United States;[67]
- Global Emissions Trading System (ETS) for international shipping, three options proposed by Norway (Germany was later added as co-sponsor), United Kingdom, and France respectively;[68]
- Penalty on Trade and Development, proposed by Bahamas;[69] and

---

64 This proposal is to establish a global reduction target for international shipping, set by either UNFCCC or IMO. Emissions above the target line would be offset largely by purchasing approved emission reduction credits. The offsetting activities would be financed by a contribution paid by ships on every tonne of bunker fuel purchased.

65 This proposal aims to levy a uniform emissions charge on all vessels calling at a port based on the amount of fuel consumed by the respective vessel on its voyage to that port (not bunker suppliers). The CBDR principle could be achieved through a self-administered national or regional fund and/or some international mechanism.

66 According to this proposal, all new ships, except for those which meet pre-set EEDI thresholds, and existing ships are required to make payment contributions based on the amount of the bunker fuel consumed/purchased and the degree to which the ship's efficiency falls short of a specific standard. Funds collected go to an International GHG Fund and its Parties decide how to allocate the revenue either to long-term in-sector reduction or to a fund to be established under the *UNFCCC*.

67 Subject all ships to mandatory energy efficiency standards. As one means of complying with the standard, an efficiency-credit trading program would be established, and these standards would become more stringent over time. Currently this proposal becomes an optional addition to a phased approach energy efficiency proposal newly-submitted by the United States.

68 This proposal aims to set a sector-wide cap on net emissions from international shipping. A number of allowances (Ship Emission Units) corresponding to the cap would be released into the market each year via a global auctioning process. The units could then be traded.

69 This proposal holds that the imposition of any costs should be proportionate to the contribution by international shipping to global CO_2 emissions. The reduction will apply to individual ships and not Member States, and developing States will not be faced with a penalty upon trade and development.

- Rebate Mechanism (RM) for a market-based instrument for international shipping, proposed by the International Union for Conservation of Nature (IUCN)[70]

Although based on the nature of MBM proposals, the 'Port State Levy' and 'Penalty on Trade and Development' could also be types of levies, Greece put these two MBM proposals into the category of 'all other proposals'. After grouping these MBM proposals, Greece narrowed the list of proposed MBMs into two groups, namely the International GHG Fund and an ETS. To justify its argument, Greece first put in abeyance the hybrid MBM proposals with the EEDI as the benchmark due to the fact that the EEDI cannot be applied to existing ships as well as for other technical reasons.[71] Then Greece eliminated the category of 'all other proposals'. It considered the Port State Levy proposal could not be implemented in a cost-effective manner due to technical uncertainty in monitoring emissions, and that ships may opt for routes through ports that lack monitoring and enforcement mechanisms to avoid the levy.[72] The Penalty on Trade and Development proposal was also asserted by Greece to be infeasible due to technical difficulty in the universal application of the Energy Efficiency Operational Indicator (EEOI).[73] It also argued that the rebate mechanism is a 'cumbersome' measurement of a country's share of global emissions based on the country's global imports, imposes a high administrative cost, and would potentially lead to unfairness and fraud.[74] Based on the above analysis, Greece

---

70 This proposal aims to compensate developing countries for the financial impact of an MBM. It could be either applied to any maritime MBM which generates revenue (add-on option) or integrated with the International Maritime Emission Reduction Scheme (IMERS) (integrated option).

71 At the time of the analysis by Greece, there were three hybrid MBM proposals that include the EEDI, namely Japan's LIS proposal, WSC's VES proposal and the US's SECT proposal. The LIS and VES proposals proposed by Japan and WSC respectively later merged into one MBM proposal, which is Efficiency Incentive Scheme (EIS) as listed in the text. However, the nature of this new MBM proposal (EIS) with the EEDI as the benchmark remains, so Greece's analysis on previous three hybrid MBM proposals still applies to current two hybrid MBM proposals that include the EEDI.

72 *Grouping and Evaluation of Proposed MBMs*, submitted by Greece, Intersessional Meeting of the Working Group on GHG Emissions from Ships 3rd Session, Agenda Item 3, IMO Doc GHG-WG 3/3 (24 February 2011) paras. 43–45.

73 Ibid. para. 53.

74 *Grouping and Evaluation of Proposed MBMs*, submitted by Greece, Intersessional Meeting of the Working Group on GHG Emissions from Ships 3rd Session, Agenda Item 3, IMO Doc GHG-WG 3/3 (24 February 2011) paras. 46–50.

believed that the only feasible options were the levy scheme (GHG Fund) and an ETS. While Germany concluded that an ETS would be more cost effective than the GHG Fund, Greece asserted that a GHG Fund could achieve the same environmental result at half the price of an ETS or even less.[75] Its argument was underpinned by analysis from many perspectives such as the certainty in relation to a cap compared with that of a price, administrative burden, carbon leakage, evasion and fraud and experience from other contexts.[76] Therefore, the view of Greece that only the GHG Fund and the ETS should be further analysed has been supported by a number of countries.[77]

Greece's comments on current MBM proposals, in particular on their deficiencies from technical, operational and financial perspectives, have been underpinned by research[78] and supported by many countries.[79] However, these proposals were comparatively rough at that time and might be improved through further development. For this reason, Greece's methodology in narrowing the list of MBM proposals might be helpful in finding the most appropriate MBM. However, a re-examination is necessary because many of these improved proposals were submitted to the IMO at the 64th MEPC meeting and more IMO commissioned research is to be completed in the near future. Meanwhile, it is worthwhile to mention that Greece's views have been supported by many developed countries rather than developing countries. This is probably because the CBDR principle was not incorporated into its proposals.[80] At the 62nd MEPC meeting, Greece underscored that an MBM should be consistent with the *LOSC* and address the concerns from developing countries in

75 *MBM Proposals: A Way Ahead*, submitted by Greece, MEPC 62nd Session, Agenda Item 5, IMO Doc MEPC 62/5/7 (6 May 2011) para. 11.

76 *Grouping and Evaluation of Proposed MBMs*, submitted by Greece, Intersessional Meeting of the Working Group on GHG Emissions from Ships 3rd Session, Agenda Item 3, IMO Doc GHG-WG 3/3 (24 February 2011) pp. 10–15.

77 *Report of the Marine Environment Protection Committee on Its Sixty-Third Session*, MEPC 63rd Session, Agenda Item 23, IMO Doc MEPC 63/23 (14 March 2012) p. 40, para. 5.25.2.

78 See, e.g., Harilaos N. Psaraftis, 'Market-Based Measures for Greenhouse Gas Emissions from Ships: A Review' (2012) 11(2) *WMU Journal of Maritime Affairs* 211.

79 *Report of the Marine Environment Protection Committee on Its Sixty-Third Session*, MEPC 63rd Session, Agenda Item 23, IMO Doc MEPC 63/23 (14 March 2012) para. 5.25.2.

80 *Report of the Marine Environment Protection Committee on Its Sixty-First Session*, MEPC 61st Session, Agenda Item 24, IMO Doc MEPC 61/24 (6 October 2010) annex 3. At the 61st MEPC meeting, many developing countries lodged their statements to the MEPC meeting report supporting the application of the CBDR principle to MBM proposals.

order to ensure a global system.[81] Nevertheless, the interests of developing countries are not well reflected in its preferred MBM proposal.

It is concluded that as a main flag State under the list in *UNFCCC* Annex I, Greece recognises the need to reduce GHG emissions from international shipping and strongly supports the IMO in regulating this issue. It emphasises the application of the NMFT principle to the GHG issue and asserts that an absolute emission reduction is impossible due to projected growth in international trade. Greece actively participated in the discussions and negotiations of the technical and operational measures within the IMO and contributed much to the improvement of the EEDI. With regard to the proposed MBMs, Greece was originally opposed to any form of MBM due to concern about the potential obstacles to trade. However, as discussions on MBMs have developed, Greece has changed its position and now supports the MBM on an International GHG Fund.

6.3.1.2 Japan

Japan is a traditional maritime power and an important flag State. As of 1 January 2015, Japan ranked 12th in the world among the flags of registration with the largest registered deadweight tonnage.[82] The dwt of the ships flying the flag of Japan was 22,419,000, which covered 1.28 per cent of the world share.[83] In the same year, Japan owned the second largest fleet in the world with 230,675,179 dwt, which accounted for 13.30 per cent of the world fleet.[84] Of these Japanese owned fleets, 80.71 per cent of them (in dwt) flew the flags of foreign States.[85]

Japan is one of the most advanced *UNFCCC* Annex I States. It has pioneered most energy-efficient shipping technologies. Japan's share of world shipbuilding output has fallen from around 34 per cent in 1999 to 18 per cent in 2011 due to worsening global economic conditions.[86] However, Japan still received

---

81 *Report of the Third Intersessional Meeting of the Working Group on Greenhouse Gas Emissions from Ships*, MEPC 62nd Session, Agenda Item 5, IMO Doc MEPC 62/5/1 (8 April 2011) Annex 1, p. 4.

82 UNCTAD, above n. 30, 42.

83 Ibid.

84 Ibid. 36.

85 Ibid.

86 During this period, China and South Korea both increased their shares of world shipbuilding output and reached 39 per cent and 31 per cent respectively. Council Working Party on Shipbuilding, 'Peer Review of Japanese Government Support Measures to the Shipbuilding Sector' (2013) <http://www.oecd.org/sti/ind/C-WP6%282012%2926-FINAL-ENG.pdf> accessed 21 August 2013, p. 23.

many international orders for building larger and more complicated vessels with more added value, compared with its Chinese and Indian counterparts. Japan's shipping industry is competitive in the international high-level or energy efficient shipbuilding market.[87] In recent years, based on the expectation of more stringent IMO rules on shipbuilding, international ship purchasers have tended to buy ships with better energy efficient technologies. At June 2012, the shares of the global shipbuilding order book (in Gross Tonnage) by China, Republic of Korea and Japan were 38.72 per cent, 33.77 per cent and 16 per cent respectively.[88] However, the shares of the global future shipbuilding order book (2014+, in Gross Tonnage) by China, South Korea and Japan are 21.41 per cent, 49.68 per cent and 15.24 per cent respectively.[89] These statistics indicate that after the adoption of the energy efficiency measures by the IMO in July 2011, the shipbuilding orders that China receives have dropped significantly, while Japan's orders are comparatively stable and Korea's orders have increased significantly. This change reveals that energy efficient technologies have already played an important role in the international shipbuilding market, and also partially justifies why China strongly opposed the adoption of the EEDI and SEEMP.[90]

Japan actively participated in and contributed to almost all discussions in relation to the GHG issue within the IMO due to its leading expertise in world shipbuilding. Its views on the regulatory principles for addressing GHG emissions from international shipping were expressed at the 58th and 59th MEPC meetings. First, Japan supported the role of the IMO in regulating the GHG issue and asserted that there should be adherence to the NMFT principle.[91] It supported the nine fundamental principles agreed at the 57th MEPC meeting.[92] However, in view of strong opposition from many countries on the

87 Ibid. 30.

88 Ibid. 29.

89 Ibid.

90 A statement lodged by China to the 62nd MEPC meeting report provides that 'the Chinese delegation opposes the adoption of this amendment and [is] in no position to acknowledge and accept the amendment'. *Report of the Marine Environment Protection Committee on Its Sixty-Second Session*, MEPC 62nd Session, Agenda Item 24, IMO Doc MEPC 62/24 (26 July 2011) annex 20, p. 2.

91 *Report of the Marine Environment Protection Committee on its 58th Session*, IMO Doc MEPC 58/23 (16 October 2008) annex 9, p. 19.

92 The 59th MEPC meeting concluded that the use of the word 'agree' would not imply approval, adoption or decision. *Report of the Marine Environment Protection Committee on its 59th Session*, Statement by the Delegation of China on GHG Issues, IMO Doc MEPC 59/24/Add.1 Annex 13 (2009) para. 4.48, p. 35. At the 57th MEPC meeting, the nine

second principle (the NMFT principle), Japan, co-sponsored by other States, suggested an improved expression of this principle in order to reach consensus. It proposed that the future IMO framework should be 'binding and equally applicable to all ships' rather than 'binding and equally applicable to all flag States'.[93] However, this proposal was not accepted by those delegations not supporting the second principle.[94] This was probably because this proposal still applied the NMFT principle, and thus was opposed by major developing States, in particular major shipbuilding developing States. Although these developing States can flag their ships with FOC States, various MOUs on port State control will make it difficult to operate and trade with substandard ships.

Second, Japan respects the CBDR principle applied in the *UNFCCC*, and argued that the CBDR principle could be reflected in other ways including through technical cooperation in the regulation of the GHG issue.[95] Compared with the positions of many other developed countries,[96] Japan's view reveals its willingness to cooperate and compromise. Unfortunately, the unsuccessful record within the IMO relating to technical cooperation,[97] in particular the transfer of technology from developed countries to developing countries, make it unlikely that this proposal will be accepted by developing countries. It appears that an innovative approach regarding the transfer of technology is needed. Third, Japan took the view that the regulatory package to be established by the IMO should consist of the EEDI, the SEEMP, and an MBM.[98]

---

fundamental principles were agreed by 'an overwhelming majority' but the second principle was opposed by some States. *Report of the Marine Environment Protection Committee on Its Fifty-Seventh Session*, MEPC 57th Session, Agenda Item 21, IMO Doc MEPC 57/21 (7 April 2008) para. 4.77.

93 *Identifying Consensus on IMO Principles on Addressing Greenhouse Gas Emissions from International Shipping*, submitted by Australia, Canada, Denmark, Germany, Japan, Marshall Islands, Norway, Panama and the United States, MEPC 58th Session, Agenda Item 4, IMO Doc MEPC 58/4/16 (1 August 2008).

94 *Report of the Marine Environment Protection Committee on Its Fifty-Seventh Session*, MEPC 57th Session, Agenda Item 21, IMO Doc MEPC 57/21 (7 April 2008) para. 4.76.

95 *Report of the Marine Environment Protection Committee on its 58th Session*, IMO Doc MEPC 58/23 (16 October 2008) annex 9, p. 19.

96 See, e.g., the US asserted that the CBDR principle 'has no place in the IMO' and is inconsistent with the actions taken by the IMO. Ibid. 10.

97 To date technical cooperation within the IMO, even within the WTO's TRIPS (Agreement on Trade-Related Aspects of Intellectual Property Rights), is regarded as not successful. See ch. 4, 4.3.3.3, ch. 7, 7.4.2.

98 *Consideration of Appropriate Targets for Reducing CO_2 Emissions from International Shipping*, submitted by Japan, MEPC 59th Session, Agenda Item 4, IMO Doc MEPC 59/4/35 (8 May 2009) para. 17.

It asserted that imposing a cap on the total GHG emissions from international shipping is not appropriate, but setting reduction targets is helpful for emission reductions.[99] These views aim at maintaining the growth of the global shipping industry while reducing GHG emissions from shipping.

With regard to the proposed technical and operational measures, Japan's contribution involved many aspects of the EEDI and SEEMP. Before the adoption of the energy efficiency measures in 2011, Japan proposed regulatory frameworks for mandatory application of the EEDI and SEEMP, a draft text for the amended Annex VI to *MARPOL 73/78*, draft texts for relevant guidelines for implementing these measures and draft texts for some MEPC Resolutions. It also developed a work plan with a timetable for the development of EEDI frameworks for ships not covered by the draft regulations, and provided a form of the International Energy Efficiency Certificate (IEE Certificate) for discussion within the IMO. Most of these proposals have been adopted by the IMO. After the adoption of the 2011 amendments to Annex VI, Japan submitted technical proposals to the IMO to expand the coverage of the EEDI. It also provided reports of the Correspondence Group on Energy Efficiency Measures for Ships where Japan served as a chair. In response to the US's proposal of enhancing technical and operational measures raised at the 65th MEPC meeting in 2013,[100] Japan, co-sponsored by five other Annex I States, proposed to establish an Energy Efficiency Data Collection System at the 66th MEPC meeting in 2014. It argued that this system would better track the energy efficiency gains in the shipping sector and facilitate emissions reduction through establishing efficiency baselines and identifying the potential for increased efficiency gains.[101]

99 Ibid. paras. 18–19.

100 *Proposal of the United States to Enhance Energy Efficiency in International Shipping*, submitted by the United States, MEPC 65th Session, Agenda Item 4, IMO Doc MEPC 65/4/19 (8 March 2013). In this proposal the US suggested to establish attained energy efficiency standards for new and existing ships through a phased approach. This proposal aroused the interest of many countries and the 65th MEPC meeting in May 2013 thus agreed to suspend the scheduled discussion on MBMS.

101 *Further Technical and Operational Measures for Enhancing Energy Efficiency of International Shipping: Establishing an Energy Efficiency Data Collection System*, submitted by Belgium, Croatia, Denmak, Japan, Norway and the United States, MEPC 66th Session, Agenda Item 4.1, IMO Doc MEPC 66/4/19 (24 January 2014) para. 4. These proponents suggested that this proposed Energy Efficiency Data Collection System could be built through amending Annex VI to MARPOL 73/78.

However, this proposal was not adopted due to the opposition of many developing countries.[102]

Through submitting various proposals to the IMO, Japan expressed its technical and regulatory preferences relating to the GHG issue, diversified the debate and expedited the regulatory process within the IMO. Firstly, Japan respects the views from the shipping industry, and asserted that the EEDI should promote efforts by shipowners, shipbuilders, suppliers and others by reflecting as accurately as possible the energy efficiency when the ship is in actual use.[103] Accordingly, at the 59th MEPC meeting, based on the views from its shipping industry, Japan proposed a two-stage verification procedure for the EEDI, namely the preliminary verification at the design stage, and the verification of the Attained EEDI at sea trial. This approach was later adopted by the IMO. Secondly, Japan insisted that the EEDI should be adopted as a new part of Annex VI to *MARPOL 73/78*.[104] It explained that Annex VI can provide a similar legal basis for the mandatory EEDI scheme with its 'well-established and workable practices'; and it is also the quickest way to make the mechanism work due to the tacit acceptance procedure.[105] Meanwhile adding a new part to Annex VI addressing this issue under the *MARPOL 73/78* would differentiate this GHG issue with the issues of SO_X and NO_X.[106] This strategy has proved to be effective in expediting the adoption of the EEDI. However, it also imposes challenges on the future implementation of the EEDI because it was adopted by majority-voting and no consensus was reached.

Thirdly, given that the Chairman and Vice Chairman of the MEPC, China, South Africa and some other countries submitted their proposals on the draft

102 *Report of the Marine Environment Protection Committee on Its Sixty-sixth Session*, MEPC 66th Session, Agenda Item 21, IMO Doc MEPC 66/21 (25 April 2014) paras. 4.1.2–4.1.3. For example, India asserted that the immediate priority should be the full and effective implementation of the adopted energy efficiency measures.

103 *Development of An Index for CO_2 Emissions Per Unit Shipping Capacity in Actual Operational Conditions*, submitted by Japan, MEPC 57th Session, Agenda Item 4, IMO Doc MEPC 57/4/11 (25 January 2008) para. 11.3.

104 *Mandatory EEDI Requirements—Draft Text for Adding A New Part to MARPOL Annex VI for Regulation of the Energy Efficiency of Ships*, submitted by Japan, Norway and the United States, MEPC 62th Session, Agenda Item 4, IMO Doc MEPC 60/4/35 (15 January 2010) para. 6.

105 A detailed analysis on the tacit acceptance procedure is provided in Chapter 4, 4.3.2. Ibid. para. 7.

106 Ibid.

MEPC Resolution on the transfer of technology to the IMO,[107] Japan also submitted its own draft MEPC Resolution on the issue in cooperation with some other countries. In its draft Resolution, Japan underscored that any technology transfer on this work should be on the basis of respect for intellectual property rights, voluntary transfer, and mutually agreed terms and conditions.[108] This proposal did not reflect the CBDR principle fully, ignored the financing issue and imposed no concrete obligations on any State, and was thus opposed by many developing countries. In May 2013 a MEPC Resolution on the transfer of technology was adopted by the IMO, which in the preamble recognised the CBDR and NMFT principles.[109] Although the expressions utilise the words 'being cognizant' to replace the proposed 'acknowledging' by other countries, it was encouraging for most developing countries to expect more beneficial measures in facilitating the transfer of technologies as regulated in the amended Annex VI to *MARPOL 73/78*.[110] As a response to this adoption, Japan, co-sponsored by Australia and the US, lodged a statement to the meeting, through which it clarified that the CBDR principle applies in the *UNFCCC* while the NMFT principle applies in the IMO and under the *MARPOL 73/78*.[111] This statement indicates that Japan did not welcome the application of the CBDR principle to this issue from any perspective, although it asserted earlier at the 58th MEPC meeting that this principle could be reflected in certain ways. At the 66th MEPC meeting in April 2014, Japan further asserted that the

107 These individuals and countries submitted their draft MEPC Resolution on the transfer of technology as a means of facilitating the implementation of regulation 23 of MARPOL Annex VI. *Report of the Marine Environment Protection Committee on Its Sixty-Fourth Session*, MEPC 64th Session, Agenda Item 23, IMO Doc MEPC 64/23 (11 October 2012) para. 4.6.

108 *Draft MEPC Resolution on Capacity-Building, Technical Assistance and Transfer of Technology Related to Energy Efficiency Measures for Ships*, submitted by Australia, Canada, Germany, Japan, the Marshall Islands, Norway and the United States, MEPC 64th Session, Agenda Item 4, IMO Doc MEPC 64/4/44 (10 August 2012) annex, reg. 5.

109 *Report of the Marine Environment Protection Committee on Its Sixty-Fifth Session*, MEPC 65th Session, Agenda Item 22, IMO Doc MEPC 65/22 (24 May 2013) Annex 4, Resolution MEPC.229(65), p. 1. The preamble paragraphs of the Resolution provide that 'BEING COGNIZANT of' the NMFT principle, and 'BEING COGNIZANT ALSO of' the CBDR principle.

110 *MARPOL 73/78* Annex VI reg. 23.2. This regulation provides that 'the Administration of a Party shall co-operate actively with other Parties, subject to its national law, regulations and policies, to promote the development and transfer of technology and exchange of information to States which request technical assistance, particularly developing States'

111 *Report of the Marine Environment Protection Committee on Its Sixty-Fifth Session*, MEPC 65th Session, Agenda Item 22, IMO Doc MEPC 65/22 (24 May 2013) annex 5, p. 3.

transfer of technology should not be regarded as a condition for developing countries to fulfil their obligations in Annex VI.[112] It appears that the work on the transfer of technologies from developed countries to developing countries, in particular the future funding of these transfers, will remain a challenging issue to be addressed.

In terms of MBMs being discussed within the IMO, Japan has been supportive of adopting an MBM for the reduction of GHG emissions from international shipping. However, Japan's views on its preferred MBMs have been amended several times. At the 59th MEPC meeting in 2009, Japan supported the MBM proposal on the International GHG Fund proposed by Denmark, but suggested that under the scheme a ship should pay its contributions to the Fund directly through established electronic accounts for individual ships instead of paying this contribution via a fuel supplier so as to reduce administrative costs.[113] This view, however, was modified at the 60th MEPC meeting in 2010 when Japan put forward a Leveraged Incentive Scheme (LIS). The LIS was based on the International GHG Fund, but treated the performance of a ship as a benchmark for partial allocation of contributions. In other words, the contribution is collected from all ships with a fixed amount per tonne of purchased fuel, but a part of these contributions collected could be refunded to ships labelled as 'good performance ships'.[114]

At the 63rd MEPC meeting in 2011, Japan provided a cost analysis on the application of efficiency improvement measures in the maritime fleet, and concluded that the current EEDI and SEEMP measures could be improved

112 *Implementation of Resolution MEPC.229(65) on Promotion of Technical Co-operation and Transfer of Technology Relating to the Improvement of Energy Efficiency of Ships*, submitted by Belgium, Japan, the United Kingdom and the United States, MEPC 66th Session, Agenda Item 4, IMO Doc MEPC 66/4/31 (7 February 2014). This joint proposal was made in response to another proposal submitted by China, India and some other developing countries. See *Implementation of Resolution MEPC.229(65) on the 'Promotion of Technical Co-operation and Transfer of Technology Relating to the Improvement of Energy Efficiency of Ships'*, submitted by Angola, Argentina, Brazil, China, Cuba, India, Malaysia, Nigeria and South Africa, MEPC 66th Session, Agenda Item 4, IMO Doc MEPC 66/4/17 (24 January 2014).

113 *Consideration of A Market-Based Mechanism to Improve the Energy Efficiency of Ships Based on the International GHG Fund*, submitted by Japan, MEPC 59th Session, Agenda Item 4, IMO Doc MEPC 59/4/34 (8 May 2009) paras. 5–6.

114 *Consideration of A Market-Based Mechanism: Leveraged Incentive Scheme to Improve the Energy Efficiency of Ships Based on the International GHG Fund*, submitted by Japan, MEPC 60th Session, Agenda Item 4, IMO Doc MEPC 60/4/37 (15 January 2010) p. 2.

through an MBM focusing on in-sector GHG emission reductions.[115] At the second Intersessional Meeting of the Working Group on GHG Emissions from Ships in 2011, in response to calls from the IMO Secretariat for further innovation and consideration of MBMs, the LIS proposed by Japan merged with the Vessel Efficiency System proposed by the World Shipping Council (WSC) due to their similarities.[116] The new consolidated MBM proposal was called the 'Efficiency Incentive Scheme' (EIS), which entails only ships which do not meet the EEDI thresholds making payment contributions to the GHG Fund. This scheme was further amended by the co-sponsors in 2012, together with a draft legal text.

As discussed earlier, at the 62nd MEPC meeting in 2011, Greece suggested that after reducing the number of current MBM proposals, only an ETS and the International GHG Fund are worthy of further consideration. Upon Greece's proposal, Japan asserted that it was premature to limit the number of MBM proposals simply based on the analysis of the draft legal texts provided for current MBM proposals.[117] This opinion, however, did not receive much support from IMO member States.

In summary, as a *UNFCCC* Annex I State, Japan has been supportive of the IMO's work in regulating GHG emissions from international shipping. Based on its strong expertise on shipping-related matters, Japan contributed to the technical and operational regulation of the GHG issue. It stuck to the application of the NMFT principle to this issue and asserted that the CBDR principle could only be reflected in the regulation of this issue through other ways such as technical cooperation. Nevertheless, during the course of drafting the MEPC resolution on the transfer of technology, Japan again underscored the NMFT principle, which reveals its reluctance to give any effect to the CBDR principle in this GHG emissions issue. With regard to the discussions on MBMs within the IMO, Japan opined that it was necessary to adopt a MBM, and it has modified its view on its preferred MBM several times. Currently Japan's preferred MBM proposal is the EIS co-sponsored by Japan and the WSC.

115 *Cost Analysis on the Application of Efficiency Improvement Measures in the Maritime Fleet*, submitted by Japan, MEPC 63rd Session, Agenda Item 5, IMO Doc MEPC 63/INF.13 (22 December 2011).

116 These two MBM proposals both focus on in-sector reduction and do not set a global capping on the total amount of CO_2 emissions from international shipping.

117 *Report of the Marine Environment Protection Committee on Its Sixty-Third Session*, MEPC 63rd Session, Agenda Item 23, IMO Doc MEPC 63/23 (14 March 2012) annex 15, p. 2.

6.3.2 *The UNFCCC Non-Annex I Flag States*

This section examines three countries as examples of the responses from the *UNFCCC* non-Annex I flag States to the issue of GHG emissions from international shipping. These three States are Panama, China and Vanuatu, which as discussed earlier, represent a major FOC State, a major developing flag State, and another developing State respectively.

6.3.2.1 Panama

Panama is the southernmost country of Central America and is situated on the isthmus connecting North and South America. The revenues from the Panama Canal tolls are a significant contribution to its Gross Domestic Product (GDP).[118] Another well-known income of Panama relates to its role as a FOC State. Partially due to the facilitation from American shipowners, Panama started its open registry in 1916 shortly after its independence in 1903.[119] Panama has grown into a leading FOC State after many years' development. As of 1 January 2015, Panama ranked first in the world among the flags of registration with the largest registered dwt.[120] The dwt of the ships flying the flag of Panama was 352,192,000, which covered 20.13 per cent of the world share.[121] Based on the dwt, 99.86 per cent of the ships flying the Panamanian flag were owned by foreigners, whereas the ships owned by Panamanian nationals are negligible.[122]

FOC States are not active participants in the discussions and negotiations on this GHG issue within the IMO. As far as Panama is concerned, to date it has only submitted two proposals addressing this issue to the IMO. At the 58th MEPC meeting in 2008, Panama, together with some other countries, submitted a proposal on the regulatory principles for addressing this matter.[123] In this document, Panama generally supported the nine fundamental principles that should be applied for future IMO regulations on this GHG issue. Meanwhile, to relieve opposition from many developing countries on the second principle,

---

118 Gale Sarah Fister, 'Panama: the Canal and Beyond' (2012) 26(12) *PM Network* 17, 17.

119 Llácer, above n. 7, 514–515.

120 UNCTAD, above n. 30, 42.

121 Ibid.

122 As at 1 January 2012, the shipowners who registered their ships with Panama are mainly from Japan (45.96 per cent), South Korea (10.10 per cent), China (7.19 per cent), Greece (5.59 per cent), the US (1.20 per cent), and Germany (1.11 per cent). UNCTAD, above n. 18, 48, Figure 2.7.

123 *Identifying Consensus on IMO Principles on Addressing Greenhouse Gas Emissions from International Shipping*, submitted by Australia, Canada, Denmark, Germany, Japan, Marshall Islands, Norway, Panama and the United States, MEPC 58th Session, Agenda Item 4, IMO Doc MEPC 58/4/16 (1 August 2008).

Panama suggested that 'binding and equally applicable to all flag States' should be replaced by 'binding and equally applicable to all ships'.[124] This document indicates that Panama, as a developing country, supports the application of the NMFT principle to this GHG issue. This position is consistent with the views from another FOC State: the Marshall Islands.[125] While most developing countries underscore the application of the CBDR principle to this issue, Panama's position lies in the fact that the increased cost for ships complying with IMO regulations will not have a direct negative impact on its income. Under any circumstance it is the foreign shipowner that bears the increased cost since FOC States rarely have their own fleet. Meanwhile, these open registry States do not need to be concerned about a decrease in their client lists due to the growing demand for registering with FOC States by foreign shipowners. Indeed not only developed countries but also many developing countries, including China, India, Saudi Arabia, flag many of their ships with FOC States.[126] Even some FOC States, Cyprus as an example, flag some of their vessels with other FOC States.[127]

With regard to the proposed technical and operational measures (EEDI and SEEMP), Panama has not submitted any technical proposals to the IMO which is probably due to its lack of regulatory interests or technical expertise. However, at the 62nd MEPC meeting in 2011, Panama, as well as Liberia, the Marshall Islands, Malta and other FOC States, voted for the adoption of the EEDI and SEEMP by the IMO. Of the top ten FOC States, nine States all voted in favour except that Saint Vincent and the Grenadines abstained.[128] Indeed both Article 16(2)(d) of the *MARPOL 73/78* and Rule 27 of the *Rules of Procedures of the MEPC* provide that an amendment to an Annex to *MARPOL 73/78* shall be adopted by a two-thirds majority of the members entitled to vote, present and voting. For this reason, the positions of these FOC States influence the fate of the proposed regulations. Similarly, the *International Convention for the Control and Management of Ships' Ballast Water and Sediments* (*BWM Convention*)[129] provides that this convention shall enter into force only when 'not less than

124 Ibid. para. 5.

125 *Report of the Marine Environment Protection Committee on its 58th Session*, IMO Doc MEPC 58/23 (16 October 2008) Annex 9, p. 21. The Marshall Islands is one of the few FOC States which have submitted their proposals on the GHG emissions issue to the IMO.

126 UNCTAD, above n. 8, 41.

127 Ibid.

128 *Report of the Marine Environment Protection Committee on Its Sixty-Second Session*, MEPC 62nd Session, Agenda Item 24, IMO Doc MEPC 62/24 (26 July 2011) p. 57.

129 *International Convention for the Control and Management of Ships' Ballast Water and Sediments*, opened for signature 13 February 2004, IMO Doc. BWM/CONF/36 (not yet in force) ('*BWM Convention*').

thirty States, the combined merchant fleets of which constitute not less than thirty-five per cent of the gross tonnage of the world's merchant shipping' have signed it.[130] Under this circumstance, the role of FOC States is decisive due to the significant amount of gross tonnages flying with their flags.

In terms of proposed MBMs, at the 62nd MEPC meeting Panama asserted that a new instrument on MBMs should be adopted within the framework of the IMO.[131] This view is different from some developing countries, such as China and India which treated the regulation of MBMs as being beyond the competence of the IMO. Nevertheless, Panama did not justify its comment in its proposal to the IMO. Panama also predicted that no MBM proposals could be adopted within less than eight to ten years,[132] thus it would be better to take the first steps at an early stage rather than waiting for a perfect solution.[133] It underscored that a cap should not be imposed on the shipping sector so as to avoid penalising global trade and growth.[134] Concerning current MBM proposals, Panama opposed the narrowing-down approach suggested by Greece as discussed earlier, but supported Japan in that all current MBM proposals should be treated as options for the long term.[135] Accordingly it proposed that these MBM proposals should be 'stratified' based on their feasibility.[136] These views generally are more concerned with the interests from the shipping industry and international trade, but are also consistent with the interests of developing countries.

Of the current MBM proposals, Panama preferred the Penalty on Trade and Development proposed by another FOC State, Bahamas.[137] As discussed in Chapter 4, this MBM proposal stresses that the imposition of any financial penalty or cost should be proportionate to the contribution by international shipping to global CO_2 emissions, and developing States should not be faced with a penalty upon trade and development. This proposal incorporates phased mandatory CO_2 cut targets for all ships dependent upon their age. Panama provided three reasons for its support of this MBM proposal. Firstly, this scheme proposes technical and operational measures that are common to all the other

130 *BWM Convention* art. 18(1).

131 *Comments on the Report of the Third Intersessional Meeting of the Working Group on Greenhouse Gas Emissions from Ships*, submitted by Panama, MEPC 62nd Session, Agenda Item 5, IMO Doc MEPC 62/5/28 (20 May 2011) para. 5.

132 Ibid.

133 Ibid. paras. 12,15.

134 Ibid. para. 8.

135 Ibid. para. 14.1.

136 Ibid. para. 15.

137 Ibid. para. 14.2.

proposals. Secondly, it can be progressed relatively quickly without the need for considering the IMO's competence. Finally, it contains valuable and reliable information useful for the IMO and *UNFCCC* in their impact assessments.[138] These views indicate that feasibility is one of the key factors for Panama in its judgment on MBM proposals. It will be more acceptable to link a MBM to currently-adopted technical and operational measures to avoid lengthy discussions and negotiations for establishing a totally different measure. The debate on the IMO's mandate and competence in MBMs has not reached a consensus between various developed and developing countries. It is thus important that all MBM proposals should be feasible. However, apart from feasibility, administrative burden is also important although it does not apply to the MBM relating to Penalty on Trade and Development.[139] The uncertainty in relation to the Bahamas proposal mainly lies in the technical difficulties associated with the 'data collection'.[140]

It is concluded that as the largest FOC State in terms of registered deadweight tonnage, Panama supported the leading role of the IMO in regulating GHG emissions from international shipping. It underscored the application of the NMFT principle to this GHG issue and voted for the adoption of the energy efficiency measures by the IMO. As to proposed MBMs, Panama recognised the urgency of addressing this issue and suggested that all current proposals should be grouped based on their feasibility. Currently Panama's preferred MBM proposal is the Penalty on Trade and Development proposed by the Bahamas.

6.3.2.2 China

China's shipping industry, in particular its shipbuilding sector, has achieved a significant proportion of the world market in many categories.[141] Meanwhile, China is also one of the major flag States in the world. As at 1 January 2015, China ranked ninth in the world among the flags of registration with the largest

---

138 Ibid. para. 14.2.

139 *Full Report of the Work Undertaken by the Expert Group on Feasibility Study and Impact Assessment of Possible Market-based Measures*, MEPC 61st Session, Agenda Item 5, IMO Doc MEPC 61/INF.2 (13 August 2010) Executive summary, p. 16.

140 Based on Bahamas's MBM proposal, in the data collection phase the collection of reliable statistics on a ship's CO_2 emissions could be achieved either through the review of paper records of fuel usage or simply by recording emissions from the funnel using a suitable sensor. However, it is believed that relevant technologies are currently not available. See ch. 4, 4.3.4.2.

141 See ch. 5, 5.4.1. This section only discusses the mainland China, excluding Chinese Hong Kong, Macau, and Chinese Taipei.

registered dwt.[142] The dwt of the ships flying the flag of China was 75,676,000, which accounted for 4.33 per cent of the world share.[143] In the same year, China owned the third largest fleet in the world with 157,557,210 dwt, which covered 9.08 per cent of the world fleet.[144] Of these Chinese owned fleets, 53.15 per cent of them (in dwt) flew the flags of foreign States.[145] Against this backdrop, China actively participated in the IMO discussions on the reduction of GHG emissions from international shipping. The response of the Chinese government to this GHG emissions issue reflects the interests of China being a developing country and an important shipping nation. Its views are examined on five aspects.

6.3.2.2.1 *The IMO's Mandate and Competence to Regulate GHG Emissions from Ships and Applicable Principles*

It is generally accepted that the varying interpretation of Article 2(2) of the *Kyoto Protocol* by various countries has been the core obstacle to the regulation of the GHG issue by the IMO.[146] In other words, it is still open to debate whether the IMO has a mandate from the *Kyoto Protocol* to regulate the GHG issue, or whether the CBDR principle should be applied to this issue. As discussed in Chapters 2 and 4, the origin of the IMO's mandate in regulating this matter determines what kind of regulatory principles apply to this regulation and what type of measures can be taken by the IMO. Given the significance of this matter, China has expressed its positions by submitting a number of proposals and statements to the IMO since the 52nd MEPC meeting in 2004. Through submitting these documents, China has attempted to address three concerns, namely: what is the scope of the IMO's mandate and competence in regulating the GHG issue; why the CBDR principle should be applied to the GHG issue; and, how the CBDR principle could be applied.

From the perspective of China, the scope of the IMO's competence in regulating the GHG issue should be limited to technology or methodology-related matters.[147] In China's view, the proposed MBMs under discussion are beyond the competence of the IMO.[148] Therefore, China supported the role of the IMO

142 UNCTAD, above n. 30, 42.

143 Ibid.

144 Ibid. 36.

145 Ibid.

146 See ch. 4, 4.2.

147 *Report of the Marine Environment Protection Committee on Its Fifty-Sixth Session*, MEPC 56th Session, Agenda Item 23, IMO Doc MEPC 56/23 (30 July 2007) para. 4.58, p. 37; *Report of the Marine Environment Protection Committee on its 59th Session*, Statement by the Delegation of China on GHG Issues, IMO Doc MEPC 59/24/Add.1 Annex 13 (2009) p. 2.

148 *Report of the Marine Environment Protection Committee on Its Sixtieth Session*, MEPC 60th Session, Agenda Item 22, IMO Doc MEPC 60/22 (12 April 2010) annex 4, p. 2.

in regulating technical issues but asserted that MBMs on this issue should be decided by the *UNFCCC* if they are to be regulated in the future.[149] Although this view has been supported by a number of developing countries,[150] China did not provide a legal basis for its assertion in its submitted documents. As discussed in Chapters 2 and 4, although the *IMO Convention* provides the IMO with an economic purpose, in practice this function has never been allowed to be exercised by the IMO. Meanwhile, China took the view that the GHG issue should be addressed through consultations by all parties in the spirit of cooperation and opposed any kind of unilateral or arbitrary action.[151] This view is consistent with the position of China's shipping industry.[152]

Based on documents submitted by China to the IMO, the rationale underpinning the application of the CBDR principle to the GHG issue consists of five elements. First, the IMO received its mandate to regulate the GHG issue from Article 2(2) of the *Kyoto Protocol*, and this is also its only mandate in regulating the GHG issue.[153] Therefore, the fundamental principles that the *UNFCCC* and its *Kyoto Protocol* have set for regulating climate change, including the CBDR principle, should also apply to the IMO in addressing GHG emissions from international shipping.[154]

Second, the CBDR principle is not just a principle drawn from the *UNFCCC* and its *Kyoto Protocol*; rather, it represents the fundamental consensus of the international community in tackling climate change.[155] Thus all relevant international organisations should give due respect to the CBDR principle when they contribute to addressing climate change, and the IMO is no exception.[156]

Third, Article 2(2) of the *Kyoto Protocol* requests *UNFCCC* Annex I States to work through the IMO to pursue the limitation or reduction of GHG emissions from marine bunker fuels, which has been 'recognised' by the IMO in

149 Ibid.

150 For example, this view was also held by Brazil, Venezuela and Malaysia. *Report of the Marine Environment Protection Committee on Its Sixty-First Session*, MEPC 61st Session, Agenda Item 24, IMO Doc MEPC 61/24 (6 October 2010) annex 3, pp. 5–7.

151 *Report of the Marine Environment Protection Committee on its 58th Session*, IMO Doc MEPC 58/23 (16 October 2008) annex 9, p. 1.

152 See ch. 5, 5.4.1.

153 *Report of the Marine Environment Protection Committee on its 59th Session*, Statement by the Delegation of China on GHG Issues, IMO Doc MEPC 59/24/Add.1 Annex 13 (2009) p. 1.

154 Ibid.; *Report of the Marine Environment Protection Committee on its 58th Session*, IMO Doc MEPC 58/23 (16 October 2008) p. 1.

155 *Report of the Outcome of the First Intersessional Meeting of the Working Group on Greenhouse Gas Emissions from Ships*, MEPC 58th Session, Agenda Item 4, IMO Doc MEPC 58/4 (4 July 2008) annex 4, p. 3.

156 Ibid.

its Resolution A.963(23) adopted at the 23rd Session of the IMO Assembly on 5 December 2003.[157] Indeed during the course of deliberations on this resolution, the MEPC recommended that the resolution

> should be based on a common policy applicable to all ships, rather than based on the provisions of the Kyoto Protocol which stated that the reduction of greenhouse gas emissions is under the responsibility of the Annex I countries of the Protocol.[158]

However, this recommendation was rejected by the IMO Assembly, which in China's opinion proved that the above assertion by the MEPC was 'wrong'.[159] Given that the Legal Division of the IMO took the view that the IMO's mandate in regulating the GHG issue was not from Article 2(2) of the *Kyoto Protocol* but from the *LOSC* and the *IMO Convention*,[160] China responded to this view in two ways. On the one hand, China argued that Article 2(2) shall only be interpreted by the Conference of the Parties (COP) and the COP serving as the Meeting of the Parties to the Kyoto Protocol (CMP), which are the competent bodies in relation to the Protocol rather than any other body.[161] On the other hand, China agreed that Articles 1 and 64 of the *IMO Convention* give the IMO competence in regulating the GHG issue but underscored that the *Kyoto Protocol*

---

157 *Application of the Principle of "Common but Differentiated Responsibilities" to the Reduction of Greenhouse Gas Emissions from International Shipping*, submitted by China and India, MEPC 58th Session, Agenda Item 4, IMO Doc MEPC 58/4/32 (15 August 2008) para. 2.

The preamble of IMO Resolution A.963(23) provides that, 'BEING AWARE ALSO that the Kyoto Protocol, which was adopted by the Conference of the Parties to the UNFCCC in December 1997 and has not yet entered into force, requires the countries listed in Annex I to the UNFCCC to pursue the limitation or reduction of GHG emissions from marine bunker fuels, working through IMO (article 2.2)'. *IMO Policies and Practices Related to the Reduction of Greenhouse Gas Emissions from Ships*, IMO Assembly 23rd Session, Agenda Item 19, IMO Doc Res A. 963(23) (5 December 2003) p. 2.

158 *Report of the Marine Environment Protection Committee on Its Fifty-Second Session*, MEPC 52nd Session, Agenda Item 24, IMO Doc MEPC 52/24 (18 October 2004) para. 4.44, p. 27.

159 Ibid.

160 International Maritime Organisation (IMO), 'Main Events in IMO's Work on Limitation and Reduction of Greenhouse Gas Emissions from International Shipping' (2011) <http://www.imo.org/MediaCentre/resources/Pages/Greenhouse%20gas%20emissions.aspx> accessed 1 September 2013, para. 121, p. 28.

161 *Report of the Marine Environment Protection Committee on its 58th Session*, IMO Doc MEPC 58/23 (16 October 2008) annex 9, p. 2.

is still 'the most direct and authoritative' origin of such authorisation.[162] From the perspective of international law, China's rebuttal of the interpretation of Article 2(2) of the *Kyoto Protocol* by the Legal Division of the IMO is persuasive in that the IMO is not the competent organisation for such interpretation.[163] Indeed, the *UNFCCC* Secretariat made a statement in 2010 supporting the application of both the CBDR principle and the NMFT principle to the regulation of shipping GHG emissions by the IMO.[164] However, China's argument on the relationship between the *Kyoto Protocol* and the *IMO Convention* in authorising the IMO's regulatory work lacks sufficient legal basis. This is because there is no hierarchy between the *Kyoto Protocol* and the *IMO Convention*, and it is thus unlikely that there can be any determination of which rules should prevail if there is a conflict between the two treaties. For this reason, it is not persuasive for China to claim that the *Kyoto Protocol* is 'the most direct and authoritative' for the IMO's work in regulating this GHG issue.[165] Meanwhile, China also pointed out that the CBDR principle does not apply to the IMO' s dealing with matters such as the *Montreal Protocol on Substances that Deplete the Ozone Layer*.[166] That was because there is not a provision such as Article 2(2) of the *Kyoto Protocol* in that treaty.[167] This argument is reasonable in that the different origins of the IMO's mandate in regulating a type of shipping issue will directly lead to differing regulatory principles.

Fourth, to apply the NMFT principle and exclude the application of the CBDR principle to the GHG issue would be unfair for developing countries. On the one hand, the largest share of GHG emissions from international shipping

---

162 *Report of the Marine Environment Protection Committee on Its Sixtieth Session*, MEPC 60th Session, Agenda Item 22, IMO Doc MEPC 60/22 (12 April 2010) annex 4, p. 2.

163 Under international law, competent organisations to interpret a treaty include the treaty Parties, an *ad hoc* tribunal or the International Court conferred by the treaty, and the organs of the competent international organisation. Ian Brownlie, *Principles of Public International Law* (Oxford University Press, 7th ed., 2008) 602.

164 *Report of the Marine Environment Protection Committee on Its Sixty-First Session*, MEPC 61st Session, Agenda Item 24, IMO Doc MEPC 61/24 (6 October 2010) annex 6, p. 2. The statement provides that '[w]e have to commit ourselves to work on a solution which respects both principles, and allows each treaty regime to retain the integrity of its principles and practices'

165 However, it can be argued that the mandate that the IMO gets from the *Kyoto Protocol* is more specific than it gets from the *IMO Convention*. See ch. 4, 4.2.

166 *Montreal Protocol on Substances that Deplete the Ozone Layer*, opened for signature 16 September 1987, 26 ILM 1550 (entered into force 1 January 1989).

167 *Report of the Marine Environment Protection Committee on Its Fifty-Second Session*, MEPC 52nd Session, Agenda Item 24, IMO Doc MEPC 52/24 (18 October 2004) para. 4.44.

can be attributed to the historical development of the shipping industry in developed countries.[168] On the other hand, developed countries control the majority of the world deadweight tonnage.[169] For these reasons, the application of the NMFT principle would place the technologically disadvantaged developing countries in a worse position for development due to their lack of 'survival emissions'.[170]

Finally, as a response to a criticism that the application of the CBDR principle to the GHG issue would possibly make most ships exempt from the global reduction regulations due to the existence of the FOC, China asserted that this concern could be addressed. In China's view, the beneficially-owned tonnage could be targeted in a way that was utilised by the Review of Maritime Transport by UNCTAD, which is based on the data supplied by Lloyd's Register-Fairplay.[171] China suggested that the nationality of ships (flag State) be defined as the nationality of shipowners for the purpose of applying the CBDR principle in the context of GHG emissions from international shipping.[172] It seems that in this way the application of the CBDR principle would not make the ships, which are owned by the nationals of developed States but are flying the flags of developing States, exempt from compulsory reduction commitments. However, shipowners may be companies or other business entities in law. It is thus possible that the nationals of developed States register their companies in developing States investing in ships so as to avoid the stringent regulations.

168 *Application of the Principle of "Common but Differentiated Responsibilities" to the Reduction of Greenhouse Gas Emissions from International Shipping*, submitted by China and India, MEPC 58th Session, Agenda Item 4, IMO Doc MEPC 58/4/32 (15 August 2008) para. 4.

169 Ibid. para. 5.

170 Ibid. para. 4. In this context, the 'survival emissions' refer to the heavy reliance of many developing countries on necessary emissions associated with their shipping industry. The enforcement of stringent IMO regulations for GHG emissions reductions may throw some substandard ships built in developing countries out of the international market, and thus make the economic situations in these developing countries worse. See also Mark J. Mwandosya, *Survival Emissions: A Perspective from the South on Global Climate Change Negotiations* (Dar es Salaam University Press and the Centre for Energy, Environment, Science and Technology, Tanzania, 2000).

171 *Application of the Principle of "Common but Differentiated Responsibilities" to the Reduction of Greenhouse Gas Emissions from International Shipping*, submitted by China and India, MEPC 58th Session, Agenda Item 4, IMO Doc MEPC 58/4/32 (15 August 2008) para. 5.

172 Ibid. China asserted that, 'the definition of countries of genuine control or the nationality being defined as the country of domicile for ship owners is the methodology used in UNCTAD's Review'. It believed that this methodology could also be used for the GHG issue to draw a clear line between the ships of Annex I countries and those of non-Annex I countries.

These five reasons adequately underpin the application of the CBDR principle to the GHG issue. However, China has not fully justified why Article 2(2) of the *Kyoto Protocol* is the only mandate in regulating the GHG issue that the IMO has received so far. In other words, the reason why the IMO's mandate in regulating the GHG issue is not given by the *LOSC* and *IMO Convention* remains unaddressed, although as discussed earlier China also agreed that the *IMO Convention* gives the IMO this competence in regulating the GHG matter.

With regard to the approach of the application of the CBDR principle to this GHG issue, China asserted that this principle should be applied to all three categories of reduction measures, namely technical, operational measures and MBMs. As discussed in Chapter 2, in a broad sense the 'differentiated responsibility' element of the CBDR principle consists of three categories: differentiated central obligations, differentiated implementation arrangements, and the granting of assistance including financial and technological assistance.[173] China made proposals in relation to all these three scenarios to the energy efficiency measures being discussed within the IMO, although two of these proposals have not received positive responses by other IMO member States. At the 61st MEPC meeting in 2010, China proposed an additional paragraph to the draft regulatory text on the GHG issue, which provided that, '[t]he application of EEDI should be mandatory [for] developed countries and voluntary [for] developing countries'.[174] This proposal reflects China's interpretation on applying the CBDR principle to this GHG issue. That is, to impose differentiated central obligations on various States. However, this proposal was not considered by the Working Group on Energy Efficiency Measures to Ships, a group commissioned by the IMO. At the 62nd MEPC meeting in 2011, China, co-sponsored by Saudi Arabia and South Africa, proposed a draft text to the IMO. This proposed article provided that,

> The regulations of EEDI and SEEMP shall apply to ships of developing countries five years after the date of their entry into force.
>
> Or
>
> The regulations of EEDI and SEEMP shall be phased in over a period of eight years for ships built for developing countries and during the period

173 Lavanya Rajamani, *Differential Treatment in International Environmental Law* (Oxford University Press, 2006) 191.

174 *Report of the Working Group on Energy Efficiency Measures for Ships*, MEPC 61st Session, IMO Doc MEPC 61/WP.10 (30 September 2010) para. 4.31.

of phasing in, developing countries shall only apply 50% of the required EEDI reduction rate.[175]

This proposed article offered a phased-in approach for developing countries in implementing the EEDI and SEEMP. This approach belongs to the 'differentiated implementation arrangement' element of the CBDR principle. These two proposed provisions have different implications: the first proposal applies to ships based on their registration from developing countries while the second applies to shipowners from developing countries. Due to the concern for the FOC, the second proposal seems more feasible. However, neither of these proposals was accepted by most countries. Under these circumstances, after the adoption of the 2011 amendments of Annex VI to *MARPOL 73/78*, China turned to the last option of partially incorporating the CBDR principle into the energy efficiency measures. The recognition of the CBDR principle was eventually written into the MEPC resolution on technical cooperation and transfer of technology. Given this outcome, China expressed that it 'would like to actively participate in related discussion under the guidance of this principle'.[176] However, whether the CBDR principle can be reflected in the implementation of this resolution is still in doubt.[177]

China has been a persistent opponent of MBMs being applied to the GHG issue. However, China has suggested that, if a MBM is to be adopted, the CBDR principle should apply in a manner that 'no extra financial responsibility' will be imposed on developing countries.[178] To achieve this goal, China has proposed two principles. First, the basic principles and key elements of the MBMs should be determined by the *UNFCCC*. Second, any funds generated from any MBM should only be provided to the shipping sector in developing countries.[179] Through comparing China's claims with the current MBM proposals, it seems that it will not be straightforward to meet China's proposal.

175 *Comments on the Proposed Mandatory Energy Efficiency Regulations*, submitted by China, Saudi Arabia and South Africa, MEPC 62nd Session, Agenda Item 5, IMO Doc MEPC 62/5/10 (5 May 2011) para. 14.

176 *Report of the Marine Environment Protection Committee on Its Sixty-Fifth Session*, MEPC 65th Session, Agenda Item 22, IMO Doc MEPC 65/22 (24 May 2013) annex 5, p. 4.

177 See ch. 4, 4.3.3.3, ch. 7, 7.4.2.

178 *Report of the Marine Environment Protection Committee on Its Sixty-First Session*, MEPC 61st Session, Agenda Item 24, IMO Doc MEPC 61/24 (6 October 2010) annex 3, p. 3.

179 Ibid.

6.3.2.2.2 *Technical Aspects of the Energy Efficiency Measures*

As asserted by China's shipping industry, shipping industries in developing countries including China are generally technologically disadvantaged.[180] Nevertheless, China still contributed to the technical improvement of the energy efficiency measures. Before the adoption of the EEDI and SEEMP by the IMO in 2011, China's involvement in this technical work consisted of two aspects. In the first place, China, through conducting sea trial tests and technical research, provided technical proposals for improving the EEDI formula, the EEDI reference line, and the calculation methods of the EEDI reference line. Through conducting this work, China asserted at the 62nd MEPC meeting in 2011 that there were still technical uncertainties in relation to the proposed EEDI and SEEMP,[181] and suggested that more time should be given to address these problems before the adoption of these measures. Indeed as early as the 59th MEPC meeting in 2009, China lodged a statement at the meeting which advocated that various countries be patient with technical and technological matters on the grounds that the EEDI is technically complicated and time consuming, and thus requires more in-depth study.[182] Nevertheless, this proposal was not accepted and the amendment was adopted at that meeting.

In the second place, China underscored that the weak voices from developing countries should be respected. Developed countries should help developing countries to strengthen their capability, so that these developing countries could effectively participate in the IMO discussions and put forward their own proposals.[183] An example is the IMO commissioned study on estimated CO_2 emission reductions associated with the EEDI and SEEMP in 2011. China asserted that this study was not reliable due to its uncertainties in future emission projections, inaccuracy of the database used, and other deficiencies.[184] Since most assessments were conducted by institutions from

---

180 See ch. 5, 5.4.1.

181 For example, China asserted that the CO_2 technologies were still at an early stage of development, and guidelines were still under discussion and development. *Comments on the Proposed Mandatory Energy Efficiency Regulations*, submitted by China, Saudi Arabia and South Africa, MEPC 62nd Session, Agenda Item 5, IMO Doc MEPC 62/5/10 (5 May 2011) para. 7.

182 *Report on the Outcome of the Second Intersessional Meeting of the Working Group on Greenhouse Gas Emissions from Ships*, MEPC 59th Sesssion, Agenda Item 4, IMO Doc MEPC 59/4/2 (8 April 2009) annex 1, p. 1.

183 *Report of the Marine Environment Protection Committee on Its Sixtieth Session*, MEPC 60th Session, Agenda Item 22, IMO Doc MEPC 60/22 (12 April 2010) annex 4, p. 3.

184 *Report of the Marine Environment Protection Committee on Its Sixty-Third Session*, MEPC 63rd Session, Agenda Item 23, IMO Doc MEPC 63/23 (14 March 2012) para. 4.41.

developed countries, China believed that more participation of developing countries in this regulatory process would better balance the interests of both developed countries and developing countries. Accordingly, China proposed that the requirement for reduction rates and applicable target years for the EEDI requirements be lowered. Specifically, with regard to the proposed reduction rate being 10 per cent in the first phase, China proposed it should be 0 (zero) and a period of five years should be given for preparations as phase 0.[185] As a compromise, eventually a two-year phase 0 was provided and a 10 per cent reduction at phase 1 maintained. This outcome, however, is significant for the shipping industries in developing countries and enables them to conduct technological preparations.

After the adoption of the energy efficiency measures, China submitted proposals to the IMO in terms of seeking an interpretation of some technical terms, expanding the coverage of the EEDI, and improving EEDI guidelines. Meanwhile, as some States criticised the SEEMP for its ineffectiveness, China took the view that experience was required of the SEEMP before any amendments could be considered.[186] This view is logical in that the SEEMP was newly adopted and only entered into force on 1 January 2013. Only after a measure has been implemented for a period, can certain deficiencies be better identified for further amendment.

6.3.2.2.3 *The Adoption of the Amendments to Annex VI to MARPOL 73/78*

There was fierce debate among IMO members on a number of issues during the course of negotiations on the proposed energy efficiency measures. Of these issues, the form of the legal instrument, circulation procedure, and voting mechanism are worthy of special mention in this section due to their significance to the issue under discussion.

Three options were available to the MEPC regarding the form of the instrument to be adopted for regulating energy efficiency measures. They are an amendment of Annex VI to *MARPOL 73/78*, Annex VII to *MARPOL 73/78*, and a new international convention.[187] China's position was that proposed energy

185 *Comments and Proposals on the Draft Regulation Text on Energy Efficiency for Ships*, submitted by China, MEPC 61st Session, Agenda Item 5, IMO Doc MEPC 61/5/26 (6 August 2010) para. 9.2.

186 *Report of the Marine Environment Protection Committee on Its Sixty-Fourth Session*, MEPC 64th Session, Agenda Item 23, IMO Doc MEPC 64/23 (11 October 2012) para. 4.97.

187 See Consideration of the Appropriate Instrument for A Mandatory Regime to Address GHG Emissions, submitted by the United Kindom, MEPC 58th Session, Agenda Item 4, IMO Doc MEPC 58/4/15 (1 August 2008).

efficiency regulation should remain under voluntary application or only be compulsory for developed countries.[188] If the regulation was to be compulsory, China supported either a new treaty or a new *MARPOL* protocol.[189] China provided three factors to underpin its argument, and in particular, why it opposed an amendment of Annex VI to *MARPOL 73/78* as the legal format for regulating this GHG issue. First, CO_2 is not a pollutant and should not be included in the Annex of *MARPOL 73/78* where severe air pollutants are addressed.[190] Besides, CO_2 is a GHG with cumulative effects, which does not match the definition of 'harmful instances' as regulated under Article 1 of *MARPOL 73/78*.[191] In other words, regulating this GHG emissions issue under the Annex VI to *MARPOL 73/78* is inconsistent with *MARPOL*'s objectives.[192] Second, regulating the energy efficiency measures in Annex VI to *MARPOL 73/78* would cause 'tremendous domestic legal obstacles' for some countries, and make future implementation and enforcement of these measures questionable.[193] Third, Annex VI also contains regulations on other air pollutants, which might represent barriers for countries to adopt mandatory energy efficiency measures and thus is not the appropriate legal instrument.[194] For these reasons, China asserted that the proposed energy efficiency measures should not be regulated under Annex VI to *MARPOL 73/78*, although this was the quickest path to implementing such requirements as mandatory measures.

GHGs are currently not regulated as a type of pollution in Chinese domestic law. The *Air Pollution Prevention and Control Law of China* (adopted in 1987 and

---

188 *Report of the Marine Environment Protection Committee on Its Sixty-First Session*, MEPC 61st Session, Agenda Item 24, IMO Doc MEPC 61/24 (6 October 2010) annex 3, p. 2.

189 Ibid.

190 *Report of the Marine Environment Protection Committee on Its Sixtieth Session*, MEPC 60th Session, Agenda Item 22, IMO Doc MEPC 60/22 (12 April 2010) Annex 4, p. 2; *Report of the Marine Environment Protection Committee on Its Sixty-First Session*, MEPC 61st Session, Agenda Item 24, IMO Doc MEPC 61/24 (6 October 2010) annex 3, p. 2.

191 *Comments on the Proposed Mandatory Energy Efficiency Regulations*, submitted by China, Saudi Arabia and South Africa, MEPC 62nd Session, Agenda Item 5, IMO Doc MEPC 62/5/10 (5 May 2011) para. 5.

192 Article 1(1) of the *MARPOL 73/78* provides that, the objective of *MARPOL* is 'in order to prevent the pollution of the marine environment by the discharge of harmful substances or effluents containing such substances in contravention of the Convention'. Ibid.

193 *Report of the Marine Environment Protection Committee on Its Sixtieth Session*, MEPC 60th Session, Agenda Item 22, IMO Doc MEPC 60/22 (12 April 2010) annex 4, p. 2.

194 *Comments on the Proposed Mandatory Energy Efficiency Regulations*, submitted by China, Saudi Arabia and South Africa, MEPC 62nd Session, Agenda Item 5, IMO Doc MEPC 62/5/10 (5 May 2011) para. 5.

amended in 1995, 2000 and 2015) does not regulate GHGs as pollutants. It is predicted that GHGs will not be regulated as a type of pollution in the short term due to pressure from various national industries. This opinion was supported by some *UNFCCC* non-Annex I States when the issue was discussed within the IMO.[195] Nevertheless, given the fact that energy efficiency measures have been adopted, these member States will need to update their domestic law so as to comply with the IMO instrument. For China, since its classification society has released its *Rules for Green Ships* which has incorporated the energy efficiency requirement,[196] it seems that compliance with these IMO rules by China's shipping industry would not be a significant concern.

The circulation procedure of the proposed mandatory energy efficiency regulations within the IMO is another issue which was opposed by China. Based on the agenda of the 62nd MEPC meeting in 2011, this meeting would consider, with a view to adoption, draft amendments of Annex VI to *MARPOL 73/78*.[197] A drafting group was also scheduled to be established for modifying the draft.[198] This agenda is consistent with *MARPOL 73/78*, which sets out a two-phased amendment adoption procedure, namely submitting the draft to an appropriate body by the IMO for consideration,[199] and the adoption by the IMO.[200] However, at this meeting the MEPC decided to circulate and adopt the draft at the same session. From the perspective of China, this decision to have the proposed amendments circulated in spite of there being no approval by the MEPC,

195 This view has been supported by Brazil, India, Saudi Arabia and the Bolivarian Republic of Venezuela. For example, the delegation of Saudi Arabia stated that '[MARPOL] was for the prevention of pollution into the marine environment. Greenhouse gas emissions are not classified by the UN as pollutants'. *Report of the Marine Environment Protection Committee on Its Sixty-Second Session*, MEPC 62nd Session, Agenda Item 24, IMO Doc MEPC 62/24 (26 July 2011) annex 20, p. 3.

196 See ch. 5, 5.4.1.

197 *Adoption of the Agenda: Annotations and Provisional Timetable*, note by the Secretariat, MEPC 62nd Session, Agenda Item 1, IMO Doc MEPC 62/1/1 (19 November 2010) para. 6.1.

198 Ibid. para. 6.2.

199 *MARPOL 73/78* art. 16(2)(b). This provision provides that 'any amendment proposed and circulated as above shall be submitted to an appropriate body by the Organization for consideration'. As a response to China's request on the meaning of 'the Organization' in this provision, the Legal Office of the IMO interpreted it as 'the IMO Secretariat' in the context of GHG emissions from international shipping. *Report of the Marine Environment Protection Committee on Its Sixty-Second Session*, MEPC 62nd Session, Agenda Item 24, IMO Doc MEPC 62/24 (26 July 2011) para. 6.49.

200 *MARPOL 73/78* art. 16(2)(d). This provision provides that 'amendments shall be adopted by a two-thirds majority of only the Parties to the Convention present and voting'

violated the IMO's customary procedure,[201] undermined the founding basis of the IMO, and also conflicted with the spirit of collectivism of the United Nations in dealing with major global issues.[202] While China's response to this procedural flaw was understandable, the adoption of this regulation appears to have been unavoidable due to the support for this regulation by the majority of member countries of Annex VI to *MARPOL 73/78*.

Both *MARPOL 73/78* and the MEPC adopt the majority-voting mechanism.[203] Being aware of the vital role of the current voting mechanism in regulating GHG emissions from ships, China, as well as many other developing countries, has attempted to replace this majority-voting mechanism with a consensus-based arrangement as to future GHG regulation. China took the view that a two-thirds majority vote should not be used to settle every issue, and the regulation of GHG emissions from ships should be made by consensus.[204] From China's perspective, climate change is a common challenge faced by the global community and should be addressed through multilateral cooperation mechanisms.[205] As 'the essence of the multilateralism lies in the degree of flexibility by which the common ground would be reached and the resolution would be accepted by all parties',[206] to date all agreements in relation to climate change, including the *UNFCCC*, its *Kyoto Protocol* and their relevant decisions, have been reached by consensus. GHG emissions from ships are a part of the global climate change issue and thus should also be addressed by unanimous agreement.

---

201 *Comments on the Note by the Secretariat on Amendments to* MARPOL *Annex VI—Inclusion of Regulations on Energy Efficiency for Ships*, submitted by Argentina, Brazil, Chile, China, Ecuador, India, Nicaragua, Peru, the Philippines, South Africa and Venezuela, MEPC 62nd Session, Agenda Item 6, IMO Doc MEPC 62/6/15 (6 May 2011) para. 4.

202 *Comments on the Proposed Mandatory Energy Efficiency Regulations*, submitted by China, Saudi Arabia and South Africa, MEPC 62nd Session, Agenda Item 5, IMO Doc MEPC 62/5/10 (5 May 2011) para. 6.

203 *MARPOL 73/78* art. 16(2)(d); International Maritime Organization (IMO), *Basic Documents Volume I* (International Maritime Organization, 2010) 113. Rule 27 of the Rules of Procedures of the MEPC provides that decisions of the committee and of its subsidiary bodies are made by a majority of the members present and voting rather than by a consensus. Whereas Article 16(2)(d) of the MARPOL 73/78 requires the adoption by a two-thirds majority of only the Parties to the Convention present and voting.

204 *Comments on the Proposed Mandatory Energy Efficiency Regulations*, submitted by China, Saudi Arabia and South Africa, MEPC 62nd Session, Agenda Item 5, IMO Doc MEPC 62/5/10 (5 May 2011) para. 12.

205 *Report of the Marine Environment Protection Committee on Its Sixty-Third Session*, MEPC 63rd Session, Agenda Item 23, IMO Doc MEPC 63/23 (14 March 2012) annex 14, p. 2.

206 Ibid.

At the 61st MEPC meeting in 2010, the secretary-general of the MEPC advocated that this GHG issue should be addressed by consensus.[207] However, it proved that a consensus could not be reached on the adoption of the energy efficiency measures within the IMO and thus the majority-voting mechanism was applied to this matter. In terms of the regulation of this GHG issue, there is growing diversity in the regulatory interests of developing countries,[208] as well as those between developed and developing countries. For this reason, it seems less likely that the MEPC would change this voting mechanism in deciding this GHG issue, although some developing countries are currently advocating for this change.[209]

Given that the energy efficiency measures were adopted by a majority vote within the IMO in 2011, China opposed the adoption of these measures and lodged a statement to the report of the 62nd MEPC meeting in 2011. In this statement China asserted that the CBDR principle was not reflected in the amendments to *MARPOL* Annex VI 'in a full and objective manner', and thus it opposed this amendment and was 'in no position to acknowledge and accept the amendment'.[210] However, the nature of this statement is not a reservation but rather an 'objection' that is 'communicated to the Organization' before an amendment is deemed to be accepted. This right of relevant Parties is provided by *MARPOL 73/78*. Article 16(2)(f)(iii) of *MARPOL 73/78* provides,

> an amendment to an appendix to an Annex to the Convention shall be deemed to have been accepted at the end of a period to be determined by the appropriate body at the time of its adoption, which period shall be not less than ten months, unless within that period *an objection is com-*

---

207 *Comments on the Proposed Mandatory Energy Efficiency Regulations*, submitted by China, Saudi Arabia and South Africa, MEPC 62nd Session, Agenda Item 5, IMO Doc MEPC 62/5/10 (5 May 2011) para. 12. At the opening of the 61st MEPC meeting, the secretary-general stated that 'given the seriousness of the contemplated measures and the need to ensure their wide and effective implementation, I see no way to make decisions on them other than by consensus'

208 Generally the regulatory interests on combating climate change are different as to major developing countries (e.g., China, India, South Africa), major flag of convenience States (e.g., Panama, Liberia), and other developing countries. A detailed discussion on this classification is provided at the conclusion of this chapter.

209 See, e.g., *Further Work on GHG Emissions from Ships*, submitted by Brazil, China, India, Peru, Saudi Arabia and South Africa, MEPC 64th Session, Agenda Item 5, IMO Doc MEPC 64/5/9 (27 July 2012) para. 8.1.

210 *Report of the Marine Environment Protection Committee on Its Sixty-Second Session*, MEPC 62nd Session, Agenda Item 24, IMO Doc MEPC 62/24 (26 July 2011) annex 20, pp. 1–2.

> *municated to the Organization* by not less than 50 per cent of the gross tonnage of the world's merchant fleet whichever condition is fulfilled. [emphasis added]

Since this provision does not stipulate what form such an 'objection' should take, the Chinese statement could be interpreted as an objection to this provision. However, this will not be enough to prevent the application of the regulation to China. In order to be exempted from this regulation, China could have made a 'declaration' that it did not accept this amendment. Article 16(2)(g)(ii) of the *MARPOL 73/78* provides that,

> ... the amendment deemed to have been accepted in accordance with the foregoing conditions shall enter into force six months after its acceptance for all the Parties *with the exception of* those which, before the date, have *made a declaration that they do not accept it* or a declaration under subparagraph (f)(ii), that their express approval is necessary. [emphasis added]

It is customary that a declaration is communicated in writing to the IMO and then circulated to the Parties by the Secretariat.[211] Therefore, there is a procedural difference between the above 'objection' and the 'declaration'. As discussed in Chapter 5, China's shipbuilding sector has a growing role in China's shipping industry and China's economy. In order to meet the requirements from its current and potential customers, China's shipbuilding sector will also need to meet the energy efficiency measures adopted by the IMO. Probably, for this reason, China did not object to the 2011 amendments to Annex VI by means of a declaration as regulated under Article 16(2)(g)(ii) of the *MARPOL 73/78*, although the latter allows it to do so to avoid the application of the amendments to China.

6.3.2.2.4 *Technical Cooperation and Transfer of Technology*

China has been a strong supporter of including effective transfer of technology from developed countries to developing countries into the regulation of GHG emissions reduction from ships. From China's point of view, the wide application and transfer of energy efficiency technologies will significantly strengthen

211 *MARPOL 73/78* art. 16(9). This provision provides that '[a]ny declaration of acceptance or of objection to an amendment under the present article shall be notified in writing to the Secretary-General of the Organization. The latter shall bring such notification and the date of its receipt to the notice of the Parties to the Convention'.

the capability of developing countries so as to help them effectively implement energy efficiency measures adopted by the IMO.[212] The *UNFCCC*, the *Bali Action Plan* and the *Cancun Agreements* all stipulate the transfer of technology requirements,[213] and this mechanism should also be included in the amended Annex VI to *MARPOL 73/78*.[214]

Before the adoption of the 2011 amendments of Annex VI, China, co-sponsored by Saudi Arabia and South Africa, proposed a draft legal provision entitled 'promotion of technical assistance and capacity building'. This proposed regulation provides that,

> In order to promote the reduction of GHG emissions from international shipping, *transparency of technology shall be increased* in the implementation of technical measures of the EEDI. All new ship designs and technology which reduce the attained EEDI value of a ship *shall be open to the public*. Developed countries *shall transfer their technology and provide financial support to* developing countries for their capacity building so as to enhance their ability to satisfy these new requirements.[215] [emphasis added]

This proposal underscores the transparency of energy efficiency technologies, the transfer of technology and financial support from developed countries to developing countries. Indeed these proposals, if adopted, would address the main concern from developing countries in complying with the proposed

---

212 *Comments on the Proposed Mandatory Energy Efficiency Regulations*, submitted by China, Saudi Arabia and South Africa, MEPC 62nd Session, Agenda Item 5, IMO Doc MEPC 62/5/10 (5 May 2011) para. 9.

213 Ibid. See also *UNFCCC* art. 4(5); *Bali Action Plan*, Decision 1/CP.13, Report of the Conference of the Parties on its Thirteenth Session, Doc FCCC/CP/2007/6/Add.1 (14 March 2008) art. 2(d); *The Cancun Agreements*, Decisions 1–2/CMP.6, Report of the Conference of the Parties serving as the Meeting of the Parties to the Kyoto Protocol on its Sixth Session, FCCC/KP/CMP/2010/12/Add.1 (15 March 2011); Decision 1/CP. 16, Report of the Conference of the Parties on its Sixteenth Session, FCCC/CP/2010/7/Add.1 (15 March 2011); *Bali Action Plan*, Decision 1/CP.13, Report of the Conference of the Parties on its Thirteenth Session, Doc FCCC/CP/2007/6/Add.1 (14 March 2008) Decision 1/CP.16 art. 1.

214 *Comments on the Proposed Mandatory Energy Efficiency Regulations*, submitted by China, Saudi Arabia and South Africa, MEPC 62nd Session, Agenda Item 5, IMO Doc MEPC 62/5/10 (5 May 2011) para. 9.

215 Ibid. para. 15; See also *Report of the Marine Environment Protection Committee on Its Sixty-First Session*, MEPC 61st Session, Agenda Item 24, IMO Doc MEPC 61/24 (6 October 2010) para. 5.45.

energy efficiency measures. However, in developed countries there are various domestic regulations on intellectual property protection,[216] and most energy efficient technologies are owned by private shipping companies. Therefore, it would be difficult for developed countries to accept this proposal. Indeed, the finally adopted Regulation 23 of the amended Annex VI to *MARPOL 73/78* ignored these expressions in relation to the transparency of technologies and financial support, and instead added that any transfer of technology should be subject to national laws, regulations and policies.[217]

After the adoption of energy efficiency measures, China advocated the adoption of an MEPC resolution on technical cooperation and transfer of technology relating to the improvement of energy efficiency of ships. It took the view that without such a resolution the MEPC could not make any smooth progress in reducing GHG emissions from ships,[218] and this resolution should be adopted prior to entry into force of the amendments to Annex VI.[219] Meanwhile, China grouped technology transfer transactions into five categories,[220] and emphasised that the role of the public sector should be strengthened although most of the technological transfer occurs in the private sector.[221] It provided

216 James Harrison, 'Recent Developments and Continuing Challenges in the Regulation of Greenhouse Gas Emissions from International Shipping' (2012) *University of Edinburgh Research Paper Series* <http://ssrn.com/abstract=2037038> accessed 15 September 2013, p. 17.

217 *MARPOL 73/78 Annex VI (2001 amendments)* reg. 23.2.

218 *Report of the Marine Environment Protection Committee on Its Sixty-Third Session*, MEPC 63rd Session, Agenda Item 23, IMO Doc MEPC 63/23 (14 March 2012) annex 13, p. 2.

219 *Promotion of Technical Cooperation and Transfer of Technology Relating to the Improvement of Energy Efficiency of Ships*, submitted by Angola, China, Jamaica, South Africa and Venezuela, MEPC 64th Session, Agenda Item 4, IMO Doc MEPC 64/4/24 (27 July 2012) para. 1.

220 Ibid. para. 5. These categories include 'the assignment, sale and licensing of all forms of industrial property'; 'the provision of know-how and technical expertise'; 'the provision of technological knowledge necessary for the installation, operation and functioning of plant and equipment; the provision of technological knowledge necessary to acquire, install and use machinery equipment, intermediate goods and/or raw materials which have been acquired by purchase, lease or other means; and the provision of technological contents of industrial and technical cooperation arrangements.'

221 Ibid. para. 7. China asserted that since the transfer of technology cannot be conducted free of charge, developed countries should provide various political, legal and policy incentives for their private sectors to facilitate this work. Indeed this proposal is consistent with Article 66(2) of the Agreement on Trade-Related Aspects of Intellectual Property Rights (TRIPS). See *Agreement on Trade-Related Aspects of Intellectual Property Rights (TRIPS)*,

approaches for facilitating technology transfer, and asserted that the terms of reference should at least include:

- 'Nature and composition and location of the team managing the process;
- Nature of technology transfer projects to be supported;
- Identification and cataloguing of available energy efficiency technologies;
- Procedures for managing requests for technology transfer;
- Sources of funding and financing; and
- Evaluation and reporting.'[222]

Based on these arguments, China, supported by some developing countries, proposed a draft MEPC resolution on this issue to the IMO. In this proposal, China underscored the CBDR principle and suggested establishing an Ad Hoc Expert Working Group on Technology Transfer with a mandate to facilitate this work.[223] As summarised by the chairman of the MEPC, the main debate on the proposed MEPC resolution involved three issues, namely: the CBDR principle, technology transfer and funding.[224] These three issues were recognised in the adopted MEPC resolution on technical cooperation and transfer of technology in May 2013. As discussed in Chapter 4, although there were no concrete obligations on any State, it was a breakthrough for developing countries to have the recognition of the CBDR principle in the preamble paragraphs of this resolution. Regarding this achievement, China asserted that the CBDR principle provided 'a sound foundation and guidance for further discussion on GHG emissions under [the] IMO', and thus it would like to actively participate in the IMO discussions under the guidance of this principle.[225] Nevertheless, China was also concerned about the provision on intellectual property protection which from the perspective of China would seriously impair the transfer of technology.[226]

Annex 1C of the *Marrakesh Agreement Establishing the World Trade Organization*, opened for signature 15 April 1994, 33 ILM 1125 (entered into force 1 January 1995) art. 66(2).

222 Ibid. para. 12.

223 *Draft MEPC Resolution on Promotion of Technical Cooperation and Technology Transfer Relating to the Improvement of Energy Efficiency of Ships*, submitted by Angola, Argentina, Brazil, China, India, Jamaica, Nigeria, Peru, South Africa and Venezuela, MEPC 64th Session, Agenda Item 4, IMO Doc MEPC 64/4/30 (27 July 2012) annex.

224 *Report of the Marine Environment Protection Committee on Its Sixty-Fourth Session*, MEPC 64th Session, Agenda Item 23, IMO Doc MEPC 64/23 (11 October 2012) para. 4.8.

225 *Report of the Marine Environment Protection Committee on Its Sixty-Fifth Session*, MEPC 65th Session, Agenda Item 22, IMO Doc MEPC 65/22 (24 May 2013) annex 5, p. 4.

226 Ibid.

Since this was the first time that the IMO had introduced the CBDR principle in its resolution,[227] it is believed that this achievement was largely due to the consensus approach adopted by the Working Group on Draft MEPC Resolution on Promotion of Technical Cooperation and Transfer of Technology relating to this energy efficiency issue.[228] At the 64th MEPC meeting in 2012, China and some other developing countries suggested that this consensus approach should be employed to deal with all decisions of the MEPC in relation to this GHG issue, but the discussion of this proposal has been postponed. At the 66th MEPC meeting in 2014, China, supported by many other developing countries, proposed to facilitate the implementation of MEPC resolution on technical cooperation and transfer of technology.[229] They treated the effective implementation of this resolution as 'a top priority' of the IMO's work in addressing this GHG issue, and suggested that further technical and operational measures for enhancing energy efficiency of international shipping should not commence until this resolution has been effectively implemented.[230] This request, however, was not agreed by the Committee.

6.3.2.2.5 *Market-based Measure Proposals*

As discussed earlier, China asserted that the regulation of MBMs in relation to this GHG issue is beyond the competence of the IMO, and claimed that the IMO should only study the methodology and operational feasibility while leaving the regulatory principles for the *UNFCCC*.[231] Apart from these positions, China's views on current MBM proposals mainly consist of two aspects. On the one hand, China asserted that currently MBM proposals on tackling GHG emissions from ships are premature and thus should not be adopted. It argued that there are many uncertainties in relation to the carbon market, the

227 Ibid.

228 See *Report of the Marine Environment Protection Committee on Its Sixty-Third Session*, MEPC 63rd Session, Agenda Item 23, IMO Doc MEPC 63/23 (14 March 2012) para. 4.72.1.

229 *Implementation of Resolution MEPC.229(65) on the 'Promotion of Technical Co-operation and Transfer of Technology Relating to the Improvement of Energy Efficiency of Ships'*, submitted by Angola, Argentina, Brazil, China, Cuba, India, Malaysia, Nigeria and South Africa, MEPC 66th Session, Agenda Item 4, IMO Doc MEPC 66/4/17 (24 January 2014).

230 *Further Technical and Operational Measures for Enhancing Energy Efficiency of International Shipping: Comments on Enhancing Energy Efficiency in Internaitonal Shipping*, submitted by Angola, Argentina, Brazil, China, Cuba, Malaysia, Nigeria and South Africa, MEPC 66th Session, Agenda Item 4.1, IMO Doc MEPC 66/4/25 (7 February 2014) para. 13.

231 See *Report of the Marine Environment Protection Committee on Its Sixty-First Session*, MEPC 61st Session, Agenda Item 24, IMO Doc MEPC 61/24 (6 October 2010) annex 3, p. 3.

calculation of the emissions, and the impacts of these measures on the shipping industry and world trade.[232] Meanwhile, these MBM proposals do not incorporate the CBDR principle well and may potentially distort competition.[233] With regard to specific MBM proposals, China treated the global levy on the marine bunker fuel proposal as an international tax and asserted that this MBM violates the CBDR principle, and if adopted would have a range of legal implications.[234] As to the proposed ETS, China asserted that an ETS would violate the consensus achieved within the *UNFCCC* by applying this scheme to all ships. It believed that under this scheme, resources would flow from the poor to the rich by penalising developing countries.[235] On the other hand, China also took a compromise position that if MBMs are to be adopted, the CBDR principle should be incorporated. China held the view that an MBM for international shipping should include three objectives.[236] They are encouraging and promoting *UNFCCC* Annex I States to reduce GHG emissions from ships; using the market as a means to enhance actions relating to financial support and the transfer of technologies; and contributing to the sustainable development of the shipping industries in developing countries.[237] Overall these objectives underscore the CBDR principle, in particular the capacity building of developing countries. Furthermore, China stressed that the discussions of MBMs must await the improvement of adopted energy efficiency measures.[238] The consensus approach adopted for the MEPC resolution on technical cooperation and transfer of technology should also apply to MBM-related research and discussions.[239]

---

232 *Uncertainties and Problems in Market-based Measures*, submitted by China and India, MEPC 61st Session, Agenda Item 5, IMO Doc MEPC 61/5/24 (5 August 2010) paras. 2–5.

233 Ibid. paras. 6–7.

234 *Report of the Outcome of the First Intersessional Meeting of the Working Group on Greenhouse Gas Emissions from Ships*, MEPC 58th Session, Agenda Item 4, IMO Doc MEPC 58/4 (4 July 2008) para. 5.20.

235 Ibid. para. 5.31.

236 *Report of the Third Intersessional Meeting of the Working Group on Greenhouse Gas Emissions from Ships*, MEPC 62nd Session, Agenda Item 5, IMO Doc MEPC 62/5/1 (8 April 2011) para. 2.17.

237 Ibid.

238 *Further Work on GHG Emissions from Ships*, submitted by Brazil, China, India, Peru, Saudi Arabia and South Africa, MEPC 64th Session, Agenda Item 5, IMO Doc MEPC 64/5/9 (27 July 2012) para. 8.6.

239 *Report of the Marine Environment Protection Committee on Its Sixty-Third Session*, MEPC 63rd Session, Agenda Item 23, IMO Doc MEPC 63/23 (14 March 2012) annex 13, p. 3.

6.3.2.3 Vanuatu

In contrast to maritime powers or newly-emerged shipbuilding developing nations which have important interests in regulating the GHG issue, some developing countries, in particular Least Developed Countries (LDCs) and Small Island Developing States (SIDS), do not have a significant commercial fleet or shipping capability. However, some of these countries, Vanuatu as an example, have actively participated in this regulatory process. As an island archipelago, Vanuatu is made up of around 80 relatively small islands of volcanic origin occupying an area of 12,000 square kilometres.[240] It is located in the South Pacific Ocean and has a population of 251,784 people.[241] Many SIDS, including Vanuatu, are among the most vulnerable to the adverse impacts of climate change. In particular the lowest lying Atolls, with the least capacity to effectively respond, suffer even more from climate change.[242] As such, the response of this category of developing flag States is different from the other two categories of developing flag States that are discussed in previous sections.

At the 58th MEPC meeting in 2008, Vanuatu lodged a statement to the report of the meeting. In this statement, Vanuatu asserted that some Pacific Micro States are already predicted to be submerged due to sea level rise as a result of global warming.[243] Therefore, Vanuatu is in favour of a global regulation on reducing GHG emissions from international shipping, and this position was echoed by the Cook Islands,[244] another SID in the South Pacific. Accordingly, in terms of regulatory principles Vanuatu supported the position that any future regulations on this GHG issue should be binding and equally applicable to all

240 *Comments on the Proposed Amendments to MARPOL Annex VI*, submitted by Vanuatu, MEPC 62nd Session, Agenda Item 6, IMO Doc MEPC 62/6/23 (20 May 2011) para. 6.

241 Secretariat of the Pacific Community (SPC), *Pacific Island Populations: Estimates and Projections of Demographic Indicators for Selected Years* (2011) <http://www.spc.int/sdd/> accessed 26 August 2013.

242 Pacific Islands Forum Secretariat, *Smaller Island States Leaders Statement on Climate Change* (6 September 2011) <http://www.forumsec.org.fj/resources/uploads/attachments/documents/SIS_Leaders_Outcomes_2011.pdf> accessed 26 August 2013, p. 17.

243 *Report of the Marine Environment Protection Committee on its 58th Session*, IMO Doc MEPC 58/23 (16 October 2008) annex 9, p. 21.

244 *Report of the Marine Environment Protection Committee on Its Sixtieth Session*, MEPC 60th Session, Agenda Item 22, IMO Doc MEPC 60/22 (12 April 2010) Annex 4, p. 12. At the 60th MEPC meeting, the Cook Islands lodged a statement to the report of the meeting which said that it fully supported the global effort in tackling climate change since its overriding concern is one of survival.

flag States in order to avoid evasion,[245] or in other words, the NMFT principle. This position relates to its view that climate change needs 'urgent action'.[246]

Vanuatu actively participated in the discussions within the IMO and expressed its opinions with regard to the proposed technical and operational measures,. At the 60th MEPC meeting in March 2010, Vanuatu suggested an alternative approach for the IMO to provide a short-term contribution to reducing GHG emissions from ships, namely using methane to power auxiliary machinery and explained that the technology is available.[247] However, this proposal did not arouse much attention from the member States, which was probably because this proposal indeed provided a technological option or method to meet the EEDI rather than a technical measure.[248]

At the 61st MEPC meeting in September 2010, Vanuatu submitted a proposal on possible exemptions for ships trading to the LDCs and SIDS from EEDI requirements.[249] This proposal was widely discussed within the IMO. Nevertheless, the majority of delegations present at the IMO did not support this proposal. They were concerned that the adoption of this provision could mean that 'the least efficient ships would serve these trades/routes indefinitely', and would prejudice the benefits of developing countries due to higher transportation costs resulting from this.[250] Indeed current research indicates that under mandatory global emission reduction equally applying to all ships, the cost increases to large developing countries are small while the increased international trade costs for SIDS are significant.[251] For this reason, it is important to address the special needs and circumstances of these LDCs

245 *Comments on the Proposed Amendments to MARPOL Annex VI*, submitted by Vanuatu, MEPC 62nd Session, Agenda Item 6, IMO Doc MEPC 62/6/23 (20 May 2011) para. 6.

246 Ibid. para. 7.

247 *Report of the Marine Environment Protection Committee on Its Sixtieth Session*, MEPC 60th Session, Agenda Item 22, IMO Doc MEPC 60/22 (12 April 2010) para. 4.31, p. 27.

248 Based on the nine fundamental principles agreed at the 57th MEPC meeting, any future regulation on this GHG issue should be 'based on a goal-based approach and not prescribe specific methods'. Therefore, the kinds and methods of technology to adopt are left to the shipping industry.

249 *Report of the Marine Environment Protection Committee on Its Sixty-First Session*, MEPC 61st Session, Agenda Item 24, IMO Doc MEPC 61/24 (6 October 2010) paras. 5.32–5.33.

250 Ibid. para. 5.33.4.

251 Haifeng Wang, 'Economic Costs of CO_2 Emissions Reduction for Non-Annex I Countries in International Shipping' (2010) 14(4) *Energy for Sustainable Development* 280, 285.

and SIDS. However, no steps have been taken by the IMO to address this issue, although the significance of this issue has been recognised.[252]

At the 62nd MEPC meeting in 2011, Vanuatu voted for the adoption of the EEDI and SEEMP by the IMO. Meanwhile, it also put forward two proposals to the IMO underscoring the importance of the safety of future ship design so as to ensure the safety of seafarers, ships and the environment through the application of the amendments.[253] In contrast to its active participation in the discussions on the technical and operational measures, Vanuatu has not expressed its positions or views on MBM proposals within the IMO.

Vanuatu is extremely vulnerable to the adverse impacts of climate change as a LDC and a SIDS. Accordingly, Vanuatu fully supports the IMO in regulating the GHG issue under discussion and has asserted that this regulation should be undertaken urgently. In terms of the proposed EEDI and SEEMP, Vanuatu suggested an exemption from these regulations for ships trading to the LDCs and SIDS so as to protect the economic interests of these countries. Although Vanuatu has not voiced its view on proposed MBMs, its vulnerability to climate change indicates that it would support any global solution which can reduce GHG emissions from ships in a quick and effective manner.

6.4 Port State Control

Port State jurisdiction consists of prescriptive and enforcement jurisdiction,[254] whereas port State control generally only involves the enforcement dimension of this jurisdiction excluding judicial jurisdiction. The IMO provides a definition of port State control:

> [p]ort State [c]ontrol is the inspection of foreign ships in national ports to verify that the condition of the ship and its equipment comply with the

252 *Report of the Marine Environment Protection Committee on Its Sixty-First Session*, MEPC 61st Session, Agenda Item 24, IMO Doc MEPC 61/24 (6 October 2010) para. 5.33.5.

253 *Comments on the Guidelines on the Method of Calculation of the Energy Efficiency Design Index for New Ships*, submitted by Vanuatu, MEPC 62nd Session, Agenda Item 5, IMO Doc MEPC 62/5/31 (20 May 2011); *Comments on the Proposed Amendments to MARPOL Annex VI*, submitted by Vanuatu, MEPC 62nd Session, Agenda Item 6, IMO Doc MEPC 62/6/23 (20 May 2011).

254 See ch. 2, 2.2.4. For the purpose of this book, enforcement jurisdiction is interpreted broadly to include both enforcement jurisdiction and judicial jurisdiction.

requirements of international regulations and that the ship is manned and operated in compliance with these rules.[255]

It appears that the key distinction between port State control and port State jurisdiction is that, under port State control, the port State only takes administrative measures of control rather than prosecuting the vessel for an alleged breach of its legislation.[256] However, the enforcement jurisdiction of the port State includes prosecution for offences committed in its ports or coastal State maritime zones, or outside the internal waters, territorial sea or EEZ of the port State.[257]

While flag States have been allocated primary responsibility for ensuring the compliance of ships on their registers with all applicable international and domestic regulations and standards, port States significantly complement the work of flag States in addressing substandard ships.[258] The legal basis of port State control lies in the customary international law that foreign vessels do not have a general right of access to ports.[259] To date port State control has been incorporated in many international conventions which confirm that port States are entitled to inspect ships in their ports or at off-shore terminals to verify whether they comply with these conventions. The principal conventions are *MARPOL 73/78*,[260] 1974 *International Convention for the Safety of Life at Sea* (*SOLAS*),[261] *1976 ILO Convention (No. 147) concerning Minimum Standards in*

---

255 International Maritime Organization (IMO), *Port State Control* <http://www.imo.org/blast/mainframe.asp?topic_id=159> accessed 22 September 2013. But see John Hare, 'Port State Control: Strong Medicine to Cure A Sick Industry' (1997) 26(3) *Georgia Journal of International and Comparative Law* 571, 571. Hare assets that port state control refers to 'the powers and concomitant obligations vested in, exercised by, and imposed upon a national maritime authority (or its delegate) by international convention or domestic statute or both'. This definition by Hare underscores the sources of powers that port state receives, whereas the IMO's definition highlights the concrete content of these powers.

256 Ho-Sam Bang, 'Recommendations for Policies on Port State Control and Port State Jurisdiction' (2013) 44(1) *Journal of Maritime Law and Commence* 115, 119.

257 *LOSC* art. 218; See also ibid.

258 Bang, above n. 6, 1.

259 Ibid. 720. See also Hare, above n. 255, 572.

260 *MARPOL 73/78* arts. 5,7; reg. 8A of Annex I; reg. 15 of Annex II; reg. 8 of Annex III; reg. 8 of Annex V; reg. 10.5 of Annex VI.

261 *International Convention for the Safety of Life at Sea*, opened for signature 1 November 1974, 1184 UNTS 2 (entered into force 25 May 1980) ('*SOLAS*') reg. 19 of Ch. I, reg. 6.2 of Ch. IX, and reg. 4 of Ch. XI.

Merchant Ships (*ILO Convention No.147*),[262] *1966 International Convention on Load Lines* (*Load Lines*),[263] *and 1978 International Convention on Standards of Training, Certification and Watchkeeping for Seafarers* (*STCW Convention*).[264] Accordingly, ships that do not meet the safety and anti-pollution thresholds contained in these conventions, and which pose a significant risk of harm to seafarers on board, to other ships, and to the marine environment, have been regarded as substandard or unseaworthy ships.[265] Meanwhile, IMO Resolutions A.787 (19) and A.882 (21) provide basic guidance on the conduct of port State control inspections.[266] In accordance with these resolutions, port State authorities may conduct the inspections on their own initiative, at the request or on the basis of, information about the ship provided by a third party.[267]

MARPOL 73/78 and its Annex VI,[268] as well as relevant IMO guidelines,[269] have provided a general framework regarding the role of port State control relating to GHG emissions from international shipping. Generally, port State inspection is limited to verifying whether there is a valid IEE Certificate on

---

262 *ILO Convention (No. 147) concerning Minimum Standards in Merchant Ships*, opened for signature 13 October 1976, Govt. Doc. Y 1.1/4:99–21 (entered into force 28 November 1978) ('*ILO Convention No. 147*') art. 4.

263 *International Convention on Load Lines*, opened for signature 5 April 1966, 640 UNTS 133 (entered into force 21 July 1968) (Protocol of 11 November 1988, entered into force 3 February 2000) ('*Load Lines*') art. 21.

264 *International Convention on Standards of Training, Certification and Watchkeeping for Seafarers*, opened for signature 7 July 1978, 1361 UNTS 2(entered into force 28 April 1984), as amended by the 1995 Protocol, 1969 UNTS (entered into force 1 February 1997) ('*STCW Convention*') art. X, reg. I/4.

265 *Procedures for Port State Control*, IMO Doc Res A. 787(19) (23 November 1995) Ch. 4, para. 4.1.1; See also Bang, above n. 6, 716–717. Based on these materials, the terms 'substandard' and 'unseaworthy' are often used interchangeably. But see Rajadurai, above n. 6, 92–93. Rajadurai asserted that Australian domestic regulations have distinguished these two terms. Based on section 207A of Australian *Navigation Act*, substandard has a different meaning which includes.

> '(1) A ship is, for the purposes of this Act, substandard if the ship is seaworthy, but conditions on board the ship are clearly hazardous to safety or health;
>
> (2) In determining whether a ship is substandard, regard shall be had to such matters as are prescribed'.

266 *Procedures for Port State Control*, IMO Doc Res A. 787(19) (23 November 1995); *Amendments to the Procedures for Port State Control*, IMO Doc Res A.882(21) (25 November 1999).

267 Rajadurai, above n. 6, 90.

268 *MARPOL 73/78* art. 5(2); reg. 10.5 of Annex VI.

269 See, e.g., *2009 Guidelines for Port State Control under the Revised MARPOL Annex VI*, IMO Doc Res MEPC. 181(59) (17 July 2009).

board.[270] The inspections by the port State control officer (PSCO) consist of initial inspections and more detailed inspections. Initial inspections mainly aim to check if there is a valid certificate on board, whereas more detailed inspections occur when the PSCO has clear grounds for believing that the condition of the ship or its equipment do not correspond substantially with the particulars of the certificates or the documents.[271] During the inspections, the PSCO should use professional judgment to determine whether to detain the ship until any noted deficiencies are corrected or to allow it to sail with certain deficiencies which do not pose an unreasonable threat of harm to the marine environment.[272] Additionally, the NMFT principle should be applied to port State control so that ships of non-Parties to the applicable conventions should also comply with the energy efficiency measures.

6.5 Response from Global and Regional Port States Organisations

Port State control activities have been regarded as being of a regional nature.[273] It is thus necessary to examine the response from global and regional port State organisations to GHG emissions reductions from international shipping. This section takes the International Association of Ports and Harbors (IAPH) and regional MOUs on port State control as examples to examine their responses to this GHG issue.

6.5.1 *The International Association of Ports and Harbors*

The IAPH was established on 7 November 1955, and is currently the only international organisation representing the voice of the world's port industry.[274] The IAPH has achieved consultative status as a non-governmental organisation (NGO) within the IMO. As a non-profit global alliance of ports, the IAPH represents roughly 200 ports from 85 countries, and these ports deal with more than 60 per cent of the world seaborne trade and nearly 80 per cent of the

270 *MARPOL 73/78 Annex VI (2011 amendments)* reg. 10.5. See also ch. 2, 2.2.4.

271 *2009 Guidelines for Port State Control under the Revised MARPOL Annex VI*, IMO Doc Res MEPC. 181(59) (17 July 2009) ch. 2, paras. 2.1 and 2.2.

272 Ibid. para. 2.3.

273 Bang, above n. 6, 726.

274 International Association of Ports and Harbors (IAPH), *History/NGO Consultative Status* (2011) <http://www.iaphworldports.org/AboutIAPH/HistoryNGOConsultativeStatus.aspx> accessed 23 September 2013.

world's container traffic.[275] Although the IAPH has not actively participated in the discussions and negotiations on GHG emissions reductions from international shipping within the IMO, its views are reflected in its various programs and new releases.

The IAPH recognised the growing contribution of GHG emissions from port related activities to global climate change, and asserted that ports should 'take practical and effective measures to create a clean air environment',[276] and that the development of guidance for such measures would be included in a so called 'Tool Box for Port Clean Air Programs'. This tool box aims to provide ports, members and non-members of the IAPH with quick access to information, options and tools for addressing port-related air quality and climate change related issues.[277]

The IAPH categorised port-related GHG emissions into three scope groups under the IAPH Tool Box for Port Clean Air Program (Figure 6.1). Scope 1 refers to direct GHG emissions from a port's directly-controlled stationary and mobile sources. Scope 2 is indirect GHG emissions with the import and consumption of purchased electricity by a port for its directly-controlled sources. Scope 3 refers to GHG emissions relating to the operation of port tenants, including those from international shipping. The IAPH asserted that of these three scope categories, GHG emissions under Scope 3 cover 'the vast majority of the port-wide [GHG] emissions'.[278] In other words, GHG emissions from international shipping serve as an important source of GHG emissions in relation to port related GHG emissions. This also provides a vital incentive for ports to support the IMO's work in regulating GHG emissions from ships. Furthermore, the IAPH requested ports to take early action to address this GHG issue in that early action will 'ameliorate the future effects of increased costs' associated with possible MBMs.[279]

---

275 Ibid.

276 *Resolution on Clean Air Programmes for Ports*, submitted by the International Association of Ports and Harbors (IAPH), MEPC 56th Session, Agenda Item 4, IMO Doc MEPC 56/INF. 14 (11 May 2007) annex.

277 the International Association of Port and Harbors (IAPH), *IAPH Tool Box for Port Clean Air Programs* (2008) <http://iaphtoolbox.wpci.nl/DRAFT%20IAPH%20TOOL%20BOX%20%20dea.pdf> accessed 23 September 2013, p. 1.

278 The International Association of Port and Harbors (IAPH), *IAPH Tool Box for Greenhouse Gases* (2008) <http://iaphtoolbox.wpci.nl/DRAFT%20IAPH%20GHG%20TOOL%20BOX%20dea.pdf> accessed 23 September 2013, p. 3.

279 Ibid. 1.

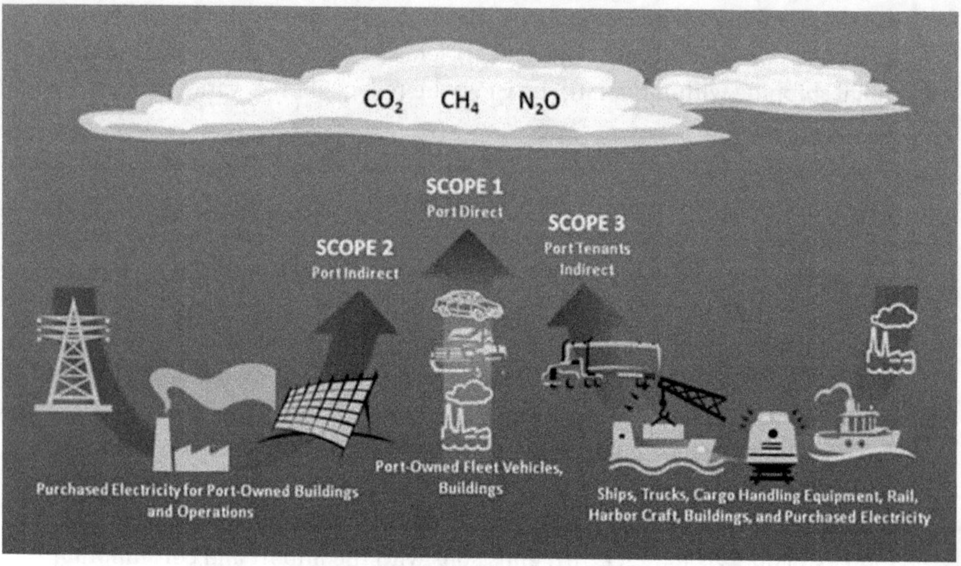

FIGURE 6.1 *GHG emissions scopes relating to port operations.*[280]

In April 2008 the IAPH requested its Port Environment Committee to work out a mechanism for assisting ports to tackle climate change. As a result, in July 2008 at the C40 World Ports Climate Conference the World Ports Climate Initiative (WPCI), which included 55 ports from the world, came into being to facilitate GHG mitigation. Also at this conference the *World Ports Climate Declaration* was adopted.[281] This declaration outlined the positions of the IAPH in regulating GHG emissions from international shipping. Based on this declaration, the IAPH expressed its support for developing technical and operational measures to address the GHG issue. On the one hand, the IAPH supported the development of clean shipping by means of improving ship fuel, ship engine and ship design.[282] On the other hand, it believed that speed reduction and environmental indexing would be effective operational measures in improving ships' energy efficiency.[283] It also urged the IMO to accelerate the incorporation of best practices into the IMO's proposed amendment

280 Ibid. 2.

281 World Ports Climate Initiative, *World Ports Climate Declaration* (July 2008) <http://www.wpci.nl/docs/Declaration.pdf> accessed 23 September 2013.

282 Ibid. p. 3, para. 2.

283 Ibid.

of Annex VI to *MARPOL 73/78*.[284] Although these proposals did not explicitly mention the EEDI and SEEMP, these views underpin the regulatory efforts of the IMO on this GHG issue.

The IAPH provided a news release stating its position on the 2011 amendments of Annex VI to *MARPOL 73/78*. According to this statement, the IAPH regarded this regulation as 'significant policy developments' by the IMO. It asserted that these measures, together with the IAPH's global air quality improvement initiatives (such as the WPCI), demonstrate that the IAPH can collaborate with the IMO 'at the port-to-port technical level and at the international regulatory level' in achieving greater GHG emissions reductions from international shipping.[285] With regard to proposed MBMs within the IMO, the IAPH has not expressed its official position. However, some of its member States have responded to this issue, which is examined in section 6.6 of this chapter.

6.5.2 *Regional Memoranda of Understanding on Port State Control*

As discussed earlier, the 1976 *ILO Convention No. 147* aimed to inspect the vessels calling at the ports of its member States. To follow up this convention, eight European countries signed a MOU in 1978 with the intent of ensuring an effective enforcement of the requirements stated under the *ILO Convention No. 147*.[286] For this reason, port State control is also asserted to originate from the 1978 MOU.[287] Regional MOUs on port State Control have become a dominant means of facilitating effective port State control at the regional level.[288] In accordance with the *1969 Vienna Convention on the Law of Treaties*, a treaty is

284 Ibid.

285 The International Association of Port and Harbors (IAPH), *Member Ports of the IAPH Welcome the International Maritime Organization's New Mandatory Measures to Reduce Greenhouse Gas Emissions from International Shipping* (22 July 2011) <http://www.iaph-worldports.org/IAPHPressReleases.aspx> accessed 23 September 2013.

286 Dr. Z. Oya Ozcayir, *Port State Control* (LLP Professional Publishing, 2001) 115. These eight European countries were Belgium, Denmark, France, Germany FR, The Netherlands, Norway, Sweden and the United Kingdom.

287 Ibid.

288 The reasons why regional MOUs have achieved rapid development include: the elimination of 'port shopping', or in other words selective visiting ports by substandard ships due to different requirements of ports; improving inspection efficiency by means of harmonisation between port States; and the reduction of the foreign ship's burden of repetitive inspections. Bang, above n. 6, 726.

'an international agreement concluded between States in written form'.[289] However, MOUs on port State Control are reached between the maritime authorities of the States concerned, and were regarded as 'an agreement based on trust'.[290] Therefore, MOUs are not legally binding, and this has also been recognised by some regional MOUs.[291] In practice an MOU, rather than a legally binding treaty, has been widely chosen by port States. This is probably because the ratification and amendment procedure of a treaty is too lengthy and less efficient than an MOU.[292]

Currently there are nine regional MOUs on port State control,[293] which if combined with the United States Coast Guard (USCG)'s port State control program,[294] cover all the regions of the world.[295] Since all these MOUs have their own organisational structures and secretariats, these MOUs can be deemed as not only agreements but also institutions on regional port State control.[296]

While regional MOUs are playing vital roles in port State control, it is worthwhile to mention that these MOUs do not give any new powers to member

---

289 *Vienna Convention on the Law of Treaties*, opened for signature 23 May 1969, 8 ILM 679 (entered into force 27 January 1980) art. 2(1)(a).

290 Ho-Sam Bang and Duck-Jong Jang, 'Recent Developments in Regional Memorandums of Understanding on Port State Control' (2012) 43(2) *Ocean Development and International Law* 170, 172.

291 See, e.g., the Tokyo MOU provides in its Preamble that 'the Memorandum is not a legally binding document and is not intended to impose any legal obligation on any of the Authorities'. Tokyo MOU, *Memorandum of Understanding on Port State Control in the Asia-Pacific Region* (29 October 2013) <http://www.tokyo-mou.org/doc/memorand.pdf> accessed 1 July 2014, Preamble. The Paris MOU has also been agreed by its participants as not legally binding. Bang and Jang, above n. 290. But see Peter B. Payoyo, *Port State Control in the Asia-Pacific: An International Legal Study of Port State Jurisdiction* (LLM Thesis, Dalhousie University 1993) 77. Payoyo asserts that the Paris MOU is technically a treaty.

292 Ozcayir, above n. 286, 116–117.

293 These nine MOUs are Paris MOU, Latin American MOU, Tokyo MOU, Caribbean MOU, Mediterranean MOU, Indian Ocean MOU, West and Central African MOU, Black Sea MOU, and Riyadh MOU.

294 The USCG is one of the five armed forces of the US and the only military organisation within the Department of Homeland Security. It is the maritime safety authority of the US, and is responsible for maritime safety, security and environmental stewardship of the US. United States Coast Guard (USCG), *Overview of the United States Coast Guard* (19 September 2013) <http://www.uscg.mil/top/about/> accessed 27 September 2013.

295 Bang, above n. 6, 726.

296 Ibid. 718.

States.[297] Rather, these MOUs provide a means of coordinating the exercise of the powers that these port States have received from various international conventions, so existing rules and standards in these conventions can be enforced in an effective and harmonised manner.[298] However, given the differing ratification status of States for specific conventions and distinct inspection mechanisms of regional MOUs, it is still difficult to ensure uniform enforcement of standards contained in various conventions.[299]

With regard to the proposed technical and operational measures and MBMs within the IMO, regional MOUs on port State control have not officially expressed their positions on these issues. The fact that these MOUs have no official status within the IMO might be one reason. Another reason is that the role of port State control/inspection in enforcing technical and operational measures has been generally limited to verifying whether there is a valid certificate on board. In this case, as regional agreements on port State control, MOUs illustrate a lack of sufficient expertise and incentive in contributing to these regulatory initiatives. However, this does not mean that these MOUs do not have any impact on the IMO's regulatory process. Indeed, the NMFT principle that runs through all MOUs on port State control was eventually incorporated in the amended Annex VI to *MARPOL 73/78* in 2011, although many developing countries opposed it.[300] This outcome, however, would not have been achieved without the current widely-accepted practice of various MOUs on port State control of which many developing countries are also members.

Regulations 20 and 21 of the 2011 amendments to Annex VI provide that the EEDI applies to new ships, new ships that undergo a major conversion, and new or existing ships that undergo a major conversion as defined in Regulation 2 of Annex VI. Regulation 2.23.3 of Annex VI stipulates that based on the date of the ship's delivery, new ship means that the delivery of the ship is on or after 1 July 2015. Since this regulation entered into force on 1 January 2013 and the major conversion of existing ships may also take substantial time, it is anticipated that the earliest date for MOU member port States to inspect the IEE Certificate of a new ship will possibly be around July 2015.[301] To date many

297 Ibid. 727.

298 Ibid.

299 See ibid. 726.

300 See ch. 4, 4.3.3.1.

301 This calculation does not take the waiver clause of Regulation 19 into account. Based on Regulation 19 of Annex VI to *MARPOL 73/78*, once a flag State approves, any ships flying the flags of this flag State could postpone the commencement date of applying the EEDI to as late as 1 July 2019 (in terms of the delivery of a ship).

MOUs, including the Paris MOU and the Tokyo MOU, have already included the IEE Certificate into their 'List of MOU Deficiency Codes',[302] which serve as the main basis for the port State control of their member port States. This prompt response by MOUs on port State control can be regarded as demonstrating their strong support for the IMO's initiatives in regulating GHG emissions from international shipping. In recent years, the effectiveness and harmonisation of regional MOUs on port State control has been strengthened through the establishment of the 'White-Grey-Black Lists',[303] the new inspection regime (NIR) based on ship risk profile,[304] the recognised organisations (RO) performance lists,[305] as well as the cooperation between different MOUs.[306] However, a gap

302 See, e.g., Paris MOU, *List of Paris MOU Deficiency Codes* (2013) <https://www.parismou.org/content/publishedmedia/2da7e155-8f2b-4afc-af3a-ea51a51a7aea/list%20of%20paris%20mou%20deficiency%20codes.pdf> accessed 24 September 2013, p. 1; Tokyo MOU, *List of Tokyo MOU Deficiency Codes* (20 August 2013) <http://www.tokyo-mou.org/doc/Tokyo%20MOU%20deficiency%20codes%20%2820%20August%202013%29.pdf> accessed 24 September 2013, p. 2.

303 Take the Paris MOU as an example: based on the total number of inspections and detentions over a 3-year rolling period of flag vessels with at least 30 inspections in the period, the 'white list' includes quality flags with a consistently low detention record; the 'grey list' indicates flags with an average port State control record; while the 'black list' covers ships with a consistent poor safety record. Paris MOU, above n. 28, 18–19.

304 The NIR was first introduced by the Paris MOU on 1 January 2011. As a risk-based targeting mechanism, the NIR ranks the risk profile of ships into high risk, standard risk and low risk based on generic and historic parameters such as company performance, the voluntary IMO member State Audit Scheme, the inspection history of the ship as well as the ship's age and ship type. The NIR generally rewards low risk ships with longer inspection intervals by port State control of up to 36 months, compared with six months in the current system. Paris MOU, *Paris MOU Reward Quality Flags* (2 November 2010) <https://www.parismou.org/Content/PublishedMedia/1039b41c-e715-4b08-8f17-080dd9c7238d/Paris%20MoU%20rewards%20quality%20flags%20%282%20November%29.pdf> accessed 25 September 2013.

On 15 February 2013, Tokyo MOU decided to adopt its own NIR from 1 January 2014. Tokyo MOU, *Tokyo MOU Will Introduce A New Inspection Regime (NIR) from 1st January 2014* (15 February 2013) <http://www.tokyo-mou.org/doc/PRESS-NIR.pdf> accessed 25 September 2013.

305 Regional MOUs also monitor the performance of classification societies acting as Recognised Organisations for flag States. Take the Paris MOU as an example: a minimum number of 60 inspections per RO are needed before the performance is taken into account for the list. The relationship between the classification society and the RO is provided in chapter 5. See ch. 5, 5.2.5.

306 To harmonise and make more uniform port State inspections, various regional MOUs on port State control have conducted joint efforts in establishing and enforcing effective

between the performance of the Paris and Tokyo MOUs and other MOUs in terms of numbers of annual inspections and the length of time that they have operated still exists,[307] and there are still many substandard ships engaged in international shipping.[308] It is arguable that current data suggest that current MOUs on port State control are successful, but it appears that it is necessary to improve their current enforcement mechanisms.[309]

6.6 Response from Main Port States

Technical and operational measures and MBMs are three routes of regulating GHG emissions from international shipping within the IMO.[310] As discussed earlier, the response from flag States to this GHG issue generally involves all these three aspects. Since port States are generally also flag States, to distinguish the port State response from flag States response to this issue, this section only examines the port State response in relation to the ports. In the context of GHG emissions from international shipping, port States are generally interested in the enforcement of technical and operational measures, and some of them may also be interested in some MBM proposals. For this reason, this section canvasses the response from port States listed in Annex I to the *UNFCCC* and those which are not listed therein generally rather than containing specific case studies as in previous sections.

---

mechanisms. For example, due to the joint effort by the Paris and Tokyo MOUs, the Tokyo MOU implemented a new coding system in 2012. Meanwhile, regional MOUs also strengthen their communications with the shipping industry. Tokyo MOU, 'Annual Report on Port State Control in the Asia-Pacific Region' (2012 2012) <http://www.tokyo-mou.org/doc/ANN12.pdf> accessed 25 September 2013.

307 Bang, above n. 256, 120.

308 For example, within the Paris MOU area, in 2012 there were 669 detentions of ships, which accounted for 3.65 per cent of all ship inspections. Compared to the year of 2003 when there were 1431 detentions covering 7.05 per cent of all ship inspections, the numbers in 2012 have decreased a bit. Whereas within the Tokyo MOU area, in 2012 there were 1421 detentions of ships accounting for 4.59 per cent of all ship inspections. Compared to the year of 2002 when there were 1307 detentions covering 6.67 per cent of all ship inspections, the numbers in 2012 were comparatively stable and the number of detentions were even higher. Paris MOU, above n. 28, 24–25; Tokyo MOU, above n. 306, p. 20.

309 Bang, above n. 256, 116.

310 See ch. 4, 4.3.3.

6.6.1 *The UNFCCC Annex I Port States*

Generally port States listed in Annex I to the *UNFCCC* are supportive of the IMO's role and work in regulating GHG emissions from international shipping. Of these countries, the European Commission, which is an observer of the IMO on behalf of the European Union (EU), has been pushing the IMO in expediting this regulatory process.[311] Other countries, such as Norway and the United States of America (US), have also been contributing to this work of the IMO.

With regard to the proposed technical and operational measures (EEDI and SEEMP), aside from the survey and certification, there was only one proposed new regulation related to port State control. An earlier version of this regulation in 2010 provided that,

> A ship to which this part of the annex applies may, when in a port or offshore terminal of another Party, be subject to inspection by officers duly authorized by that Party for the purpose of determining whether the ship is in compliance with this part of the Annex. Any such inspection is limited to verifying, when appropriate, that there is on board an International Certificate on the Energy Efficiency on the ship.[312]

Norway underscored that this proposed regulation on port State control should be reconsidered in that it partly repeats Article 5 of *MARPOL 73/78* and thus might lead to confusion and misinterpretations.[313] This view was echoed by the working group on energy efficiency for ships at the 61st MEPC meeting in 2010.[314] At the 62nd MEPC meeting in 2011, a new regulation 10.5 was adopted without too much debate, which reads:

---

311 See, e.g., at the 57th MEPC meeting, the European Commission reiterated its strong preference for global solutions to this GHG issue, and asserted that if it is not possible for the IMO to maintain the established timelines, it would retain 'the right to initiate appropriate action to protect the environment'. *Report of the Marine Environment Protection Committee on Its Fifty-Seventh Session*, MEPC 57th Session, Agenda Item 21, IMO Doc MEPC 57/21 (7 April 2008) para. 4.9.

312 *Report of the Outcome of the Intersessional Meeting of the Working Group on Energy Efficiency Measures for Ships*, MEPC 61st Session, Agenda Item 5, IMO Doc MEPC 61/5/3 (7 July 2010) annex 1, p. 6.

313 *Comments on the Draft Regulatory Text on Energy Efficiency for Ships*, submitted by Norway, MEPC 61st Session, Agenda Item 5, IMO Doc MEPC 61/5/6 (16 July 2010) para. 2.

314 *Report of the Working Group on Energy Efficiency Measures for Ships*, MEPC 61st Session, IMO Doc MEPC 61/WP.10 (30 September 2010) para. 4.15.

> In relation to chapter 4, any port State inspection shall be limited to verifying, when appropriate, that there is a valid International Energy Efficiency Certificate on board, in accordance with article 5 of the Convention.

The smooth adoption of this regulation was probably because this regulation has become a standard phrase for port State control,[315] although it indeed excludes possible unilateral actions by port States in dealing with shipping GHG emissions.[316] Additionally, at the 61st MEPC meeting, Norway also proposed to develop a new chapter in the *2009 Guidelines for Port State Control under the Revised MARPOL Annex VI* to provide basic guidance for port State control in relation to the energy efficiency regulations.[317] Due to the different nature of air pollutants (e.g., SO_X, NO_X) and GHG emissions (e.g., CO_2) and their different certificate requirements, this proposal is significant although it has not yet been addressed.

Many port States made prompt responses to the adopted energy efficiency measures by the IMO in 2011, such as including the IEE Certificate in their port inspection requirements. In December 2012, the USCG amended its Vessel Inspection Alternative Regulations to add the IEE Certificate to the list of certificates that a recognised classification society may issue on behalf of the USCG.[318] In May 2013, the USCG released the *MARPOL* Annex VI IEE Certificate Implementation Guidance, which provided interim guidance to ensure compliance with the IEE Certificate by the US flagged ships and foreign ships calling on US ports.[319] It appears that as a port State which is not a participating

---

315 See, e.g., *MARPOL 73/78* art. 5(2); Paris MOU, *Paris Memorandum of Understanding on Port State Control* (23 May 2013) <http://www.guardiacostiera.it/servizi/documents/sicurezzanavigazione/MOU,%20incl.%2029th%20Amendment.pdf> accessed 26 September 2013 ('Paris MOU') annex 9, sec 2.

316 See ch. 4, 4.3.3.3.

317 *Comments on the Draft Regulatory Text on Energy Efficiency for Ships*, submitted by Norway, MEPC 61st Session, Agenda Item 5, IMO Doc MEPC 61/5/6 (16 July 2010) para. 7.2.

318 United States Coast Guard (USCG), *Adding International Energy Efficiency (IEE) Certificate to List of Certificates A Recognised Classification Society May Issue* (10 December 2012) <https://www.federalregister.gov/articles/2012/12/10/2012-29749/adding-international-energy-efficiency-iee-certificate-to-list-of-certificates-a-recognized> accessed 27 September 2013.

319 United States Coast Guard (USCG), *MARPOL Annex VI International Energy Efficiency (IEE) Certificate Implementation Guidance* (5 May 2013) <http://www.regulations.gov/#!documentDetail;D=USCG-2012-1095-0002> accessed 27 September 2013.

member of any regional MOU,[320] the US still properly exercises its obligations under the Annex VI to *MARPOL 73/78*.[321]

In terms of proposed MBM proposals, port States under Annex I to the *UNFCCC* have revealed more interest than *UNFCCC* non-Annex I States. Among these States, Norway, the UK, France and Germany supported a global ETS,[322] while the European Commission provided the IMO with specific information on the EU-ETS.[323] However, the European Commission also asserted that if the IMO could not achieve satisfactory progress in regulating this GHG issue, the EU would possibly take unilateral actions. Namely, the EU might include GHG emissions from international shipping into its Emission Trading Scheme (ETS). Indeed, the EU has taken unilateral measures in the international aviation sector and attributed its actions to slow and unsatisfied work under the International Civil Aviation Organization (ICAO). On 1 January 2012 the EU included the emissions from the international aviation industry into the EU-ETS. In December 2012 the EU suspended this policy due to the improved performance from the ICAO, or perhaps because of strong opposition from many countries including the US, Russia, China and India. In the same year, the EU published a consultation document asking for views on how best to reduce GHG emissions from ships so as to finally include GHG emissions from international shipping in an EU-ETS.[324] Although EU's unilateral actions have been opposed by many developing countries, and some developed countries,[325]

---

320 The US is an observer to some regional MOUs, such as the Paris MOU, Tokyo MOU, Caribbean MOU, Mediterranean MOU, Black sea MOU, and Latin American MOU.

321 Another example is that the US has not ratified the 2004 International Convention for the Control and Management of Ships' Ballast Water and Sediments (BWM Convention), but it has adopted more stringent rules than the BWM Convention.

322 See ch. 4, 4.3.4.2.

323 See, e.g., *Key Design Elements for Designing A 'Cap and Trade' Greenhouse Gas Emissions Trading Scheme* (*ETS*), submitted by the European Commission, Intersessional Meeting of the Greenhouse Gas Working Group 1st Session, Agenda Item 5, IMO Doc GHG-WG 1/5/3 (30 May 2008); *Information on Experience with the European Union Emissions Trading Scheme (EU-ETS) and the Carbon Market*, submitted by the European Commission, Intersessional Meeting of the Working Group on GHG Emissions from Ships 3rd Session, Agenda Item 2, IMO Doc GHG-WG 3/INF.3 (25 February 2011).

324 Will Nichols, *EU Launches Attempt to Deliver Shipping Emissions Trading Scheme* (24 January 2012) <http://www.businessgreen.com/bg/news/2140997/eu-launches-attempt-deliver-shipping-emissions-trading-scheme> accessed 26 September 2013.

325 For example, China's shipping industry is against any unilateral actions in regulating GHG emissions from international shipping, and Greece and the UK also opposed any attempt to jeopardise IMO's work on this issue. *Report of the Marine Environment Protection*

these actions objectively have also pushed the IMO's work in accelerating the regulation of this issue.

Compared with Norway and the European Commission, the US first put forward its own MBM proposal on the SECT, but at the 64th MEPC meeting in 2012, the US proposed an establishment of attained energy efficiency standards for new and existing ships through a phased approach while the SECT became an optional addition. In other words, the US suggested suspending the MBM discussions and instead focusing on the further improvement of ship's energy efficiency through technical and operational measures. This view aims to reduce the transportation cost for the shipping industry, and thus was supported by many countries. Meanwhile, this view is also consistent with the interests of port States since these States cannot receive direct economic benefits generated by MBMs, although their environmental benefits may be secured for the long term.

6.6.2 *The UNFCCC Non-Annex I Port States*

Port States from either Annex I or non-Annex I to the *UNFCCC* have similar interests in reducing GHG emissions from international shipping so as to protect the environment of the ports. However, in terms of the means to achieve the emissions reduction from ships, port States from non-Annex I to the *UNFCCC* have slightly different positions. Singapore is a port State under non-Annex I to the *UNFCCC*,[326] and is also an important flag State.[327] At the 62nd MEPC meeting in 2011, Singapore proposed some refinements to the draft Regulation 19 of Annex VI to *MARPOL 73/78*. Under the draft amended Annex VI by the IMO Secretariat in 2010, Regulation 19 included three clauses stipulating that this chapter only applies to all ships 400 gross tonnage and above, and does not apply to domestic shipping and ships which have diesel-electric

Committee on its 58th Session, IMO Doc MEPC 58/23 (16 October 2008) annex 9, pp. 1,9,20. See also ch. 5, 5.4.1.

326 But Singapore is regarded as a developed country based on the OECD criteria. See OECD, *Geographical Distribution of Financial Flows to Developing Countries: Disbursements, Commitments, Country Indicators* (2013) <http://www.oecd-ilibrary.org/docserver/download/4313023e.pdf?expires=1380277397&id=id&accname=ocid53013930&checksum=8AACEBAA8DAD0ABE76999F69BE37B75E> accessed 27 September 2013, p. 6.

327 For example, as of 1 January 2012, Singapore ranked fifth in the world among the flags of registration with the largest registered deadweight tonnage. Meanwhile Singapore controls the 11th largest owned fleets (dwt) in the world with 712 vessels registered in Singapore and 398 registered in other flag States. UNCTAD, above n. 8, 41,44.

propulsion, turbine propulsion or hybrid propulsion systems.[328] The exclusion of any exemption for any flag State and grace period under this proposed regulation indeed revealed the regulatory interests of most *UNFCCC* Annex I States. In this way not only the NMFT principle would be thoroughly exercised but also the regulation could be implemented sooner after it enters into force so as to achieve the best environmental effect. However, due to strong opposition from many developing countries, it appeared less likely that this proposed regulation could be adopted.

Under these circumstances, Singapore proposed an extra four paragraphs to add to the proposed Regulation 19 of Chapter 4. These four paragraphs constituted two key points. One was to provide a waiver clause for all flag States to postpone their commencement date to four years after the entry into force of the regulation based on the building contract of a new ship, or six and a half years based on the delivery date of a new ship.[329] This proposal was a compromise solution. It provided a grace period for all flag States, in particular for developing flag States.[330] Meanwhile, it also applied the NMFT principle and excluded the application of the CBDR principle.[331] For this reason, this proposal was agreed by a large number of States and finally adopted by the IMO. The other point of Singapore's proposal was to give port States the entitlement to deny ships' port entry based on whether they comply with the EEDI.[332]

---

328 *Amendments to MARPOL Annex VI—Inclusion of Regulations on Energy Efficiency for Ships*, note by the Secretariat, MEPC 62nd Session, Agenda Item 6, IMO Doc MEPC 62/6/3 (21 December 2010) annex, p. 9.

329 *Amendments to MARPOL Annex VI—Inclusion of Regulations on Energy Efficiency for Ships*, submitted by Singapore, MEPC 62nd Session, Agenda Item 6, IMO Doc MEPC 62/6/21 (20 May 2011) annex, p. 1.

330 Due to differentiated economic and technical situations and enforcement capacities between developed and developing States, this waiver clause is most likely to be used by some developing flag States. See also ch. 4, 4.3.2.

331 However, Singapore supported the further study on the feasibility of applying the CBDR principle in international shipping so as to explore all possible options. *Report of the Marine Environment Protection Committee on its 58th Session*, IMO Doc MEPC 58/23 (16 October 2008) annex 9, p. 12.

332 *Amendments to MARPOL Annex VI—Inclusion of Regulations on Energy Efficiency for Ships*, submitted by Singapore, MEPC 62nd Session, Agenda Item 6, IMO Doc MEPC 62/6/21 (20 May 2011) annex, p. 1. Singapore's proposed regulation in relation to the port State's entitlement provided that,

'Subject to the provisions of international law, a Party to the present Convention shall be entitled to deny entry of ships operating in accordance with the provisions of paragraph 4 of this regulation into the ports or offshore terminals under its jurisdiction, except when this is necessary for the purpose of securing the safety of the ship or saving life at sea.

Based on international law principles, port States have sovereignty over their ports or internal waters and are thus entitled to utilise this right.[333] However, Singapore's proposal was not agreed by member States of the MEPC. Indeed, the denial of ships' port entry solely based on their noncompliance with the EEDI is inconsistent with various regional MOUs, conflicts with other international regulations on places of refuge for ships in distress,[334] and may also distort international trade. The purpose of Singapore's proposal on conditioning ships' port entry was to strengthen the enforcement of port State control of the IMO's energy efficiency measures. It is thus argued by Singapore that this goal can be achieved at the cost of safety or trade related benefits. It appears that innovative mechanisms or improvement in current mechanisms for port State control are necessary in order to improve the effectiveness of port State control on this issue.[335]

Of the current seven types of MBM proposals, the Port State Levy proposed by Jamaica is most related to the interests of port States. This proposal levies a uniform emissions charge on all vessels calling at their respective ports based on the amount of fuel consumed by the respective vessel on its voyage to that port.[336] Port States collect the levy on their own behalf or through regional mechanisms, and they are also responsible for allocating the respective revenues associated with the levy through a self-administered national or regional fund or an international mechanism. Under this proposal, the CBDR principle can also be achieved through access by developing countries to the fund. Chapter 4 of this book discussed the fact that this proposal has shortcomings in its measurement of emissions solely by fuels consumed, enforcement difficulties and high administrative costs. However, this proposal protects the interests of flag and port States by enabling their participation in this scheme. Meanwhile this scheme can be grouped into either in-sector or out-of-sector reduction,[337] which provides more options and would thus be attractive for

In such cases, that party shall communicate to the Organization for circulation to the Parties to the present Protocol particulars thereof for their information.'

333 R.R. Churchill and A.V. Lowe, *The Law of the Sea* (Manchester University Press, 3rd ed., 1999) 61.

334 Anthony Morrison, *Places of Refuge for Ships in Distress: Problems and Methods of Resolution* (Martinus Nijhoff Publishers, 2012) 96–106, 144, 166.

335 Further discussion on this issue is provided in Chapter 7, 7.4.4. See also Bang, above n. 256, 131–132.

336 See ch. 4, 4.3.4.2.

337 The third Intersessional Meeting of the Working Group on GHG Emissions from Ships grouped the MBMs proposals into two categories, namely 'focus on in-sector' and 'in-sector and out-of-sector'. *Report of the Third Intersessional Meeting of the Working Group*

some flag and port States. As discussed in previous chapters, Greece and Republic of Korea, and their shipping industries, as well as the International Chamber of Shipping, have expressed that their preferred MBM is a levy or fund based scheme.[338]

6.7 Conclusion

Flag States and port States are two vital stakeholders in the global regulation of GHG emissions from international shipping, and they are involved in both the legislative and implementing process of this issue. Of the flag States under Annex I to the *UNFCCC*, both Greece and Japan have strongly supported the leading role of the IMO in regulating this GHG issue and contributed to the adoption of energy efficiency measures by the IMO. They have both insisted that the NMFT principle should apply to this issue. Although Japan asserted that the CBDR principle could be reflected in the energy efficiency measures, this view was supplanted by its subsequent statement. Currently both Greece and Japan welcome the adoption of an MBM to reduce GHG emissions from ships, but they have different preferred MBMs. It is concluded that flag States under *UNFCCC* Annex I have similar positions towards the adoption of technical and operational measures by the IMO. Their attitudes towards proposed MBMs are also positive, although they have different preferences on the form MBMs should take.

In comparison with *UNFCCC* Annex I flag States, non-Annex I flag States have a more diverse response towards this GHG issue due to their differing regulatory interests. Based on the case studies, Panama supported the leading role of the IMO in tackling this issue and underscored the urgency of this work. Meanwhile, Panama emphasised the application of the NMFT principle to this matter, and welcomed the MBMs. Of the current MBM proposals, Panama preferred the Penalty on Trade and Development proposed by the Bahamas. As the largest developing country, China supported the IMO in regulating the shipping GHG emissions. However, China has insisted on the application of the CBDR principle to all three measures for regulating the GHG issue and is dubious about the IMO's competence in regulating the MBMs. Although China opposed the adoption of the energy efficiency measures by the IMO in 2011,

---

on Greenhouse Gas Emissions from Ships, MEPC 62nd Session, Agenda Item 5, IMO Doc MEPC 62/5/1 (8 April 2011) annex 3.

338 But Greece interpreted the 'levy' as only referring to the International GHG Fund. See ch. 6, 6.3.1.1.

China did not object to this regulation due to its desire for expanding its international shipbuilding market. China believes that MBMs are premature at this stage, but it has also accepted the compromise position that if a MBM is to be adopted, the CBDR principle should be incorporated. Vanuatu is a LDC and SIDS which is extremely vulnerable to the adverse impacts of climate change. It supported the IMO's regulatory initiatives but suggested that an exemption for ships trading to the LDCs and SIDS be provided from these regulations. Due to similar economic and shipping situations, Panama, China and Vanuatu can be grouped into three categories representing the regulatory interests of major FOC States, major developing flag States, and other developing States respectively. Therefore, it can be deduced that major developing flag States and some other developing States are the main supporters of the application of the CBDR principle to the regulation of this GHG issue and these States pay more attention to their needs in capacity building and technology transfer rather than the regulation itself. Nevertheless, major FOC States support the NMFT principle and tend to welcome most of the relevant regulatory measures.

With regard to the response from port States to GHG emissions from international shipping, global and regional port States organisations, as well as various port States, have all recognised the importance of regulating this issue and asserted that ports should take practical and effective measures to address this problem. Among them, the IAPH has taken some initiatives in tackling this matter, which can be regarded as a supplement to the IMO's regulatory outcomes. Various regional MOUs on port State control have also added the IEE Certificate to their 'List of MOU Deficiency Codes' in support of the IMO's work. Currently various port States generally support the reduction of GHG emissions from international shipping, and have asserted that current port State control on this GHG issue could be strengthened. However, disagreements remain as to the means to achieve this reduction. Generally port States under non-Annex I to the *UNFCCC* require more grace periods and assistance in capacity building for implementing the IMO regulations which address the GHG issue.

CHAPTER 7

The Future Development of Legal and Institutional Frameworks to Reduce Greenhouse Gas Emissions from International Shipping

7.1 Introduction

The regulatory framework for the reduction of greenhouse gas (GHG) emissions from international shipping has come into being through the continuous efforts of the international community, particularly through the International Maritime Organization (IMO). The most significant achievement is the adopted technical and operational measures in the form of 2011 amendments to Annex VI to the *International Convention for the Prevention of Pollution from Ships* (*MARPOL 73/78*).[1] During this regulatory process, the UN, the IMO, the shipping industry, and various flag and port States have responded differently according to their respective interests. Given the deficiencies in current legal and institutional frameworks for regulating GHG emissions from international shipping analysed in previous chapters, it is necessary to find ways to improve the adopted technical and operational measures, possibly including adopting market-based measures (MBMs). However, it is challenging to balance the interests of various stakeholders in the development of future regulatory efforts regarding the smooth adoption of these measures and the compliance by the global shipping industry.

This chapter consists of five parts. The first part discusses the influence and interaction of key actors in regulating and implementing GHG emissions from international shipping. The second part outlines the current international regulatory framework for the reduction of GHG emissions from international shipping. Based on these findings in the first and second parts and the analyses from previous chapters, the remaining three parts identify and examine gaps in current legal and institutional frameworks on the GHG issue and propose recommendations for addressing these deficiencies. Among them, the third and fourth parts focus on the adopted energy efficiency measures and potential MBMs respectively whereas the fifth part examines the current institutional arrangement.

1 *International Convention for the Prevention of Pollution from Ships* (MARPOL 73/78), signed 2 November 1973, 12 ILM 1319, as amended by the 1978 Protocol to the 1973 Convention, 1341 UNTS 3, 17 ILM 546 (entered into force 2 October 1983).

 | DOI 10.1163/9789004329317_008

7.2 Influence and Interaction of Various Stakeholders in Regulating and Implementing GHG Emissions from International Shipping

GHG emissions from international shipping are a type of 'conditional' pollution.[2] They contribute to global climate change and have been partially regulated by the IMO. Various stakeholders have participated in this regulatory process and contributed to the adoption of energy efficiency measures in July 2011. To identify the roles of these key actors and balance their different regulatory interests would facilitate the improvement of current energy efficiency measures, as well as the future development and adoption of MBMs.

Regarding the regulation and enforcement of GHG emissions from international shipping, key actors include flag States, port States, national shipping industries, international shipping associations, the UN, the IMO, the Intergovernmental Panel on Climate Change (IPCC) and other non-governmental organisations (NGOs). Figure 7.1 depicts the interaction of these actors in regulating and enforcing GHG emissions from international shipping. The first step in the regulatory process is that national shipping industries of various States express their views or suggestions on approaches to reducing shipping GHG emissions to their governments (flag States or port States) and relevant global shipping associations of which they are members. The driving forces of the responses by these national shipping industries are the efforts of the international community, in particular pertaining to the requirements from the *United Nations Framework Convention on Climate Change* (*UNFCCC*)[3] and its *Kyoto Protocol*.[4] Given that flag States, port States and some global shipping associations are either member parties or observers of the IMO,[5] the second step is that these IMO member parties and observers contribute to

2 See ch. 2, 2.1.2.

3 *United Nations Framework Convention on Climate Change*, opened for signature 9 May 1992, 31 ILM 848 (entered into force 21 March 1994) ('*UNFCCC*').

4 The *Kyoto Protocol* to the *UNFCCC* includes the reduction of GHG emissions from domestic shipping into national commitments of the *UNFCCC* Annex I States, and empowers the IMO to regulate GHG emissions from international shipping. See *Kyoto Protocol to the United Nations Framework Convention on Climate Change*, opened for signature 16 March 1998, 37 ILM 22 (entered into force 16 February 2005) ('*Kyoto Protocol*') arts. 2(1)(vii), 3(1), 10(b)(i); and 2(2).

5 Within the IMO, observers can be Intergovernmental Organisations (IGOs) with observer status, or Non-Governmental International Organisations (NGOs) which have been granted consultative status with IMO. Global shipping associations are generally NGOs in consultative status. IMO, *Member States, IGOs and NGOs* (2013) <http://www.imo.org/About/Membership/Pages/Default.aspx> accessed 30 October 2013.

the adoption of these proposals through either submitting proposals or participating in discussions and negotiations on behalf of their States or shipping sectors within the IMO. The third step is that regulations, such as the Energy Efficiency Design Index (EEDI) and the Ship Energy Efficiency Management Plan (SEEMP),[6] are adopted within the IMO through either a consensus or a majority-voting mechanism.

The whole process involves complicated interaction and competition between flag States, port States and global shipping associations. Nevertheless, only contracting flag States and port States are qualified to vote, whereas international shipping associations may influence the decision-making process through other means such as submitting proposals and speaking at the MEPC meetings.[7] At the time of voting for the amendments of Annex VI to *MARPOL 73/78* on 15 July 2011, there were 64 contracting States and 59 of them registered to attend the 62nd MEPC meeting.[8] However, as of 19 April 2016, the number of contracting States had increased to 86 and accounted for 95.34 per cent of the world tonnage (by gross tonnage).[9] As the IMO received its specific mandate in relation to regulation of GHG emissions from international shipping through Article 2(2) of the *Kyoto Protocol to the UNFCCC*,[10] any updates in the global climate change regime, such as the *Paris Agreement to the UNFCCC* adopted in December 2015,[11] are also likely to influence the regulatory process within the IMO.

---

6 The EEDI and SEEMP represent the main technical measure and operational measure respectively adopted by the IMO in July 2011 aiming to reduce GHG emissions from international shipping. See ch. 4, 4.3.3.

7 See International Maritime Organization (IMO), *Basic Documents Volume I* (International Maritime Organization, 2010), pp. 157–158, Rules Governing Relationship with Non-Governmental International Organizations, Rules 6–7. These rules provide that a non-governmental international organization with a consultative status has the right to submit written statements on items of the agenda of different committees, the right to be represented by an observer at relevant sessional meetings, and shall speak on relevant items of the agenda on the invitation of the Chairman and with the approval of the body concerned.

8 *Report of the Marine Environment Protection Committee on Its Sixty-Second Session*, MEPC 62nd Session, Agenda Item 24, IMO Doc MEPC 62/24 (26 July 2011) para. 6.110.

9 IMO, *Summary of Status of Conventions* (19 April 2016) <http://www.imo.org/About/Conventions/StatusOfConventions/Pages/Default.aspx> accessed 13 May 2016.

10 See ch. 4, 4.2.

11 *Paris Agreement to the United Nations Framework Convention on Climate Change* (2015), opened for signature 22 April 2016, FCCC/CP/2015/L.9 (not yet in force) ('*Paris Agreement*').

FIGURE 7.1 *Interactions of various stakeholders in regulating and enforcing GHG emissions from international shipping.*
[NOTE: R REPRESENTS 'REGULATORY PROPOSALS'; E REPRESENTS 'ENFORCEMENT REQUIREMENTS']

There are two main steps involved in the enforcement process for adopted measures. First, any State which accepts and ratifies new IMO regulations must ensure that its shipping industry complies with these regulations. To achieve this goal, these States may need to enact domestic legislation to incorporate IMO rules. Alternatively, the classification society employed by a Contracting State may adopt its rules incorporating these IMO rules and apply them to shipping companies of the State.[12] Global shipping associations will be informed of any new regulations from the IMO, and will usually provide sample guidelines for their members on implementing these regulations. Secondly, national shipping industries will adhere to their domestic legislation or classification society rules in building ships and operating shipping. The shipping industry of a State which is not a contracting Party to IMO regulations may recognise and follow the regulations on a voluntary basis. Otherwise, Regional Memoranda of Understanding (MOUs) on port State control enforcement regimes may

12 For example, China's Classification Society has released its own rules incorporating EEDI and SEEMP requirements, whereas the Chinese Government has not updated its domestic legislation due to legal barriers. See ch. 5, 5.4.1.

result in these substandard ships becoming less competitive in the global shipping market.

The regulatory and enforcement process which has occurred in relation to GHG emissions from international shipping reveals the importance of flag States and port States. Although a flag State can also be a port State, the port State's enforcement role is limited to verifying the International Energy Efficiency Certificate (IEE Certificate). Under the circumstances when a flag State is not a port State as well, the flag State has been allocated primary responsibility for ensuring the compliance of ships on their registers with all applicable international and domestic regulations and standards. It is thus reasonable to assert that generally flag States have more influence than port States as to the regulation and enforcement of this GHG issue. National shipping industries do not directly participate in the discussions and negotiations within the IMO or have the right to vote. However, it is common practice that the shipping industry of a State initiates the regulatory process for IMO regulations through its flag State, port State or global shipping associations of which it is a member. Meanwhile, the shipping industry often serves as the body for receiving the benefits or bearing the costs of the enforcement of these regulations. For this reason, no regulation can be adopted or implemented without the support of the shipping industry.[13] From the regulation and enforcement perspective, the order of importance of these stakeholders can be roughly ranked from high to low as the shipping industry, flag States and port States. Meanwhile, both *MARPOL 73/78* and the IMO's Marine Environment Protection Committee (MEPC) have adopted a two-thirds majority voting mechanism,[14] which is supplemented by a tonnage-based arrangement.[15] This voting arrangement, when combined with the diverse regulatory interests of States not under Annex I

---

13 See G.P. Pamborides, *International Shipping Law: Legislation and Enforcement* (Kluwer Law International, 1999) 145. Pamborides asserts that 'shipping is too valuable to the world's economy to jeopardise', and even powerful port States cannot exercise port State control 'without the prior consent of the [shipping] industry'.

14 *MARPOL 73/78* art. 16(2)(d). This provision provides that amendments to *MARPOL 73/78* shall be adopted by a two-thirds majority of only the Parties to the Convention present and voting. International Maritime Organization (IMO), above n. 7, 113. Rule 27 of the *Rules of Procedures of the MEPC* provides that, 'decisions of the Committee and of its subsidiary bodies shall be made and reports, resolutions and recommendations adopted by a majority of the Members entitled to vote, present and voting'.

15 *MARPOL 73/78* art. 16(2)(f)(ii)(iii). These two provisions of Article 16 provide that under some circumstances, to prevent an amendment to an Annex or to an amendment to an appendix to an Annex from being accepted, the combined merchant fleet of some opponent States should constitute not less than 50 per cent of the gross tonnage of the world's merchant fleet.

to the *UNFCCC*, would make developing flag or port States disadvantaged in the negotiations and adoption of IMO regulations.[16] These findings can be utilised to analyse the gaps in current regulatory frameworks for the GHG issue and underpin gap-filling options if there are conflicts between the interests of stakeholders.

7.3 Formation of the International Regulatory Framework for Reducing GHG Emissions from International Shipping

The international effort to regulate GHG emissions from international shipping can be traced back to 1995 when the *UNFCCC*'s Subsidiary Body on Scientific and Technological Advice (SBSTA) and the Subsidiary Body for Implementation (SBI) were requested to examine the allocation and control of emissions from international bunker fuels.[17] In 1996 the SBSTA identified five options from the eight options provided by the *UNFCCC* Secretariat as the basis for future work on the allocation of emissions from aviation and marine bunker fuels.[18] Since the IMO started its regulatory work on GHG emissions from international shipping in 1997,[19] two parallel regimes, the global climate

16 Generally Annex I States under the *UNFCCC* have similar regulatory interests in terms of regulating GHG emissions from international shipping. Nevertheless, non-Annex I States, including major Flag of Convenience (FOC) States, major developing flag States, and other developing States, have different positions towards the regulation of this GHG issue in particular towards the CBDR principle. It is thus difficult for developing States to ensure the incorporation of their interests into IMO regulations. An IMO regulation could be passed without the consent of a few major developing States under the majority-voting mechanism. See ch. 6, 6.7.

17 *Methodological Issues*, Decision 4/CP.1, Report of the Conference of the Parties on its First Session, FCCC/CP/1995/7/Add.1 (28 March – 7 April 1995) art. 1(f), p. 16.

18 These options are: no allocation; allocation to the country where the bunker fuel is sold; allocation to the country of the transporting company, the country of registration of registration of the aircraft/vessel, or the country of the operator; allocation to the country of departure or destination of the aircraft/vessel; and allocation to the country of departure or destination of the passenger/cargo. Sebastian Oberthür, 'Institutional Interaction to Address Greenhouse Gas Emissions from International Transport: ICAO, IMO and the Kyoto Protocol' (2003) 3(3) *Climate Policy* 191, 193.

19 In 1997 the IMO adopted Resolution 8 on 'CO_2 emissions from ships', which requested the IMO to undertake a study on GHG emissions from ships and consider feasible CO_2 reduction strategies. See International Maritime Organisation (IMO), 'Main Events in IMO's Work on Limitation and Reduction of Greenhouse Gas Emissions from International Shipping' (2011) <http://www.imo.org/MediaCentre/resources/Pages/Greenhouse%20gas%20emissions.aspx> accessed 5 November 2013, p. 3.

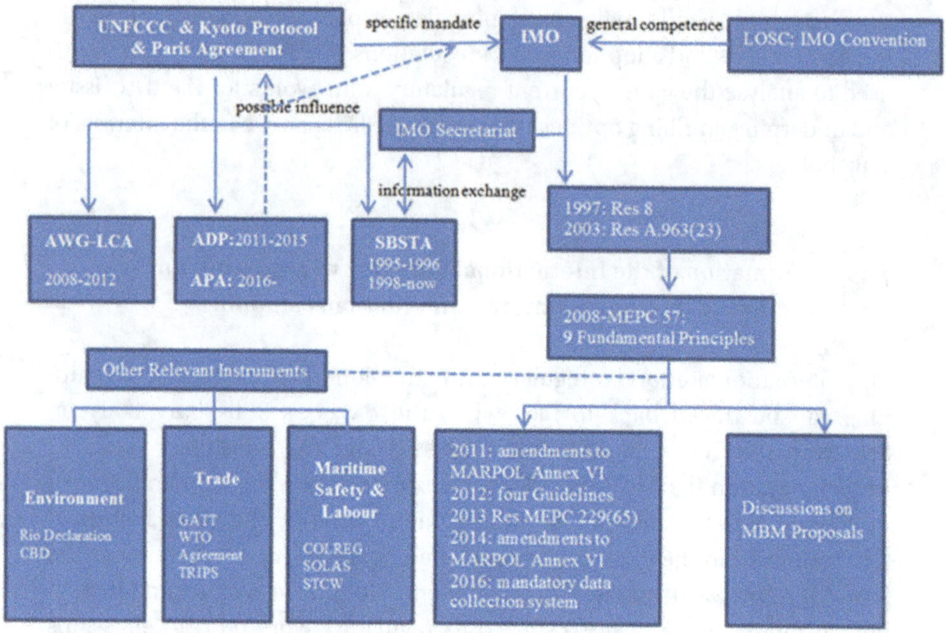

FIGURE 7.2 *International regulatory framework for reducing GHG emissions from international shipping.*

change regime and the IMO GHG emissions regime, have been contributing to the international regulatory process on this GHG issue. Figure 7.2 depicts how these two regimes interact and contribute to the development of the current international regulatory framework for the reduction of GHG emissions from international shipping.

The first regime is the global climate change regime[20] where the SBSTA worked on the allocation and control of international bunker fuels from 1995 to 1996, and since 1998 has been collaborating with the IMO.[21] The Ad Hoc Working

20 See ch. 3, 3.3.

21 Since the IMO received its specific mandate from Article 2(2) of the *Kyoto Protocol*, the regular progress report submitted by the IMO to the SBSTA could be regarded as a type of obligation that the IMO bears under the Kyoto Protocol. But Hackmann has asserted that this cooperation between the *UNFCCC* and the IMO is a 'reciprocal exchange of information and a reciprocal participation in relevant meetings', and both institutions are independent in their decisions. Bernd Hackmann, 'Analysis of the Governance Architecture to Regulate GHG Emissions from International Shipping' (2012) 12(1) *International Environmental Agreements: Politics, Law and Economics* 85, 95. See also ch. 4, 4.2.

Group on Long-term Cooperative Action (AWG-LCA), established as a subsidiary body under the *UNFCCC* in 2007, had worked on the issue of international bunker fuels before 2012. Given no substantial outcomes on the GHG emissions issue had been achieved by the AWG-LCA,[22] this working group finalised its work at the 2012 Doha Climate Change Conference as mandated. The Ad Hoc Working Group on the Durban Platform for Enhanced Action (ADP), established as a new subsidiary body under the *UNFCCC* at the Durban Climate Change Conference in 2011, had been working on negotiating a global climate change agreement that would be adopted by 2015 and would enter into force from 2020. With the adoption of the *Paris Agreement* in December 2015, the ADP was terminated as mandated and the subsequent Ad Hoc Working Group on the Paris Agreement (APA) started its work in 2016. Whether the APA's future work will involve GHG emissions from international shipping remains unclear; however, as discussed in Chapter 3, the *Paris Agreement* is still an agreement under the *UNFCCC*. Since the IMO received its specific GHG mandate from Article 2(2) of the *Kyoto Protocol*, the *Paris Agreement* does not change the origin of the IMO's GHG mandate, but the future work of the APA may clarify or limit the IMO's regulatory competence so as to influence the IMO's current regulations on the GHG issue.

The second regime is the IMO GHG emissions regime[23] where the IMO has adopted conventions, codes, resolutions and guidelines to regulate GHG emissions from international shipping. Of these various regulatory initiatives, Resolution 8 and Resolution A.963 (23)[24] were adopted by the IMO in 1997 and 2003 respectively, which have underpinned the subsequent actions of the IMO. At the 57th MEPC meeting in 2008, the IMO adopted nine fundamental principles on which to base its future regulation of the GHG emissions issue. Under the IMO GHG emissions regime to date, technical and operational measures have been adopted by the IMO to regulate the GHG emissions from

22 Although no outcome has been achieved under the AWG-LCA, some of the options could possibly be adopted by the 2015 Climate Change Agreement. For instance, at the 14th Session of the AWG-LCA in June 2011, member States put forward six options to address GHG emissions from international aviation and shipping. The second option was that the *UNFCCC* set global emissions targets of 20 per cent below 2005 levels in 2020 for international shipping on a scale consistent with the agreed 2 degree Celsius objective, and the use of MBMs may contribute to achieving this target. See *Work of the AWG-LCA Contact Group at AWG-LCA 14.2*, AWG-LCA 14th Session (2nd part), Agenda Item 3.2.4, Bonn (7–17 June 2011).

23 See ch. 4, 4.3.

24 *IMO Policies and Practices Related to the Reduction of Greenhouse Gas Emissions from Ships*, IMO Assembly 23rd Session, Agenda Item 19, IMO Doc Res A. 963(23) (5 December 2003).

international shipping in the form of amendments to Annex VI to *MARPOL 73/78*, as well as its guidelines and resolutions, whereas MBM proposals are still under discussions within the IMO.

Aside from these two parallel regimes that are tackling GHG emissions from international shipping, some international instruments on maritime safety and labour, international trade and environment, also contribute to addressing the issue. Some international environmental instruments, such as the *Rio Declaration on Environment and Development* (*Rio Declaration*),[25] and the *Convention on Biological Diversity* (*CBD*),[26] have provided general regulatory principles for the regulation of the GHG issue. With regard to the adopted technical and operational measures, maritime safety and labour related treaties, such as the *Convention on the International Regulations for Preventing Collisions at Sea* (*COLREG*),[27] *International Convention for the Safety of Life at Sea* (*SOLAS*),[28] and *International Convention on Standards of Training, Certification and Watchkeeping for Seafarers* (*STCW*),[29] will apply to the issue. Additionally, the *Agreement on Trade-Related Aspects of Intellectual Property Rights* (*TRIPS*)[30] can be applied to the transfer of technology issues incorporated in Regulation 23 of the amended Annex VI to *MARPOL 73/78*. Concerning the proposed MBMs which are currently being discussed within the IMO, whether these proposed MBMs are compatible with WTO rules is also an issue in debate.[31]

---

25 *Rio Declaration on Environment and Development*, 31 ILM 874 (14 June 1992) ('Rio Declaration').

26 *Convention on Biological Diversity*, opened for signature 5 June 1992, 31 ILM 818 (entered into force 29 December 1993) ('*CBD*').

27 *Convention on the International Regulations for Preventing Collisions at Sea*, opened for signature 20 October 1972, UKTS 77 (entered into force 15 July 1977) ('*COLREG*').

28 *International Convention for the Safety of Life at Sea*, opened for signature 1 November 1974, 1184 UNTS 2 (entered into force 25 May 1980) ('*SOLAS*').

29 *International Convention on Standards of Training, Certification and Watchkeeping for Seafarers*, opened for signature 7 July 1978, 1361 UNTS 2(entered into force 28 April 1984), as amended by the 1995 Protocol, 1969 UNTS (entered into force 1 February 1997) ('*STCW*').

30 *Agreement on Trade-Related Aspects of Intellectual Property Rights* (*TRIPS*), Annex 1C of the *Marrakesh Agreement Establishing the World Trade Organization*, opened for signature 15 April 1994, 33 ILM 1125 (entered into force 1 January 1995) ('*TRIPS*').

31 See, e.g., *Possible Incompatibility between WTO Rules and A Market-based Measure for International Shipping*, submitted by India, MEPC 62nd Session, Agenda Item 5, IMO Doc MEPC 62/5/27 (20 May 2011); *Possible Incompatibility between the WTO Rules and Market-based Measures for International Shipping*, submitted by India and Saudi Arabia, MEPC 64th Session, Agenda Item 5, IMO Doc MEPC 64/5/3 (29 June 2012).

Although the *Kyoto Protocol* was adopted in 1997, the international regulatory framework for the reduction of GHG emissions from international shipping has only been developing in recent years with the adoption of energy efficiency measures by the IMO as the main achievement. Furthermore, the lack of a consensus in adopting this regulation indicates the existence of substantially different views between countries. Therefore, it is important to identify and address the divergence so as to ensure widespread and effective compliance with these regulations by all countries.

7.4 Deficiencies in Adopted Energy Efficiency Measures and Approaches for Their Future Improvement

Although GHG emissions from international shipping have been partially regulated by the IMO in the form of amendments to Annex VI of *MARPOL 73/78*, deficiencies still exist in this regulation.[32] Based on the objective of achieving absolute GHG emissions reduction from international shipping, there are still gaps in the current legal and institutional framework for the reduction of shipping GHG emissions. This part identifies these deficiencies and gaps in adopted energy efficiency measures, and proposes solutions from four perspectives, namely expanding the coverage and strengthening the effectiveness of technical measures, strengthening the effectiveness of operational measures, improving the enforcement of energy efficiency measures by flag and port States, and regulating the mandatory data collection system in a prudential manner.

7.4.1 *Deficiencies in Current Technical and Operational Measures*

The revised Annex VI to *MARPOL 73/78* adopted in July 2011 is a significant advance in regulating GHG emissions from international shipping. In particular, the amendments make mandatory the EEDI for new ships, and the SEEMP for all ships. The IMO-commissioned research has indicated that the EEDI and SEEMP will achieve significant GHG emissions reduction from international

32 See, e.g., James Harrison, 'Recent Developments and Continuing Challenges in the Regulation of Greenhouse Gas Emissions from International Shipping' (2012) *University of Edinburgh Research Paper Series* <http://ssrn.com/abstract=2037038> accessed 5 November 2013, pp. 25–26; Md. Saiful Karim, 'IMO Mandatory Energy Efficiency Measures for International Shipping: The First Mandatory Global Greenhouse Gas Reduction Instrument for an International Industry' (2011) 7(1) *Macquarie Journal of International and Comparative Environmental Law* 111, 113.

shipping.[33] Nevertheless, some deficiencies remain and create challenges for future implementation of these measures.[34] This section identifies these deficiencies from international law and stakeholders' perspectives.

The adopted EEDI and SEEMP do not fully incorporate relevant international environmental law principles. First, these measures do not adequately reflect the principle of environmental liability for transboundary harm. Based on this principle, flag States have a duty to prevent, reduce and control transboundary harm resulting from cumulative GHG emissions from international shipping, and port States also have a duty to cooperate in mitigating transboundary environmental risks.[35] However, the new paragraph 5 of Regulation 10 of Annex VI limits port State inspection to verifying if there is a valid IEE Certificate on board. This limitation excludes possible unilateral actions or more stringent rules by port States. Given that there is no uniform enforcement of standards by various regional Memoranda of Understanding (MOUs) on port State control, port State control has often been regarded as ineffective.[36] In this sense, a comparatively flexible port State control, such as the United States Coast Guard (USCG)'s port State control program, would facilitate effective port State control in addressing this GHG issue[37] and thus better reflect environmental liability for possible transboundary harm.

Second, the EEDI and SEEMP are not fully consistent with the polluter-pays principle. Under the EEDI and SEEMP not all polluters are responsible for the environmental cost. The 2011 amendments of Annex VI only made the EEDI applicable to certain types of new ships. This accounted for 70 per cent of emissions from new ships, but existing ships are not covered.[38] Having realised

33 Zabi Bazari and Tore Longva, '*Assessment of IMO Mandated Energy Efficiency Measures for International Shipping*' (IMO Doc MEPC 63/INF.2, Annex, 31 October 2011) p. 8.

34 See Harrison, above n. 32, 2. See also ch. 4, 4.3.3.3.

35 See ch. 2, 2.3.

36 Ho-Sam Bang, 'Is Port State Control an Effective Means to Combat Vessel-Source Pollution? An Empirical Survey of the Practical Exercise by Port States of Their Powers of Control' (2008) 23(4) *The International Journal of Marine and Coastal Law* 715, 726. See also ch. 6, 6.5.3.

37 The USCG is the maritime safety authority of the US, and is responsible for maritime safety, security and environmental stewardship of the US. The US is not a formal member of any MOUs (except as an observer to some MOUs), but its unilateral port State control measures under the USCG has been generally regarded as effective and successful. Bang, above n. 36, 741.

38 IMO, above n. 19, 12. The types of ships being regulated by the EEDI in 2011 included bulk carrier, gas carrier, tanker, container ship, general cargo ships, refrigerated cargo carrier, and combination carrier.

this problem, the IMO has been refining the application scope of the EEDI. Eventually at the 66th MEPC meeting in April 2014, amendments of Annex VI to *MARPOL 73/78* were adopted to extend the application scope of the EEDI to include an extra five types of ships. They are Liquefied Natural Gas (LNG) carriers, roll-on/roll-off (ro-ro) cargo ships (vehicle carriers), ro-ro cargo ships, ro-ro passenger ships, and cruise passenger ships having non-conventional propulsion.[39] However, this extended application scope of the EEDI still does not include all types of new ships. In particular, the 2014 amendments of Annex VI exempts ships not propelled by mechanical means, platforms including Floating Production Storage and Offloading Facilities (FPSOs) and Floating Storage Units (FSUs) and drilling rigs, regardless of their forms of propulsion, as well as cargo ships having an ice-breaking capability.[40] Passenger ships other than cruise passenger ships will also remain unregulated by the EEDI. Therefore, the EEDI requirements do not make all polluters (shipowners in this context) pay for the environmental cost of their GHG emissions. Furthermore, the adopted EEDI only regulates the design of new ships, which diminishes its effectiveness as well as decreasing its accuracy as an emissions indicator.[41] It can thus be deduced that the cost-bearing mechanism under the EEDI could be improved. Indeed the SEEMP makes all ship operators engaged in international voyages liable for the preparation and implementation of the SEEMP. However, the lack of reduction target-setting and monitoring weakens

39 *Amendments to MARPOL Annex VI and the NOx Technical Code 2008*, IMO Doc Res MEPC.251(66) (4 April 2014) reg 21 ('*MARPOL 73/78 Annex VI (2014 amendments)*').

40 *MARPOL 73/78 Annex VI (2014 amendments)* reg 19(2)(3).

41 Stathis Palassis, 'Climate Change and Shipping' in Robin Warner and Clive Schofield (eds), *Climate Change and the Oceans: Gauging the Legal and Policy Currents in the Asia Pacific and Beyond* (Edward Elgar Publishing Limited, 2012) 200, 218; See also S.M.Rashidul Hasan, *Impact of EEDI on Ship Design and Hydrodynamics: A Study of the Energy Efficiency Design Index and Other Related Emission Control Indexes* (Master of Science Thesis, Chalmers University of Technology, 2011) <http://publications.lib.chalmers.se/records/fulltext/151284.pdf> accessed 2 December 2013, pp66–67. Hasan takes the view that under current EEDI regulations the EEDI is only used as a design parameter, and may cause 'the sister vessel dilemma' or 'destroy the sister vessel concept'. For example, keel lay of two sister vessels is in Phase 0 (1 January 2013 – 31 December 2014) and Phase 1 (1 January 2015 – 31 December 2019) respectively. Based on Regulation 21 of Annex VI to MARPOL 73/78, the first vessel meets the EEDI requirement, while the second vessel has to be modified to achieve the EEDI requirements of Phase 1. These two vessels are thus not sisters anymore.

the effect of this measure,[42] and actually renders polluters (ship operators in this context) exempt from liabilities.

The adopted EEDI and SEEMP lack full support from stakeholders in emissions reductions from international shipping. Under the *UNFCCC* process the Parties have mainly discussed regulatory principles in relation to GHG emissions from shipping, whereas the discussion of technical and operational measures falls into the work of the IMO. In the IMO discussions, national shipping industries from both developed and developing States have generally supported the adoption of technical and operational measures. However, the shipping industries in many developing countries have insisted that the principle of Common but Differentiated Responsibility (CBDR) be incorporated into the EEDI and SEEMP.[43] As the main regional ship designers and shipbuilders association, the Community of European Shipyards' Association (CESA) preferred the SEEMP to the EEDI in that it believed that the SEEMP would be more effective than the EEDI.[44] Clearly the adopted EEDI and SEEMP have not fully incorporated the CBDR principle, and the SEEMP is also not as effective in reducing shipping GHG emissions as CESA expected.

Flag States under Annex I to the *UNFCCC* generally support the adopted EEDI and SEEMP by the IMO. Nevertheless, many non-Annex I flag States opposed the adoption of these measures due to their differing regulatory interests. For example, some major developing countries, such as China and India, insisted on the application of the CBDR principle to these measures. They asserted that these measures should not be regulated in the form of an amendment to Annex VI of *MARPOL 73/78* due to the different nature of air pollutants (e.g., SO_X, NO_X) and GHG emissions (e.g., CO_2). Some less developed countries, such as Vanuatu, suggested that an exemption for ships trading to the Least Developed Countries (LDCs) and Small Island Developing States (SIDS) be provided from the EEDI and SEEMP. Furthermore, many developing countries proposed that effective transfer of technology and financial assistance from developed countries to developing countries be strengthened so as to enhance their capability to implement these measures. There are differing views among

---

42 See ch. 4, 4.3.3.3.

43 See, e.g., Anil Devli, *Overview of the Shipping Sector in India* (1 April 2011) <http://www.ahlers.com/images/news/2011/overview%20of%20the%20shipping%20sector%20in%20india%20%28mr%20anil%20devli%29.pdf> accessed 26 November 2013, p. 8; see also ch. 5, 5.4.

44 *Development of a CO_2 Design Index for New Ships*, submitted by the Community of European Shipyards' Association (CESA), MEPC 58th Session, Agenda Item 4, IMO Doc MEPC 58/4/12 (1 August 2008) para. 5. See also ch. 5, 5.2.1.

commentators as to whether the EEDI and SEEMP should be amended to take into account these requests from developing countries based on the CBDR principle. However, it is generally accepted that current mechanisms on financial assistance and transfer of technology in relation to this GHG issue are weak.[45] Some scholars even argue that the implementation of the EEDI might 'trigger another migration of [the] shipbuilding industry in the future' if the current transfer of technological and financial resources from developed countries to developing countries cannot be improved.[46]

In contrast to active participation of most flag States in discussing proposed technical and operational measures, most port States have not made timely responses to the regulatory initiatives by the IMO. Nevertheless, some port States agree that current port State control on the GHG issue could be strengthened so as to ensure compliance with the EEDI and SEEMP by shipping companies. While some *UNFCCC* Annex I States suggested maintaining the current port inspection mechanism,[47] Singapore, as a non-Annex I port State, proposed a more stringent port State control measure which was not adopted by the IMO.[48]

7.4.2 *Expanding the Coverage and Strengthening the Effectiveness of Technical Measures*

The EEDI is the main technical measure that the IMO has adopted in tackling GHG emissions from international shipping. However, this technical measure needs to be improved so as to reduce GHG emissions more effectively and

45 See, e.g., Harrison, above n. 32, 16–18; Derya Aydin Okur, *The Challenge of Regulating Greenhouse Gas Emissions from International Shipping and the Complicated Principle of 'Common but Differentiated Responsibilities'* (2012) <http://web.deu.edu.tr/hukuk/dergiler/dergimiz13-1/2-deryaaydinokur.pdf> accessed 26 November 2013, p. 45.

46 Jianing Zheng, Hao Hu and Lei Dai, 'How would EEDI Influence Chinese Shipbuilding Industry?' (2013) 40(5) *Maritime Policy & Management* 495, 499, 509. The authors assert that China might possibly lose its leading shipbuilding position in the world since it achieved this position with regard to three major indicators (ship deliveries, new ship orders and booked ship orders) in 2010.

47 See, e.g., *Comments on the Draft Regulatory Text on Energy Efficiency for Ships*, submitted by Norway, MEPC 61st Session, Agenda Item 5, IMO Doc MEPC 61/5/6 (16 July 2010) para. 2; *Report of the Working Group on Energy Efficiency Measures for Ships*, MEPC 61st Session, IMO Doc MEPC 61/WP.10 (30 September 2010) para. 4.15. See also ch. 6, 6.6.1.

48 See *Amendments to MARPOL Annex VI—Inclusion of Regulations on Energy Efficiency for Ships*, submitted by Singapore, MEPC 62nd Session, Agenda Item 6, IMO Doc MEPC 62/6/21 (20 May 2011) annex, p. 1.

efficiently. This improvement could be achieved through the following four means.

First, the EEDI only applies to certain types of new ships and excludes existing ships, which has limited its reduction effectiveness and this coverage should be expanded. Although the application of the EEDI is projected to be extended from seven types of new ships to eleven types of new ships by April 2014, a number of types of new ships would remain unregulated by the EEDI. In this case two strategies could be utilised to address this problem. One is to expand the coverage of the EEDI's application to include most types of new ships through technological innovation. Indeed this step-by-step approach was utilised by the IMO to counter strong opposition from developing countries and expedite the regulation process.[49] It is likely that as technologies develop, the current EEDI formula could accommodate more types of ships including existing ships, although the EEDI has been recognised as 'inappropriate' by the MEPC for application to existing ships.[50] The other strategy is to consider other enhanced technical measures based on the current EEDI. For example, the US has proposed establishing attained energy efficiency standards for new and existing ships through a phased approach. This proposal would be implemented by means of amendments of Annex VI to *MARPOL 73/78* and aroused wide attention within the MEPC at the 65th MEPC meeting in May 2013.[51] The IMO suspended the discussion of MBMs at the 65th MEPC meeting, which was probably due to the attraction of this new proposal on technical measures to many countries.

Second, the EEDI applies at the design stage of a ship and there are limits to what can be achieved at a later stage. The sole use of the EEDI as a design parameter may lead to 'the sister vessel dilemma' or the 'destroy the sister vessel concept'.[52] The EEDI would be more effective in reducing GHG emissions if the current EEDI formula could be optimised taking shipping operation and

49 See ch. 4, 4.3.3.1.

50 *Report of the Marine Environment Protection Committee on Its Sixty-Third Session*, MEPC 63rd Session, Agenda Item 23, IMO Doc MEPC 63/23 (14 March 2012) para. 5.54.

51 See *Further Details on the Proposal of the United Nations to Reduce Greenhouse Gas Emissions from International Shipping*, submitted by the United States of America, MEPC 64th Session, Agenda Item 5, IMO Doc MEPC 64/5/6 (26 July 2012); *Draft Legal Text with Respect to the Proposal of the United States to Reduce Greenhouse Gas Emissions from International Shipping*, submitted by the United States of America, MEPC 64th Session, Agenda Item 5, IMO Doc MEPC 64/5/7 (26 July 2012); *Proposal of the United States to Enhance Energy Efficiency in International Shipping*, submitted by the United States, MEPC 65th Session, Agenda Item 4, IMO Doc MEPC 65/4/19 (8 March 2013).

52 Hasan, above n. 41.

the EEDI's implementation phases, as indicated in Annex VI to *MARPOL 73/78*, into account. However, this possibility depends on the availability of relevant technologies.

Third, given that the granting of financial and technological assistance constitutes one way of implementing differentiated responsibility under the CBDR principle,[53] it is important to ensure a smooth transfer of technologies from developed countries to developing countries as required in Regulation 23 of Annex VI to *MARPOL 73/78*. In this way, the capacity of developing countries in building cost-effective ships could be strengthened, which would be helpful in ensuring globally uniform enforcement of these measures including both the EEDI and SEEMP. However, neither Regulation 23 of Annex VI nor the subsequent MEPC Resolution on Promotion of Technical Co-operation and Transfer of Technology relating to the Improvement of Energy Efficiency of Ships[54] has imposed concrete obligations to transfer such technology on any State. Rather, this Resolution underscores respect for intellectual property rights.[55] Since most energy efficient technologies are owned by private shipping companies in developed countries, it appears that a market-based approach to technology acquisition might be a better option for developing countries. Indeed, Article 66(2) of the *Agreement on Trade-Related Aspects of Intellectual Property Rights* (*TRIPS*)[56] requires developed country parties to 'provide incentives to enterprises and institutions in their territories for the purpose of promoting and encouraging technology transfer to least developed country members in order to enable them to create a sound and viable technological base'. However, research indicates that even based on comparatively lax criteria, only 22 per cent of reported initiatives by developed countries fulfilled Article 66(2).[57] Accordingly, this mechanism has been criticised for its

---

53 Rajamani, above n. 252.

54 *Promotion of Technical Co-operation and Transfer of Technology relating to the Improvement of Energy Efficiency of Ships*, IMO Doc Res MEPC.229(65) (17 May 2013) ('Resolution MEPC.229(65)').

55 *Resolution MEPC.229(65)* art. 4.

56 *Agreement on Trade-Related Aspects of Intellectual Property Rights* (*TRIPS*), Annex 1C of the *Marrakesh Agreement Establishing the World Trade Organization*, opened for signature 15 April 1994, 33 ILM 1125 (entered into force 1 January 1995).

57 The commencement year for developing countries that are party to the WTO to adopt TRIPS has been postponed from 2006 to 2013, and even to 2016 for some countries as it relates to protections for pharmaceuticals. It is thus difficult to get relevant data on the TRIPS-based transfer of technologies from developing countries. Amanda Watson, 'Does TRIPs Increase Technology Transfer to the Developing World? The Empirical Evidence' (2011) 20(3) *Information & Communications Technology Law* 253, 271,273.

lack of effectiveness in transferring technologies from developed countries to developing countries.[58] Insufficient financial incentives are one of the key factors among many reasons behind this lack of effectiveness.[59] For this reason, the establishment of a global technology acquisition fund[60] financed by developed countries, either within the shipping industry or under a broader *UNFCCC* regime, might contribute to addressing this problem. Alternatively this fund could also be linked to the MBM proposal in the IMO on the International GHG Fund. However, it remains unclear whether this proposal is feasible and cost-effective, and to what extent it would be accepted by most countries.

Fourth and finally, the amended Annex VI to *MARPOL 73/78* provides that the IMO must review the status of technological developments regularly. Based on the results of these reviews, the time period, reference line parameters for relevant ship types and reduction rates of the EEDI could possibly be upgraded.[61] Since the shipping industries in some developing countries are concerned that these upgraded technical thresholds would become a form of trade barrier for them,[62] it is important to make this reviewing process fair and equitable for developing countries. Since such reviews have been scheduled for around 2015 and 2022 by the IMO, it is important that a certain percentage of reviewers should be from major developing countries. In this way the technological disadvantages of developing countries can be taken into account in the reviewing process.

7.4.3 *Strengthening the Effectiveness of Operational Measures*

Compared with the EEDI, the SEEMP has received less attention from the international community. This is probably because of their different roles in tack-

58 See, e.g., Wei Guo, 'The TRIPs Agreement Does Little to Promote the Development of Technology Transfer to Developing Countries' (2009) 3(3) *Management Science and Engineering* 20, 22–26; Nitya Nanda and Nidhi Srivastava, 'Clean Technology Transfer and Intellectual Property Rights' (2009) 9(3) *Sustainable Development Law & Policy* 42, 46. But Navraj Singh asserts that intellectual property rights protection is only a relatively peripheral factor contributing to the ineffectiveness of climate technology transfer. Ghaleigh Navraj Singh, 'Barriers to Climate Technology Transfer—The Chimera of Intellectual Property Rights' (2011) 5(2) *Carbon & Climate Law Review: CCLR* 220, 233.

59 These reasons include the lack of financial means of developing countries, lack of intellectual property rights (IPR) protection in developing countries and the monopoly created by IPR-based market power. Nanda and Srivastava, above n. 58, 43–44; Navraj Singh, above n. 58, 229–231.

60 Nanda and Srivastava, above n. 58.

61 *MARPOL Annex VI* (2011) reg 21.6.

62 See ch. 5, 5.4.1.

ling GHG emissions from international shipping. The EEDI can reduce GHG emissions and influence global shipbuilding migration directly, and influence international trade indirectly. In contrast, the SEEMP has no impact on international trade, and its reduction potential is dependent on the performance of individual ship operators due to its lack of mandatory reduction targets and monitoring requirements. However, three approaches could be adopted to strengthen the effectiveness of the SEEMP.

A way to improve the effectiveness of the SEEMP would be to set a reduction target. However, an IMO-commissioned report asserts that it is less likely to have a target-based regulatory framework for the SEEMP 'in the foreseeable future'.[63] The SEEMP applies to all existing and new ships of 400 gross tonnage and above, and applies to various types of ships. The *Guidelines for the Development of a Ship Energy Efficiency Management Plan* (SEEMP Guidelines) in 2012 provide for procedures and measures at the stages of planning, implementation, monitoring, and self-evaluation and improvement, and incorporate best practices for the fuel-efficient operation of ships.[64] However, different types of ships may have differing 'best practices' on each international voyage based on the different purposes of the voyages. It is thus technically difficult to set these reduction targets at an agreed level for most stakeholders. Given that the main objective of the SEEMP is to minimise shipping GHG emissions by means of reducing fuel consumption, to provide some other incentives might be more effective.[65] These incentives include:

- 'high fuel and carbon prices;
- more vigorous awareness building and cultural change on board ships;
- more collaboration between industry stakeholders and a solution to the issue of split-incentives; and
- effective monitoring of SEEMP implementation via rigorous audits and reviews.'[66]

Currently monitoring of the SEEMP mainly relies on the voluntary use of the Energy Efficiency Operational Indicator (EEOI) or other performance indicators that ship operators choose. However, as a monitoring tool and a

63 Bazari and Longva, above n. 33, 7.

64 *2012 Guidelines for the Development of a Ship Energy Efficiency Management Plan (SEEMP)*, Resolution MEPC.213(63), IMO Doc MEPC 63/23 Annex 9 (2 March 2012) ('*SEEMP Guidelines*') art. 5.

65 Bazari and Longva, above n. 33, Appendix 4, 12.

66 Ibid. 15.

benchmark tool for ship and fleet efficiency performance,[67] the EEOI was initially introduced for trial purposes on a voluntary basis in 2009.[68] Given that the SEEMP was regulated as a mandatory operational measure in 2011 and the EEOI has now been generally accepted by the shipping industry,[69] it would be a natural progression for the EEOI to be mandated. It is likely that now that the SEEMP has been regulated as a mandatory measure, it will achieve more reduction potential. It is arguable that the EEOI would provide a 'more accurate and verifiable measurement of fuel consumption that could pave the way for CO_2 foot printing and data verification in the future'.[70] Indeed a penalty on trade and development proposal by the Bahamas in the IMO seeks to collect emission statistics from either the EEOI or ship funnels using a suitable sensor.[71]

Generally the SEEMP provides an approach for monitoring ship and fleet efficiency performance, and it is advantageous for ship operators to adopt new technologies and allied practices when they seek to optimise the performance of the ship. In this sense, successful financial and technological transfer from developed countries to developing countries, which has been discussed in the previous section, would also facilitate the enforcement of these measures by shipowners or ship operators from developing countries.

7.4.4 *Improving the Enforcement of Energy Efficiency Measures by Flag and Port States*

The enforcement of energy efficiency measures (EEDI and SEEMP) mainly relies on flag States and port States.[72] Flag States have primary responsibility for ensuring the compliance of ships on their registers with all applicable international and domestic regulations and standards, whereas port States

67 IMO, above n. 19, 4.

68 *Guidelines for Voluntary Use of the Ship Energy Efficiency Operational Indicator* (*EEOI*), Ref. T5/1.01, IMO Doc MEPC.1/Circ.684 (17 August 2009).

69 See ch. 5, 5.3.2.

70 Bazari and Longva, above n. 33, 7–8.

71 *Need and Purpose of an MBM: How Technical and Operational Measures Are the Only Direct and Effective Means to Deliver Cuts in CO_2 Emissions*, submitted by the Bahamas, Intersessional Meeting of the Working Group on GHG Emissions from Ships 3rd Session, Agenda Item 2, IMO Doc GHG-WG 3/2 (22 December 2010) para. 17. But see Psaraftis, above n. 109, 221. Psaraftis asserts that it is impossible to establish an EEOI baseline and a reliable EEOI for all types of ships. See also ch. 4, 4.3.4.2.

72 C. Pisani, 'Fair at Sea: The Design of A Future Legal Instrument on Marine Bunker Fuels Emissions within the Climate Change Regime' (2002) 33(1) *Ocean Development and International Law* 57, 71.

significantly complement the work of flag States in addressing substandard ships.[73] In practical terms, the main approach for flag State enforcement is through survey and certification, while port States exercise their enforcement via port State control.

Chapters 2 and 6 discussed the prescriptive and enforcement jurisdiction of flag States and port States on the GHG emissions issue. For most countries, the first step in enforcing energy efficiency measures is to incorporate these measures into their domestic law and policy or, in some countries, into rules made by classification societies. These rules would thus be applicable to the ships registered in these flag States. After the adoption of the EEDI and SEEMP in 2011, some non-Annex I flag States, such as South Korea and China, either incorporated these rules into their domestic legislation and classification rules or are currently engaged in doing this through their national regulatory process. However, it is worthwhile to note that only 49 out of 64 Parties of Annex VI to *MARPOL 73/78* voted for the amendments to Annex VI in July 2011.[74] Some significant dissenters, including Brazil, Chile, China, Kuwait and Saudi Arabia, voted against this regulation. Therefore, it is argued that 'these dissents could have a significant impact on the implementation and application of the Regulation'.[75] Furthermore Brazil and Finland objected to this amendment so these energy efficiency measures would not apply to them. It is possible that some substandard ships may seek suitable routes to avoid the regulation. In this sense, it is necessary for the IMO to continue to improve these energy efficiency measures, in particular the regulation on technological and financial transfer from developed countries to developing countries, so that more flag States might be attracted to ratify Annex VI to *MARPOL 73/78*.

Regarding those flag States which have ratified Annex VI, it is important that these States or the Recognised Organisations (RO)[76] that they nominate, verify and issue the International Energy Efficiency Certificate (IEE Certificate) stringently in accordance with relevant IMO regulations and guidelines. Additionally, to overcome the possible 'laziness' of some flag of convenience (FOC) States in complying with the EEDI and SEEMP requirements,

73 Bang, above n. 36, 1. See also Pisani, above n. 72, 66. Pisani asserts that the objective of port State control is to eradicate substandard ships by means of imposing pressure on flag States, shipowners, classification societies, and insurers to comply with their obligations under international law.

74 *Report of the Marine Environment Protection Committee on Its Sixty-Second Session*, MEPC 62nd Session, Agenda Item 24, IMO Doc MEPC 62/24 (26 July 2011) para. 6.110.

75 Harrison, above n. 32, 11.

76 See ch. 5, 5.2.5.

establishing a compulsory self-assessment scheme to assess the ability of flag States to enforce these IMO measures may also be necessary.[77] In 2005 the IMO adopted a voluntary Member State Audit Scheme, based on which the IMO would assess whether a Member State has complied with an IMO convention once such an audit was requested by that State.[78] However, this scheme was regarded as less effective due to its optional nature.[79] At the 66th MEPC meeting in April 2014, the IMO adopted amendments to Annex VI to *MARPOL 73/78* which makes the IMO Audit Scheme mandatory through adding a Chapter 5 entitled 'verification of compliance with the provisions of this annex'. Based on these amendments, the IMO shall conduct periodic audits in accordance with the audit standard as specified in *IMO Instruments Implementation Code* (*III Code*)[80] to verify compliance with and implementation of this Annex by flag States, coastal States and port States which have ratified the amendments.[81] The amendments will impose some pressure for States, in particular FOC States, to exercise their obligations and responsibilities contained in this Annex. Nevertheless, as only States which have ratified these amendments are legally bound by them, it appears vital to push more States to ratify the revised Annex VI to *MARPOL 73/78*. Furthermore, the audit covers nine categories of administrative, legal and technical issues.[82] How to ensure the smooth auditing and good cooperation from relevant party States seems challenging. However, a proposed compulsory self-assessment scheme for assessing the ability of flag States may not be necessary provided that the newly-adopted mandatory IMO Audit Scheme be effectively implemented.

---

77 Ho-Sam Bang, 'Recommendations for Policies on Port State Control and Port State Jurisdiction' (2013) 44(1) *Journal of Maritime Law and Commence* 115, 132.

78 *Framework and Procedures for the Voluntary IMO Member State Audit Scheme*, Assembly 24th Session, Agenda Item 19, IMO Doc Res A.974 (24) (21 December 2005).

79 Bang, above n. 77.

80 *IMO Instruments Implementation Code* (*III Code*), IMO Doc Res A.1070(28) (4 December 2013).

81 *MARPOL 73/78 Annex VI* (*2014 amendments*) reg 25(1).

82 *Framework and Procedures for the IMO Member State Audit Scheme*, IMO Doc Res A.1067(28) (4 December 2013) reg 7.4.2. These nine categories are: '(1) jurisdiction; (2) organization and authority; (3) legislation, rules and regulations; (4) promulgation of IMO instruments, rules and regulations; (5) enforcement arrangements; (6) control, survey, inspection, audit, verification, approval and certification functions; (7) selection, recognition, authorization, empowerment and monitoring of recognized organization, as appropriate, and of nominated surveyors; (8) investigations required to be reported to the Organization; and (9) reporting to the Organization and other Administrations'.

The role of port States in enforcing the EEDI and SEEMP has also been controversial. Chapter 6 discussed the debate within the IMO in which Singapore suggested giving port States the right to deny ships port entry based on whether they comply with the EEDI. However, Singapore's proposal was not agreed by the MEPC and the adopted regulation limits port State inspection to verifying if there is a valid IEE Certificate on board.[83] This regulation excludes possible unilateral actions by port States, and has actually become a standard phase for port State control. The purpose of this provision appears to be to establish a globally uniform port State control regime, in which the NMFT principle can be uniformly applied. However, this goal is less likely to be achieved in the foreseeable future due to the diverse financial and technological capacity of ports in different regions, in particular those regions where most ports are developing countries. The current imbalance of performance among nine regional MOUs on port State control has confirmed this situation.[84] Under these circumstances, a differentiated strategy might be helpful in improving the enforcement of energy efficiency measures by port States. This strategy would consist of two elements. On the one hand, the gaps in performance among nine regional MOUs on port State control should be narrowed. This goal could be achieved through the assistance provided by developed State PSC MOUs to developing State PSC MOUs.[85] The assistance may include strengthening the exchange of information by organising joint ministerial meetings, coordinating activities by hosting regular port State control Committee meetings, training inspectors and increasing technical and financial assistance.[86] It has also been suggested that the IMO should develop uniform MOUs on port State control management techniques.[87] On the other hand, currently some States, such as the US, exercise more stringent unilateral port State control measures than IMO rules, which in the context of the EEDI and SEEMP would overcome the ineffectiveness of some IMO rules and thus should be allowed. The number of ships trading with the US has remained stable after the US Coast Guard (USCG) adopted its own port State control program.[88] This is evidence for the

83 *MARPOL Annex VI* (2011) reg 10.5.

84 Ho-Sam Bang and Duck-Jong Jang, 'Recent Developments in Regional Memorandums of Understanding on Port State Control' (2012) 43(2) *Ocean Development and International Law* 170, 184. Bang and Jang assert that the performance of nine regional MOUs on port State control 'is of variable quality' as to their commitments to port State control related activities. See also ch. 6, 6.5.3.

85 Bang and Jang, above n. 84.

86 Ibid.

87 Ibid.

88 Bang, above n. 36, 744.

proposition that this type of unilateral action does not lead to a mass shift of shipping routes but rather improves 'the quality of shipping'.[89]

Aside from further improvement of MOUs and unilateral actions on port State control, it has been suggested that the IMO should update the *2009 Guidelines for Port State Control under the Revised MARPOL Annex VI* (*2009 Guidelines*).[90] The *2009 Guidelines* were enacted to meet the needs of port State control on air pollution (e.g., SO_X, NO_X). Since *MARPOL* Annex VI has been amended to include GHG emissions (e.g., CO_2) which are of a different nature to air pollutants, the *2009 Guidelines* should be updated to reflect the differing Certificate requirements. This proposal was first put forward by Norway. At the 61st MEPC meeting in 2010, Norway proposed developing a new chapter in the *2009 Guidelines* to provide basic guidance for port State control in relation to the energy efficiency regulations.[91] However, this issue has not yet been addressed.

7.4.5 *Regulating the Mandatory Data Collection System in a Prudential Manner*

At the 69th MEPC meeting in April 2016, the IMO approved mandatory data collection requirements for ships engaged in international shipping to report their fuel consumption. It is expected that this draft will be adopted at the 70th MEPC meeting in October 2016 in the form of the amendments to *MARPOL* Annex VI and could enter into force in 2018.[92] This move can be regarded as the response of the IMO to the unilateral EU regulation on Monitoring, Reporting and Verification (MRV) and the newly-adopted *Paris Agreement*. On the one hand, in June 2013 the European Commission developed a proposal for an EU Regulation on MRV and submitted it to the European Parliament and the Council. This proposal aimed to further the reduction of shipping GHG emissions due to the slow progress within the IMO. Consequently, the MRV Regulation 2015/757 was adopted and came into force on 1 July 2015. This regulation imposed some pressure on the ongoing discussions within the IMO. On the other hand, Article 13 of the *Paris Agreement* establishes an enhanced

89 Ibid.

90 *2009 Guidelines for Port State Control under the Revised MARPOL Annex VI*, IMO Doc Res MEPC.181(59) (17 July 2009).

91 *Comments on the Draft Regulatory Text on Energy Efficiency for Ships*, submitted by Norway, MEPC 61st Session, Agenda Item 5, IMO Doc MEPC 61/5/6 (16 July 2010) para. 7.2; See also *Report of the Working Group on Energy Efficiency Measures for Ships*, MEPC 61st Session, IMO Doc MEPC 61/WP.10 (30 September 2010) para. 4.14.

92 IMO, *IMO Takes Further Action on Climate Change* (22 April 2016) < http://www.imo.org/en/MediaCentre/PressBriefings/Pages/11-data-collection-.aspx > accessed 14 May 2016.

transparency framework for climate action and support, which provides a basis for a similar transparency MRV system in the international shipping industry. It is possible that this Agreement has expedited the regulatory process within the IMO.

The establishment of the mandatory data collection system, as the first step of the three-step approach (data collection, data analysis and decision making) within the IMO, is conducted under the IMO agenda of 'Further Technical and Operational Measures for Enhancing the Energy Efficiency of International Shipping'. The collection and analysis of the data on shipping GHG emissions would help to decide whether further reduction measures are needed, and it may also provide a foundation for the adoption of a MBM in the future. This is because most of the current MBM proposals require the measurement of GHG emissions from ships. However, some large developing States have expressed their concerns on the development of this system. For example, at the 68th MEPC meeting in May 2015, India expressed its 'deep concern' regarding the parallel development of a data collection system on a regional basis.[93] Indeed the simultaneous application of both the EU MRV and the IMO's data collection system to the EU region would possibly cause some problems relating to their enforcement. Meanwhile, as another BRICS State,[94] China asserted that it is premature to adopt the data collection system at this stage. It took the view that the concerns from the shipping industry should be addressed first, such as the collection of unnecessary data, the possible leakage of commercial sensitive information, and the irrationality and infeasibility of the establishment of mandatory operational eficiency standards.[95] Accordingly, China purported that this draft requirement should be implemented voluntarily.[96]

The concerns from the large developing States such as India and China on the mandatory data collection system reveal the regulatory interests of these States as well as the significance of harmonising the IMO's data collectin system and the EU MRV. Due to their lack of core energy efficiency technologies, the delayed implementation of the data collection requirements would be beneficial for the development of the shipping industries in developing States. In order to hold the increase of global average temperature to well

93 *Report of the Marine Environment Protection Committee on Its Sixty-Eighth Session*, MEPC 68th Session, Agenda Item 21, IMO Doc MEPC 68/21 (29 May 2015) para. 4.2.7.

94 BRICS is the acronym for an association of five large emerging national economies, namely Brazil, Russia, India, China and South Africa.

95 *Report of the Marine Environment Protection Committee on Its Sixty-Eighth Session*, MEPC 68th Session, Agenda Item 21, IMO Doc MEPC 68/21/Add.1 (29 May 2015) annex 25, pp. 6–7.

96 Ibid.

below 2 or 1.5 degree Celsius above pre-industrial levels,[97] it will be beneficial and urgent to adopt the mandatory data collection system within the IMO. Furthermore, technically it is unlikely to incorporate the differentiated treatment to the establishment of the mandatory data collection system. Under the circumstances, it is necessary to take into account the concerns and benefits of developing States in the development of this system so as to ensure its future implementation by these developing States.

7.5 Analysis of Necessity and Feasibility of Adopting a Market-Based Measure

This part provides in-depth analysis on the MBMs. It first examines whether it is necessary to adopt a MBM, and then it discusses how to select the most suitable MBM from four perspectives.

7.5.1 *Is a Market-Based Measure Necessary?*

Whether MBMs should be adopted by the IMO to tackle GHG emissions from international shipping has been controversial since this type of measure was formally put forward in the 2000 IMO GHG Study.[98] Various countries, international organisations and scholars have expressed their differing views on this issue. At the 65th MEPC meeting in May 2013, the discussion of MBMs was suspended due to a proposal of the US on furthering technical and operational measures. However, it is predicted that in the long term, certain MBMs should be adopted as a supplementary method of reducing shipping GHG emissions in addition to the current energy efficiency measures.

First, studies have indicated that using EEDI and SEEMP alone would not achieve absolute emissions reduction from international shipping.[99] Although the IMO is currently working on the improvement of the technical aspects of the EEDI, it is difficult for the IMO to achieve technical breakthroughs in a short time due to the intricacies of ship types and shipping features. Achieving an increase of no more than two degrees Celsius in the global average temperature by 2100 has become the goal of international community in tackling

---

97 *Paris Agreement* art. 2(1)(a).

98 See Kjell Skjølsvik et al, 'Study of Greenhouse Gas Emissions from Ships' (International Maritime Organization (IMO), 2000).

99 See, e.g., Bazari and Longva, above n. 33, executive summary, p. 8; Palassis, above n. 41, 220; Aydin Okur, above n. 45, 39. See also ch. 4, 4.3.3.3.

climate change.[100] However, a recent study suggests that international shipping needs to make its 'fair and proportionate contribution' so as to reach the two degrees goal, which means CO_2 emissions from international shipping need to be cut within the next decade and fall by at least 80 per cent by 2050 compared to their 1990 levels.[101] Against this background, it is important for the international shipping industry to examine the possibility of adopting MBMs for more GHG reductions rather than waiting for the effects of applying energy efficiency measures to be identified. Given the EEDI and SEEMP only entered into force on 1 January 2013 and there is a waiver clause that some flag States may choose,[102] it may take a long time for the international community to identify the precise effectiveness of these measures in terms of a reduction of GHG emissions from international shipping. While a delay in introducing MBMs to international shipping would be more costly for future implementation,[103] it would be more cost-effective for the international shipping industry to adopt MBMs at an early stage.

Second, the adoption of MBMs could incorporate international law principles in a full and objective manner. The CBDR principle and the No More Favourable Treatment (NMFT) principle are two important international law principles.[104] The debate on applying these principles to GHG emissions from international shipping has been a constant theme running through the international regulatory process in the IMO. Consequently, the energy efficiency measures were adopted by a majority vote in 2011 rather than by a consensus.

100 The two degrees Celsius goal was first put forward by the G-8 in 2009, and later agreed in the Copenhagen Accord. In 2010 this goal was formally incorporated into the *UNFCCC* process, and was later included in Article 2(1)(a) of the *Paris Agreement*. Nevertheless, the specific reduction targets and time frame for achieving this goal have not yet been agreed under the *UNFCCC* process. See Lavanya Rajamani, 'The Cancun Climate Agreements: Reading the Text, Subtext and Tea Leaves' (2011) 60(2) *The International and Comparative Law Quarterly* 499, 501; *Paris Agreement* art. 2(1)(a).

101 A. Bows-Larkin et al, 'High Seas, High Stakes: High Seas Project Final Report' (Tyndall Centre for Climate Change Research, University of Manchester, 2014) <http://www.mace.manchester.ac.uk/media/eps/schoolofmechanicalaerospaceandcivilengineering/research/centres/tyndall/pdf/High_Seas_High_Stakes_High_Seas_Project_Final_Report.pdf> accessed 8 July 2014.

102 Regulation 19 of the 2011 amended Annex VI to *MARPOL 73/78* provides that flag States may postpone their actual commencement date of the EEDI to six and a half years from 1 January 2013.

103 Rachael Dillon, 'The Growing Challenge of Climate Change for the Maritime Industry and the Role of Market-Based Measures to Reduce Emissions' (2012) 79(2) *Journal of Transportation Law, Logistics & Policy* 139, 151.

104 See ch. 2, 2.5.

The main opposition came from some developing countries which criticised that the CBDR principle was not reflected in these measures 'in a full and objective manner'.[105] Compared with adopted energy efficiency measures, proposed MBMs could potentially address the conflict between these two principles. Of the current seven types of MBM proposals submitted to the IMO, the rebate mechanism for a market-based instrument for international shipping, the port State levy, and some other proposals all incorporate elements reflecting both the CBDR and the NMFT principles. There is a higher possibility that these MBMs will be accepted by both developed and developing countries, if both these principles can be adequately incorporated into future MBMs.

MBMs are also consistent with the polluter-pays principle.[106] MBMs can be designed to internalise the external cost of GHG emissions from international shipping through a GHG Fund, different emissions trading schemes, or other means. That is to set a price on the external cost and thereby internalise it so as to address the market failure for GHG emissions.[107] Consequently, the polluters,[108] namely ship owners or ship operators, would pay for their ships' GHG emissions.[109] Indeed this mechanism provides an economic incentive for the polluters to reduce their GHG emissions. Additionally, it is arguable that a 'universally applied and uniformly regulated' global MBM would be 'fully compatible with' the rules of the World Trade Organization (WTO).[110]

105 See, e.g., *Report of the Marine Environment Protection Committee on Its Sixty-Second Session*, MEPC 62nd Session, Agenda Item 24, IMO Doc MEPC 62/24 (26 July 2011) annex 20, p. 1.

106 See ch. 2, 2.6.

107 Erika Ekström, *Market Based Measures to Regulate CO_2 Emissions from International Shipping* (2009) <http://lup.lub.lu.se/luur/download?func=downloadFile&recordOId=1557129&fileOId=1586238> accessed 27 December 2013, pp. 25–26. In this context, market failure refers to 'environmental externalities'. See OECD, *Environmental Externalities* (4 March 2003) <http://stats.oecd.org/glossary/detail.asp?ID=824> accessed 4 April 2014. Environmental externalities 'refers to the economic concept of uncompensated environmental effects of production and consumption that affect consumer utility and enterprise cost outside the market mechanism'. See also ch. 4, 4.3.4.1.

108 GHG emissions from international shipping can be regarded as a type of 'conditional' pollution, and GHG emissions have been regulated as a type of pollution in some countries such as the US, Germany and Australia. See ch. 2, 2.1.2.

109 See Harilaos N. Psaraftis, 'Market-Based Measures for Greenhouse Gas Emissions from Ships: A Review' (2012) 11(2) *WMU Journal of Maritime Affairs* 211, 213.

110 Within the IMO, the WTO secretariat and an IMO Expert Group asserted that MBMs are compatible with WTO rules, whereas India argued that the view of WTO secretariat could not represent the WTO and MBMs are conflicted with relevant rules of the WTO. *Report of the Marine Environment Protection Committee on Its Sixty-First Session*, MEPC 61st Session, Agenda Item 24, IMO Doc MEPC 61/24 (6 October 2010) annex 8, p. 3; *Possible*

Third, the adoption of MBMs reflects a majority view among the main stakeholders in the GHG issue. Generally speaking, the international shipping industry supports the adoption of MBMs by the IMO in reducing GHG emissions from international shipping. For instance, as an influential shipowners' and ship operators' association, the Round Table of International Shipping Associations asserted in 2012 that MBMs might be eventually introduced for shipping although it also commented that MBMs 'are not justified at this particular time'.[111] The Community of European Shipyards' Associations (CESA) is the main regional shipping NGO representing the interests of shipbuilders within the IMO. CESA asserted that the EEDI cannot achieve any short-term emissions reduction from ships, but it regarded MBMs as a 'more effective solution' to address the issue under discussion.[112] Additionally, except for the shipping industries in a few large developing countries such as China and India, most national industries of various countries welcome the adoption of MBMs although they have different preferences on the form MBMs should take.[113]

Flag States have made similar responses to the proposed MBMs. The position of the large developing States on MBMs is also not intransigent. For example, China opposes the adoption of MBMs by the IMO at this stage but it would also accept a compromise position provided that the CBDR principle could be incorporated in the proposed MBMs to be adopted in the future.[114] Due to their less important role in the proposed MBMs, most port States have not expressed their opposition to them. Developing port States are generally more concerned about the assistance that they can obtain in strengthening their capacity building for implementing IMO regulations on the GHG emissions issue.

Fourth, the adoption of MBMs by the international shipping industry is consistent with the practice in the international aviation industry, and would

---

Incompatibility between the WTO Rules and Market-based Measures for International Shipping, submitted by India and Saudi Arabia, MEPC 64th Session, Agenda Item 5, IMO Doc MEPC 64/5/3 (29 June 2012).

111 Round Table of International Shipping Associations, *Round Table Associations Position Paper on GHG+MBMs* (22 February 2012) <https://www.bimco.org/About/Press/Press_Releases/2012/2012_02_22_Round_Table_MBM.aspx> accessed 27 December 2013.

112 *Phase-in Implementation of the Energy Efficiency Design Index for Standard and Complex Ship Types*, submitted by the Community of European Shipyards' Associations (CESA), MEPC 59th Session, Agenda Item 4, IMO Doc MEPC 59/4/38 (20 May 2009) para. 3.

113 See ch. 5, 5.3–5.4.

114 *Report of the Marine Environment Protection Committee on Its Sixty-First Session*, MEPC 61st Session, Agenda Item 24, IMO Doc MEPC 61/24 (6 October 2010) annex 3, p. 3.

raise revenue for climate finance.[115] Article 2(2) of the *Kyoto Protocol* delegates the regulation of GHG emissions in their respective sectors to the IMO and the International Civil Aviation Organization (ICAO). Similar to the IMO's three-pillar reduction strategy (technical, operational, and market-based measures), the ICAO has agreed its four-pillar reduction strategy, namely aircraft technology, operational improvement, sustainable alternative fuels, and MBMs.[116] In particular, at the 38th ICAO Assembly held from 24 September to 4 October 2013, a consensus agreement was reached on the development of a global MBM scheme for international aviation. This MBM scheme is to be decided by the 39th ICAO Assembly in 2016, and to be implemented from 2020.[117] Given that both the international aviation and shipping industries received their GHG mandate from the *Kyoto Protocol* and that they are facing similar regulatory barriers,[118] the successful practice of the ICAO in adopting a future MBM will provide useful experience for the IMO. Meanwhile, MBM-generated revenues from international aviation and shipping industries could be utilised for climate finance in other sectors through the *UNFCCC* process.[119] Additionally,

115 To date there has been no universally agreed definition of 'climate change finance' or 'climate finance' as it is often called. Generally speaking, climate finance refers to 'financial flows for reducing emissions, i.e. mitigation', and 'measures for adapting to the consequences of climate change', in particular funding for tackling climate change in developing countries. However, many scholars assert that climate finance should also include 'resources for activities supporting low-carbon development and energy infrastructure transitions', including 'capacity building, research and development, and technology transfer'. In this section, 'climate finance' should be defined from a narrow sense, namely it refers to 'financial flows for reducing [GHG] emissions'. Luis Gomez-Echeverri, 'The Changing Geopolitics of Climate Change Finance' (2013) 13(5) *Climate Policy* 632, 635.

116 *Information Relevant to Emissions from Fuel Used for International Aviation and Maritime Transport*, UNFCCC SBSTA 39th Session, Warsaw, Doc FCCC/SBSTA/2013/MISC.20 (10 November 2013) p. 3.

117 At the 38th ICAO Assembly, EU's Emissions Trading System (ETS) was rejected as an option for MBM to be adopted by ICAO by 2016.

118 For example, how to reconcile the conflict between the IMO and ICAO's non-discrimination principle with the *UNFCCC*'s CBDR principle has been at the heart of these debates as to the regulation of these GHG issues within the IMO and the ICAO. A detailed analysis of the similarities and differences between the international aviation and shipping industries is provided by: Michael Keen, Ian Parry and Jon Strand, 'Market-Based Instruments for International Aviation and Shipping as a Source of Climate Finance' (2012) <http://elibrary.worldbank.org/doi/pdf/10.1596/1813-9450-5950> accessed 27 December 2013, pp. 10–16.

119 Ibid. 52. However, it is often argued by the aviation and shipping industries that these industries should not be targeted as a source of such revenue in a disproportionate

a recent report concludes that the economic impacts of MBMs for international shipping on developing countries are likely to be small, and undesirable economic impacts can be addressed through a combination of appropriate measures.[120]

Although there is mounting recognition that MBMs should be adopted to reduce GHG emissions from international shipping, opposition from some developing countries remains. Nevertheless, since the main reasons for their opposition are the uncertainties surrounding the form of MBMs and their economic impacts, lack of the CBDR principle, and the lack of competence of the IMO to regulate MBMs,[121] these barriers can potentially be addressed through well-designed mechanisms and better institutional arrangements.

7.5.2 *How to Select the Most Suitable Market-Based Measure?*

As discussed early in this chapter, the adoption of MBMs is a necessary step in achieving absolute GHG emissions reductions from international shipping in the long term. The key question faced by the international community is what type of MBMs to adopt. Although this choice is mostly a political decision, other factors such as the cost of regulation and the compatibility of the MBMs with international law principles will also play a role in the final decision.[122] This section considers options for the most suitable MBMs to address the GHG emissions issue through grouping and analysing current MBM proposals, and proposing relevant mechanisms for selected MBMs.

---

manner. See *Information Relevant to Emissions from Fuel Used for International Aviation and Maritime Transport*, UNFCCC SBSTA 39th Session, Warsaw, Doc FCCC/SBSTA/2013/MISC.20 (10 November 2013) p. 6.

120 Annela Anger et al, 'Research to Assess Impacts on Developing Countries of Measures to Address Emissions in the International Aviation and Shipping Sectors' (2013) <http://www.climatestrategies.org/research/our-reports/category/69/376.html> accessed 28 December 2013, p. 4. See also United Nations Conference on Trade and Development (UNCTAD), 'Review of Maritime Transport 2013' (2013) <http://unctad.org/en/PublicationsLibrary/rmt2013_en.pdf> accessed 11 January 2014, p. 108.

121 See, e.g., *Uncertainties and Problems in Market-based Measures*, submitted by China and India, MEPC 61st Session, Agenda Item 5, IMO Doc MEPC 61/5/24 (5 August 2010) pp. 2–3; *Market-based Measures—Inequitable Burden on Developing Countries*, submitted by India, MEPC 61st Session, Agenda Item 5, IMO Doc MEPC 61/5/19 (2 August 2010) p. 3; Dillon, above n. 103, 151–152.

122 See Fredrik Carlsson and Henrik Hammar, 'Incentive-Based Regulation of CO_2 Emissions from International Aviation' (2002) 8(6) *Journal of Air Transport Management* 365, 365.

7.5.2.1 Grouping of Proposed Market-Based Measures

To date, various MBM proposals have been discussed and debated within the IMO and further modified by countries and NGOs. Currently seven MBM options are available.[123] They are:

- GHG Fund, one option was proposed by Cyprus, Denmark, the Marshall Islands, Nigeria and the International Parcel Tankers Association (IPTA), and the other option was a Speed-based GHG Fund proposed by the Clean Shipping Coalition (CSC);[124]
- Port State Levy (PSL), proposed by Jamaica;[125]
- Efficiency Incentive Scheme (EIS), proposed by Japan and the World Shipping Council (WSC);[126]
- Ship Efficiency and Credit Trading (SECT), proposed by the United States;[127]

---

123 A table on these seven types of MBM proposals, as well as relevant base documents, is provided at section 4.3.4.2 of chapter 4.

124 This proposal is to establish a global reduction target for international shipping, set by either UNFCCC or IMO. Emissions above the target line would be offset largely by purchasing approved emission reduction credits. The offsetting activities would be financed by a contribution paid by ships on every tonne of bunker fuel purchased. The other option proposed by the CSC is to establish a speed-based GHG Fund or Compensation Fund to include regulated slow steaming in the design and impact assessment of any MBM proposals. It would set average target speeds for different types and sizes of ships in order to meet the agreed emissions reduction target set by the IMO for an MBM. Additional speed levies or contributions would be payable for ships having higher average speeds. Revenues could be used to purchase offsets.

125 This proposal aims to levy a uniform emissions charge on all vessels calling at a port, based on the amount of fuel consumed by the vessel on its voyage to that port (not bunker suppliers). The CBDR principle could be achieved through a self-administered national or regional fund and/or some international mechanism.

126 According to this proposal, all new ships, except for those which meet pre-set EEDI thresholds, and existing ships are required to make payment contributions based on the amount of the bunker fuel consumed/purchased and the degree to which the ship's efficiency falls short of a specific standard. Funds collected go to an International GHG Fund and its Parties decide how to allocate the revenue either to long-term in-sector reduction or to a Fund to be established under UNFCCC.

127 Subject all ships to mandatory energy-efficiency standards. As one means of complying with the standard, an efficiency-credit trading programme would be established, and these standards would become more stringent over time. Currently this proposal becomes an optional addition to a phased approach energy-efficiency proposal newly submitted by the United States.

- Global Emissions Trading System (ETS) for international shipping, three options proposed by Norway (Germany was later added as a co-sponsor), United Kingdom, and France, respectively;[128]
- Penalty on Trade and Development (Bahamas), proposed by Bahamas;[129] and
- Rebate Mechanism (RM) for a market-based instrument for international shipping, proposed by the International Union for Conservation of Nature (IUCN). It consists of integrated RM and add-on RM.[130]

There are different ways of grouping MBM proposals based on diverse criteria. For the purpose of this chapter, two types of groupings are provided. These groupings are used to assist the analysis of these MBM proposals in the following sections.

These MBM proposals can be grouped into three categories, namely environmental fee-related MBM proposals, tradable permit scheme-related MBM proposals, and hybrid MBM proposals.[131] Of these seven MBM options, GHG Fund, PSL, Bahamas's Penalty on Trade and Development, and the integrated RM belong to the category of environmental fee-related MBMs. They provide the polluter with an incentive to reduce GHG emissions in order to pay lower fees which take the form of a contribution, a levy, or a penalty. The three types of ETS are tradable permit scheme-related MBMs, which seek to reduce GHG emissions through setting a global cap/reduction target and allocating

128 This proposal aims to set a sector-wide cap on net emissions from international shipping. A number of allowances (Ship Emission Units) corresponding to the cap would be released into the market each year via a global auctioning process. The units could then be traded.

129 This proposal holds that the imposition of any costs should be proportionate to the contribution by international shipping to global CO_2 emissions. The reduction will apply to individual ships and not Member States, and developing States will not be faced with a penalty on trade and development. Currently Bahamas has modified this MBM proposal into a technical and operational proposal, but this option as a MBM still remains.

130 This proposal aims to compensate developing countries for the financial impact of a MBM. It could be either applied to any maritime MBM which generates revenue (add-on option) or integrated with the International Maritime Emission Reduction Scheme (IMERS) (integrated option).

131 According to the theory of environmental law and policy, MBMs can be classified into three groups, namely environmental fees (contribution), tradable permit (allowance) schemes, and liability rules. *Scientific Study on International Shipping and Market-based Instruments*, MEPC 60th Session, Agenda Item 4, IMO Doc MEPC 60/INF.21 (15 January 2010) annex, p. 14. See also ch. 4, 4.3.4.2.

emissions allowances. The EIS and SECT can be regarded as hybrid MBMs with the EEDI as a benchmark, whereas the add-on RM is a hybrid MBM built into any other MBM.

Based on the areas in which the reduction of GHG emissions from ships will mainly take place, these MBM proposals could be classified into two categories: focus on in-sector, and focus on both in-sector and out-of-sector. This type of categorisation was agreed at the third intersessional meeting of the working group on GHG emissions from ships in 2011. It was noted that this grouping aims to 'simplify future assessment and facilitate the decision making process of MEPC'.[132] The grouping of current MBM proposals based on this criterion is provided in Table 7.1. Furthermore, the proponents of these MBM proposals, as well as some other countries, have identified the strengths and weaknesses of these MBM proposals in each group as identified in Table 7.1.[133] There has been no clear tendency within the IMO as to which group of MBMs suits international shipping most but it has been suggested that in-sector reduction MBMs should only be treated as 'a transitory policy' while a MBM that covers all sectors should be adopted in the longer term.[134] From the perspective of reduction effectiveness, an IMO-commissioned report has revealed that the majority of reductions estimated for ten MBM proposals are achieved by 'in-sector and out-of-sector' MBM proposals.[135]

---

132 *Report of the Third Intersessional Meeting of the Working Group on Greenhouse Gas Emissions from Ships*, MEPC 62nd Session, Agenda Item 5, IMO Doc MEPC 62/5/1 (8 April 2011) para. 3.38.

133 The strengths and weaknesses of MBM proposals as assessed by their proponents are summarised in the MEPC report. See ibid. annex 4, p. 1. Additionally, some countries, such as South Korea, also provided their assessment on two groups of MBM proposals under this classification. See *The Evaluation on the Relative Strengths and Weaknesses of the Reduction Mechanisms Employed by the MBM Proposals*, submitted by the Republic of Korea, Intersessional Meeting of the Working Group on GHG Emissions from Ships 3rd Session, Agenda Item 3, IMO Doc GHG-WG 3/3/1 (25 February 2011).

134 Meriem Hamdi-Cherif, Céline Guivarch and Philippe Quirion, 'Sectoral Targets for Developing Countries: Combining 'Common but Differentiated Responsibilities' with 'Meaningful Participation'' (2011) 11(1) *Climate Policy* 731, 744.

135 *Full Report of the Work Undertaken by the Expert Group on Feasibility Study and Impact Assessment of Possible Market-based Measures*, MEPC 61st Session, Agenda Item 5, IMO Doc MEPC 61/INF.2 (13 August 2010) para. 1.20.

TABLE 7.1 *Grouping of the MBM proposals based on the reduction sectors*[136]

MBM Proposals	GHG Fund	ETS	EIS	SECT	PSL	Bahamas	RM (integrated)	RM[1] (add-on)
Focus on In-Sector	(Yes[2])		Yes	Yes	Yes	Yes		Yes
In-Sector & Out-of-Sector	Yes	Yes			(Yes[3])		Yes	Yes

NOTE: '1' REPRESENTS THAT THE ADD-ON RM CAN BE APPLIED TO BOTH GROUPS BUT CANNOT BE USED WITH ALL MBM PROPOSALS; '2' REPRESENTS THAT THE SPEED-BASED GHG FUND PROPOSED BY THE CSC CAN BE USED AS AN IN-SECTOR MBM; '3' REPRESENTS THE POSSIBLE USE OF REVENUES FOR OUT-OF-SECTOR REDUCTIONS, BUT THIS IS NOT CLEARLY DEFINED IN DOCUMENT MEPC 60/4/40.

7.5.2.2 Criteria and Methodology for Selecting Market-Based Measures

It is difficult to select the most suitable MBMs for further reducing GHG emissions from international shipping. Having described the current MBM proposals and their characteristics, this section continues to address two relevant issues: the setting of criteria, and the methodology for utilising these criteria to analyse the suitability of current MBM options for further reducing GHG emissions from international shipping.

The IMO has commissioned some studies[137] and organised a number of discussions and debate on the proposed MBMs. At the 60th MEPC meeting in 2010, the Committee agreed, by majority, to the Terms of Reference (TOR) for the

136 This table is based on the table from a MEPC report with minor changes. *Report of the Third Intersessional Meeting of the Working Group on Greenhouse Gas Emissions from Ships*, MEPC 62nd Session, Agenda Item 5, IMO Doc MEPC 62/5/1 (8 April 2011) annex 3, p. 1.

137 The IMO-commissioned studies in relation to MBMs include the 2000 IMO GHG Study, 2009 Second IMO GHG Study, 2009 Scientific Study on International Shipping and Market-Based Instruments, 2010 Feasibility Study and Impact Assessment Report, and an ongoing study on possible impacts on consumers and industries in developing countries which is to be finalised in 2014. See Skjølsvik et al, above n. 98; ø. Buhaug et al, 'Second IMO GHG Study 2009' (International Maritime Organization (IMO), 2009); *Scientific Study on International Shipping and Market-based Instruments*, MEPC 60th Session, Agenda Item 4, IMO Doc MEPC 60/INF.21 (15 January 2010); *Full Report of the Work Undertaken*

Expert Group on Feasibility Study and Impact Assessment of Possible MBMs. These TOR provide the criteria for the Expert Group to assess the feasibility and impact of these MBM proposals. These nine criteria ('nine criteria') are:

(1) 'the environmental effectiveness, e.g., the extent to which the proposed MBM is effective in contributing to the reduction of greenhouse gas emissions from international shipping;
(2) the cost-effectiveness of the proposed MBM and its potential impact(s) on trade and sustainable development;
(3) the proposed MBM's potential to provide incentives to technological change and innovation-and the accommodation of current emission reduction and energy efficiency technologies;
(4) the practical feasibility of implementing the proposed MBM;
(5) the need for technology transfer to, and capacity-building within, developing countries, in particular the least developed countries (LDCs) and the small island developing States (SIDS), in relation to implementation and enforcement of the proposed MBM, including the potential to mobilize climate change finance for mitigation and adaptation actions;
(6) the MBM proposal's relation with other relevant conventions such as UNFCCC, Kyoto Protocol and WTO, as well as its compatibility with customary international law, as depicted in UNCLOS;
(7) the potential additional administrative burden, and the legal aspects for National Administrations by implementing and enforcing the proposed MBM;
(8) the potential additional workload, economic burden and operational impact for individual ships, the shipping industry and the maritime sector as a whole, of implementing the proposed MBM; and
(9) the MBM's compatibility with the existing enforcement and control provisions under the IMO legal framework.'[138]

The delegations of China, Saudi Arabia, Brazil, South Africa, India and Venezuela made statements on the establishment of the expert group on MBMs and these nine criteria. They reserved their rights to not agree with the conclusions of the expert group and one of their main reasons was that the

---

by the Expert Group on Feasibility Study and Impact Assessment of Possible Market-based Measures, MEPC 61st Session, Agenda Item 5, IMO Doc MEPC 61/INF.2 (13 August 2010).

138 *Full Report of the Work Undertaken by the Expert Group on Feasibility Study and Impact Assessment of Possible Market-based Measures*, MEPC 61st Session, Agenda Item 5, IMO Doc MEPC 61/INF.2 (13 August 2010) para. 1.2.

CBDR principle was not clearly stated in these criteria.[139] However, based on a broad interpretation of the CBDR principle, the fifth criterion above could be regarded as one type of differentiated responsibility. Clearly many developing countries, in particular major developing countries, were not satisfied with these criteria. Indeed at the 57th MEPC meeting, the Committee agreed 'by an overwhelming majority' to take the following nine principles as its reference for further debate on GHG emissions from international shipping including proposed MBMs.[140] These nine principles ('nine principles') are:

(1) 'effective in contributing to the reduction of total global greenhouse gas emissions;
(2) binding and equally applicable to all flag States in order to avoid evasion;
(3) cost-effective;
(4) able to limit, or at least, effectively minimize competitive distortion;
(5) based on sustainable environmental development without penalizing global trade and growth;
(6) based on goal-based approach and not prescribe specific methods;
(7) supportive of promoting and facilitating technical innovation and R&D in the entire shipping sector;
(8) accommodating to leading technologies in the field of energy efficiency; and
(9) practical, transparent, fraud free and easy to administer.'[141]

The second of these principles was opposed by many developing countries, such as China, India, Brazil, South Africa, Venezuela and Barbados. The second principle is the incorporation of the NMFT principle, but it excludes the CBDR principle in this context. Although the chairman of the MEPC proposed to use 'ships' to replace 'flag States' in the second principle, this suggestion was not accepted by those States not supporting principle 2.[142] These nine principles were later condensed into four by the 2009 Second IMO GHG Study, namely equal applicability to all flag States, minimisation of competitive distortion, environmental effectiveness and cost-effectiveness, and

139 *Report of the Marine Environment Protection Committee on Its Sixtieth Session*, MEPC 60th Session, Agenda Item 22, IMO Doc MEPC 60/22 (12 April 2010) annex 9.

140 *Report of the Marine Environment Protection Committee on Its Fifty-Seventh Session*, MEPC 57th Session, Agenda Item 21, IMO Doc MEPC 57/21 (7 April 2008) para. 4.77.

141 Ibid. para. 4.73.

142 Ibid. paras. 4.75–4.76.

non-prescriptive.[143] Since the draft of nine criteria was based on the nine principles, these nine principles can be used as background information in interpreting the nine criteria.

Based on the spirit of the nine principles, the nine criteria as endorsed by the IMO can be summarised into five in order to simplify understanding of the analysis to be conducted in the following section. These five criteria are:

(1) environmental effectiveness and cost-effectiveness;
(2) the incentive to technological change;
(3) practical feasibility of implementation;
(4) compatibility with international law and IMO legal framework; and
(5) financial and technological transfer.

The first criterion is condensed from the IMO's Criteria 1, 2, 7 and 8. The 'cost' in this proposed first criterion includes both the costs of the emission-reduction measures and the administrative costs and economic burden associated with MBMs. The second and third criteria are condensed from the IMO's Criteria 3 and 4 respectively. The fourth criterion is drawn from the IMO's Criteria 6 and 9, whereas the fifth criterion is condensed from the IMO's fifth criterion. If we take the second principle of the nine principles reached at the 57th MEPC meeting into account, the fifth criterion above actually excludes the full application of the CBDR principle to this GHG emissions issue. In particular, these two means of differentiated responsibility—differentiated central obligations and differentiated implementation—are excluded from any future MBMs.

Given that the ICAO has reached a consensus agreement on developing a global MBM scheme for international aviation, a general examination of their criteria for assessing MBMs is helpful for the IMO to follow due to their similar mandate from Article 2(2) of the *Kyoto Protocol*. At the 37th ICAO Assembly meeting in 2010, ICAO adopted Resolution A37–19 which provides 15 guiding principles for the design and implementation of MBMs for international aviation. Although the preamble of this resolution acknowledges both the CBDR principle and the non-discrimination principle, the CBDR principle was not explicitly stated in these guiding principles. Principle n stipulates that 'it is strongly recommended that' the MBM-generated revenues should be applied

143 Buhaug et al, above n. 137, 73. Based on the 2009 Second IMO GHG Study, 'equal applicability to all flag States' was drawn from the second principle, 'minimization of competitive distortion' was drawn from the fourth principle, 'environmental effectiveness and cost-effectiveness' was drawn from the fifth principle, while 'non-prescriptive' was drawn from the sixth principle.

'in the first instance to mitigating the environmental impact of aircraft engine emissions, including mitigation and adaptation, as well as assistance to and support for developing countries'.[144] However, three years later, an updated 16 guiding principles for MBMs of international aviation were adopted by Resolution A38–18 of the ICAO in October 2013. An added Principle p provides that 'MBMs should take into account the CBDR principle and the principle of non-discrimination and equal and fair opportunities'.[145] Also in this Resolution a consensus agreement on setting a global MBM for international aviation was confirmed.[146] It can be deduced from these references that the incorporation of the CBDR principle in principles related to future MBMs in international aviation played some role in facilitating the achievement of this consensus although what type of MBMs will be adopted in the international aviation sector still remains unclear. Therefore, it is arguable that to achieve a similar consensus in the international shipping sector the criteria for selecting MBMs for reducing GHG emissions from international shipping should be:

(1) environmental effectiveness and cost-effectiveness;
(2) the incentive to technological change;
(3) practical feasibility of implementation;
(4) compatibility with international law and IMO legal framework; and
(5) incorporation of the CBDR and NMFT principles.

Having established the criteria for selecting MBMs, there are two methods of utilising these criteria to analyse the proposed MBMs. One is to examine each MBM option against these criteria, and to compare the outcomes. This approach was adopted by the 2010 IMO Feasibility Study and Impact Assessment Report, but it simply assesses each MBM proposal in accordance with the earlier mentioned 'nine criteria'. Consequently, no detailed comparison and policy recommendations were provided, and the conclusion that 'all proposals could be implemented' is not substantiated.[147] The other method

144 *Consolidated Statement of Continuing ICAO Policies and Practices Related to Environmental Protection-Climate Change*, Assembly 37th Session, ICAO Doc Res A37–19 (8 October 2010) annex, prin n.

145 *Consolidated Statement of Continuing ICAO Policies and Practices Related to Environmental Protection-Climate Change*, Assembly 38th Session, ICAO Doc Res A38–18 (4 October 2013) annex, prin p.

146 Ibid. regs 17–19.

147 *Full Report of the Work Undertaken by the Expert Group on Feasibility Study and Impact Assessment of Possible Market-based Measures*, MEPC 61st Session, Agenda Item 5, IMO

is to narrow the list of current MBM proposals based on certain criteria, such as the practical feasibility of implementation. The most appropriate MBM option would appear after infeasible MBM options are removed from the list. This approach was put forward by Greece in one of its proposals to the IMO in 2011. Based on this approach, Greece grouped the MBM proposals into four categories, and only the GHG Fund and ETS were left after it removed other infeasible options.[148] Greece asserted that its preferred MBM option was the GHG Fund after it compared these two MBM schemes carefully.[149] Although Greece's approach was not agreed by all countries,[150] this narrowing-down method appears to be a more practical way of finding the most suitable MBM options. The next section will utilise the narrowing-down approach to analyse current MBM proposals.

7.5.2.3 Selection of the Most Suitable Market-Based Measure

Studies on the pros and cons of current MBM proposals have been conducted by various States and research institutions.[151] However, no MBM proposal has been widely accepted by most countries. Countries' preferences for different MBM options vary widely. While Chapter 4 provides a general assessment of the current seven types of MBM proposals, this section re-examines these MBM options based on five criteria and the narrowing-down methodology discussed in the previous section.

Doc MEPC 61/INF.2 (13 August 2010) para. 1.61. Some studies have indicated that these MBM proposals with the EEDI as the benchmark are not feasible. See ch. 4, 4.3.4.2.

148 *Grouping and Evaluation of Proposed MBMs*, submitted by Greece, Intersessional Meeting of the Working Group on GHG Emissions from Ships 3rd Session, Agenda Item 3, IMO Doc GHG-WG 3/3 (24 February 2011) p. 10. See also ch. 6, 6.3.1.1.

149 Ibid. 15.

150 For example, Japan did not support Greece's narrowing-down approach in finding the most suitable MBMs. *Report of the Marine Environment Protection Committee on Its Sixty-Third Session*, MEPC 63rd Session, Agenda Item 23, IMO Doc MEPC 63/23 (14 March 2012) annex 15, p. 2. See also ch. 6, 6.3.1.2.

151 See, e.g., *Grouping and Evaluation of Proposed MBMs*, submitted by Greece, Intersessional Meeting of the Working Group on GHG Emissions from Ships 3rd Session, Agenda Item 3, IMO Doc GHG-WG 3/3 (24 February 2011); *The Evaluation on the Relative Strengths and Weaknesses of the Reduction Mechanisms Employed by the MBM Proposals*, submitted by the Republic of Korea, Intersessional Meeting of the Working Group on GHG Emissions from Ships 3rd Session, Agenda Item 3, IMO Doc GHG-WG 3/3/1 (25 February 2011); *Full Report of the Work Undertaken by the Expert Group on Feasibility Study and Impact Assessment of Possible Market-based Measures*, MEPC 61st Session, Agenda Item 5, IMO Doc MEPC 61/INF.2 (13 August 2010).

7.5.2.3.1 *Step 1: Remove Hybrid MBMs with the EEDI as a Benchmark*

The EIS and SECT are two hybrid MBMs with the EEDI as a benchmark. As discussed in Chapter 4, both of these MBM proposals have been modified by their proponents several times. Regarding the SECT, the main problem with this proposal is that all ships, including existing ships, would be subject to mandatory energy efficiency standards which embed the EEDI within their formulation. Under this scheme, a 'good EEDI' ship sells credits to a 'bad EEDI' ship.[152] However, the EEDI has been widely regarded as only applying to new ships. Just as asserted by the International Association of Dry Cargo Shipowners (INTERCARGO), the EEDI 'does not apply to, and hence it cannot and should not be used for, existing ships'.[153] It is thus infeasible for the SECT MBM to be practically applied to the whole shipping industry, which is also not consistent with the third criterion of selecting MBMs, i.e., practical feasibility of implementation. At the 64th MEPC meeting in 2012, the US modified its SECT MBM and made it an enhanced energy efficiency measure, or in other words, an enhanced technical and operational measure, rather than a MBM. Based on this new proposal, the SECT becomes an optional addition to a phased approach where it provides 'standards that encourage feasible improvement in technical and operational energy efficiency'.[154] Therefore the SECT as an independent MBM does not meet the five criteria outlined above and should be removed from this selective process.

Under the EIS scheme co-sponsored by Japan and the WSC in 2011, the EEDI applied to both new and existing ships,[155] which is infeasible due to the limitation of the EEDI's application scope. In 2012, as a response to these concerns on the EEDI's application to existing ships, Japan and the WSC updated the EIS. Based on the updated EIS, the EEDI only applies to new ships whereas existing ships are required to continue to pay the contribution.[156] However, those existing ships which have undergone a major conversion should be treated as new ships in accordance with Regulation 5.4.3 of revised Annex VI to

152 Psaraftis, above n. 109, 217.

153 *Application of the EEDI to Existing Ships*, submitted by INTERCARGO, MEPC 63rd Session, Agenda Item 5, IMO Doc MEPC 63/5/12 (6 January 2012) para. 19.

154 *Proposal of the United States to Enhance Energy Efficiency in International Shipping*, submitted by the United States, MEPC 65th Session, Agenda Item 4, IMO Doc MEPC 65/4/19 (8 March 2013) para. 1.

155 *Efficiency Incentive Scheme (ETS)*, submitted by Japan and the World Shipping Council, MEPC 63nd Session, Agenda Item 5, IMO Doc MEPC 63/5/3 (25 November 2011) para. 2.

156 *Draft Legal Text on the Modified Efficiency Incentive Scheme (EIS)*, submitted by Japan, MEPC 64th Session, Agenda Item 5, IMO Doc MEPC 64/5/2 (28 June 2012) paras. 13–14.

MARPOL 73/78.[157] In this case it appears that there is no technical difficulty with the EEDI. However, the revenues generated by this scheme are mainly allocated to cover the costs and expenses of administration, adaptation projects under the *UNFCCC*, R&D projects with the aim of reducing the shipping industry's CO_2 emissions, and the IMO's technical cooperation program.[158] The failure to incorporate the CBDR principle into this scheme is not consistent with the fifth criterion for selecting MBMs (the incorporation of the CBDR and NMFT principles).

These two hybrid MBMs are also not 'cost-effective'. Given that the EEDI already applies to new ships under the 2011 amendments of Annex VI to *MARPOL 73/78*, the adoption of either of these two hybrid MBMs would lead to a 'great cost to society'.[159] This is because if either of these MBMs were adopted, new ships would be doubly regulated and impacted in two ways: firstly they would be directly impacted by a technical measure, and secondly they would be indirectly impacted by a hybrid MBM.[160] Although existing ships under the EIS would not be impacted by a hybrid MBM, a combination of these factors would 'accelerate the marginalization of the majority of older ships' in the international shipping industry.[161] This consequence is not consistent with the first criterion of selecting MBMs (the incentive to technological change). Therefore, it appears reasonable to remove these two hybrid MBMs from the list of most suitable MBMs for the international shipping industry on the basis that they are practically infeasible.

7.5.2.3.2 *Step 2: Eliminate a Penalty on Trade and Development MBM by the Bahamas*

The Penalty on Trade and Development MBM raised by the Bahamas in 2010 is an incomplete MBM, or a 'do-nothing' proposal.[162] Essentially the Bahamas suggested some regulatory principles for designing MBMs. For example, the imposition of any costs should be proportionate to the contribution by international shipping to global CO_2 emissions; care must be taken to avoid restricting world trade; and developing States should not be faced with a pen-

157 Ibid. para. 15.

158 Ibid. annex art. 8.3.

159 *Grouping and Evaluation of Proposed MBMs*, submitted by Greece, Intersessional Meeting of the Working Group on GHG Emissions from Ships 3rd Session, Agenda Item 3, IMO Doc GHG-WG 3/3 (24 February 2011) para. 38.

160 Ibid.

161 Ibid.

162 Psaraftis, above n. 109, 221.

alty upon their trade and development. However, these principles could also constitute a MBM in that they suggest a scheme based on the contribution paid by international shipping for the purpose of reducing global CO_2 emissions. This scheme might work on the ground that fuel price would serve as 'a key driver' for the reduction of GHG emissions as fuel prices rise.[163]

In 2011 the Bahamas submitted an updated proposal and its draft regulation. It proposed a phased reduction program for the MBM based on the age of new and existing ships.[164] The Bahamas also argued that 'technical and operational measures are the only direct and effective means to deliver cuts in CO_2 emissions'.[165] This amendment to the Bahamas proposal does not develop any further mechanism in terms of being a fully developed MBM. The fact that the Bahamas original MBM was incomplete and has not been developed as a fully-fledged MBM proposal justifies its exclusion from the list of suitable MBM options at least at this stage.

7.5.2.3.3 *Step 3: Put on Hold the ETS Proposals*

To date some studies have been conducted on the effectiveness of an ETS,[166] as well as a comparative analysis of a global ETS and an international GHG Fund.[167] The examined aspects of these two MBMs include the certainty in cap or price, administrative burden, carbon leakage, evasion and fraud, and experience in other contexts. Although no clear preference has been given in most of these studies, it seems that a global ETS and an international GHG

163 Ibid. 214.

164 *Mandatory CO_2 Emission Cut Targets through Technical and Operational Measures*, submitted by the Bahamas, MEPC 62nd Session, Agenda Item 5, IMO Doc MEPC 62/5/13 (6 May 2011); See also *Draft Regulations To Be Included in MARPOL Annex VI for the Control of CO_2 Emissions from Ships*, note by the Bahamas, MEPC 63rd Session, Agenda Item 5, IMO Doc MEPC 63/5/1 (24 November 2011).

165 *Need and Purpose of an MBM: How Technical and Operational Measures Are the Only Direct and Effective Means to Deliver Cuts in CO_2 Emissions*, submitted by the Bahamas, Intersessional Meeting of the Working Group on GHG Emissions from Ships 3 rd Session, Agenda Item 2, IMO Doc GHG-WG 3/2 (22 December 2010).

166 See, e.g., Garyfalia Nikolakaki, 'Economic Incentives for Maritime Shipping Relating to Climate Protection' (2013) 12(1) *WMU Journal of Maritime Affairs* 17; Ekström, above n. 107; Anca Cristea et al, 'Trade and the Greenhouse Gas Emissions from International Freight Transport' (2013) 65(1) *Journal of Environmental Economics and Management* 153; Kevin Anderson and Alice Bows, 'Executing A Scharnow Turn: Reconciling Shipping Emissions with International Commitments on Climate Change' (2012) 3(6) *Carbon Management* 615.

167 See, e.g., Psaraftis, above n. 109, 223–227.

Fund are the MBM proposals which are favoured by most commentators. The two schemes (a global ETS and an international GHG Fund) both provide the incentive to technological change (the first criterion),[168] leave room for potential incorporation of the CBDR and NMFT principles (the fifth criterion),[169] and can be compatible with international law and IMO legal framework (the fourth criterion).[170] Therefore, this section only examines the ETS against two of the five criteria outlined above, i.e., environmental effectiveness and cost-effectiveness (the first criterion), and practical feasibility of implementation (the third criterion).

Firstly, it is practically infeasible for a global ETS to be implemented due to the lack of support from its main stakeholders. The relationship between a global ETS and its main stakeholders is complex. While the design of an ETS can influence decisions of stakeholders, the interaction of stakeholders also impacts the effectiveness of the ETS.[171] Chapters 5 and 6 have identified the responses from main stakeholders on the proposed ETS MBMs. The international shipping industry, in particular shipowners and ship operators, prefer a levy or compensation fund-based MBM and oppose any ETS proposals in that a global ETS is 'unworkable' for the shipping industry.[172] Indeed, this opposition mainly concerns the setting of a cap within the ETS, which from the point of view of the shipping industry would distort international trade and impede the benign development of the industry.[173] Furthermore, it is argued that a global emissions cap is 'extremely unlikely at least in the short run' due to expected opposition from developing countries.[174] The responses from national shipping

168 Both the two schemes set the contribution or a sector-wide cap on net emissions from international shipping based on actual performance of ships, which provides an incentive for ships to reduce GHG emissions through technological upgrade.

169 Both the two schemes suggest that the CBDR principle could be reflected in these schemes in certain ways. This issue is further discussed in this section.

170 To date the divergent views on this criterion mainly lie in two aspects, i.e., whether the CBDR principle incorporated by the UNFCCC has been fully reflected in these MBM options, and whether the IMO has the mandate and competence in regulating these MBMs. These issues are discussed in chapters 2,4,7.

171 Zhihong Yang et al, 'An Analysis of Greenhouse Gas Emission Trading System from the Perspective of Stakeholders' (2010) 2 *Procedia Environmental Sciences* 82, 82.

172 See, e.g., Round Table of International Shipping Associations, above n. 111.

173 Ibid.

174 Hamdi-Cherif, Guivarch and Quirion, above n. 733–734. Hamdi-Cherif, Guivarch and Quirion assert that five factors make it unlikely for a global emissions cap to be accepted by developing countries. They are the lack of incorporation of the CBDR principle, possible constraints to economic growth and sustainable development, political protests from

industries of various countries are more complex. Most national shipping industries in developing countries have not expressed their views on preferred MBMs. While shipping industries in Greece and South Korea oppose an ETS, Australia and the UK's shipping industries support it.[175] It can be deduced that the overwhelming majority of the global shipping industry opposes the future adoption of an ETS by the international shipping industry. Compared with the shipping industry, flag and port States have more diverse views on their preferred MBMs. Based on the case studies in Chapter 6, Greece and Japan's preferred MBMs are GHG Fund and EIS respectively, whereas Panama prefers the Bahamas' proposal, China dislikes any MBM, and it appears that Vanuatu tends towards accepting any MBM.

There is no consensus on a preferred MBM among the main stakeholders in the GHG emissions issue. However, there is an overwhelming trend in global shipping industry for opposing any ETS, which if combined with the different weighting of these stakeholders as discussed in the first part of this chapter,[176] would make the ETS less likely to be feasible for future implementation. In other words, it is most likely that the proposed ETS proposals would violate the third criterion of selecting MBMs (practical feasibility).

Secondly, current ETS proposals have incorporated both the CBDR and the NMFT principles, but the approaches to this incorporation would make this regulation costly and less effective. The main approach for Norway's ETS proposal to incorporate the CBDR principle is to provide two exemptions from applying the scheme, namely ships below certain sizes and ships on international

developing countries, revenue management issues, and the lack of support from the US. But these authors also recognise that a global emission cap should be an ultimate goal in the long term. Although this analysis is not specifically aimed at the shipping industry, the opposition from developing countries on this emissions cap issue is similar in these two cases.

175 It is also worthwhile to mention that the global ETS for international shipping proposed by the shipping industries of Australia, the UK, Sweden and Belgium in 2009 is a bit different from the three ETS proposals proposed by Norway, UK and France. For example, the ETS proposed by these national industries recognises the CBDR principle but does not provide specific means to incorporate the principle. Instead, it simply mentions that this principle needs to be reflected in an ETS. Whereas the three ETS proposals discussed in this section incorporate the CBDR principle. See Australian Shipowners Association et al, above n. 249, para. 21.

176 That is, from regulation and enforcement perspectives, the importance of these stakeholders can be roughly ranked from high to low as the shipping industry, flag States and port States. In other words, the weighting of the shipping industry is higher than that of flag and port States in this maritime context.

voyages to SIDS and/or LDCs.[177] To ensure that the criteria for exemption are always fulfilled, this ETS scheme also has a time limitation (e.g., five years), and a new application is required to prolong the exemption after this required period expires.[178] However, the second exemption would probably make some shipowners and ship operators opt for certain shipping routes through the SIDS or LCDS so as to get emission exemptions.[179] This evasion would significantly diminish the effectiveness of this MBM. Furthermore, it is argued that if a global ETS only applies to the shipping industry rather than other transport modes (such as aviation, rail, road and inland waters), the shipping industry would become 'more costly' and trade may shift to other modes of transport.[180] The fact that some global trade can only be conducted by ships does not justify the ETS's sole application to the shipping industry. Currently it remains uncertain whether a global ETS would apply to other transport modes. While both of these two potential problems constitute carbon leakage[181] and would possibly lead to evasion and fraud,[182] these ETS proposals are clearly not consistent with the first criterion of selecting MBMs (environmental effectiveness and cost-effectiveness). However, it is suggested that these problems could be resolved if the CBDR principle were to be incorporated in other ways such as a phased application of an ETS.[183]

It is thus arguable that current MBM proposals for an ETS would violate the first and third criteria of selecting MBMs. Indeed, the violation of the first criterion could be addressed through further modification of the scheme. However,

177 *A Further Outline of A Global Emission Trading System* (ETS) *for International Shipping*, submitted by Norway, MEPC 60th Session, Agenda Item 4, IMO Doc MEPC 60/4/22 (15 January 2010) annex 2, p. 12.

178 Ibid.

179 See ch. 4, 4.3.4.2.

180 Nikolakaki, above n. 166, 36.

181 Carbon leakage generally refers to differentiated carbon policies and their subsequent impacts on GHG emissions. See ch. 4, 4.3.4.2.

182 Psaraftis, above n. 109, 226. Psaraftis asserts that the ETS provides 'substantial' potential for evasion and fraud, and some fraud cases have been reported within the EU ETS and elsewhere.

183 A. Miola, M. Marra and B. Ciuffo, 'Designing A Climate Change Policy for the International Maritime Transport Sector: Market-Based Measures and Technological Options for Global and Regional Policy Actions' (2011) 39(9) *Energy Policy* 5490, 5492. Miola, Marra and Ciuffo propose a three-stage implementation of an international maritime ETS. At stage 1, an ETS set by the IMO and the UNFCCC would be open for voluntary participation by States and ports; at Stage 2, the ETS would cover the traffic of all ports of Annex I States; at Stage 3, the ETS would cover traffic in all ports on a global level. But see Cristea et al, above n. 166, 170. These authors assert that it is difficult for the CBDR principle to be incorporated into a cap-and-trade approach.

the setting of a cap on total emissions from the shipping industry, which as a core feature of an ETS constitutes the primary reason for violating the third criterion in the international shipping context, can never be addressed. For these reasons, it is suggested that the MBM proposal on a global ETS should be put on hold.

7.5.2.3.4 *Step 4: Balance Three Environmental Fee-Related MBM Proposals*

Environmental fee-related MBM proposals have been favoured by the international shipping industry and many national shipping industries. Of the three MBM proposals left from the previous selection, the GHG Fund and the PSL can be used either for 'in-sector' reduction or 'in-sector and out-of-sector' reduction whereas the integrated RM can only be used for 'in-sector and out-of-sector' reduction (see Table 7.1).

To further narrow the list of most-suitable MBM proposals, the Speed-based GHG Fund should be removed first for its conflict with several of the five proposed criteria for selecting MBMs. The Speed-based GHG Fund, as an in-sector MBM, requests an additional speed levy or speed compensation fund contribution based on shipping emissions generated at these higher than average speeds.[184] However, speed limits or slow steaming may affect the supply chain, competition in the shipping market, and safety.[185] Indeed speed reduction has become one of the widely-used operational measures in relation to the SEEMP.[186] It is thus dubious whether a speed-based GHG Fund is a suitable MBM, let alone the technical complexity involved in finalising an average speed for various types and sizes of ships. Furthermore, as acknowledged by the Clean Shipping Coalition (CSC), this MBM proposal lacks sufficient incentives for technological innovation. Thus the CSC has suggested that this could be addressed by providing an extra incentive beyond the emissions-related speed levy.[187] For example, ships with a better EEDI might pay less. However, if the scheme is linked to the EEDI, it becomes an MBM with the EEDI as a benchmark, as discussed in the first step of this section. Thus it will not be applicable to all ships in the international shipping industry. The above two

---

184 *Reducing Global Ship Emissions Using a Speed-Related GHG or Compensation Fund*, submitted by Clean Shipping Coalition (CSC), MEPC 64th Session, Agenda Item 5, IMO Doc MEPC 64/5/8 (27 July 2012) para. 12.

185 H. Lindstad, B. E. Asbjørnslett and A. H. Strømman, 'Reductions in Greenhouse Gas Emissions and Cost by Shipping at Lower Speeds' (2011) 39(6) *Energy Policy* 3456, 3463.

186 Bazari and Longva, above n. 33, 7. See also ch. 4, 4.3.3.2.

187 *Reducing Global Ship Emissions Using a Speed-Related GHG or Compensation Fund*, submitted by Clean Shipping Coalition (CSC), MEPC 64th Session, Agenda Item 5, IMO Doc MEPC 64/5/8 (27 July 2012) para. 14.

factors make this scheme inconsistent with the second (the incentive to technological change) and third (practical feasibility of implementation) criteria for selecting MBMs. Additionally, the proposal does not describe the use of revenues, and it seems also difficult to incorporate the CBDR principle into this scheme. This is because the speed-based GHG Fund only applies to the shipping sector, and it would be very challenging for the IMO to apply both the CBDR and NMFT principles within the same sector.

The GHG Fund, the PSL and the integrated RM are all levy-related MBMs based on the amount of bunker fuels that ships purchase or consume. Although it is not accurate to measure a ship's emissions solely based on the fuels that it purchases or consumes, GHG emissions emitted by bunker fuels cover the overwhelming majority of all sources of these emissions from ships engaged in international voyages.[188] Indeed, both the GHG Fund and PSL recognise that the CBDR principle could be reflected by the differentiated use of revenues rather than full incorporation of this principle. It is thus important to compare these two measures against other criteria, in particular the first criterion of selecting MBMs (environmental effectiveness and cost-effectiveness). According to the IMO feasibility study and impact assessment report in 2010, under a range of scenarios, the total reduction achievable from a GHG Fund is 13–40 per cent while the number for PSL is 2–8 per cent.[189] Meanwhile, the cost of an MBM is estimated to be 50 \$/tonne CO_2 abated for a GHG Fund, but 770 \$/tonne CO_2 abated for a PSL.[190] On the basis of this assessment, the GHG Fund is much more environmentally effective and cost-effective than the PSL. In view of this difference, it would be reasonable to put on hold the PSL at this stage.

The next step, also the final step, is to compare the GHG Fund and an integrated RM and choose one of them as the most suitable MBM proposal for addressing the issue under discussion. Generally speaking, these two schemes are very similar except for two main differences. One is that the integrated RM better incorporates the CBDR principle so as to ensure 'no net incidence' on developing countries. Based on its disbursement mechanism, a bank (fund) established under an integrated RM would collect the market-driven levy from ships and first rebate it to developing countries.[191] Whereas under a GHG Fund,

188 See ch. 1, 1.2.2.1.

189 *Full Report of the Work Undertaken by the Expert Group on Feasibility Study and Impact Assessment of Possible Market-based Measures*, MEPC 61st Session, Agenda Item 5, IMO Doc MEPC 61/INF.2 (13 August 2010) para. 1.20.

190 Ibid. para. 1.29.

191 *Further Information on a Rebate Mechanism for a Market-Based Measure for International Shipping*, submitted by the International Union for Conservation of Nature (IUCN), MEPC 61st Session, Agenda Item 5, IMO Doc MEPC 61/5/33 (6 August 2010) para. 12.

the CBDR principle is only reflected by allocating revenues to mitigation and adaptation activities in developing countries and in particular in LDCs, SIDS, and the Land Locked Developing Countries (LLDCs).[192] However, this lack of full incorporation of the CBDR principle could be addressed by including an add-on RM.

The other main difference between these two schemes is their different registers or institutional arrangements. The GHG Fund provides two options for the payment of the contribution.[193] Under the first option, the ships must buy fuel at a registered bunker fuel supplier which is required to collect information on all fuels sold on a ship specific basis, and to collect and transfer GHG contributions to the International GHG Fund Administrator.[194] The Administrator maintains a global registry of registered bunker fuel suppliers and of GHG Contributions received, where each ship has its own account.[195] The second option allows the contribution to be paid to the GHG Fund directly by shipowners. Under the integrated RM, ships report the fuel bunkered to a central Emissions Registry (ER) which holds an emission account for each ship, and a predetermined global bank (Bank/Fund) provides a payment account for each ship.[196] A ship is required to report fuel bunkered to ER and pay the levy to the Bank directly. The Bank will disburse the revenues after it gets the feedback from the ER on certification and enforcement by flag and port States.[197] While the compliance by shipowners with the payment to the GHG Fund under the second option mainly relies on the documentation and national administrations, the establishment of registered bunker fuel suppliers under the first option, together with the International GHG Fund Administrator would to a significant extent reduce the possibility of evasion and fraud. However, under the integrated RM, the fact that ships report and pay to different institutions individually increases the risks of evasion and fraud significantly. The cost of the integrated RM was not determined by the 2010 IMO feasibility study and

---

192 *An International Fund for Greenhouse Gas Emissions from Ships*, submitted by Denmark, MEPC 59th Session, Agenda Item 4, IMO Doc MEPC 59/4/5 (9 April 2009) para. 8.

193 See *The International Greenhouse Gas Fund*, submitted by Cyprus, Denmark, the Marshall Islands and Nigeria, Intersessional Meeting of the Working Group on GHG Emissions from Ships 3rd Session, Agenda Item 3, IMO Doc GHG-WG 3/3/4 (25 February 2011) annex.

194 *An International Fund for Greenhouse Gas Emissions from Ships*, submitted by Denmark, MEPC 59th Session, Agenda Item 4, IMO Doc MEPC 59/4/5 (9 April 2009) para. 6.

195 Ibid.

196 *Further Information on a Rebate Mechanism for a Market-Based Measure for International Shipping*, submitted by the International Union for Conservation of Nature (IUCN), MEPC 61st Session, Agenda Item 5, IMO Doc MEPC 61/5/33 (6 August 2010) para. 12.

197 Ibid.

impact assessment report.[198] However, it is predicted that the cost of this integrated RM would be higher than the GHG Fund in that there are two independent international institutions (ER and Bank) under the integrated RM while there is only one such institution under the GHG Fund (Administrator). It is concluded that the GHG Fund would be more environmentally effective and cost-effective than the integrated RM.

Through comparing these two MBM proposals, it is argued that based on current MBM proposals the most suitable MBM for international shipping is an add-on RM built into a global GHG Fund. It is likely that there would be a concern on the part of stakeholders that the administrative costs of this new MBM proposal would be higher due to the possible increased number of administrative bodies. However, it is likely that through the setting of proper mechanisms, this proposal could be environmentally effective, cost-effective and feasible, and its incorporation of economic incentives and the CBDR and NMFT principles would be attractive for developing countries. Since all MBM proposals need to comply with international law and the IMO legal framework, and this issue could be further clarified by the participation of relevant international organisations, the fourth criterion of selecting MBMs (compatibility with international law and IMO legal framework) is not discussed in this section.

7.5.2.4 Proposed Mechanisms for the Selected Market-Based Measure

Given that an add-on Rebate Mechanism (RM) built into a global GHG Fund is proposed as the most suitable MBM for furthering the reduction of GHG emissions from international shipping, the question of how to properly establish mechanisms for this MBM is significant for its future implementation. This section briefly introduces this measure and discusses five elements of this scheme. The institutional arrangements in relation to this scheme are discussed in the next section.

7.5.2.4.1 *Overview of the Add-on RM Built into a Global GHG Fund*

Ships over certain tonnages (e.g., 400 Gross Tonnage (GT)) must pay GHG contributions based on the bunker fuel they purchase. This payment could be paid either by Registered Bunker Fuel Suppliers or the shipowners to the International GHG Fund. The contributions collected by the GHG Fund should first be rebated to developing countries based on their share of global seaborne imports by value. This share should be reviewed and adjusted annually to reflect the true situations of these developing countries and thus ensure 'no net incidence' on developing countries. The remaining revenues should be

198 This integrated RM was submitted to the IMO on 6 August 2010, but the 2010 IMO feasibility study and impact assessment report was formally released on 13 August 2010.

allocated through the mechanisms under the *UNFCCC* process. This scheme would establish a global reduction target for international shipping, set by either the *UNFCCC* or the IMO. Therefore, these revenues would then be used to offset excessive GHG emissions above the target line (see Figure 7.3). Figure 7.3 reveals that GHG emissions under a business as usual scenario will increase significantly due to the predicted growth in international trade. Since the effectiveness of technical and operational measures is not sufficient to achieve combined GHG emissions reductions for the sector, the GHG Fund could be utilised to offset excessive emissions so as to achieve the reduction target. Any additional funds remaining would be available for mitigation and adaptation activities via the *UNFCCC* and R&D, and financial and technological transfer within the IMO framework.

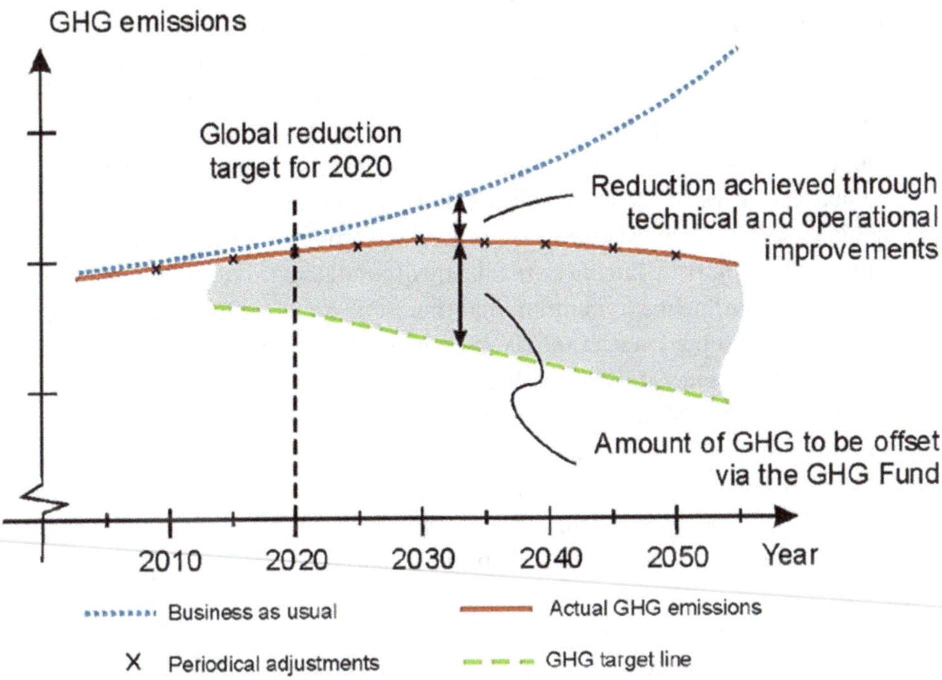

FIGURE 7.3 *Meeting a global reduction target through offsetting mechanism in an international GHG fund for GHG emissions from ships.*[199]

199 *An International Fund for Greenhouse Gas Emissions from Ships*, submitted by Cyprus, Denmark, the Marshall Islands, Nigeria and the International Parcel Tankers Association (IPTA), MEPC 60th Session, Agenda Item 4, IMO Doc MEPC 60/4/8 (18 December 2009) para. 39.

7.5.2.4.2 *The Fulfilment of the CBDR Principle*

Each developing country Party to the *UNFCCC* would be entitled to obtain an unconditional payment (rebate) equal to the cost incurred due to an add-on RM built into a global GHG Fund. The amount of rebate would be calculated annually in proportion to a key. The proposed key is a country's share of global seaborne imports by value. Under this scheme, once the contributions are collected by the International GHG Fund, these revenues should be disbursed in two steps. In the first step, any economic costs incurred by a developing country Party participating in this scheme is paid/rebated unconditionally, however, a developing country could decide to forego a part of or the entire rebate. In the second step, the remaining revenue is disbursed through the operating entity of the financial mechanism of the *UNFCCC*.[200]

The first step of revenue disbursement makes consumers in developed countries exclusively responsible for the net revenue, so that there would be 'no net incidence' on developing countries. This consequence is the same as that of the differentiated central obligations category of the CBDR principle. Developing countries would not bear any mandatory obligations under this scheme although the NMFT principle also applies in this process. Furthermore, the second step of revenue allocation provides an extra bonus for the most vulnerable countries, which can be regarded as the granting of financial and technological assistance under the CBDR principle. Through these means, both the CBDR and the NMFT principles are fully incorporated in this proposed MBM.

It is also worthwhile to mention that the list of countries in Annex I to the *UNFCCC* was adopted in 1992 and has been one of the main criteria for identifying developed States and developing States. It is anticipated that this list will be updated and maintained in the 2015 international climate change agreement. As discussed earlier, there are divergent regulatory interests and economic situations among developing countries. Therefore, it will be necessary to review and update the list for developing countries while also reviewing the share of these countries in global seaborne imports.

7.5.2.4.3 *Application Threshold of the Scheme*

This scheme is a merger of two current MBM proposals. Concerning the application threshold of this scheme, the proposers of the International GHG Fund

200 The information in this paragraph comes from a proposal submitted by the IUCN. *A Rebate Mechanism for A Market-based Instrument for International Shipping*, submitted by the International Union for Conservation of Nature (IUCN), MEPC 60th Session, Agenda Item 4, IMO Doc MEPC 60/4/55 (29 January 2010).

have suggested that it apply to ships over 400 GT,[201] while the IUCN have proposed a much higher threshold, such as over 4,000 GT.[202] The reason for the IUCN suggesting a higher threshold for an MBM is to favour these SIDS and LDCs since the ports of these developing countries can only receive smaller ships.[203] It has also been argued that a higher threshold could encourage the modal shift from some land-transport to more energy efficient coastal shipping in many countries, as well as facilitating the implementation of this scheme by reducing the number of ships subject to it.[204] Since the threshold of 400 GT has become common for IMO-regulated treaties, it is reasonable to maintain this threshold if future regulation on this proposed MBM rests solely with the IMO. However, there would be fewer barriers to achieving such a convention which stipulates a higher application threshold for the scheme if this scheme is reached through an independent international convention under the auspices of the IMO and/or other competent international bodies. This convention, if adopted, would be beneficial for ensuring the interests of the most vulnerable developing countries.

7.5.2.4.4 *The Setting of Reduction Targets*

Reduction targets have been a topic for discussion and debate within the IMO since the 59th MEPC meeting in 2009.[205] As an 'integral and obvious part of any emissions reduction plan',[206] reduction targets are different from a reduction cap. While imposing a cap on the total GHG emissions from international

---

201 *The International Greenhouse Gas Fund,* submitted by Cyprus, Denmark, the Marshall Islands and Nigeria, Intersessional Meeting of the Working Group on GHG Emissions from Ships 3rd Session, Agenda Item 3, IMO Doc GHG-WG 3/3/4 (25 February 2011) annex 3, art. 5.

202 *A Rebate Mechanism for A Market-based Instrument for International Shipping,* submitted by the International Union for Conservation of Nature (IUCN), MEPC 60th Session, Agenda Item 4, IMO Doc MEPC 60/4/55 (29 January 2010) para. 33.

203 Ibid.

204 Ibid. para. 34.

205 See, e.g., *IMO Must Act Decisively to Reduce GHG Emissions from Shipping If It Is to Retain Its Competence in Technical and Political Matters Related to Shipping and GHGs,* submitted by Friends of the Earth International, Greenpeace International and World Wild Fund for Nature, MEPC 59th Session, Agenda Item 4, IMO Doc MEPC 59/4/47 (22 May 2009) paras. 11–14; *Comments on MEPC 60/4/48, "An International Fund for Greenhouse Gas Emissions from Ships",* submitted by Japan, MEPC 60th Session, Agenda Item 4, IMO Doc MEPC 60/4/51 (29 January 2010) para. 4.

206 *IMO Must Act Decisively to Reduce GHG Emissions from Shipping If It Is to Retain Its Competence in Technical and Political Matters Related to Shipping and GHGs,* submitted by Friends of the Earth International, Greenpeace International and World Wild Fund

shipping has been opposed by the shipping industries and many States, the setting of reduction targets compatible with the selected MBM has been widely accepted.[207] A phased reduction target has been regulated for the EEDI in the revised Annex VI to *MARPOL 73/78*.[208] However, how to set a global reduction target for the selected add-on RM built into a global GHG Fund is not so straightforward. Two issues need to be properly addressed, namely, who sets the target and what the target is.

First, should the reduction target be set by the *UNFCCC* or the IMO? Given that the GHG Fund proposal provides two options, a decision needs to be made in the future. Taking into account the different expertise and mandate of these two organisations, it would be reasonable to propose that the *UNFCCC* decides the reduction target for MBMs involving 'in-sector and out-of-sector' reductions, while the IMO decides the target for in-sector MBMs. Based on this criterion, the reduction target of this proposed MBM would be decided by the *UNFCCC* process, including the scheduled global climate change agreement to be adopted in 2015.

Second, what will the reduction target be? Regarding this question, the draft text of the Convention on the International GHG Fund only provides that 'international shipping shall reduce its emissions of CO_2 by x% [in 20xx] compared to [20xx]'.[209] While the proposers of the GHG Fund suggested that this target should be set based on emissions levels in 2007,[210] some NGOs proposed that this target should reflect the emissions reductions target in developed countries.[211] The setting of reduction targets is a complex and technical matter.

---

for Nature, MEPC 59th Session, Agenda Item 4, IMO Doc MEPC 59/4/47 (22 May 2009) para. 11.

207 See, e.g., *Consideration of Appropriate Targets for Reducing CO_2 Emissions from International Shipping*, submitted by Japan, MEPC 59th Session, Agenda Item 4, IMO Doc MEPC 59/4/35 (8 May 2009) para. 18.3, 19.

208 *MARPOL Annex VI* (2011) reg 21.

209 *The International Greenhouse Gas Fund*, submitted by Cyprus, Denmark, the Marshall Islands and Nigeria, Intersessional Meeting of the Working Group on GHG Emissions from Ships 3rd Session, Agenda Item 3, IMO Doc GHG-WG 3/3/4 (25 February 2011) annex, art. 3.

210 *An International Fund for Greenhouse Gas Emissions from Ships*, submitted by Cyprus, Denmark, the Marshall Islands, Nigeria and the International Parcel Tankers Association (IPTA), MEPC 60th Session, Agenda Item 4, IMO Doc MEPC 60/4/8 (18 December 2009) para. 40.

211 *IMO Must Act Decisively to Reduce GHG Emissions from Shipping If It Is to Retain Its Competence in Technical and Political Matters Related to Shipping and GHGs*, submitted by Friends of the Earth International, Greenpeace International and World Wild Fund

However, it has been suggested that three elements should be taken into account in deciding an appropriate global reduction target for this proposed MBM.[212] They are the prospects for the growth in the global economy before 2020, the long life time of ships, and the reduction potential and reduction efforts of other sectors in particular the international aviation sector which is currently working on a MBM to be adopted by 2016 and to be implemented by 2020.[213]

7.5.2.4.5 *The Timing for Adopting the Scheme and its Legal Instrument*

The discussions in Chapter 5 indicated that the international shipping industry believe that MBMs are 'not justified at this particular time' although it agrees that MBMs are necessary for achieving absolute emissions reduction from ships.[214] Given that there are ongoing discussions on furthering the improvement of current energy efficiency measures within the IMO, and the effectiveness of technical and operational measures has not been assessed yet, it is likely that a proposed MBM will be not adopted in the short term. The 2015 *Paris Agreement* does not touch upon the reduction of GHG emissions from international shipping. The provision on the levy scheme appeared in the 'Negotiating Text' for the Paris Climate Agreement in February 2015, but it was eventually removed from the Agreement. However, the *Paris Agreement* creates the concept of Nationally Determined Contributions (NDCs) as a type of self-differentiated commitment for emissions reduction.[215] It is possible that this practice may enable the international shipping industry to adopt a similar Intended IMO Determined Contribution in the near future. This possibility, which if combined with the possible adoption of a global MBM by the international aviation sector in 2016, may justify the consideration of a MBM by the international shipping sector in or after 2017. As to the legal instrument, it is

---

for Nature, MEPC 59th Session, Agenda Item 4, IMO Doc MEPC 59/4/47 (22 May 2009) para. 12. These NGOs assert that,

'The shipping industry is by any measure a mature and well-developed industry, and as such, its targets must reflect those of developed countries. This means: international shipping GHG emissions must be reduced to at least 40% below the 1990 levels by 2020 and at least 80% below the 1990 levels by 2050.'

212 *An International Fund for Greenhouse Gas Emissions from Ships*, submitted by Cyprus, Denmark, the Marshall Islands, Nigeria and the International Parcel Tankers Association (IPTA), MEPC 60th Session, Agenda Item 4, IMO Doc MEPC 60/4/8 (18 December 2009) para. 41.

213 Ibid.

214 Round Table of International Shipping Associations, above n. 111.

215 See ch. 3, 3.2.2.3.

likely that the add-on RM built into a global GHG Fund MBM would be adopted by means of an international convention under the auspices of the IMO and the *UNFCCC* if it finds support among States.

7.6 Institutional Fragmentation, Imbalance of Interests, and Proposed Reforms

Similar to relevant laws, regulations and policies, institutional arrangements are also an integral part of a regulatory framework. However, not all international environmental institutions have been regarded as effective and legitimate.[216] It is arguable that under the current regulatory framework the institutional governance of this GHG issue is fragmented and ineffective, and the interests of the *UNFCCC* Annex I States and non-Annex I States are not balanced. The two factors constitute the main barriers for the international community to regulate this GHG issue.[217] This part examines these two deficiencies and proposes institutional and legal reforms to address them.

7.6.1 *Institutional Fragmentation in Global Reduction of Shipping GHG Emissions*

Generally 'fragmentation' refers to the 'isolation and disconnect between regimes and institutions' which may lead to overlapping and conflicting legal and policy mandates.[218] Fragmentation has been regarded as 'leading to inefficiencies, a lack of synergy... inconsistent or contradictory standards',[219] but it has also been described as reflecting an 'unprecedented normative and

216 Steinar Andresen and Ellen Hey, 'The Effectiveness and Legitimacy of International Environmental Institutions' (2005) 5(3) *International Environmental Agreements: Politics, Law and Economics* 211, 223.

217 See, e.g., Per Kågeson, 'Applying the Principle of Common but Differentiated Responsibility to the Mitigation of Greenhouse Gases from International Shipping' (Centre for Transport Studies, Stockholm, 2011) <http://www.transguide.org/SWoPEc/CTS2011-5.pdf> accessed 28 December 2013, p. 5; Aydin Okur, above n. 45, 28.

218 Karen N Scott, 'International Environmental Governance: Managing Fragmentation through Institutional Connection' (2011) 12(1) *Melbourne Journal of International Law* 177, 178.

219 Philippe Roch and Franz Xaver Perrez, 'International Environmental Governance: The Strive Towards a Comprehensive, Coherent, Effective and Efficient International Environmental Regime' (2005) 16(1) *Colorado Journal of International Environmental Law and Policy* 1, 16.

institutional expansion of international law'.[220] Accordingly, fragmentation is often referred to in the analyses of international law or international environmental law. Institutional fragmentation generally has a crucial impact on the effectiveness and performance of international environmental institutions.[221] However, this impact might be either positive or negative depending on the degree of fragmentation.[222]

In the context of GHG emissions from international shipping, institutional fragmentation is one of the main reasons for the lack of consensus of different countries in adopting applicable regulatory principles. The *UNFCCC* and the IMO are the main institutions governing this GHG issue, and they represent the global climate change regime and the IMO GHG emissions regime respectively. Additionally, the European Union (EU) and the WTO are also involved in the regulatory process in relation to MBMs. Generally speaking, the institutional fragmentation of this GHG reductions issue consists of the following three aspects.

First, the precise roles of the *UNFCCC* and the IMO in regulating GHG emissions from international shipping are ambiguous, and the coordination of negotiation within these two fora has been weak. Article 2(2) of the *Kyoto Protocol* delegates the regulation of this GHG issue to the IMO, but it does not explicitly provide the types of measures that the IMO can adopt, in particular whether the IMO can regulate the issue by prescribing MBMs. Also there are divergent interpretations of this article as to whether it means that only Annex I States have reduction obligations in the IMO's subsequent regulations,[223] or whether the CBDR principle should apply to this GHG issue or not. Although the *UNFCCC* secretariat has supported the application of the CBDR principle to this issue, the Conference of Parties (COP), which is the competent body for interpretation,[224] has not clarified its views. Meanwhile, while the IMO has been discussing the regulation of this GHG issue, the *UNFCCC* has also been

220 Gerhard Hafner, 'Pros and Cons Ensuing from Fragmentation of International Law' (2004) 25(4) *Michigan Journal of International Law* 849, 856.

221 Hackmann, above n. 21, 93.

222 Ibid.

223 See, e.g., W.B. Fitzgerald, O.J.A. Howitt and I.J. Smith, 'Greenhouse Gas Emissions from the International Maritime Transport of New Zealand's Imports and Exports' (2011) 39(3) *Energy Policy* 1521, 1523. Fitzgerald et al. assert that the first point of Article 2(2) is consistent with the CBDR principle. But see Kågeson, above n. 217, 27. Kågeson opines that this article should not be interpreted as meaning that the outcome of IMO's decision-making process solely applies to Annex I States.

224 Ian Brownlie, *Principles of Public International Law* (Oxford University Press, 7th ed., 2008) 630. Under international law, competent organisations to interpret a treaty include the

working on this matter under its SBSTA, AWG-LCA and the ADP. Given that regulatory principles and methodology issues, including reduction targets and measures, have been generally discussed within the *UNFCCC* process, IMO member States are unlikely to commit themselves to mandatory instruments before decisions on regulatory principles and targets are taken under the *UNFCCC* process.[225] As a result, these issues were discussed under both the *UNFCCC* and IMO processes. The duplication of discussions and processes, together with the lack of ongoing mechanisms for consultation between the *UNFCCC* and the IMO, makes the global regulation of this shipping GHG emissions issue lengthy and ineffective. Furthermore, negotiators within the *UNFCCC* are generally climate change experts who do not fully understand the specificities of international shipping.[226] This fragmentation between the two fora requires better coordination.

Second, possible unilateral actions by the EU on regulating GHG emissions from international shipping are consistent with international law, but might diminish the authority of the IMO's current work. It is thus important for the two institutions to have better coordination and collaboration. The EU and its member States are Parties to both the *UNFCCC* and the *Kyoto Protocol*, so they should be bound by Article 2(2) of the *Kyoto Protocol*. According to the *Vienna Convention on the Law of Treaties*,[227] the term 'working through' in Article 2(2) of the *Kyoto Protocol* should not be interpreted as 'exclusively working through' based on its 'ordinary meaning'. Therefore it is legitimate for the EU to take unilateral measures under the *Kyoto Protocol*. A number of studies have supported the EU's unilateral actions under international law.[228] It is argued that

---

treaty Parties, 'an *ad hoc* tribunal or the International Court' which has had jurisdiction conferred on it by the treaty, and the 'organs' of the competent international organisation.

225 David Ellul, *Greenhouse Gas Emissions from International Shipping* (Certificate of Advanced Studies in Environmental Diplomacy Thesis, Univeristy of Geneva, 2008) 24.

226 Ibid.

227 *Vienna Convention on the Law of Treaties*, opened for signature 23 May 1969, 8 ILM 679 (entered into force 27 January 1980) art. 31(1).

228 See, e.g., Moniek Heerings, *The Legality of A Future EU Emission Trading Scheme for Shipping* (Master of Laws in Law of the Sea Thesis, Master Thesis, University of Tromsø 2012) <http://www.ub.uit.no:8080/munin/bitstream/handle/10037/5125/thesis.pdf?sequence=2> accessed 31 December 2013; Aoife O'Leary, David Holyoake and Marta Ballesteros, 'Legal Implications of EU Action on GHG Emissions from the International Maritime Sector' (2011); Per Kågeson, 'Linking CO_2 Emissions from International Shipping to the EU ETS' (2 July 2007) <http://www.natureassociates.se/pdf/nya/CO2%20shipping%20final.pdf> accessed 31 December 2013.

unilateral measures of the EU in tackling GHG emissions from international shipping will be consistent with international law provided they are regulated

> in accordance with the principles of non-discrimination, good faith and non-abuse of right, and designed in ways that minimise impact on the right of innocent passage and freedom of high seas and respect the sovereignty of other countries.[229]

To date the EU has attributed its unilateral actions in regulating GHG emissions to the slow and unsatisfactory regulatory process of international authorities. On 1 January 2012 the EU included the emissions from the international aviation industry into the EU-ETS due to slow progress within ICAO. In December 2012 the EU suspended this policy due to improved performance by ICAO, or perhaps because of strong opposition from many countries, including the US, Russia, China and India.[230] In the same year, the EU published a consultation document asking for views on how best to reduce GHG emissions from ships so as to finally include GHG emissions from international shipping into an EU ETS.[231] Once shipping GHG emissions are included in the EU-ETS, the co-existence of two regulatory mechanisms, namely the EU-ETS and potential IMO MBMs, will make implementation and compliance by developing States shipping industries more difficult. Additionally, unilateral actions, such as unilateral levy or taxation, would probably 'harm local tourism, commerce, and the competitiveness of national carriers, raise import prices and reduce the demand for exports'.[232] It is also possible that ship operators may change their usual shipping routes so as to avoid the unilateral measures, which will diminish the effectiveness of these measures. Currently the European Commission (EC) is an observer to the IMO on behalf of the EU but there is no consultation or coordination mechanism between the two. Due to the limited authority

229 O'Leary, Holyoake and Ballesteros, above n. 228, iii.

230 But the EU's legislation on aviation emissions was ruled by the European Court of Justice as 'fully compliant with international law and relevant bilateral agreements' on 21 December 2011 in a legal case brought by some US airlines and their trade association against the inclusion of aviation in the EU ETS. Sherry P. Broder and Jon M. Van Dyke, 'The Urgency of Reducing Air Pollution from Global Shipping' in Aldo E. Chircop et al. (eds), *The Regulation of International Shipping: International and Comparative Perspectives: Essays in Honor of Edgar Gold* (Martinus Nijhoff Publishers, 2012) 249, 286.

231 Will Nichols, *EU Launches Attempt to Deliver Shipping Emissions Trading Scheme* (24 January 2012) <http://www.businessgreen.com/bg/news/2140997/eu-launches-attempt-deliver-shipping-emissions-trading-scheme> accessed 31 December 2013.

232 Keen, Parry and Strand, above n. 118, 53.

accorded to an observer,[233] more efforts should be made to strengthen the coordination between the IMO and EU.

Third, it is argued that current MBM proposals in the IMO would have a negative influence on international trade, and there is concern that these measures might violate WTO rules. However, the WTO has not contributed to this regulatory process. In 2012 the IMO Council instructed the Secretariat to seek comments from the WTO on this issue, which was triggered by a proposal submitted by India and Saudi Arabia asserting the incompatibility between WTO rules and MBMs for international shipping.[234] Meanwhile, the Secretary-General of the IMO also wrote to the Director General of the WTO in November 2012 requesting the views of the WTO on this matter. However, the WTO has not responded. Rather it provided a neutral document which sets out the most relevant WTO disciplines to the types of MBMs that the IMO is considering on the ground that the WTO Secretariat is not authorised to interpret WTO rules.[235] Therefore the problem remains unaddressed. Indeed whether the seven types of proposed MBMs comply with WTO rules needs to be analysed on a case-by-case basis. However, there is no doubt that these measures involve international trade and some of them involve different sectors rather than just the shipping sector.[236] It is thus necessary for the WTO to be more actively involved in regulating MBMs for international shipping, so that a consensus on this compatibility issue can be achieved.

In 2011 the IMO partially regulated the GHG issue from the technical and operational perspectives. Nevertheless, it took the IMO 14 years to develop this regulation since it adopted Resolution 8 on 'CO_2 emissions from ships' in 1997, and the regulation was adopted by a majority vote rather than by a consensus which has imposed challenges on its future implementation. Future steps

---

233 See IMO, above n. 7.

234 See *Possible Incompatibility between the WTO Rules and Market-based Measures for International Shipping*, submitted by India and Saudi Arabia, MEPC 64th Session, Agenda Item 5, IMO Doc MEPC 64/5/3 (29 June 2012).

235 *World Trade Organization's Views on Document MEPC 64/5/4 Submitted by India and Saudi Arabia*, note by the Secretary-General, MEPC 65th Session, Agenda Item 5, IMO Doc MEPC 65/INF.18 (21 February 2013) para. 5.

236 The third Intersessional Meeting of the Working Group on GHG Emissions from Ships grouped the MBM proposals into two categories, namely 'focus on in-sector' and 'in-sector and out-of-sector'. Based on this grouping, current MBM proposals involving out-of-sector emission reductions are the International GHG Fund, the Emissions Trading Scheme, the Port State Levy and the Rebate Mechanism. *Report of the Third Intersessional Meeting of the Working Group on Greenhouse Gas Emissions from Ships*, MEPC 62nd Session, Agenda Item 5, IMO Doc MEPC 62/5/1 (8 April 2011) annex 3.

in regulating the GHG issue, in particular the regulation of possible globally uniform MBMs, could be expedited if the institutional fragmentation which characterises current relationships between the IMO and other relevant treaty bodies and organisations such as the *UNFCCC*, EU and WTO could be better managed.[237]

7.6.2 *Optimising Institutional Arrangements*

Given that institutional fragmentation exists in current international regulation of GHG emissions from international shipping, how to effectively manage this fragmentation is significant for the smooth implementation of technical and operational measures, as well as the future adoption and enforcement of a selected MBM. This section provides some approaches for improving institutional arrangements for the GHG emissions issue.

7.6.2.1 Technical and Operational Measures

It has been suggested that the institutional fragmentation existing in global regulation of the GHG issue has significantly delayed the regulatory process.[238] The vagueness and overlap of the mandates of the *UNFCCC* and the IMO has been at the centre of most of these discussions. Technical and operational measures have been adopted after 14 years of discussions and negotiations within the IMO, and the IMO is the sole regulator of these measures. However, there is still room for improving the institutional arrangements for these technical and operational measures. Currently the MEPC is improving the regulation on the transfer of technologies from developed countries to developing countries. However, neither Regulation 23 of the revised Annex VI to *MARPOL 73/78* nor the subsequent Resolution on the transfer of technologies in May 2013[239] has provided a good solution to addressing this issue. As discussed earlier, a market-based approach to technology acquisition might be a better option for developing countries. Strengthening the communication and coordination between the IMO, *UNFCCC* and WTO may assist in resolving this issue. This is particularly the case because the TRIPS has accumulated experience in relation to technological transfer via certain incentives, and the financing

237 Scott asserts that there are two approaches for managing the risks of fragmentation and maximising its potential: one is to draw on the international rules and principles relating to the interpretation and application of treaties, and the other is to utilise environmental governance mechanisms. Scott, above n. 218, 181–182.

238 See, e.g., Oberthür, above n. 18, 202–203; Hackmann, above n. 21, 96.

239 Resolution MEPC.229(65).

mechanism under the *UNFCCC*, or perhaps a future MBM, would possibly provide funds for a market-based approach to technology acquisition.

7.6.2.2 Market-Based Measure

For the proposed add-on RM built into a global GHG Fund, institutional arrangements could be established in the following two ways.

First, which institutions have the mandate in regulating this proposed add-on RM built into a global GHG Fund? To date the IMO has been the sole forum in which the regulation of MBMs for the international shipping industry has been discussed. However, the competence of the IMO in regulating MBMs has not been widely accepted. Many developing countries opposed the IMO's role in regulating MBMs based on its technical competence from the *IMO Convention* and the lack of a clear mandate from the *Kyoto Protocol*.[240] The proposed add-on RM built into a global GHG Fund is an MBM which involves in-sector and out-of-sector emissions reduction, as well as international trade. It is thus less feasible for the IMO to exclusively regulate this proposed MBM. Furthermore, it has been suggested that under this scheme a global reduction target should be decided by the *UNFCCC* process although the *Paris Agreement* to the *UNFCCC* does not touch upon this GHG issue. Therefore, it is arguable that this scheme could be established jointly by the *UNFCCC* and the IMO. However, it is also crucial to make a clear mandate division between these two international organisations. For example, the *UNFCCC* could set the global reduction target and coordinate the work among different sectors, while the IMO could be responsible for technical issues in relation to the shipping sector. This could be achieved through a future convention on this proposed MBM and better coordination of the negotiations between these two fora.

The 63rd MEPC meeting in 2012 invited international organisations, such as the Food and Agriculture Organization of the United Nations (FAO), the United Nations Conference on Trade and Development (UNCTAD) and the United Nations World Food Programme (WFP), to 'provide appropriate information' on the GHG emissions issue.[241] While it is useful for these organisations to provide necessary information, some of these organisations as stakeholders in this GHG issue could have played more important roles. For example, currently the EU is attempting to find ways to include GHG emissions from international shipping into an EU ETS as it did for the aviation sector. This possibility is potentially harmful to the uniform global regulation of

240 See ch. 4, 4.2.

241 *Report of the Marine Environment Protection Committee on Its Sixty-Third Session*, MEPC 63rd Session, Agenda Item 23, IMO Doc MEPC 63/23 (14 March 2012) para. 5.13.7.

the shipping GHG issue, it is thus important for the IMO and *UNFCCC* to have better coordination and communication with the EU in the regulatory process for a future MBM. Meanwhile, due to the nature of this scheme being a trade-related MBM, the involvement of the WTO in either the working group or relevant discussions on regulating this scheme is also necessary. In this way, the compatibility of this proposed MBM could be clarified by the WTO so as to clear possible legal barriers and facilitate the smooth adoption and implementation of this measure in the future.

Second, which institutions should be established to exercise the functions of this proposed MBM scheme? Based on current MBM proposals on the GHG Fund and the add-on RM, there should be at least one central body (International GHG Fund) established under a convention on the international GHG Fund, and another body (RM administrator) established to exercise the role of rebating revenues to developing countries. As the international GHG Fund must be independent of the UN and the IMO,[242] Denmark has suggested establishing the Assembly and the Secretariat for this Fund so as to fulfil the relevant responsibilities. The proposers of the GHG Fund provided two options to address the question of who should be entrusted with administering the GHG Fund: a new division created within the IMO Secretariat, or a completely new body.[243] It may be advisable to establish a new body to administer this GHG Fund, which could be established within the *UNFCCC*, due to the different nature of this GHG issue from other pollution related matters.

The main role of the RM element built into the international GHG Fund is to calculate and rebate the share of developing countries according to global seaborne imports by value. It is expected that the administrative burden could be significantly reduced if the proposed new MBM scheme merges these two bodies into one, namely, incorporates the functions of a RM administrator into the

242 *An International Fund for Greenhouse Gas Emissions from Ships*, submitted by Denmark, MEPC 59th Session, Agenda Item 4, IMO Doc MEPC 59/4/5 (9 April 2009) para. 38. Based on the successful experience from the existing International Oil Pollution Compensation Funds (IOPC Funds) mechanism, Denmark asserted that the international GHG Fund should also be independent of the UN and the IMO in that 'these two organizations are not geared to manage the tasks and responsibilities of an international fund of this nature'.

243 *An International Fund for Greenhouse Gas Emissions from Ships*, submitted by Cyprus, Denmark, the Marshall Islands, Nigeria and the International Parcel Tankers Association (IPTA), MEPC 60th Session, Agenda Item 4, IMO Doc MEPC 60/4/8 (18 December 2009) paras. 24–30. In this proposal, the proposers argued that the IOPC Funds might not be a suitable body to administer the proposed GHG Fund due to the complex functionality of the proposed Fund.

GHG Fund. Then the functions of the add-on RM could be fulfilled within the GHG Fund through the administration of a new body within the *UNFCCC*.

While this section proposes a MBM merged from two current MBM proposals, it is also possible that an in-sector MBM could be adopted by the shipping industry in the future provided that a similar MBM is adopted by the international aviation sector in 2016. If that occurs, the institutional arrangements would be different from those envisaged under the proposed add-on RM built into a global international GHG Fund. In that circumstance it is possible that the IMO would be the sole regulator of the GHG emissions issue due to the nature of the MBM and the mandate of the IMO as discussed in Chapter 4.[244] As things currently stand, it appears important for various stakeholders to raise their awareness and knowledge of current MBM options, facilitate the political will and cooperate closely with various States in adopting an MBM suitable for both developed and developing countries in the long term.[245]

7.6.3 *Imbalance of Interests between the UNFCCC Annex I States and Non-Annex I States and Proposed Reforms*

To date no substantial outcomes in relation to global shipping emissions reductions have been achieved within the global climate change regime, whereas the main outcome within the IMO GHG emissions regime is the amended Annex VI to *MARPOL 73/78*, as well as related guidelines and relevant resolutions. As discussed earlier, the adopted EEDI and SEEMP lack full support from the relevant stakeholders in GHG emissions reduction from international shipping. It is arguable that a disproportionate burden in reducing shipping GHG emissions from a technical perspective has been imposed on developing countries, mainly *UNFCCC* non-Annex I States. This is because the adopted EEDI and SEEMP requirements equally apply to the ships flying the flags of both developed and developing countries while the development of the shipping industries in developing countries fall behind those in developed countries. The CBDR principle has not been fully reflected in these energy efficiency

244 See Ellul, above n. 225, 24. Ellul asserts that the IMO is the most suitable body for regulating MBMs in that 'the existing IMO framework would provide the necessary implementation and verification mechanisms by requiring flag States and allowing port States and coastal States to enforce such international measures'.

245 See, e.g., the ICAO Assembly Resolution A38–18 requires 'the active engagement and cooperation of States and the industry' in drafting and discussing a future MBM which should reflect the 'collective commitments' of the international aviation industry. *Consolidated Statement of Continuing ICAO Policies and Practices Related to Environmental Protection-Climate Change*, Assembly 38th Session, ICAO Doc Res A38–18 (4 October 2013) preamble.

measures. This imbalance of interests between the *UNFCCC* Annex I States (developed States) and Non-Annex I States (developing States) imposes challenges on the future implementation of these measures.[246]

The analyses in previous chapters reveal that applying both the CBDR and NMFT principles to the global regulation of GHG emissions from international shipping can properly balance the interests of both developed and developing States. The incorporation of the CBDR principle in the regulatory measures adopted by the IMO on this GHG issue complies with the mandate that the IMO received from the *Kyoto Protocol*, and it would also be feasible for the IMO to incorporate both the CBDR and the NMFT principles into its regulation of this issue.[247] In ICAO Assembly Resolution A38–18 which was adopted in October 2013, the CBDR principle has been explicitly incorporated and treated as one of the guiding principles for the design and implementation of MBMs for international aviation.[248] Although some developed countries, such as Australia, Japan and the US, have reservations on some of the ICAO provisions, it is unlikely that the final ICAO MBMs will completely ignore the CBDR principle. Taking the ICAO experience as an example, it appears likely that the CBDR principle will also be incorporated in the MBMs to be adopted by the international shipping industry in reducing shipping GHG emissions. This is also because most stakeholders in this GHG issue, in particular the shipping industries of some developed flag and port States,[249] as well as some developed countries,[250] have expressed their willingness to incorporate the CBDR

246 See ch. 4, 4.3.3.3.

247 See ch. 4, 4.2.

248 See *Information Relevant to Emissions from Fuel Used for International Aviation and Maritime Transport*, UNFCCC SBSTA 39th Session, Warsaw, Doc FCCC/SBSTA/2013/MISC.20 (10 November 2013) Appendix, ICAO Assembly Resolution A38–18, preambular para. 10, paras. 6,7,16b,20,21,annex Guiding principle p.

249 For example, in 2009 the shipping industries in the UK, Australia, Belgium and Sweden co-released their MBM proposal which recognises that the CBDR principle needs to be reflected in a global ETS. Australian Shipowners Association et al, *A Global Cap-and-Trade System to Reduce Carbon Emissions from International Shipping* (2009) <http://www.asa.com.au/wp-content/uploads/Joint-Industry-ETS-Discussion-PapervFINAL1.pdf> accessed 1 January 2014, p. 8. See also ch. 5, 5.3.

250 For example, three ETS proposals submitted by Norway, UK, and France to the IMO have incorporated the CBDR principle by providing some exemptions to ships on international voyages to small island developing countries, and the least developed countries. See *A Further Outline of A Global Emission Trading System (ETS) for International Shipping*, submitted by Norway, MEPC 60th Session, Agenda Item 4, IMO Doc MEPC 60/4/22 (15 January 2010) annex 2, p. 12.

principle in possible future MBMs. The strong influence of the shipping industry in the regulatory process of the IMO, as discussed early in this chapter, will be an important factor in the final outcome on MBMs.[251]

Applying the CBDR principle to the regulation of this GHG issue does not mean that developing countries would be exempt from obligations in reducing shipping GHG emissions. Differentiated responsibility consists of differentiated central obligations, differentiated implementation arrangements, and the granting of assistance, including financial and technological assistance.[252] The appropriate forms of the differentiated responsibilities vary with the types of measures that the IMO would adopt. In the technical and operational measures that the IMO has adopted, developing countries and developed countries have the same obligations to comply with these measures. There is some granting of financial and technological assistance but this needs to be strengthened.

It is suggested that the current grouping of developed and developing States based on the Annexes to the *UNFCCC* should be reformed so as to better balance the interests of developed and developing States in furthering the reduction of GHG emissions from international shipping. Specifically, under the CBDR principle the status of non-Annex I States to the *UNFCCC* (developing countries) should be reviewed and updated regularly to suit the changing situations of these countries. From an economic perspective, two options are available to States for achieving absolute GHG emissions reductions, namely to reduce their Gross Domestic Product (GDP), and to reduce their emissions intensity.[253] Since emission intensity is defined as emissions per GDP,[254] to reduce GHG emissions from international shipping under both options means to slow down economic development which is particularly difficult for developing countries. The fact that most current pollution from international shipping can be attributed to the historical contribution of ships from developed countries and there are differentiated capacities of developed and developing countries, justifies the application of the CBDR principle in regulating this issue. Under the CBDR principle, developing countries that are parties to the

251 Compared with MBM proposals, the application of the CBDR principle to energy efficiency measures was not supported by the shipping industries from developed countries. The majority voting mechanism within the MEPC and the divergent interests of developing countries in this regard secured the adoption of these measures without fully incorporating the CBDR principle. See ch. 5, 5.3–5.4.

252 Lavanya Rajamani, *Differential Treatment in International Environmental Law* (Oxford University Press, 2006) 191.

253 Sven Bode, 'Long-Term Greenhouse Gas Emission Reductions—What's Possible, What's Necessary?' (2006) 34(9) *Energy Policy* 971, 971–972.

254 Ibid. 971.

UNFCCC have a responsibility to take precautionary measures to anticipate, prevent or minimise the causes of climate change and mitigate its adverse effects on a voluntary basis.[255] However, as their economy develops, some developing countries' capacity and emissions have expanded significantly and they should thus be treated differently from other developing countries under the *UNFCCC* or IMO process. As discussed in Chapters 5 and 6, major developing countries, major Flag of Convenience (FOC) countries and other developing countries all have different regulatory interests and capacities. Some Organisation for Economic Co-operation and Development (OECD) member countries, such as the Republic of Korea and Israel, are still non-Annex I countries under the *UNFCCC*. Singapore, as a non-Annex I country, is often recognised as a developed country.

From this perspective, it appears necessary to review and update the country list of Annex I to the *UNFCCC* and/or create a new list for industrial developing countries so as to reflect their updated per capita emissions and the economic capacity of these countries. As the obligations of these countries under the Annexes differ, it would encourage more developed countries, such as the US and Japan, to support the application of the CBDR principle to the regulatory process of either the IMO or the *UNFCCC*.[256] It is also anticipated that in this way the current weak situations of developing countries in the amended Annex VI to *MARPOL 73/78*, including the arrangement for financial and technological transfer, could thus be strengthened so that 'another migration of shipbuilding industry in the future' would not be triggered.[257] Similarly, probably there would be less opposition from developed States as to the adoption and implementation of a future MBM.

7.7 Conclusion

Stakeholders play a vital role in the formation and improvement of the regulatory framework of GHG emissions from international shipping, and the order of importance of these stakeholders from the regulation and enforcement perspectives can be roughly ranked from high to low as the shipping industry, flag States and port States. To date the global climate change regime under the

255 *UNFCCC* art. 3.3.

256 To date the US has not ratified the Kyoto Protocol, and some developed countries, such as Japan and Canada, have pulled out of the second commitment period of the Kyoto Protocol.

257 Zheng, Hu and Dai, above n. 46.

UNFCCC process and the IMO GHG emissions regime have formed and been contributing to the development of this regulatory framework. Based on the analyses of previous chapters, this chapter has identified three main deficiencies in current legal and institutional frameworks for reducing GHG emissions from international shipping. They are deficiencies in adopted energy efficiency measures, lack of MBMs, institutional fragmentation and imbalance of interests between developed and developing States. It has been further suggested that these deficiencies could be improved through seven means:

- to improve the EEDI through expanding its scope of application, improving the EEDI formula, and establishing a market-based approach for technological transfer;
- to strengthen the effectiveness of the SEEMP through providing other incentives for ship operators, making the EEOI a mandatory tool and the granting of more effective financial and technological transfer from developed countries to developing countries;
- to improve flag State control through attracting more flag States to ratify Annex VI to *MARPOL 73/78* and ensure the smooth enforcement of the mandatory IMO Audit Scheme by flag States, coastal States and port States which are parties to Annex VI; and to strengthen port State control through narrowing the gaps in performance among nine regional MOUs on port State control, allowing certain unilateral actions on port State control and updating the *2009 Guidelines for Port State Control under the Revised MARPOL Annex VI*;
- to regulate the mandatory data collection system in a prudential manner;
- to adopt an MBM based on an add-on RM built into a global GHG Fund;
- to optimise institutional arrangements for these adopted technical and operational measures, and a selected MBM to be adopted in the future; and
- to apply both the CBDR and NMFT principles to the further regualtion of GHG emissions from international shipping.

CHAPTER 8

Conclusion

> [We] seek solutions that benefit the environment. A differentiated deal is better for the environment than no deal at all, and time is short...
>
> Statement by World Wide Fund for Nature (WWF)[1]

8.1 Introduction

The process to regulate the reduction of GHG emissions from international shipping is comprehensive and controversial. It involves political, historical, economic and technical considerations and cuts across international maritime law, international environmental law, and international trade law. It has been necessary to find political compromises and develop innovative regulatory efforts. The purpose of this book is to identify gaps in the current regulatory framework for the reduction of GHG emissions from international shipping, and to identify options and recommendations for improving this framework. This book synthesised applicable international environmental law principles and examined the responses of the main stakeholders.

This concluding chapter consists of three parts. The first part examines the application of selected international environmental law principles to the GHG emissions issue. The second part identifies the responses to this issue from the UN, the IMO, the shipping industry, flag States and port States. The gaps and gap-filling recommendations are provided in the last part.

8.2 Applicable International Environmental Law Principles

Chapter 2 examined the applicability of international environmental law principles to the reduction of GHG emissions from international shipping. It was argued that GHG emissions from international shipping, in particular CO_2, are a type of 'conditional' pollution. From this perspective, GHG emissions are pollutants on the basis that they engender 'deleterious effects' or lead to 'significant' environmental impact. This view is consistent with the definition of

1 *Report of the Marine Environment Protection Committee on its 58th Session*, IMO Doc MEPC 58/23 (16 October 2008) annex 9, p. 25.

 | DOI 10.1163/9789004329317_009

marine pollution in the *United Nations Convention on the Law of the Sea* (LOSC) and with the national legislation of a number of countries. GHG emissions from international shipping, being a type of pollution, trigger the application of many marine pollution-related treaties. This argument and the principles relating to flag State, coastal State and port State jurisdiction also underpin the application of international environmental law principles to GHG emissions from international shipping.

In customary international law, States are obliged to prevent, reduce and control transboundary harm resulting from activities under their jurisdiction or control where transboundary harm occurs. States also have a duty to cooperate in mitigating transboundary environmental risks and emergencies, through notification, consultation, negotiation, and in appropriate cases, environmental impact assessment. It was argued that GHG emissions from international shipping might lead to transboundary harm under four scenarios, namely:

- The harm is caused to the high seas and the deep seabed, or international airspace which is the airspace above the high seas and exclusive economic zones (EEZ) of coastal States;
- The harm is caused between the flag State and the coastal State or port State;
- The harm is caused between the flag State and a third State; or
- The harm is caused between two flag States.

On this basis, the duties associated with transboundary harm would apply in the context of GHG emissions from international shipping. These include a flag State's primary prescriptive and enforcement jurisdiction and responsibility to prevent, reduce and control transboundary harm resulting from GHG emissions from the ships entitled to fly its flag. To achieve this goal, flag States need to adopt national legislations on the reduction of such emissions, taking into account the amended Annex VI to the *International Convention for the Prevention of Pollution from Ships* (*MARPOL 73/78*) irrespective of whether they have ratified this amendment. Flag States need to conduct regular surveys, issue or empower other parties to issue the International Energy Efficiency Certificate to ships flying their flags, as well as impose administrative penalties or institute proceedings in relation to offences. Furthermore, coastal States and port States also have a duty to cooperate in mitigating transboundary environmental risks arising from excessive GHG emissions from international shipping.

The precautionary principle is a customary international law principle that has been incorporated into many international treaties. Although GHG emissions from ships have been recognised as harmful, there is not yet scientific proof that they have caused specific impacts. The application of the precautionary principle to this context would justify the action of States in taking proactive steps to tackle shipping GHG emissions. In contrast to the precautionary principle, the polluter-pays principle aims to address three relevant questions, namely: who is the polluter? what should the polluter pay for? and, how to pay? It was argued that in the context of the GHG emissions issue the polluter should include ship owners, ship operators, and flag States under certain circumstances. The cost should be put in a global context through adopting uniform measures, whereas the means of payment could include various technical, operational, and market-based measures (MBMs). In particular, MBMs which involve the global emissions reduction of different sectors may better reflect the polluter-pays principle.

The Common but Differentiated Responsibility (CBDR) principle is an important international environmental law principle, and has been widely incorporated into global climate change regulations. The No More Favourable Treatment (NMFT) principle is a principle which has been consistently applied to all IMO treaty instruments. The reduction of GHG emissions from international shipping involves measures taken by both the global climate change regime and IMO regulations. While the NMFT principle underscores the uniform application of applicable standards to all ships calling at ports of port States, the CBDR principle emphasises differentiated treatment towards developed States and developing States in the course of tackling climate change. It is thus controversial whether the two principles should be applied to the GHG emissions issue because in many respects they conflict. Chapter 2 examined divergent views on these principles and asserted that both principles should be applied to the issue. However, it is difficult to devise methods to incorporate the CBDR and NMFT principles into the IMO's regulation of GHG emissions from international shipping. While it is easier to apply the NMFT principle, the main challenge is in the application of the CBDR principle. It was suggested that differentiated treatment should be broadly interpreted to cover differentiated central obligations, differentiated implementation arrangements, and the granting of assistance including financial and technological assistance. Accordingly, the CBDR principle could be applied to the GHG issue in different ways depending on the nature of various measures for addressing this issue.

The findings in Chapter 2 were then discussed in Chapters 3 to 6 in relation to the responses from the main stakeholders. These responses, together with the findings in Chapter 2, underpinned the identification of gaps and gap-filling options in relation to the regulatory framework for the reduction of GHG emissions from international shipping as discussed in Chapter 7.

8.3 Findings on Main Stakeholders in GHG Emissions from International Shipping

The main stakeholders of GHG emissions from international shipping, including the UN, the IMO, the shipping industry, and various flag States and port States, play significant roles in the regulation and enforcement of global regulatory initiatives. Chapters 3 to 6 assessed the responses from these stakeholders to the GHG emissions issue.

The UN has provided legal and institutional responses to the reduction of GHG emissions from international shipping. Chapter 3 focused on these responses. This GHG emissions issue has been discussed under both the international climate change regime (the United Nations Framework Convention on Climate Change (*UNFCCC*) process) and the IMO GHG Emissions regime. Regarding the international climate change regime, the *UNFCCC*'s Subsidiary Body on Scientific and Technological Advice (SBSTA) started to address this problem in 1995 before the IMO received its GHG mandate from Article 2(2) of the *Kyoto Protocol* in 1997. While the SBSTA failed to reach consensus in adopting the allocation of emissions from marine bunker fuels, the subsequent Ad-Hoc Working Group on Long-term Cooperative Action (AWG-LCA) under the *UNFCCC* had not achieved any breakthrough in terms of regulatory principles on this issue before it was terminated at the Doha Climate Change Conference in 2012. Currently regulatory measures to reduce shipping GHG emissions mainly rely on the work of the IMO.

The *Paris Agreement to the UNFCCC* adopted in December 2015 does not address GHG emissions from international shipping. Nevertheless, this Agreement has some imporant implications for this shipping emissions issue. In particular, it is projected that the subsequent conferences of the *UNFCCC* and its *Paris Agreement* will continue to discuss this GHG matter; during the period of post-2020 and beyond the CBDR principle will still be applicable to the IMO's regulation on shipping GHG emissions although the *Kyoto Protocol* that provided the IMO with a GHG mandate will expire then; the evolved CBDR principle under the *Paris Agreement* will play a role in the IMO's further regulation on the reduction of shipping GHG emissions. It appears that the interaction between the *UNFCCC* process and the IMO GHG regime will continue.

A number of UN institutions have been established to address climate change. Among them, the IPCC has emphasised the necessity and urgency of tackling GHG emissions from shipping by releasing five assessment reports. The United Nations Environment Programme (UNEP) and World Meteorological Organization (WMO) raised public awareness on the need to tackle the issue and implement the outcomes within the international climate change regime;

the *UNFCCC* and its *Kyoto Protocol* and their conferences of the parties of the *UNFCCC* (COPs) and COPs serving as the Meeting of the Parties to the *Kyoto Protocol* (CMPs), as well as the *UNFCCC*'s SBSTA and AWG-LCA, have provided crucial platforms for different countries to negotiate the regulatory principles and reduction targets in relation to global regulation on marine bunker fuels, the main source of emissions from international shipping.

Chapter 4 examined the regulatory initiatives achieved within the IMO. It was argued that the *Convention on the International Maritime Organization (IMO Convention)* and the *LOSC* provide the IMO with general competence to regulate GHG emissions from ships, while the *Kyoto Protocol* gives the IMO a specific mandate to regulate this matter. These competences enable the IMO to apply both the CBDR and NMFT principles to address GHG emissions from international shipping, which recalled the discussion of the two principles in Chapter 2. It was also argued that the IMO has an exclusive role in regulating technical and operational measures and non-exclusive role in regulating MBMs, with regard to reducing GHG emissions from international shipping. This chapter continued to examine the amendments to Annex VI to the *International Convention for the Prevention of Pollution from Ships* (*MARPOL 73/78*). These amendments partially regulate this GHG emissions issue by making mandatory the Energy Efficiency Design Index (EEDI) for new ships, and the Ship Energy Efficiency Management Plan (SEEMP) for all ships. It was argued that these adopted technical and operational measures are a significant advance in regulating GHG emissions from ships. These measures were a breakthrough in the lengthy deadlock on negotiations between various countries on shipping GHG emissions within the IMO, and also confirmed the leading role of the IMO in regulating this issue. Scenario modelling has demonstrated that the estimated CO_2 emissions reduction due to combined EEDI and SEEMP will lead to significant emissions reduction, if projected growth in world trade is not taken into account. However, some deficiencies also exist in these technical and operational measures. Their effectiveness needs to be improved and strengthened, and the lack of sufficient support from major developing countries also imposes challenges for their future implementation. Furthermore, it was anticipated that MBMs would in time be adopted by the IMO and/or other competent international institutions as a supplement for the EEDI and SEEMP to reduce GHG emissions from ships. To date seven types of MBM proposals have been submitted to the IMO for further discussion and debate. However, no MBM proposals have been widely accepted.

The shipping industry plays a crucial role in the reduction of GHG emissions from international shipping. The industry puts forward suggestions and provides feedback for the introduction of a new instrument, and develops initiatives to implement the instrument after it is adopted. Chapter 5 canvassed the response of the shipping industry to this GHG emissions issue. At the international and regional level, global shipping organisations support the leading role of the IMO in regulating GHG emissions from ships, and agree that both technical and operational measures would help to reduce GHG emissions. It was argued that a GHG Fund or levy-related MBM is more acceptable to international and regional organisations. Although most of these organisations assert that the NMFT principle should be solely applied to this GHG emissions issue, the international shipowners association accepts the incorporation of the CBDR principle into proposed MBMs.

At the national level, shipping industries from various countries generally welcome the EEDI and SEEMP except that some insist that the CBDR principle should be incorporated into these measures. The divergence of these shipping industries mainly lies in their differing views on the proposed MBMs. Case studies indicated that *UNFCCC* Annex I States support the adoption of a MBM but disagree on their preferred MBMs. For example, Australia and the UK support a global Emissions Trading System (ETS) for international shipping and accept the application of the CBDR principle in this regard while Greece prefers a GHG Fund or levy relevant MBM. Within the *UNFCCC* non-Annex I States, the Korean shipping industry prefers a GHG Fund or levy-related MBM, whereas China and India believe that it is still premature to adopt any MBM. It was thus argued that the development status of a developing country, in particular its technological capability, determines the willingness of its shipping industry to accept an MBM. Emerging non-Annex I economies possessing better technologies, such as South Korea and Singapore, tend to accept an MBM more easily.

Flag States and port States are two vital stakeholders in the GHG emissions reduction from international shipping issue, and they are involved in both the legislative and implementing processes around this issue. Chapter 6 identified the responses of flag States and port States. Case studies on Greece and Japan revealed that flag States under the *UNFCCC* Annex I have similar positions towards the adoption of technical and operational measures by the IMO. Their attitudes to the proposed MBMs are also positive, although they have different preferences on the form MBMs should take. In comparison with the *UNFCCC* Annex I flag States, non-Annex I flag States have more diverse responses towards this GHG issue due to their differing regulatory interests. Case studies on Panama, China and Vanuatu indicated that major developing flag States and some other developing States are the main supporters of applying the CBDR principle to the regulation of this GHG issue and these States pay more attention to their needs in capacity building and technology transfer rather than the regulation itself. Generally they prefer technical and operational measures rather than MBMs. Nevertheless, major 'flag of convenience' (FOC) States support the NMFT principle and tend to welcome most relevant regulatory measures.

The role of port States in implementing the adopted EEDI and SEEMP has been limited to verifying the International Energy Efficiency Certificate (IEE Certificate). However, many port States have voiced their views. They have recognised the importance of regulating this issue and asserted that ports should take practical and effective measures to address this problem. While the International Association of Ports and Harbors (IAPH) has taken initiatives in tackling this matter, regional Memoranda of Understanding (MOUs) on port State control have also added the IEE Certificate to their 'List of MOU Deficiency Codes' to support the IMO's work. However, it was argued in Chapter 6 that current port State control on this GHG issue should be strengthened. This is because the current MOUs on port State control are not sufficient in achieving effective reduction of GHG emissions from ships, and disagreements remain among these port States as to the means to achieve this reduction. For instance, some port States under non-Annex I to the *UNFCCC* require more grace periods and assistance in capacity building for implementing the IMO regulations on addressing the GHG issue while other port States disagree with this view.

8.4 Gaps and Gap-Filling Recommendations

The global regulation of GHG emissions from international shipping is a process in which various stakeholders interact and contribute to the formation and improvement of the regulatory framework. The order of importance of these stakeholders from the regulation and enforcement perspectives can be roughly ranked from high to low as the shipping industry, flag States and port States. To date two parallel regimes, namely the global climate change regime under the *UNFCCC* process and the IMO GHG emissions regime, have contributed to the regulation of this GHG issue. Currently GHG emissions from international shipping have been partially regulated through technical and operational measures in the form of a revised Annex VI to *MARPOL 73/78*, and seven types of MBM proposals have been submitted to the IMO for intensive discussion and debate.

Based on the analyses in previous chapters, Chapter 7 identified three main deficiencies in the current legal and institutional framework for reducing GHG emissions from international shipping. Firstly, the adopted technical and operational measures do not fully incorporate international environmental law principles and lack full support from the main stakeholders in GHG emissions reduction from international shipping. Secondly, the absence of MBMs in the current regulatory framework for GHG emissions reduction from ships cannot achieve absolute reductions in the long term and does not reflect the widely held views in support of MBMs among the main stakeholders of the GHG issue. Thirdly, the institutional fragmentation and the imbalance of interests between the *UNFCCC* Annex I States and non-Annex I States exist in current regulatory framework for the reduction of GHG emissions from international shipping. The two factors constitute the main barriers for the international community to regulate this GHG issue.

It was further argued in Chapter 7 that the gaps existing in the current regulatory framework of this GHG emissions issue could be addressed in seven ways. Firstly, to improve the EEDI in three respects, namely: to expand the application scope of the EEDI through technological innovation and other enhanced technical measures based on current EEDI; to improve the EEDI formula; and, to establish a market-based approach for technological transfer. Secondly, to strengthen the effectiveness of the SEEMP through three approaches, which are providing other incentives for ship operators, making the EEOI a mandatory tool, and the granting of more effective financial and technological transfer from developed countries to developing countries. Thirdly, to improve flag State control through attracting more flag States to ratify Annex VI to *MARPOL 73/78* and ensure the smooth enforcement of the mandatory IMO Audit Scheme by flag States, coastal States and port States which are parties to Annex VI. In addition, to strengthen port State control through narrowing the gaps in performance among nine regional MOUs on port State control, allowing certain unilateral actions on port State control and updating the *2009 Guidelines for Port State Control under the Revised MARPOL Annex VI*. Fourthly, to regulate the mandatory data collection system in a prudential manner, taking into account the benefits of and concerns from developing States. Fifthly, to adopt an MBM based on an add-on Rebate Mechanism (RM) built into a global GHG Fund. Sixthly, to optimise institutional arrangements for these adopted technical and operational measures, and a selected MBM to be adopted in the future. In this respect, it is important to strengthen communication and coordination between the IMO, *UNFCCC* and WTO so as to address the institutional fragmentation existing in the regulation and implementation of energy efficiency measures. Whereas under a proposed add-on RM built into a global GHG Fund, it is crucial to clarify the institutional mandates of the *UNFCCC* and the IMO, and merge unnecessary administrative bodies to make this scheme cost-effective. Seventh and finally, to apply both the CBDR and NMFT principles to the further regualtion of GHG emissions from international shipping.

It is anticipated that the path toward the improvement of the current regulatory framework of GHG emissions from international shipping, including the enhancement of current energy efficiency measures and the ultimate selection of an MBM for international shipping, would be long. Limiting an increase of 2 or 1.5 degrees Celsius in the global average temperature by 2100 has become the goal of the *Paris Agreement to the UNFCCC*.[2] However, a report by Asian Development Bank reveals that an increase of 2 degrees Celsius by 2050 is 'almost unavoidable'.[3] Given the tight schedule of achieving this goal, the reduction of GHG emissions from international shipping as an important contribution to achieving that target has drawn mounting attention from the international community. To date the IMO has been refining the adopted technical and operational measures. At the 66th MEPC meeting in April 2014, the amendments to Annex VI of *MARPOL 73/78* were adopted to expand the EEDI application to include an extra five types of ships.[4] Meanwhile the Working Group on Further Technical and Operational Measures for Enhancing Energy Efficiency of International Shipping was established to facilitate the enhancement of further technical and operational measures. However, more needs to be done to tackle the rising GHG emissions from international shipping. As more developed countries and global shipping organisations come to accept the application of both the CBDR and NMFT principles to the GHG issue, in particular the MBM proposals, it seems that finding ways to incorporate both principles into the issue under discussion will be the next step.

---

2 *Paris Agreement* art. 2(1)(a).

3 Michael Westphal, Gordon Hughes and Jorn Brommelhorster (eds), *Economics of Climate Change in East Asia* (Asian Development Bank, 2013) executive summary, xvi.

4 *Amendments to MARPOL Annex VI and the NOx Technical Code 2008*, IMO Doc Res MEPC.251(66) (4 April 2014) reg. 21. These five added ships are LNG carriers, ro-ro cargo ships (vehicle carriers), ro-ro cargo ships, ro-ro passenger ships and cruise passenger ships having non-conventional propulsion.

Bibliography

1 Articles/Books/Reports

Anderson, Kevin and Alice Bows, 'Executing A Scharnow Turn: Reconciling Shipping Emissions with International Commitments on Climate Change' (2012) 3(6) *Carbon Management* 615

Andresen, Steinar, 'The Effectiveness of UN Environmental Institutions' (2007) 7(4) *International Environmental Agreements: Politics, Law and Economics* 317

Andresen, Steinar and Ellen Hey, 'The Effectiveness and Legitimacy of International Environmental Institutions' (2005) 5(3) *International Environmental Agreements: Politics, Law and Economics* 211

Anger, Annela et al., 'Research to Assess Impacts on Developing Countries of Measures to Address Emissions in the International Aviation and Shipping Sectors' (2013) <http://www.climatestrategies.org/research/our-reports/category/69/376.html>

Australian Government: Department of Infrastructure and Transport, 'Australian Sea Freight 2013–14' (2015) <http://bitre.gov.au/publications/2015/files/asf_2013_14.pdf>

Australian Shipowners Association, *Proposed Carbon Tax Considerations and Implications for the Australian Shipping Industry* (10 May 2011) <http://www.asa.com.au/wp-content/uploads/ASA-Submission-Carbon-Price-May-2011.pdf>

Australian Shipowners Association, *Reforming Australia's Shipping: Stakeholder Discussion Paper* (31 January 2011) <http://www.asa.com.au/wp-content/uploads/ASA-Submssion-Shipping-Reform-Discussion-Paper.pdf>

Australian Shipowners Association et al., *A Global Cap-and-Trade System to Reduce Carbon Emissions from International Shipping* (2009) <http://www.asa.com.au/wp-content/uploads/Joint-Industry-ETS-Discussion-PapervFINAL1.pdf>

Aydin Okur, Derya, *The Challenge of Regulating Greenhouse Gas Emissions from International Shipping and the Complicated Principle of 'Common but Differentiated Responsibilities'* (2012) <http://web.deu.edu.tr/hukuk/dergiler/dergimiz13-1/2-deryaaydinokur.pdf>

Bafundo, Nina E., 'Compliance with the Ozone Treaty: Weak States and the Principle of Common but Differentiated Responsibility' (2006) 21(3) *American University International Law Review* 461

Bali Action Plan, Decision 1/CP.13, Report of the Conference of the Parties on its Thirteenth Session, Doc FCCC/CP/2007/6/Add.1 (14 March 2008)

Balme, Richard, 'China's Climate Change Policy: Governing at the Core of Globalization' (2011) 5(1) *Carbon & Climate Law Review* 44

Bang, Ho-Sam, 'Is Port State Control an Effective Means to Combat Vessel-Source Pollution? An Empirical Survey of the Practical Exercise by Port States of Their Powers of Control' (2008) 23(4) *The International Journal of Marine and Coastal Law* 715

Bang, Ho-Sam, 'Recommendations for Policies on Port State Control and Port State Jurisdiction' (2013) 44(1) *Journal of Maritime Law and Commence* 115

Bang, Ho-Sam and Duck-Jong Jang, 'Recent Developments in Regional Memorandums of Understanding on Port State Control' (2012) 43(2) *Ocean Development and International Law* 170

Barrett, Jill, 'The Negotiation and Drafting of the Climate Change Convention' in Robin Churchill and David Freestone (eds), *International Law and Global Climate Change* (1991) 183

Baumert, Kevin A., 'Participation of Developing Countries in the International Climate Change Regime: Lessons for the Future' (2006) 38(2) *The George Washington International Law Review* 365

Bazari, Zabi and Tore Longva, '*Assessment of IMO Mandated Energy Efficiency Measures for International Shipping*' (IMO Doc MEPC 63/INF.2, Annex, 31 October 2011)

BIMCO, *GHG and Market-based Measures: BIMCO's Position* (October 2011) <https://www.bimco.org/en/About/Viewpoint/07_Greenhouse_Gases.aspx>

Birnie, Patricia W., Alan E. Boyle and Catherine Redgwell, *International Law and the Environment* (Oxford University Press, 3rd ed., 2009)

Boczek, B.A., *Flags of Convenience: An International Legal Study* (Oxford University Press, 1962)

Bodansky, D. M., 'The Emerging Climate Change Regime' (1995) 20(1) *ANNUAL REVIEW OF ENERGY AND THE ENVIRONMENT* 425

Bodansky, Daniel, 'The History of the Global Climate Change Regime' in Urs Luterbacher and Detlef F. Sprinz (eds), *International relations and global climate change* (Cambridge, MA: MIT Press, 2001) 23

Bode, Sven, 'Long-Term Greenhouse Gas Emission Reductions—What's Possible, What's Necessary?' (2006) 34(9) *Energy Policy* 971

Bolin, Bert, *A History of the Science and Politics of Climate Change: the Role of the Intergovernmental Panel on Climate Change* (Cambridge University Press, 2007)

Boone, Laura, 'Reducing Air Pollution from Marine Vessels to Mitigate Arctic Warming: Is it Time to Target Black Carbon?' (2012) (1) *Carbon & Climate Law Review* 13

Bows-Larkin, A. et al., 'High Seas, High Stakes: High Seas Project Final Report' (Tyndall Centre for Climate Change Research, University of Manchester, 2014) <http://www.mace.manchester.ac.uk/media/eps/schoolofmechanicalaerospaceandcivilengineering/research/centres/tyndall/pdf/High_Seas_High_Stakes_High_Seas_Project_Final_Report.pdf>

Bratspies, Rebecca M. and Russell A. Miller (eds), *Transboundary Harm in International Law: Lessons from the Trail Smelter Arbitration* (Cambridge University Press, 2006)

Broder, Sherry P. and Jon M. Van Dyke, 'The Urgency of Reducing Air Pollution from Global Shipping' in Aldo E. Chircop et al. (eds), *The Regulation of International Shipping: International and Comparative Perspectives : Essays in Honor of Edgar Gold* (Martinus Nijhoff Publishers, 2012) 249

Brownlie, Ian, *Principles of Public International Law* (Oxford University Press, 7th ed., 2008)

Brunner, Ronald D., 'Science and the Climate Change Regime' (2001) 34(1) *Policy Sciences* 1

Buhaug, ø. et al., 'Second IMO GHG Study 2009' (International Maritime Organization (IMO), 2009)

Carlson, Scott N., 'The Montreal Protocol's Environmental Subsidies and GATT: A Needed Reconciliation' (1994) 29(2) *Texas International Law Journal* 211

Carlsson, Fredrik and Henrik Hammar, 'Incentive-Based Regulation of CO_2 Emissions from International Aviation' (2002) 8(6) *Journal of Air Transport Management* 365

Carr, Yvette, 'The International Legal Issues Relating to the Facilitation of Sub-seabed CO_2 Sequestration Projects in Australia' (2007) 14 *Australian International Law Journal* 137

Choudhary, S.R., 'Common but Differentiated Responsibility in International Environmental Law from Stockholm to Rio' in Konrad Ginther, Erik Denters and Paul J.I.M. de Waart (eds), *Sustainable Development and Good Governance* (1995) 322

Churchill, Robin and David Freestone (eds), *International Law and Global Climate Change* (Graham & Trotman/M. Nijhoff, 1991)

Churchill, R.R. and A.V. Lowe, *The Law of the Sea* (Manchester University Press, 3rd ed., 1999)

Churchill, R.R., 'The Meaning of the "Genuine Link" Requirement in relation to the Nationality of Ships. A Study Prepared for the International Transport Workers' Federation' (2000) <http://www.itfglobal.org/seafarers/icons-site/images/ITF-Oct2000.pdf>

Clark, R.B., *Marine Pollution* (Oxford University Press, 5th ed., 2001)

Cogliati-Bantz, Vincent P., 'Disentangling the "Genuine Link": Enquiries in Sea, Air and Space Law' (2010) 79(3) *Nordic Journal of International Law* 383

Committee on Climate Change, 'International Aviation & Shipping Review: Scope of Carbon Budgets—Statutory Advice on Inclusion of International Aviation and Shipping' (1 April 2012) <http://www.theccc.org.uk/publication/international-aviation-shipping-review/>

Communications from Parties Included in Annex I to the Convention: Guidelines, Schedule and Process for Consideration, SBSTA Fourth Session, Agenda Item 5(a), Doc FCCC/SBSTA/1996/9/Add.1 (24 October 1996)

Cooperative Sectoral Approaches and Sector-Specific Actions in order to Enhance the Implementation of Article 4, Paragraph 1(c), of the Convention, AWG-LCA 14th Session (Third Part), Panama City (1–7 October 2011)

Copenhagen Accord, Decision 2/CP.15, Report of the Conference of the Parties on its Fifteenth Session, FCCC/CP/2009/11/Add.1 (30 March 2010)

Corbett, James et al., 'Morality from Ship Emissions: A Global Assessment' (2007) 41(24) *Environmental Science & Technology* 8512

Corbett, James J., Paul S. Fischbeck and Spyros N. Pandis, 'Global Nitrogen and Sulfur Inventories for Oceangoing Ships' (1999) 104(3) *Journal of Geophysical Research* 3457

Council Working Party on Shipbuilding, 'Peer Review of Japanese Government Support Measures to the Shipbuilding Sector' (2013) <http://www.oecd.org/sti/ind/C-WP6%282012%2926-FINAL-ENG.pdf>

Crist, P., 'Greenhouse Gas Emissions Reduction Potential from International Shipping' (OECD/ITF, 2009)

Cristea, Anca et al., 'Trade and the Greenhouse Gas Emissions from International Freight Transport' (2013) 65(1) *Journal of Environmental Economics and Management* 153

Cullet, Philippe, 'Common but Differentiated Responsibilities' in Malgosia Fitzmaurice, David M. Ong and Panos Merkouris (eds), *Research Handbook on International Environmental Law* (2010) 161

Cullet, Philippe, 'Differential Treatment in International Law: Towards a New Paradigm of Inter-state Relations' (1999) 10(3) *European Journal of International Law* 549

Currie, Duncan E. J. and Kateryna Wowk, 'Climate Change and CO_2 in the Oceans and Global Oceans Governance' (2009) 3(4) *Carbon & Climate Law Review* 387

Dai, Xinyuan, 'Global Regime and National Change' (2010) 10(6) *Climate Policy* 622

deFur, Peter L. and Michelle Kaszuba, 'Implementing the Precautionary Principle' (2002) 288(1–2) *The Science of The Total Environment* 155

Dickson, Barnabas, 'The Precautionary Principle in CITES: A Critical Assessment' (1999) 39(2) *Natural Resources Journal* 211

Dickson, Barney, Rosie Cooney and Ebscohost, *Biodiversity and the Precautionary Principle: Risk Uncertainty and Practice in Conservation and Sustainable Use* (Earthscan, 2005)

Dietz, Thomas and Paul C. Stern, 'Exploring New Tools for Environmental Protection' in Thomas Dietz and Paul C. Stern (eds), *New Tools for Environmental Protection: Education, Information, and Voluntary Measures* (National Academies Press, 2002)

Dillon, Rachael, 'The Growing Challenge of Climate Change for the Maritime Industry and the Role of Market-Based Measures to Reduce Emissions' (2012) 79(2) *Journal of Transportation Law, Logistics & Policy* 139

Draft Articles on Prevention of Transboundary Harm from Hazardous Activities, ILC Report GAOR A/56/10 (2001)

Dubash, Navroz K. and Lavanya Rajamani, 'Beyond Copenhagen: Next Steps' (2010) 10(6) *Climate Policy* 593

Durrant, Nicola, 'Carbon Capture and Storage Laws in Australia: Project Facilitation or a Precautionary Approach?' (2010) 18(4) *Environmental Liability Journal* 148

Egiyan, G.S., "Flag of Convenience' or 'Open Registration' of Ships' (1990) 14(2) *Marine Policy* 106

Egiyan, G.S., 'The Principle of Genuine Link and the 1986 UN Convention on the Registration of Ships' (1988) 12(3) *Marine Policy* 314

Ekström, Erika, *Market Based Measures to Regulate* CO_2 *Emissions from International Shipping* (2009) <http://lup.lub.lu.se/luur/download?func=downloadFile&recordOId=1557129&fileOId=1586238>

Ellul, David, *Greenhouse Gas Emissions from International Shipping* (Certificate of Advanced Studies in Environmental Diplomacy Thesis, Univeristy of Geneva, 2008)

Establishment of An Ad Hoc Working Group on the Durban Platform for Enhanced Action, UNFCCC Decision 1/CP.17, Doc. FCCC/CP/2011/9/Add.1 (2011)

Eyring, Veronika et al., 'Transport Impacts on Atmosphere and Climate: Shipping' (2010) 44(37) *Atmospheric Environment* 4735

Faure, Michael and Ying Song (eds), *China and International Environmental Liability: Legal Remedies for Transboundary Pollution* (Edward Elgar Publishing, 2008)

Fisher, Elizabeth, 'Is the Precautionary Principle Justiciable?' (2001) 13(3) *Journal of Environmental Law* 315

Fisher, Elizabeth, 'Precaution, Precaution Everywhere: Developing a" Common Understanding" of the Precautionary Principle in the European Community' (2002) 9(1) *Maastricht journal of European and comparative law* 21

Fitzgerald, W.B., O.J.A. Howitt and I. J. Smith, 'Greenhouse Gas Emissions from the International Maritime Transport of New Zealand's Imports and Exports' (2011) 39(3) *Energy Policy* 1521

Forrest, Craig, 'Shipping and the Marine Environment in Australia' in Warwick Gullett, Clive Schofield and Joanna Vince (eds), *Marine Resources Management* (LexisNexis Butterworths, 2011) 123

Freestone, David, 'Climate Change and the Oceans' (2009) 3(4) *Carbon & Climate Law Review* 383

French, Duncan, 'Developing States and International Environmental Law: the Importance of Differentiated Responsibilities' (2000) (49) *International & Comparative Law Quarterly* 35

Full Report of the Work Undertaken by the Expert Group on Feasibility Study and Impact Assessment of Possible Market-based Measures, MEPC 61st Session, Agenda Item 5, IMO Doc MEPC 61/INF.2 (13 August 2010)

Gaines, Sanford E., 'The Polluter-Pays Principle: from Economic Equity to Environmental Ethos' (1991) 26(3) *Texas International Law Journal* 463

Gallagher, Kevin P., 'International Trade and Air Pollution: Estimating the Economic Costs of Air Emissions from Waterborne Commerce Vessels in the United States' (2005) 77(2) *Journal of Environmental Management* 99

Gehring, T. and S. Oberthür, 'Interplay: Exploring Institutional Interaction' in Oran R. Young, Leslie A. King and Heike Schroeder (eds), *Institutions and Environmental Change: Principal Findings, Applications, and Research Frontiers* (MA: MIT Press, 2008)

Giziakis, Costas and Anastasia Christodoulou, 'Environmental Awareness and Practice concerning Maritime Air Emissions: the Case of the Greek Shipping Industry' (2012) 39(3) *Maritime Policy & Management* 353

Gomez-Echeverri, Luis, 'The Changing Geopolitics of Climate Change Finance' (2013) 13(5) *Climate Policy* 632

Government of India, Ministry of Shipping, 'Annual Report 2012–2013' (2013) <http://www.shipping.nic.in/showfile.php?lid=1195>

Government of India, Ministry of Shipping, 'Maritime Agenda: 2010–2020' (January 2011) <http://www.shipping.nic.in/showfile.php?lid=261>

Grammenos, Costas T. and Chong Ju Choi, 'The Greek Shipping Industry: Regulatory Change and Evolving Organizational Forms' (1999) 29(1) *International Studies of Management & Organization* 34

Green, Bryan A., 'Lessons from the Montreal Protocol: Guidance for the Next International Climate Change Agreement' (2009) 39(1) *Environmental Law* 253

Griffin, Andrew, 'MARPOL 73/78 and Vessel Pollution: A Glass Half Full or Half Empty?' (1994) 1(2) *Indiana Journal of Global Legal Studies* 489

Grubb, Michael, 'Cancun: the Art of the Possible' (2011) 11(2) *Climate Policy* 847

Gullett, Warwick, 'Environmental Impact Assessment and the Precautionary Principle: Legislating Caution in Environmental Protection' (1998) 5(3) (Sept 1998) *Australian Journal of Environmental Management* 146

Gullett, Warwick, 'Environmental Protection and the Precautionary Principle: a Response to Scientific Uncertainty in Environmental Management' (1997) 14(1) *Environmental and Planning Law Journal* 52

Gullett, Warwick, 'The Precautionary Principle in Australia: Policy, Law and Potential Precautionary EIAs' (2000) 11(2) *Risk: Health, Safety & Environment* 93

Gullett, Warwick, Chris Paterson and Elizabeth Fisher, 'Substantive Precautionary Decision-Making: the Australian Fisheries Management Authority's "Lawful Pursuit" of the Precautionary Principle' (2001) 7(2) *The Australian Journal of Natural Resources Law and Policy* 95

Guo, Wei, 'The TRIPS Agreement Does Little to Promote the Development of Technology Transfer to Developing Countries' (2009) 3(3) *Management Science and Engineering* 20

Hackmann, Bernd, 'Analysis of the Governance Architecture to Regulate GHG Emissions from International Shipping' (2012) 12(1) *International Environmental Agreements: Politics, Law and Economics* 85

Hafner, Gerhard, 'Pros and Cons Ensuing from Fragmentation of International Law' (2004) 25(4) *Michigan Journal of International Law* 849

Hale, R. et al., 'Predicted Levels of Future Ocean Acidification and Temperature Rise could Alter Community Structure and Biodiversity in Marine Benthic Communities' (2011) 120(5) *Oikos* 661

Hamdi-Cherif, Meriem, Céline Guivarch and Philippe Quirion, 'Sectoral Targets for Developing Countries: Combining 'Common but Differentiated Responsibilities' with 'Meaningful Participation'' (2011) 11(1) *Climate Policy* 731

Haralambides, H.E., 'Structure and Operations in the Liner Shipping Industry' in David A. Hensher and Kenneth J. Button (eds), *Handbook of Transport Modelling* (Elsevier Ltd, 2007) 607

Harding, Ronnie and Elizabeth Fisher, *Perspectives on the Precautionary Principle* (Federation Press, 1999)

Hare, John, 'Port State Control: Strong Medicine to Cure A Sick Industry' (1997) 26(3) *Georgia Journal of International and Comparative Law* 571

Harremoës, Poul, *The Precautionary Principle in the 20th Century: Late Lessons from Early Warnings* (Earthscan Publications, 2002)

Harris, Paul, 'Common but Differentiated Responsibility: the Kyoto Protocol and United States Policy' (1999) 27(7) *N.Y.U. Environmental Law Journal* 27

Harrison, James, 'Recent Developments and Continuing Challenges in the Regulation of Greenhouse Gas Emissions from International Shipping' (2012) *University of Edinburgh Research Paper Series* <http://ssrn.com/abstract=2037038>

Hassan, Daud, *Protecting the Marine Environment from Land Based Sources of Pollution* (Ashagate, 2006)

Hasan, S.M. Rashidul, *Impact of EEDI on Ship Design and Hydrodynamics: A Study of the Energy Efficiency Design Index and Other Related Emission Control Indexes* (Master of Science Thesis, Chalmers University of Technology, 2011) <http://publications.lib.chalmers.se/records/fulltext/151284.pdf>

Heerings, Moniek, *The Legality of A Future EU Emission Trading Scheme for Shipping* (Master of Laws in Law of the Sea Thesis, Master Thesis, University of Tromsø 2012) <http://www.ub.uit.no:8080/munin/bitstream/handle/10037/5125/thesis.pdf?sequence=2>

Hodgkinson, D.I. and R. Garner, *Global Climate Change: Australian Law and Policy* (LexisNexis Butterworths, 2008)

Hoffman, Jennifer, Ana Fonseca and Carlos Drews(eds), 'Cetaceans and Other Marine Biodiversity of the Eastern Tropical Pacific: Options for Adapting to Climate Change'

(Report from a workshop held at San Jose, Costa Rica, 2009) <http://wwf.panda.org/?uNewsID=166824>

Horner, Christopher, *The Politically Incorrect Guide to Global Warming and Environmentalism* (Regnery Publishing, 2007)

House of Commons, Energy and Climate Change Committee, 'The EU Emissions Trading System: Tenth Report of Session 2010–12' (2012) <http://www.publications.parliament.uk/pa/cm201012/cmselect/cmenergy/1476/1476.pdf#page=50&zoom=auto,0,186>

Hulme, Mike, *Why We Disagree about Climate Change* (Cambridge University Press, 2009)

Hulme, Milke, 'The Idea of Climate Change' (2010) 19(3) *GAIA: Ecological Perspectives for Science & Society* 171

Implications of the United Nations Convention on the Law of the Sea for the International Maritime Organization, IMO Doc LEG/MISC.6 (10 September 2008)

Intergovernmental Panel on Climate Change (IPCC), 'Fifth Assessment Report (AR5)' (2014) <http://www.ipcc.ch/>

Intergovernmental Panel on Climate Change (IPCC), 'Fifth Assessment Report: Working Group I Report' (2013) <http://www.climatechange2013.org/images/uploads/WGIAR5-SPM_Approved27Sep2013.pdf>

Intergovernmental Panel on Climate Change (IPCC), 'Fifth Assessment Report: Working Group II Report' (2014) <http://ipcc-wg2.gov/AR5/images/uploads/IPCC_WG2AR5_SPM_Approved.pdf>

Intergovernmental Panel on Climate Change (IPCC), 'Fifth Assessment Report: Working Group III Report' (2014) <http://report.mitigation2014.org/spm/ipcc_wg3_ar5_summary-for-policymakers_approved.pdf>

Intergovernmental Panel on Climate Change (IPCC), 'Fourth Assessment Report' (2007) <http://www.ipcc.ch/publications_and_data/ar4/syr/en/contents.html>

Intergovernmental Panel on Climate Change (IPCC), 'IPCC Guidelines for National Greenhouse Gas Inventories' (2006) <http://www.ipcc-nggip.iges.or.jp/public/2006gl/vol2.html>

Intergovernmental Panel on Climate Change (IPCC), 'Third Assessment Report' (2001) <http://www.grida.no/publications/other/ipcc_tar/>

International Law Commission Report (2004) Commentary to Principle 4, 15–17

International Center for Climate Governance, *International Climate Policy* (No. 39, January 2016) <http://www.iccgov.org/wp-content/uploads/2016/01/ICCG-International-Climate-Policy-Magazine-39.pdf>

International Maritime Organisation (IMO), 'Main Events in IMO's Work on Limitation and Reduction of Greenhouse Gas Emissions from International Shipping' (2011) <http://www.imo.org/MediaCentre/resources/Pages/Greenhouse%20gas%20emissions.aspx>

International Maritime Organization (IMO), *Basic Documents Volume I* (International Maritime Organization, 2010)

International Monetary Fund (IMF), 'World Economic Outlook, April 2011' (2011) <http://www.ioha2012.net/?page_id=1945>

Jiang, Liping and Jørgen T. Lauridsen, 'Price Formation of Dry Bulk Carriers in the Chinese Shipbuilding Industry' (2012) 39(3) *Maritime Policy & Management* 339

John, A. Church and J. White Neil, 'A 20th Century Acceleration in Global Sea-level Rise' (2006) 33(L01602) *Geophysical Research Letters* 1

Joint Group of Experts on Scientific Aspects of Marine Pollution (GESAMP), 'Report of the First Session (London, UN Doc.GESAMP I/11, 1969)' (1969)

Juda, Lawrence, 'IMCO and the Regulation of Ocean Pollution from Ships' (1977) 26(3) *The International and Comparative Law Quarterly* 558

Kågeson, Per, 'Applying the Principle of Common but Differentiated Responsibility to the Mitigation of Greenhouse Gases from International Shipping' (Centre for Transport Studies, Stockholm, 2011) <http://www.transguide.org/SWoPEc/CTS2011-5.pdf>

Kågeson, Per, 'Linking CO_2 Emissions from International Shipping to the EU ETS' (2007) <http://www.natureassociates.se/pdf/nya/CO2%20shipping%20final.pdf>

Karim, Md. Saiful, 'IMO Mandatory Energy Efficiency Measures for International Shipping: The First Mandatory Global Greenhouse Gas Reduction Instrument for an International Industry' (2011) 7(1) *Macquarie Journal of International and Comparative Environmental Law* 111

Karim, Md. Saiful and Shawkat Alam, 'Climate Change and Reduction of Emissions of Greenhouse Gases from Ships: An Appraisal' (2011) 1(1) *Asian Journal of International Law* 131

Kashubsky, Mikhail, *Offshore Petroleum Security: Analysis of Offshore Security Threats, Target Attractiveness, and the International Legal Framework for the Protection and Security of Offshore Petroleum Installations* (PhD Thesis, University of Wollongong, 2011)

Kasoulides, George, 'The 1986 United Nations Convention on the Conditions for Registration of Vessels and the Question of Open Registry' (1989) 20(6) *Ocean Development and International Law* 543

Keen, Michael, Ian Parry and Jon Strand, 'Market-Based Instruments for International Aviation and Shipping as a Source of Climate Finance' (2012) <http://elibrary.worldbank.org/doi/pdf/10.1596/1813-9450-5950>

Kiss, Alexandre Charles and Dinah Shelton, *International Environmental Law* (Transnational Publishers, 3rd ed., 2004)

Koivurova, Timo, 'International Legal Avenues to Address the Plight of Victims of Climate Change: Problems and Prospects' (2007) 22(2) *Journal of Environmental Law and Litigation* 267

Kollamthodi, Sujith et al., 'Greenhouse Gas Emissions from Shipping: Trends, Projections and Abatement Potential: Final Report' (The Shadow Committee on Climate Change, 2008)

Korea Offshore & Shipbuilding Association (KOSHIPA), 'Review of 2012: Innovative Tech & Creative Production Processes' (2013) <http://www.koshipa.or.kr/eng/koshipa/koshipa3/introduction05.htm>

Kuntz-Duriseti, Kristin, 'Evaluating the Economic Value of the Precautionary Principle: Using Cost Benefit Analysis to Place a Value on Precaution' (2004) 7(4) *Environmental Science & Policy* 291

Lee, Sang-Yoon and Young-Tae Chang, 'Shipping Companies' Awareness and Preparedness for Greenhouse Gas Regulations: A Korean Case' in Theo Notteboom (ed), *Current Issues in Shipping, Ports and Logistics* (2011) 25

Legality of the Threat or Use of Nuclear Weapons (Advisory Opinion) (1996) ICJ Reports 241, para. 29

Lindstad, H., B. E. Asbjørnslett and A. H. Strømman, 'Reductions in Greenhouse Gas Emissions and Cost by Shipping at Lower Speeds' (2011) 39(6) *Energy Policy* 3456

Llácer, Francisco J. Montero, 'Open Registers: Past, Present and Future' (2003) 27(6) *Marine Policy* 513

Logan, Cheryl, 'A Review of Ocean Acidification and America's Response' (2010) 60(10) *BioScience* 819

MacInnis, Kenneth A., 'The Polluter Pays Principle: Preventing Ship-Source Pollution in the Arctic' in Aldo E. Chircop et al. (eds), *The Regulation of International Shipping: International and Comparative Perspectives* (2012) 143

Marr, Simon, *The Precautionary Principle in the Law of the Sea: Modern Decision Making in International Law* (Martinus Nijhoff, 2003)

Matsui, Yoshiro, 'Some Aspects of the Principle of "Common but Differentiated Responsibilities"' (2002) 2(2) *International Environmental Agreements : Politics, Law and Economics* 151

Mattson, Gini, 'MARPOL 73/78 and Annex I: An Assessment of Its Effectiveness' (2006) 9(2) *Journal of International Wildlife Law & Policy* 175

McCarthy, Elena, 'Ocean Noise, Scientific Uncertainty, and the Paradox of the Precautionary Principle' (2007) 10(3) *Journal of International Wildlife Law & Policy* 233

Meng, Qing-nan, *Land-based Marine Pollution: International Law Development* (Graham & Trotman, 1987)

Methodological Issues, Decision 4/CP.1, Report of the Conference of the Parties on its First Session, FCCC/CP/1995/7/Add.1 (28 March–7 April 1995)

Michaelowa, Axel and Karsten Krause, 'International Maritime Transport and Climate Policy' (2000) 35(3) *Intereconomics* 127

Ministry of Finance, Government of India, 'Position Paper on the Ports Sector in India' (December 2009) <http://pppinindia.com/pdf/ppp_position_paper_ports_122k9.pdf>

Miola, A., M. Marra and B. Ciuffo, 'Designing A Climate Change Policy for the International Maritime Transport Sector: Market-Based Measures and Technological Options for Global and Regional Policy Actions' (2011) 39(9) *Energy Policy* 5490

Moffat, Jodie, 'Arranging Deckchairs on the Titanic: Climate Change, Greenhouse Gas Emissions and International Shipping' (2010) 24(2) *Australian and New Zealand Maritime Law Journal* 104

Molenaar, Erik Jaap, *Coastal State Jurisdiction over Vessel-Source Pollution* (Kluwer Law International, 1998)

Morris, Julian, *Rethinking Risk and the Precautionary Principle* (Butterworth-Heinemann, 2000)

Morrison, Anthony, *Places of Refuge for Ships in Distress: Problems and Methods of Resolution* (Martinus Nijhoff Publishers, 2012)

Moser, Susanne, Heide Hackmann and Françoise Caillods, 'Global Environmental Change Changes Everything: Key Messages and Recommendations' in ISSC/UNESCO (ed), *World Social Science Report 2013: Changing Global Environments* (OECD Publishing and Unesco Publishing, 2013)

Mwandosya, Mark J., *Survival Emissions: A Perspective from the South on Global Climate Change Negotiations* (Dar es Salaam University Press and the Centre for Energy, Environment, Science and Technology, Tanzania, 2000)

Nanda, Nitya and Nidhi Srivastava, 'Clean Technology Transfer and Intellectual Property Rights' (2009) 9(3) *Sustainable Development Law & Policy* 42

Nash, Jonathan Remy, 'Too Much Market? Conflict between Tradable Pollution Allowances and the "Polluter Pays" Principle' (2000) 24(2) *Harward Environmental Law Review* 465

Nast, Timothy, 'The Reponse of the International Shipping Industry to Global Climate Change' (2013) 44(1) *Journal of Maritime Law and Commence* 29

National Development and Reform Commission of China, *Enhanced Actions on Climate Change: China's Intended Nationally Determined Contributions* (30 June 2015) <http://www4.unfccc.int/submissions/indc/Submission%20Pages/submissions.aspx>

Navraj Singh, Ghaleigh, 'Barriers to Climate Technology Transfer—The Chimera of Intellectual Property Rights' (2011) 5(2) *Carbon & Climate Law Review : CCLR* 220

Ngwadla, Xolisa, Achala C. Abeysinghe and Adéyêmi Freitas, *The 2015 Climate Agreement: Lessons from the Bali Road Map* (2012) <http://www.eurocapacity.org/downloads/2015ClimateAgreement.pdf>

Nicholls, Robert J. and Anny Cazenave, 'Sea-Level Rise and Its Impact on Coastal Zones' (2010) 328 (18 June 2010) *Science* 1517

Nierenberg, Nicolas, Walter R. Tschinkel and Victoria J. Tschinkel, 'Early Climate Change Consensus at the National Academy: The Origins and Making of Changing Climate' (2010) 40(3) *Historical Studies in the Natural Sciences* 318

Nikolakaki, Garyfalia, 'Economic Incentives for Maritime Shipping Relating to Climate Protection' (2013) 12(1) *WMU Journal of Maritime Affairs* 17

O'Leary, Aoife, David Holyoake and Marta Ballesteros, 'Legal Implications of EU Action on GHG Emissions from the International Maritime Sector' (2011)

O'Riordan, Timothy and James Cameron, *Interpreting the Precautionary Principle* (Earthscan Publications, 1994)

Oberthür, Sebastian, 'The Climate Change Regime: Interactions with ICAO, IMO, and the EU Burden-Sharing Agreement' in Sebastian Oberthür and Thomas Gehring (eds), *Institutional Interaction in Global Environmental Governance* (The MIT Press, 2006) 53

Oberthür, Sebastian, 'Institutional Interaction to Address Greenhouse Gas Emissions from International Transport: ICAO, IMO and the Kyoto Protocol' (2003) 3(3) *Climate Policy* 191

Oceans Beyond Piracy, 'The Economics Cost of Somali Piracy 2012' (2013) <http://oceansbeyondpiracy.org/sites/default/files/attachments/View%20Full%20Report_1.pdf>

Organisation for Economic Co-operation and Development (OECD), *The Polluter Pays Principle* (OECD Publishing, 1975)

OECD, *Geographical Distribution of Financial Flows to Developing Countries: Disbursements, Commitments, Country Indicators* (2013) <http://www.oecd-ilibrary.org/docserver/download/4313023e.pdf?expires=1380277397&id=id&accname=ocid53013930&checksum=8AACEBAA8DAD0ABE76999F69BE37B75E>

Ong, D. M., 'International Legal Efforts to Address Human-induced Global Climate Change' in M. Fitzmaurice, D. M. Ong and Panos Merkouris (eds), *Research Handbook on International Environmental Law* (Edward Elgar Publishing Limited, 2010) 450

Outcomes of the United Nations Climate Change Conferences held in Lima in December 2014 and in Geneva in February 2015, Note by the Secretariat, MEPC 68th Session, Agenda Item 5, IMO Doc MEPC 68/5 (18 February 2015)

Orr, James et al., 'Anthropogenic Ocean Acidification over the Twenty-first Century and Its Impact on Calcifying Organisms' (2005) 437 *Nature* 681

Ozcayir, Dr. Z. Oya, *Port State Control* (LLP Professional Publishing, 2001)

Palassis, Stathis, 'Climate Change and Shipping' in Robin Warner and Clive Schofield (eds), *Climate Change and the Oceans: Gauging the Legal and Policy Currents in the Asia Pacific and Beyond* (Edward Elgar Publishing Limited, 2012) 200

Pamborides, G. P., *International Shipping Law: Legislation and Enforcement* (Kluwer Law International, 1999)

Paris MOU, '2012 Annual Report on Port State Control' (2012) <https://www.parismou.org/Publications/Annual_report/>

Parker, Larry and John Blodgett, '"Carbon Leakage" and Trade: Issues and Approaches' (19 December 2008) <http://www.fas.org/sgp/crs/misc/R40100.pdf>

Payoyo, Peter B., *Port State Control in the Asia-Pacific: An International Legal Study of Port State Jurisdiction* (LLM Thesis, Dalhousie University 1993)

Peel, Jacqueline, 'Issues in Climate Change Litigation' (2011) 5(1) *Carbon & Climate Law Review* 15

Peel, Jacqueline, *The Precautionary Principle in Practice: Environmental Decision-making and Scientific Uncertainty* (Federation Press, 2005)

Pisani, C., 'Fair at Sea: The Design of A Future Legal Instrument on Marine Bunker Fuels Emissions within the Climate Change Regime' (2002) 33(1) *Ocean Development and International Law* 57

Popovski, Vesselin and Kieran G. Mundy, 'Defining Climate-Change Victims' (2012) 7(1) *Sustainability Science* 5

Porter, Michael E. et al., 'Shipbuilding Cluster in the Republic of Korea' (Harvard Business School, 2010) <http://www.isc.hbs.edu/pdf/Student_Projects/Korea_Shipbuilding_2010.pdf>

Proposal to Develop an 'Intended IMO Determined Contribution' on CO_2 Reduction for International Shipping, submitted by International Chamer of Shipping (ICS), MEPC 69th Session, Agenda Item 7, IMO Doc MEPC 69/7/1 (12 February 2016)

Psaraftis, Harilaos N., 'Market-Based Measures for Greenhouse Gas Emissions from Ships: A Review' (2012) 11(2) *WMU Journal of Maritime Affairs* 211

Purdy, Ray and Richard Macrory, *Geological Carbon Sequestration: Critical Legal Issues* (January 2004) Tyndall Centre for Climate Change Research <http://www.tyndall.ac.uk/sites/default/files/wp45.pdf>

Pyhålå, Minna, Anne Brusendorff and Hanna Paulomåki, 'The Precautionary Principle' in Malgosia Fitzmaurice, David M. Ong and Panos Merkouris (eds), *Research Handbook on International Environmental Law* (2010) 203

Qin, Dahe, *Opening Remark at Working Group I—Twelfth Session* (23 September 2013) <http://www.ipcc.ch/meetings/session36/speeches/op_wg1_p12_Dahe_Qin.pdf>

QinetiQ, Lloyd's Register, and University of Strathclyde, 'Global Marine Trends 2030' (2013) <https://www.dropbox.com/sh/ysc3kkspzsxs6de/n9hnjB3CQf/GMT2030%20LowRes.pdf>

Rajadurai, Amborse, 'Regulation of Shipping: The Vital Role of Port State Control' (2004) 18 *Australian and New Zealand Maritime Law Journal* 83

Rajamani, Lavanya, 'The Cancun Climate Agreements: Reading the Text, Subtext and Tea Leaves' (2011) 60(2) *The International and Comparative Law Quarterly* 499

Rajamani, Lavanya, 'The Climate Regime in Evolution: The Disagreements that Survive the Cancun Agreements' (2011) 5(2) *Carbon & Climate Law Review* 136

Rajamani, Lavanya, *Differential Treatment in International Environmental Law* (Oxford University Press, 2006)

Rajamani, Lavanya, 'The Durban Platform for Enhanced Action and the Future of the Climate Regime' (2012) 61(2) *International and Comparative Law Quarterly* 501

Rajamani, Lavanya, 'The Nature, Promise, and Limits of Differentiated Treatment' (2005) 16(1) *Yearbook of International Environmental Law* 81

Rajamani, Lavanya, 'The Principle of Common but Differentiated Responsibility and the Balance of Commitments under the Climate Regime' (2000) 9(2) *Review of European Community & International Environmental Law* 120

Rao, Pemmaraju Sreenivasa, '*First Report on the Legal Regime for Allocation of Loss in Case of Transboundary Harm Arising out of Hazardous Activities*, International Law Commission, 55th Session, 5 May–6 June and 7 July–8 August, 2003, UN Doc A/CN.4/531 (21 March 2003)' (2003)

Rayfuse, Rosemary, Mark G. Lawrence and Kristina M. Gjerde, 'Ocean Fertilisation and Climate Change: The Need to Regulate Emerging High Seas Uses' (2008) 23(2) *The International Journal of Marine and Coastal Law* 297

Shrestha, Maxim, 'COP 21 and the Paris Agreement: Achievement or Half Measure?' (5 February 2016) (No. 028) *RSIS Commentary* 2

Report of the Ad Hoc Working Group on Long-Term Cooperative Action under the Convention on Its Seventh Session, Held in Bangkok from 28 September to 9 October 2009, and Barcelona from 2 to 6 November 2009, Doc FCCC/AWGLCA/2009/14 (20 November 2009)

Report of the Conference of the Parties on Its Eighteenth Session, Held in Doha from 26 November to 8 December 2012, Doc FCCC/CP/2012/8 (28 February 2013)

Report of the Conference of the Parties on Its Nineteenth Session, Held in Warsaw from 11 to 23 November 2013, Doc FCCC/CP/2013/10 (31 January 2014)

Report of the Conference of the Parties on Its Seventeenth Session, Held in Durban from 28 November to 11 December 2011, Doc FCCC/CP/2011/9/Add.1 (15 March 2012)

Report of the Conference of the Parties Serving as the Meeting of the Parties to the Kyoto Protocol on Its Ninth Session, Held in Warsaw from 11 to 23 November 2013, Doc FCCC/KP/CMP/2013/9 (31 January 2014)

Report of the Conference of the Parties Serving as the Meeting of the Parties to the Kyoto Protocol on Its Seventh Session, Held in Durban from 28 November to 11 December 2011, Doc FCCC/KP/CMP/2011/10/Add. 1 (15 March 2012)

Report of the Conference of the Parties Serving as the Meeting of the Parties to the Kyoto Protol on Its Eighth Session, Held in Doha from 26 November to 8 December 2012, Doc FCCC/KP/CMP/2012/13 (28 February 2013)

Report of the Expert Workshop on the Update of GHG Emissions Estimate for International Shipping (Update-EW), note by the Secretariat, MEPC 65th Session, Agenda Item 5, IMO Doc MEPC 65/5/2 (4 March 2013)

Report of the Marine Environment Protection Committee on its 58th Session, IMO Doc MEPC 58/23 (16 October 2008)

Report of the Marine Environment Protection Committee on its 59th Session, Statement by the Delegation of China on GHG Issues, IMO Doc MEPC 59/24/Add.1 Annex 13 (2009)

Report of the Marine Environment Protection Committee on Its Fifty–Second Session, MEPC 52nd Session, Agenda Item 24, IMO Doc MEPC 52/24 (18 October 2004)

Report of the Marine Environment Protection Committee on Its Fifty-Seventh Session, MEPC 57th Session, Agenda Item 21, IMO Doc MEPC 57/21 (7 April 2008)

Report of the Marine Environment Protection Committee on Its Fifty–Sixth Session, MEPC 56th Session, Agenda Item 23, IMO Doc MEPC 56/23 (30 July 2007)

Report of the Marine Environment Protection Committee on Its Forty–Ninth Session, MEPC 49th Session, Agenda Item 22, IMO Doc MEPC 49/22 (8 August 2003)

Report of the Marine Environment Protection Committee on Its Sixtieth Session, MEPC 60th Session, Agenda Item 22, IMO Doc MEPC 60/22 (12 April 2010)

Report of the Marine Environment Protection Committee on Its Sixty–Fifth Session, MEPC 65th Session, Agenda Item 22, IMO Doc MEPC 65/22 (24 May 2013)

Report of the Marine Environment Protection Committee on Its Sixty–First Session, MEPC 61st Session, Agenda Item 24, IMO Doc MEPC 61/24 (6 October 2010)

Report of the Marine Environment Protection Committee on Its Sixty–Fourth Session, MEPC 64th Session, Agenda Item 23, IMO Doc MEPC 64/23 (11 October 2012)

Report of the Marine Environment Protection Committee on Its Sixty–Second Session, MEPC 62nd Session, Agenda Item 24, IMO Doc MEPC 62/24 (26 July 2011)

Report of the Marine Environment Protection Committee on Its Sixty–Seventh Session, MEPC 67th Session, Agenda Item 20, IMO Doc MEPC 67/20 (31 October 2014)

Report of the Marine Environment Protection Committee on Its Sixty–Eighth Session, MEPC 68th Session, Agenda Item 21, IMO Doc MEPC 68/21 (29 May 2015)

Report of the Marine Environment Protection Committee on Its Sixty–Sixth Session, MEPC 66th Session, Agenda Item 21, IMO Doc MEPC 66/21 (25 April 2014)

Report of the Marine Environment Protection Committee on Its Sixty–Third Session, MEPC 63rd Session, Agenda Item 23, IMO Doc MEPC 63/23 (14 March 2012)

Report of the Marine Environmental Protection Committee on Its Fifty–Fifth Session, MEPC 55th Session, Agenda Item 23, IMO Doc MEPC 55/23 (16 October 2006)

Report of the Outcome of the First Intersessional Meeting of the Working Group on Greenhouse Gas Emissions from Ships, MEPC 58th Session, Agenda Item 4, IMO Doc MEPC 58/4 (4 July 2008)

Report of the Outcome of the Intersessional Meeting of the Working Group on Energy Efficiency Measures for Ships, MEPC 61st Session, Agenda Item 5, IMO Doc MEPC 61/5/3 (7 July 2010)

Report of the Third Intersessional Meeting of the Working Group on Greenhouse Gas Emissions from Ships, MEPC 62nd Session, Agenda Item 5, IMO Doc MEPC 62/5/1 (8 April 2011)

Report of the United Nations Conference on Environment and Development, UN Doc.A/CONF.151/26 (1992)

Report of the Working Group on Energy Efficiency Measures for Ships, MEPC 61st Session, IMO Doc MEPC 61/WP.10 (30 September 2010)

Report of the Working Group on International Liability, in International Law Commission Report (1996) GAOR A/51/10, Annex 1, 235

Report on the Outcome of the Second Intersessional Meeting of the Working Group on Greenhouse Gas Emissions from Ships, MEPC 59th Sesssion, Agenda Item 4, IMO Doc MEPC 59/4/2 (8 April 2009)

Roch, Philippe and Franz Xaver Perrez, 'International Environmental Governance: The Strive Towards a Comprehensive, Coherent, Effective and Efficient International Environmental Regime' (2005) 16(1) *Colorado Journal of International Environmental Law and Policy* 1

Rothwell, Donald and Tim Stephens, *The International Law of the Sea* (Hart Publishing, 2010)

Round Table of International Shipping Associations, *Round Table Associations Position Paper on GHG+MBMs* (22 February 2012) <https://www.bimco.org/About/Press/Press_Releases/2012/2012_02_22_Round_Table_MBM.aspx>

Russell, V.S., 'Pollution: Concept and Definition' (1974) 6(3) *Biological Conservation* 157

Sands, Philippe, *Principles of International Environmental Law* (Cambridge University Press, 2nd ed., 2003)

Sarah Fister, Gale, 'Panama: the Canal and Beyond' (2012) 26(12) *PM Network* 17

Saxena, Rajiv, 'Overseas Shipping Made Cheaper' (2010) 42(7) *Industrial Engineer: IE* 24

Schofield, Clive, 'Shifting Limits? Sea Level Rise and Options to Secure Maritime Jurisdictional Claims' (2009) 3(4) *Carbon & Climate Law Review* 405

Schwartz, Priscilla, 'The Polluter-Pays Principle' in Malgosia Fitzmaurice, David M. Ong and Panos Merkouris (eds), *Research Handbook on International Environmental Law* (2010) 243

Scientific Study on International Shipping and Market-based Instruments, MEPC 60th Session, Agenda Item 4, IMO Doc MEPC 60/INF.21 (15 January 2010)

Scott, Karen N, 'International Environmental Governance: Managing Fragmentation through Institutional Connection' (2011) 12(1) *Melbourne Journal of International Law* 177

Selin, Henrik and Noelle Eckley, 'Science, Politics, and Persistent Organic Pollutants: The Role of Scientific Assessments in International Environmental Co-operation' (2003) 3(1) *International Environmental Agreements* 17

Semedo, Alvaro et al., 'Wave-Induced Wind in the Marine Boundary Layer' (2009) 66(8) *JOURNAL OF THE ATMOSPHERIC SCIENCES* 2256

Shaw, Bill, 'The Global Environment: A Proposal to Eliminate Marine Oil Pollution' (1987) 27(1) *Natural Resources Journal* 157

Shaw, Malcolm N., *International Law* (Cambridge University Press, 6th ed., 2008)

Skjølsvik, Kjell et al., 'Study of Greenhouse Gas Emissions from Ships' (International Maritime Organization (IMO), 2000)

Steinemann, Anne, 'Improving Alternatives for Environmental Impact Assessment' (2001) 21(1) *Environmental Impact Assessment Review* 3

Stopford, Martin, *Martin Economics* (Routledge, 3rd ed., 2009)

Sullivan, Timothy J., 'Environment and Enforcement: Regulation and the Social Definition of Pollution' (1984) 12(2) *Ecology Law Quarterly* 423

Summerskill, Michael, *Oil Rigs: Law and Insurance* (Stevens & Sons, 1979)

Tan, Alan Khee-Jin, *Vessel-Source Marine Pollution: the Law and Politics of International Regulation* (Cambridge University Press, 2006)

The Berlin Mandate, Report of the Conference of the Parties on its First Session, FCCC/CP/1995/7/Add.1 (28 March–7 April 1995)

The Bonn Agreements on the Implementation of the Buenos Aires Plan of Action, Decision 5/CP.6, Report of the Conference of the Parties on the Second Part of its Sixth Session, FCCC/CP/2001/5 (25 September 2001)

The Buenos Aires Plan of Action, Decision 1/CP.4, Report of the Conference of the Parties on its Fourth Session, FCCC/CP/1998/16/Add.1 (25 January 1999)

The Cancun Agreements, Decisions 1–2/CMP.6, Report of the Conference of the Parties serving as the Meeting of the Parties to the Kyoto Protocol on its Sixth Session, FCCC/KP/CMP/2010/12/Add.1 (15 March 2011); Decision 1/CP.16, Report of the Conference of the Parties on its Sixteenth Session, FCCC/CP/2010/7/Add.1 (15 March 2011)

The Geneva Ministerial Declaration, Report of the Conference of the Parties on its Second Session, FCCC/CP/1996/15/Add.1 (8–19 July 1996)

The Marrakesh Accords, Decisions 2–14/CP.7, Report of the Conference of the Parties on its Seventh Session, FCCC/CP/2001/13/Add.1 (21 January 2002)

The International Geosphere-Biosphere Programme, The Intergovernmental Oceanographic Commission and The Scientific Committee on Oceanic Research, 'Ocean Acifidication Summary for Policymakers—Third Symposium on the Ocean in a High-CO_2 World' (International Geosphere-Biosphere Programme, Stockholm, Sweden, 2013)

Tickner, Joel A. and Ken Geiser, 'The Precautionary Principle Stimulus for Solutions and Alternatives-based Environmental Policy' (2004) 24(7–8) *Environmental Impact Assessment Review* 801

Tobey, James A. and Henri Smets, 'The Polluter-Pays Principle in the Context of Agriculture and the Environment' (1996) 19(1) *WORLD ECONOMY* 63

Tokyo MOU, 'Annual Report on Port State Control in the Asia-Pacific Region' (2012 2012) <http://www.tokyo-mou.org/doc/ANN12.pdf>

UK Chamber of Shipping, 'Shipping's Carbon Emissions--Design and Implementation of Market-based Measure: Part 1: A Cap-and-Trade Emissions Trading System' (2011) <http://www.ukchamberofshipping.com/media/filer/2012/10/10/shippings_carbon_emissions_-_cap-and-trade_2011.pdf>

UK Chamber of Shipping, 'Shipping's Carbon Emissions--Design and Implementation of Market-based Measures: Part 2: An International GHG Contribution Fund' (2011) <http://www.ukchamberofshipping.com/media/filer/2012/10/10/shippings_carbon_emissions_-_levy_2011.pdf>

United Nations Conference on Trade and Development (UNCTAD), 'Review of Maritime Transport 2012' (2012) <http://unctad.org/en/PublicationsLibrary/rmt2012_en.pdf>

United Nations Conference on Trade and Development (UNCTAD), 'Review of Maritime Transport 2013' (2013) <http://unctad.org/en/PublicationsLibrary/rmt2013_en.pdf>

United Nations Conference on Trade and Development (UNCTAD), 'Review of Maritime Transport 2015' (2015) <http://unctad.org/en/PublicationsLibrary/rmt2015_en.pdf>

United Nations Environment Programme (UNEP), 'Annual Report 2010' (2011) <www.unep.org/annualreport>

United Nations Environment Programme (UNEP), *Key Achievements of the Montreal Protocol To Date* <http://ozone.unep.org/Publications/MP_Key_Achievements-E.pdf>

United Nations Environment Programme (UNEP), 'The Emissions Gap Report 2012: A UNEP Synthesis Report' (November 2012) <http://www.unep.org/pdf/2012gapreport.pdf>

Van der Putten, Wim H., Mirka Macel and Marcel E. Visser, 'Predicting Species Distribution and Abundance Responses to Climate Change: Why It is Essential to Include Biotic Interactions across Trophic Levels' (2010) 365 *Philosophical Transactions of the Royal Society B: Biological Sciences* 2025

Van Dyke, J. M., 'Applying the Precautionary Principle to Ocean Shipments of Radioactive Materials' (1996) 27(4) *Ocean Development and International Law* 379

Verzijl, J.H.W, *International Law in Historical Perspective, Part V, Nationality and Other Matters Relating to Individuals* (A.W. Sijthoff-Leiden, 1972)

Wang, Haifeng, 'Economic Costs of CO_2 Emissions Reduction for Non-Annex I Countries in International Shipping' (2010) 14(4) *Energy for Sustainable Development* 280

Wang, Haifeng, 'GHG Emissions from the International Goods Movement by Ships and the Adaptation Funding Distribution' in Zongwei Luo (ed), *Green Finance and Sustainability: Environmentally-Aware Business Models and Technologies* (Business Science Reference, 2011) 274

Wang, Tao and Jim Watson, 'China's Carbon Emissions and International Trade: Implications for Post-2012 Policy' (2008) 8(6) *Climate Policy* 577

Warner, Robin, 'Marine Snow Storms: Assessing the Environmental Risks of Ocean Fertilization' (2009) 3(4) *Carbon & Climate Law Review* 426

Watson, Amanda, 'Does TRIPs Increase Technology Transfer to the Developing World? The Empirical Evidence' (2011) 20(3) *Information & Communications Technology Law* 253

Weisbach, David, 'Negligence, Strict Liability, and Responsibility for Climate Change' (2012) 97(2) *Iowa Law Review* 521

Weisslitz, Michael, 'Rethinking the Equitable Principle of Common but Differentiated Responsibility: Differential versus Absolute Norms of Compliance and Contribution in the Global Climate Change Context' (2002) 13(2) *Colorado Journal of International Environmental Law and Policy* 473

Westphal, Michael, Gordon Hughes and Jorn Brommelhorster (eds), *Economics of Climate Change in East Asia* (Asian Development Bank, 2013)

Work of the AWG-LCA Contact Group at AWG-LCA 14.2, AWG-LCA 14th Session (2nd part), Agenda Item 3.2.4, Bonn (7–17 June 2011)

World Commission on Environment and Development (WCED), 'Our Common Future' (Oxford University Press, 1987)

Xue, Hanqin, *Transboundary Damage in International Law* (Cambridge University Press, 2003)

Yamin, Farhana and Joanna Depledge, *The International Climate Change Regime: A Guide to Rules, Institutions and Procedures* (Cambridge University Press, 2004)

Yang, Zhihong et al., 'An Analysis of Greenhouse Gas Emission Trading System from the Perspective of Stakeholders' (2010) 2 *Procedia Environmental Sciences* 82

Young, O.R., *The Institutional Dimensions of Environmental Change: Fit, Interplay, and Scale* (MA: MIT Press, 2002)

Zhang, Zhongxiang, 'Multilateral Trade Measures in a Post-2012 Climate Change Regime? What Can Be Taken from the Montreal Protocol and the WTO?' (2009) 37(12) *Energy Policy* 5105

Zheng, Jianing, Hao Hu and Lei Dai, 'How would EEDI Influence Chinese Shipbuilding Industry?' (2013) 40(5) *Maritime Policy & Management* 495

中国工业和信息化部[Ministry of Industry and Information Technology of the People's Republic of China], '《2012年全国造船完工量超过 6000

万载重吨新接订单同比下降四成》[In 2002 National Accomplished Shipbuilding Output Exceeds 60000000 DWT and New Orders Decreased by 40%]' (14 January 2013) <http://www.miit.gov.cn/n11293472/n11293832/n11294132/n12858417/n12858628/15121853.html>

中国工业和信息化部[Ministry of Industry and Information Technology of the People's Republic of China], '《船舶工业"十二五"发展规划》[Development Plan for the Ship Industry during the 12th Five-Year Plan (2011–2015)]' (2012) <http://www.china.com.cn/policy/txt/2012-03/12/content_24876042.htm>

刘能治[Liu Nengye], '《航运减排国际法新动向》[New Development of International Law on the Reduction of Greenhouse Gas Emissions from Shipping]' (2011) 34(9) 世界海运[*World Maritime Shipping*] 11

朱建华[Zhu Jianhua], '《我国船舶运输温室气体减排对策探讨》Approaches for the Reduction of Greenhouse Gas Emissions from Ships in China' (2010) 1 水运科学研究 *Research on Waterborne Transportation* 1

李志文[Li Zhiwen], '《船舶温室气体减排国际立法的新发展及其启示》[New Development of International Regulation in Reducing Greenhouse Gas Emissions from Ships and Enlightenments to China]' (2012) 152(6) 法商研究 *Journal of Studies in Law and Business* 141

解玉真等[Xie Yuzhen et al.], '《EEDI对中国造船及航运业的影响》[The Impacts of the EEDI on the Chinese Shipbuilding and Shipping Industries]' (2011) 11 中国海事*China Maritime* 23

2 Cases

Anvil Hill Project Watch Association Inc v Minister for the Environment and Water Resources (2007) FCA 1481

Corfu Channel Case (*U.K. v Albania*) (1949) ICJ Rep. 4

Gabcikovo-Nagymaros Case (*Hungary v Slovakia*) (1997) ICJ Reports 7

Lac Lanoux Arbitration (*France v Spain*) (1957) 12 UN RIAA 285

Lotus Case (*France v Turkey*) (1927) PCIJ Series A, No. 10

Massachusetts v Environmental Protection Agency 549 U.S. 497 (2007), 127 S.Ct. 1438

Massachusetts v. Environmental Protection Agency, 549 U.S. 497 (2007)

The MOX Plant Case (*Ireland v United Kingdom*) (2001) 47 ILM 405; ITLOS, Order of 3 December 2001 on Provisional Measures

Nuclear Tests Case (*Australia v France*) (*Interim Protection*) (1974) ICJ Reports 253

Pulp Mills on the River Uruguay (*Argentina v Uruguay*) (2006) ICJ Reports

Request for an Examination of the Situation in Accodance with Paragraph 63 of the Court's Judgment of 20 December 1974 in the Nuclear Tests (*New Zealand v France*) (1995) ICJ Reports 288

Southern Bluefin Tuna Case (*Australia & New Zealand v Japan*) (1999) Case Nos.3 and 4 ITLOS

Trail Smelter Case (*United States of America v Canada*) (*Reports of International Arbitral Awards*) (1938 & 1941) 3 UN RIAA 1905

3 Legislation

Air Pollution Prevention and Control Law of the People's Republic of China, adopted in 1987 and amended in 1995, 2000 and 2015 respectively, entered into force on 1 January 2016

Australian Clean Energy Bill 2011 (*Explanatory Memorandum*), adopted by the Parliament of the Commonwealth of Australia and House of Representatives (2010–2011)

Australian Clean Energy Bill 2011 (*Explanatory Memorandum*), Policy Context

Clean Air Act of the United States of America, 17 December 1963, 42 USC 7401–7626

Clean Air Act of the United States of America, Pub L No 108–201, Stat, 42 USC §7401 et seq. (1970, as amended in 1977 and 1990)

Environmental Protection Law of the People's Republic of China, adopted at the 11th Meeting of the Standing Committee of the Seventh National People's Congress on 26 December 1989, and amended on 24 April 2014

4 Treaties

African Convention on the Conservation of Nature and Natural Resource, opened for signature 15 September 1968, 1001 UNTS 4 (entered into force 9 October 1969)

Agreement Governing the Activities of States on the Moon and Other Celestial Bodies, opened for signature 5 December 1979, 18 ILM 1434 (entered into force 11 July 1984)

Agreement on Trade-Related Aspects of Intellectual Property Rights (*TRIPS*), Annex 1C of the *Marrakesh Agreement Establishing the World Trade Organization*, opened for signature 15 April 1994, 33 ILM 1125 (entered into force 1 January 1995)

Amendments to MARPOL Annex VI and the NOx Technical Code 2008, IMO Doc Res MEPC.251(66) (4 April 2014)

Association of South Ease Asian Nations Agreement on the Conservation of Nature and Natural Resources, opened for signature 9 July 1985, 15 EPL 64 (not yet in force)

Charter of the United Nations and Statute of the International Court of Justice, signed 26 June 1945, 59 STAT 1031 (entered into force 24 October 1945)

Convention for the Prevention of Marine Pollution from Land-based Sources, opened for signature 4 June 1974, 13 ILM 352 (entered into force 6 May 1978)

Convention for the Protection of the Mediterranean Sea against Pollution, signed 16 February 1976, 15 ILM 300 (entered into force 12 February 1978)

Convention for the Protection of the Marine Environment of the North-East Atlantic, opened for signature 22 September 1992, 32 ILM 1068 (entered into force 25 March 1998)

Convention on Biological Diversity, opened for signature 5 June 1992, 31 ILM 818 (entered into force 29 December 1993)

Convention on Long-Range Transboundary Air Pollution, opened for signature 13 November 1979, 18 ILM 1442 (entered into force 16 March 1983)

Convention on the Conservation and Management of Highly Migratory Fish Stocks in the Western and Central Pacific Ocean, opened for signature 5 September 2000, 40 ILM 278 (entered into force 19 June 2004)

Convention on the Inter-Governmental Maritime Consultative Organization, opened for signature 6 March 1948, 289 UNTS 3 (entered into force 17 March 1958), amended and renamed as *Convention on the International Maritime Organization*, opened for signature 14 November 1975, 9 UTS 61 (entered into force 22 May 1982)

Convention on the International Regulations for Preventing Collisions at Sea, opened for signature 20 October 1972, UKTS 77 (entered into force 15 July 1977)

Convention on the Prevention of Marine Pollution by Dumping of Wastes and Other Matter, opened for signature 29 December 1972, 18 ILM 510 (entered into force 30 August 1975)

Convention on the Protection and Use of Transboundary Watercourses and International Lakes, opened for signature 17 March 1992, 31 ILM 1312 (entered into force 6 October 1996)

Convention on the Transboundary Effects of Industrial Accidents, opened for signature 17 March 1992, 31 ILM 1330 (entered into force 19 April 2000)

Doha Amendment to the Kyoto Protocol, adopted 8 December 2012, Decision 1/CMP.8, C.N.718.2012.TREATIES-XXVII.7.c (not yet in force)

General Agreement on Tariff and Trade, opened for signature 30 October 1947, 55 UNTS 194 (in force provisionally since 1 January 1948 under the 1947 Protocol of Application, 55 UNTS 308)

IAEA Vienna Convention on Civil Liability for Nuclear Damage, opened for signature 29 May 1963, 1063 UNTS 265 (entered into force 12 November 1977)

ILO Convention (No. 147) concerning Minimum Standards in Merchant Ships, opened for signature 13 October 1976, Govt. Doc. Y 1.1/4:99–21 (entered into force 28 November 1978)

IMO International Convention on Civil Liability for Bunker Oil Pollution Damage, opened for signature 27 March 2001, 40 ILM 1493 (entered into force 21 November 2008)

International Convention for the Control and Management of Ships' Ballast Water and Sediments, opened for signature 13 February 2004, IMO Doc. BWM/CONF/36 (not yet in force)

International Convention for the Prevention of Pollution from Ships (MARPOL *73/78*), signed 2 November 1973, 12 ILM 1319, as amended by the 1978 Protocol to the 1973 Convention, 1341 UNTS 3, 17 ILM 546 (entered into force 2 October 1983)

International Convention for the Safety of Life at Sea, opened for signature 1 November 1974, 1184 UNTS 2 (entered into force 25 May 1980)

International Convention for the Unification of Certain Rules relating to Bills of Lading for the Carriage of Goods by Sea, opened for signature 25 August 1924, 120 LNTS 155 (entered into force 2 June 1931)

International Convention on Civil Liability for Oil Pollution Damage, opened for signature 27 November 1992, IMO LEG/CONF.9.15 (entered into force 30 March 1996)

International Convention on Civil Liability for Oil Pollution Damage, opened for signature 29 November 1969, 973 UNTS 3 (entered into force 19 June 1975)

International Convention on Liability and Compensation for Damage in Connection with the Carriage of Hazardous and Noxious Substances by Sea, opened for signature 3 May 1996, 25 ILM 1406 (not yet in force)

International Convention on Load Lines, opened for signature 5 April 1966, 640 UNTS 133 (entered into force 21 July 1968) (Protocol of 11 November 1988, entered into force 3 February 2000)

International Convention on Oil Pollution Preparedness, Response and Cooperation, opened for signature 30 November 1990 (entered into force 13 May 1995)

International Convention on Salvage, opened for signature 28 April 1989, 1953 UNTS 165 (entered into force 14 July 1996)

International Convention on Standards of Training, Certification and Watchkeeping for Seafarers, opened for signature 7 July 1978, 1361 UNTS 2(entered into force 28 April 1984), as amended by the 1995 Protocol, 1969 UNTS (entered into force 1 February 1997)

International Convention on the Establishment of an International Fund for Compensation for Oil Pollution Damage, opened for signature 18 December 1971, 11 ILM 284 (entered into force 16 October 1978)

International Convention on the Establishment of an International Fund for Compensation for Oil Pollution Damage, opened for signature 27 November 1992, 87 UKTScm 3433 (entered into force 30 May 1996)

International Plant Protection Convention, opened for signature 6 December 1951, 150 UNTS 67 (entered into force 3 April 1952)

Kyoto Protocol to the United Nations Framework Convention on Climate Change, opened for signature 16 March 1998, 37 ILM 22 (entered into force 16 February 2005)

Midwest Greenhouse Gas Reduction Accord, signed on 15 November 2007, <http://www.c2es.org/us-states-regions/regional-climate-initiatives/mggra>

Montreal Protocol on Substances that Deplete the Ozone Layer, opened for signature 16 September 1987, 26 ILM 1550 (entered into force 1 January 1989)

OECD Convention on Third Party Liability in the Field of Nuclear Energy, opened for signature 29 July 1960, 956 UNTS 251 (entered into force 1 April 1968)

Paris Agreement to the United Nations Framework Convention on Climate Change (2015), opened for signature 22 April 2016, FCCC/CP/2015/L.9 (not yet in force)

Protocol on Liability and Compensation for Damage Resulting from Transboundary Movements of Hazardous Wastes and Their Disposal, opened for signature 10 December 1999, EMuT 989:22/B (not yet in force)

Protocol to the Contracting Parties to the Convention on the Prevention of Marine Pollution by Dumping of Wastes and Other Matter, opened for signature 7 November 1996, 36 ILM 1 (entered into force 24 March 2006)

Protocol to the Convention on the Prevention of Marine Pollution by Dumping of Wastes and Other Matter, opened for signature 7 November 1996, EMuT 972:96/D (entered into force 24 March 2006)

The 1990 Amendment to the Montreal Protocol on Substances that Deplete the Ozone Layer (London), opened for signature 29 June 1990, 30 ILM 537 (entered into force 10 August 1992)

The 1992 Amendment to the Montreal Protocol on Substances that Deplete the Ozone Layer (Copenhagen), opened for signature 25 November 1992, 32 ILM 874 (entered into force 14 June 1994)

The 1997 Amendment to the Montreal Protocol on Substances that Deplete the Ozone Layer (Montreal), opened for signature 25 September 1997, UNEP/OzL.Pro.9/12 (entered into force 10 November 1999)

The 1999 Amendment to the Montreal Protocol on Substances that Deplete the Ozone Layer (Beijing), opened for signature 17 December 1999, UNEP/OzL.Pro.11/10 (entered into force 25 February 2002)

Treaty Banning Nuclear Weapon Tests in the Atmosphere, in Outer Space and under Water, opened for signature 5 August 1963, 480 UNTS 43 (entered into force 10 October 1963)

Treaty on Principles Governing the Activities of States in the Exploration and Use of Outer Space, including the Moon and Other Celestial Bodies, opened for signature 27 January 1967, 610 UNTS 205 (entered into force 10 October 1967)

United Nations Convention on Conditions for the Registration of Ships, opened for signature 7 February 1986, 26 ILM 1229 (not yet in force)

United Nations Convention on the Law of the Sea, opened for signature 10 December 1982, 1833 UNTS 3 (entered into force 16 November 1994)

United Nations Framework Convention on Climate Change, opened for signature 9 May 1992, 31 ILM 848 (entered into force 21 March 1994)

Vienna Convention for the Protection of the Ozone Layer, opened for signature 22 March 1985, 26 ILM 1529 (entered into force 22 September 1988)

Vienna Convention on the Law of Treaties, opened for signature 23 May 1969, 8 ILM 679 (entered into force 27 January 1980)

5 Other

2009 Guidelines for Port State Control under the Revised MARPOL Annex VI, IMO Doc Res MEPC. 181(59) (17 July 2009)

2012 Guidelines for the Development of a Ship Energy Efficiency Management Plan (SEEMP), Resolution MEPC. 213(63), IMO Doc MEPC 63/23 Annex 9 (2 March 2012)

2012 Guidelines on Survey and Certification of the Energy Efficiency Design Index (EEDI), Resolution MEPC. 214(63), IMO Doc MEPC 63/23/Add.1 Annex 10 (2 March 2012)

2012 Guidelines on the Method of Calculation of the Attained Energy Efficiency Design Index (EEDI) for New Ships, Resolution MEPC.212(63), IMO Doc MEPC 63/23 Annex 8 (2 March 2012)

A Cross-industry Goal-based Approach to Reduction of GHG Emissions from New Ships, submitted by the ICS, BIMCO, CESA, INTERCARGO, INTERTANKO and OCIMF, MEPC 57th Session, Agenda Item 4, IMO Doc MEPC 57/4/8 (23 January 2008)

Adoption of the Agenda: Annotations and Provisional Timetable, note by the Secretariat, MEPC 62nd Session, Agenda Item 1, IMO Doc MEPC 62/1/1 (19 November 2010)

Agenda 21, A/CONF.151/26 (Vols. I, II, III) (13 June 1992)

Amendment to Draft Compromise MEPC Resolution on Promotion of Technical Co-operation and Transfer of Technology relating to the Improvement of Energy Efficiency of Ships, submitted by South Africa, MEPC 65th Session, Agenda Item 4, IMO Doc MEPC 65/4/33/Corr.1 (16 May 2013)

Amendments to MARPOL Annex VI—Inclusion of Regulations on Energy Efficiency for Ships, submitted by the International Chamber of Shipping (ICS), MEPC 62nd Session, Agenda Item 6, IMO Doc MEPC 62/6/24 (20 May 2011)

Amendments to MARPOL Annex VI—Inclusion of Regulations on Energy Efficiency for Ships, note by the Secretariat, MEPC 62nd Session, Agenda Item 6, IMO Doc MEPC 62/6/3 (21 December 2010)

Amendments to MARPOL Annex VI—Inclusion of Regulations on Energy Efficiency for Ships, submitted by Singapore, MEPC 62nd Session, Agenda Item 6, IMO Doc MEPC 62/6/21 (20 May 2011)

Amendments to the Procedures for Port State Control, IMO Doc Res A.882(21) (25 November 1999)

Application of EEDI to Ships other than Those Operating with Conventional Machinery and Power Distribution Arrangements, submitted by ICS, CLIA, INTERFERRY and the Marshall Islands, Intersessional Meeting of the Greenhouse Gas Working Group 2nd Session, Agenda Item 2, IMO Doc GHG-WG 2/2/19 (6 February 2009)

Application of Power Correction Factor F_j for Enhanced Safety, submitted by INTERFERRY, MEPC 60th Session, Agenda Item 4, IMO Doc MEPC 60/4/20 (15 January 2010)

Application of the EEDI to Existing Ships, submitted by INTERCARGO, MEPC 63rd Session, Agenda Item 5, IMO Doc MEPC 63/5/12 (6 January 2012)

Application of the Principle of "Common but Differentiated Responsibilities" to the Reduction of Greenhouse Gas Emissions from International Shipping, submitted by China and India, MEPC 58th Session, Agenda Item 4, IMO Doc MEPC 58/4/32 (15 August 2008)

Benefits and Possible Adverse Impacts of Market-based Instruments, submitted by World Wide Fund for Nature (WWF), MEPC 58th Session, Agenda Item 4, IMO Doc MEPC 58/4/39 (15 August 2008)

Bergen Ministerial Declaration on Sustainable Development, 20 EPL 200 (15 May 1990)

Charter of Economic Rights and Duties of States, UNGA Res.3281(XXIX), 29th Sess, Agenda Item 48, A/RES/29/3281 (12 December 1974)

CO_2 Reduction Requires Efficient Instruments Based on Sound Technical Solutions, submitted by the Community of European Shipyards' Association (CESA), Intersessional Meeting of the Greenhouse Gas Working Group 2nd Session, Agenda Item 2, IMO Doc GHG-WG 2/2/22 (6 February 2009)

Comments and Proposals on the Draft Regulation Text on Energy Efficiency for Ships, submitted by China, MEPC 61st Session, Agenda Item 5, IMO Doc MEPC 61/5/26 (6 August 2010)

Comments on MEPC 59/4/8 and MEPC 59/4/9 relating to the Energy Efficiency Design Index, the Ship Energy Management Plan and Possible Market-based Instruments, submitted by INTERTANKO, MEPC 59th Session, Agenda Item 4, IMO Doc MEPC 59/4/43 (22 May 2009)

Comments on MEPC 60/4/48, "An International Fund for Greenhouse Gas Emissions from Ships", submitted by Japan, MEPC 60th Session, Agenda Item 4, IMO Doc MEPC 60/4/51 (29 January 2010)

Comments on the Draft Regulatory Text on Energy Efficiency for Ships, submitted by Norway, MEPC 61st Session, Agenda Item 5, IMO Doc MEPC 61/5/6 (16 July 2010)

Comments on the EEDI Baseline Formula, submitted by Greece, MEPC 60th Session, Agenda Item 4, IMO Doc MEPC 60/4/15 (15 January 2010)

Comments on the Guidelines on the Method of Calculation of the Energy Efficiency Design Index for New Ships, submitted by Vanuatu, MEPC 62nd Session, Agenda Item 5, IMO Doc MEPC 62/5/31 (20 May 2011)

Comments on the Note by the Secretariat on Amendments to MARPOL Annex VI—Inclusion of Regulations on Energy Efficiency for Ships, submitted by Argentina, Brazil, Chile, China, Ecuador, India, Nicaragua, Peru, the Philippines, South Africa and Venezuela, MEPC 62nd Session, Agenda Item 6, IMO Doc MEPC 62/6/15 (6 May 2011)

Comments on the Outcomes of the United Nations Climate Change Conference Held in Copenhagen, Denmark, submitted by OCIMF and INTERTANKO, MEPC 60th Session, Agenda Item 4, IMO Doc MEPC 60/4/46 (28 January 2010)

Comments on the Proposed Amendments to MARPOL Annex VI, submitted by Vanuatu, MEPC 62nd Session, Agenda Item 6, IMO Doc MEPC 62/6/23 (20 May 2011)

Comments on the Proposed Mandatory Energy Efficiency Regulations, submitted by China, Saudi Arabia and South Africa, MEPC 62nd Session, Agenda Item 5, IMO Doc MEPC 62/5/10 (5 May 2011)

Comments on the Report of the Third Intersessional Meeting of the Working Group on Greenhouse Gas Emissions from Ships, submitted by Panama, MEPC 62nd Session, Agenda Item 5, IMO Doc MEPC 62/5/28 (20 May 2011)

Comments related to the Outcome of Informal Consultations Conducted by the Chairman and Proposals for Further Progress, submitted by Oil Companies International Marine Forum (OCIMF), MEPC 59th Session, Agenda Item 4, IMO Doc MEPC 59/4/46 (22 May 2009)

Comments related to Trial Calculations of the EEDI for Subgroups of Ro-ro Cargo Ships, submitted by INTERFERRY and CESA, MEPC 60th Session, Agenda Item 4, IMO Doc MEPC 60/4/48 (29 January 2010)

Consideration of A Market-Based Mechanism to Improve the Energy Efficiency of Ships Based on the International GHG Fund, submitted by Japan, MEPC 59th Session, Agenda Item 4, IMO Doc MEPC 59/4/34 (8 May 2009)

Consideration of A Market-Based Mechanism: Leveraged Incentive Scheme to Improve the Energy Efficiency of Ships Based on the International GHG Fund, submitted by Japan, MEPC 60th Session, Agenda Item 4, IMO Doc MEPC 60/4/37 (15 January 2010)

Consideration of Adoption of Three Principles for Market-based Instruments, submitted by Cruise Lines International Association (CLIA), MEPC 59th Session, Agenda Item 4, IMO Doc MEPC 59/4/32 (8 May 2009)

Consideration of Appropriate Targets for Reducing CO_2 Emissions from International Shipping, submitted by Japan, MEPC 59th Session, Agenda Item 4, IMO Doc MEPC 59/4/35 (8 May 2009)

Consideration of Commitments for Subsequent Periods for Parties Included in Annex I to the Convention under Article 3, Para. 9 of the Kyoto Protocol, Decision 1/CMP.1, Doc FCCC/KP/CMP/2005/8/Add.1 (2006)

Consideration of Safety Issues Related To EEDI, submitted by Greece, MEPC 61st Session, Agenda Item 5, IMO Doc MEPC 61/5/23 (5 August 2010)

Consideration of the Appropriate Instrument for A Mandatory Regime to Address GHG Emissions, submitted by the United Kindom, MEPC 58th Session, Agenda Item 4, IMO Doc MEPC 58/4/15 (1 August 2008)

Considerations of the Application of the EEDI Reference Lines to LNG Vessels, submitted by Society of International Gas Tanker and Terminal Operators (SIGTTO), MEPC 62nd Session, Agenda Item 6, IMO Doc MEPC 62/6/20 (20 May 2011)

Consolidated Statement of Continuing ICAO Policies and Practices Related to Environmental Protection-Climate Change, Assembly 37th Session, ICAO Doc Res A37–19 (8 October 2010)

Consolidated Statement of Continuing ICAO Policies and Practices Related to Environmental Protection-Climate Change, Assembly 38th Session, ICAO Doc Res A38–18 (4 October 2013)

Control of Greenhouse Gas Emissions from International Maritime Transport, submitted by the International Chamber of Shipping (ICS), MEPC 60th Session, Agenda Item 4, IMO Doc MEPC 60/4/13 (15 January 2010)

Cooperation between States in the Field of the Environment, UNGA Res 2995 XXVII (15 December 1972)

Cost Analysis on the Application of Efficiency Improvement Measures in the Maritime Fleet, submitted by Japan, MEPC 63rd Session, Agenda Item 5, IMO Doc MEPC 63/INF.13 (22 December 2011)

Decision 1/CP.21 Aoption of the Paris Agreement, Doc FCCC/CP/2015/10/Add.1 (29 January 2016)

Declaration on Sustainable Development and Plan of Implementation of the World Summit on Sustainable Development, A/CONF.199/CRP/7 and A/CONF/L/6/Rev.2 (4 September 2002)

Development of a CO_2 Design Index for New Ships, submitted by the Community of European Shipyards' Association (CESA), MEPC 58th Session, Agenda Item 4, IMO Doc MEPC 58/4/12 (1 August 2008)

Development of An Index for CO_2 Emissions Per Unit Shipping Capacity in Actual Operational Conditions, submitted by Japan, MEPC 57th Session, Agenda Item 4, IMO Doc MEPC 57/4/11 (25 January 2008)

Draft Legal Text on the Modified Efficiency Incentive Scheme (EIS), submitted by Japan, MEPC 64th Session, Agenda Item 5, IMO Doc MEPC 64/5/2 (28 June 2012)

Draft Legal Text on Uses of Financing Generated from A Maritime MBM, submitted by the World Wide Fund for Nature (WWF), MEPC 64th Session, Agenda Item 5, IMO Doc MEPC 64/5/10 (27 July 2012)

Draft Legal Text with Respect to the Proposal of the United States to Reduce Greenhouse Gas Emissions from International Shipping, submitted by the United States of America, MEPC 64th Session, Agenda Item 5, IMO Doc MEPC 64/5/7 (26 July 2012)

Draft MEPC Resolution on Capacity-Building, Technical Assistance and Transfer of Technology Related to Energy Efficiency Measures for Ships, submitted by Australia, Canada, Germany, Japan, the Marshall Islands, Norway and the United States, MEPC 64th Session, Agenda Item 4, IMO Doc MEPC 64/4/44 (10 August 2012)

Draft MEPC Resolution on Promotion of Technical Cooperation and Technology Transfer Relating to the Improvement of Energy Efficiency of Ships, submitted by Angola,

Argentina, Brazil, China, India, Jamaica, Nigeria, Peru, South Africa and Venezuela, MEPC 64th Session, Agenda Item 4, IMO Doc MEPC 64/4/30 (27 July 2012)

Draft Principles on the Allocation of Loss in the Case of Transboundary Harm Arising Out of Hazardous Activities, Annex, UN Doc A/RES/61/36 (18 December 2006)

Draft Regulations To Be Included in MARPOL Annex VI for the Control of CO_2 Emissions from Ships, note by the Bahamas, MEPC 63rd Session, Agenda Item 5, IMO Doc MEPC 63/5/1 (24 November 2011)

Efficiency Incentive Scheme (ETS), submitted by Japan and the World Shipping Council, MEPC 63nd Session, Agenda Item 5, IMO Doc MEPC 63/5/3 (25 November 2011)

Emission "Caps" and Reduction Targets, submitted by the World Shipping Council (WSC), MEPC 60th Session, Agenda Item 4, IMO Doc MEPC 60/4/28 (15 January 2010)

Energy Efficiency Design Index for Propulsion Redundancy, submitted by INTERTANKO, MEPC 60th Session, Agenda Item 4, IMO Doc MEPC 60/4/4 (18 December 2009)

Energy Efficiency Design Index for Tankers, submitted by INTERTANKO, MEPC 60th Session, Agenda Item 4, IMO Doc MEPC 60/4/3 (18 December 2009)

Ensuring No Net Incidence on Developing Countries from A Global Maritime Market-Based Mechanism, submitted by World Wide Fund for Nature (WWF), MEPC 63rd Session, Agenda Item 5, IMO Doc MEPC 63/5/6 (22 December 2011)

Ensuring No Net Incidence on Developing Countries from a Global Maritime Market-Based Mechanism, submmited by World Wide Fund for Nature (WWF), IMO Doc. MEPC 63/5/6 (22 December 2011)

Environment and Economics: Guiding Principles Concerning International Economic Aspects of Environmental Policies, OECD Doc. No.C(72)128, 1972 WL 24710 (26 May 1972)

Establishment of An 'EEDI Database', submitted by Liberia, IACS, and INTERCARGO, MEPC 66th Session, Agenda Item 4, IMO Doc MEPC 66/4/13 (24 January 2014)

Establishment of An 'EEDI Database', submitted by the International Association of Classification Societies (IACS), MEPC 65th Session, Agenda Item 4, IMO Doc MEPC 65/4/31 (8 March 2013)

Follow-Up to the Recommendations of the Twenty-Ninth Session of the Committee on Fisheries, Rome, Italy, 31 January–4 February 2011, COFI 30th Session, FAO Doc COFI/2012/Inf.5 (9–13 July 2012)

Framework and Procedures for the IMO Member State Audit Scheme, IMO Doc Res A.1067(28) (4 December 2013)

Framework and Procedures for the Voluntary IMO Member State Audit Scheme, Assembly 24th Session, Agenda Item 19, IMO Doc Res A.974 (24) (21 December 2005)

Further Details on the Proposal of the United Nations to Reduce Greenhouse Gas Emissions from International Shipping, submitted by the United States of America, MEPC 64th Session, Agenda Item 5, IMO Doc MEPC 64/5/6 (26 July 2012)

Further Information on a Rebate Mechanism for a Market-Based Measure for International Shipping, submitted by the International Union for Conservation of Nature (IUCN), MEPC 61st Session, Agenda Item 5, IMO Doc MEPC 61/5/33 (6 August 2010)

A Further Outline of A Global Emission Trading System (ETS) for International Shipping, submitted by Norway, MEPC 60th Session, Agenda Item 4, IMO Doc MEPC 60/4/22 (15 January 2010)

Further Prospects for EEDI Improvement, submitted by Greece, MEPC 62nd Session, Agenda Item 5, IMO Doc MEPC 62/5/6 (5 May 2011)

Further Technical and Operational Measures for Enhancing Energy Efficiency of International Shipping: Comments on Enhancing Energy Efficiency in Internaitonal Shipping, submitted by Angola, Argentina, Brazil, China, Cuba, Malaysia, Nigeria and South Africa, MEPC 66th Session, Agenda Item 4.1, IMO Doc MEPC 66/4/25 (7 February 2014)

Further Technical and Operational Measures for Enhancing Energy Efficiency of International Shipping: Establishing an Energy Efficiency Data Collection System, submitted by Belgium, Croatia, Denmak, Japan, Norway and the United States, MEPC 66th Session, Agenda Item 4.1, IMO Doc MEPC 66/4/19 (24 January 2014)

Further Work on GHG Emissions from Ships, submitted by Brazil, China, India, Peru, Saudi Arabia and South Africa, MEPC 64th Session, Agenda Item 5, IMO Doc MEPC 64/5/9 (27 July 2012)

Future IMO Regulation regarding Greenhouse Gas Emissions from International Shipping, submitted by Denmark, Marshall Islands, BIMCO, ICS, INTERCARGO, INTERTANKO and OCIMF, MEPC 57th Session, Agenda Item 4, IMO Doc MEPC 57/4/2 (21 December 2008)

Grouping and Evaluation of Proposed MBMs, submitted by Greece, Intersessional Meeting of the Working Group on GHG Emissions from Ships 3rd Session, Agenda Item 3, IMO Doc GHG-WG 3/3 (24 February 2011)

Guidance on the Treatment of Innovative Energy Efficiency Technologies for Calculation and Verification of the Attained EEDI, MEPC 64th Session, Agenda Item 4, IMO Doc MEPC 64/4/39 (9 August 2012)

Guidelines for Calculation of Reference Lines for Use with the Energy Efficiency Design Index (EEDI), Resolution MEPC.215(63), IMO Doc MEPC 63/23/Add.1 Annex 11 (2 March 2012)

Guidelines for the Implementation of the Ship Operational Index—Ship Efficiency Management Tool, submitted by INTERTANKO, OCIMF and BIMCO, MEPC 58th Session, Agenda Item 4, IMO Doc MEPC 58/4/13 (1 August 2008)

Guidelines for Voluntary Use of the Ship Energy Efficiency Operational Indicator (EEOI), Ref. T5/1.01, IMO Doc MEPC.1/Circ.684 (17 August 2009)

Identifying Consensus on IMO Principles on Addressing Greenhouse Gas Emissions from International Shipping, submitted by Australia, Canada, Denmark, Germany, Japan, Marshall Islands, Norway, Panama and the United States, MEPC 58th Session, Agenda Item 4, IMO Doc MEPC 58/4/16 (1 August 2008)

II Yearbook International Law Commission (1980) Pt1, 160, 138–139

IMO Instruments Implementation Code (III Code), IMO Doc Res A.1070(28) (4 December 2013)

IMO Must Act Decisively to Reduce GHG Emissions from Shipping If It Is to Retain Its Competence in Technical and Political Matters Related to Shipping and GHGs, submitted by Friends of the Earth International, Greenpeace International and World Wild Fund for Nature, MEPC 59th Session, Agenda Item 4, IMO Doc MEPC 59/4/47 (22 May 2009)

IMO Policies and Practices Related to the Reduction of Greenhouse Gas Emissions from Ships, IMO Assembly 23rd Session, Agenda Item 19, IMO Doc Res A.963(23) (5 December 2003)

Implementation of Resolution MEPC.229(65) on Promotion of Technical Co-operation and Transfer of Technology Relating to the Improvement of Energy Efficiency of Ships, submitted by Belgium, Japan, the United Kingdom and the United States, MEPC 66th Session, Agenda Item 4, IMO Doc MEPC 66/4/31 (7 February 2014)

Implementation of Resolution MEPC.229(65) on the 'Promotion of Technical Co-operation and Transfer of Technology Relating to the Improvement of Energy Efficiency of Ships', submitted by Angola, Argentina, Brazil, China, Cuba, India, Malaysia, Nigeria and South Africa, MEPC 66th Session, Agenda Item 4, IMO Doc MEPC 66/4/17 (24 January 2014)

The Importance of Using Effective Anti-fouling Coatings in relation to Greenhouse Gas Emissions from Shipping, submitted by the International Paint and Printing Ink Council (IPPIC), MEPC 60th Session, Agenda Item 4, IMO Doc MEPC 60/4/21 (15 January 2010)

Information on Experience with the European Union Emissions Trading Scheme (EU-ETS) and the Carbon Market, submitted by the European Commission, Intersessional Meeting of the Working Group on GHG Emissions from Ships 3rd Session, Agenda Item 2, IMO Doc GHG-WG 3/INF.3 (25 February 2011)

Information Relevant to Emissions from Fuel Used for International Aviation and Maritime Transport, UNFCCC SBSTA 39th Session, Warsaw, Doc FCCC/SBSTA/2013/MISC.20 (10 November 2013)

An International Fund for Greenhouse Gas Emissions from Ships, submitted by Cyprus, Denmark, the Marshall Islands, Nigeria and the International Parcel Tankers Association (IPTA), MEPC 60th Session, Agenda Item 4, IMO Doc MEPC 60/4/8 (18 December 2009)

An International Fund for Greenhouse Gas Emissions from Ships, submitted by Denmark, MEPC 59th Session, Agenda Item 4, IMO Doc MEPC 59/4/5 (9 April 2009)

An International Fund for Greenhouse Gas Emissions from Ships, submitted by Greece, MEPC 60th Session, Agenda Item 4, IMO Doc MEPC 60/4/49 (29 January 2010)

The International Greenhouse Gas Fund—Strengths and Weaknesses, submitted by Cyprus, Denmark, the Marshall Islands, Liberia, Nigeria, the Republic of Korea and the International Parcel Tankers Association (IPTA), MEPC 62nd Session, Agenda Item 5, IMO Doc MEPC 62/5/33 (20 May 2011)

The International Greenhouse Gas Fund, submitted by Cyprus, Denmark, the Marshall Islands and Nigeria, Intersessional Meeting of the Working Group on GHG Emissions from Ships 3rd Session, Agenda Item 3, IMO Doc GHG-WG 3/3/4 (25 February 2011)

Introduction of A Cubic Capacity Correction Factor into the EEDI Formula, submitted by the International Parcel Tankers Association (IPTA), MEPC 62th Session, Agenda Item 6, IMO Doc MEPC 62/6/13 (5 May 2011)

Key Design Elements for Designing A 'Cap and Trade' Greenhouse Gas Emissions Trading Scheme (ETS), submitted by the European Commission, Intersessional Meeting of the Greenhouse Gas Working Group 1st Session, Agenda Item 5, IMO Doc GHG-WG 1/5/3 (30 May 2008)

Legal Aspects of the Organization's Work on Greenhouse Gas Emissions in the Context of the Kyoto Protocol, note by the Secretariat, IMO Doc MEPC 58/4/20 (1 August 2008)

A Mandatory CO_2 Design Index for New Ships, submitted by Denmark, Marshall Islands, BIMCO, ICS, INTERCARGO, INTERTANKO and OCIMF, MEPC 57th Session, Agenda Item 4, IMO Doc MEPC 57/4/3 (21 December 2007)

Mandatory CO_2 Emission Cut Targets through Technical and Operational Measures, submitted by the Bahamas, MEPC 62nd Session, Agenda Item 5, IMO Doc MEPC 62/5/13 (6 May 2011)

Mandatory EEDI Requirements—Draft Text for Adding A New Part to MARPOL Annex VI for Regulation of the Energy Efficiency of Ships, submitted by Japan, Norway and the United States, MEPC 62th Session, Agenda Item 4, IMO Doc MEPC 60/4/35 (15 January 2010)

Market-based Measures—Inequitable Burden on Developing Countries, submitted by India, MEPC 61st Session, Agenda Item 5, IMO Doc MEPC 61/5/19 (2 August 2010)

Market Based Measures—Impact on India's Shipping Trade, submitted by India, IMO Doc MEPC 63/5/8 (23 December 2011)

MBM Proposals: A Way Ahead, submitted by Greece, MEPC 62nd Session, Agenda Item 5, IMO Doc MEPC 62/5/7 (6 May 2011)

Minimum Propulsion Power to Maintain the Manoeuvrability in Adverse Conditions, submitted by Greece, MEPC 64th Session, Agenda Item 4, IMO Doc MEPC 64/4/37 (9 August 2012)

Need and Purpose of an MBM: How Technical and Operational Measures Are the Only Direct and Effective Means to Deliver Cuts in CO_2 Emissions, submitted by the Bahamas, Intersessional Meeting of the Working Group on GHG Emissions from Ships 3 rd Session, Agenda Item 2, IMO Doc GHG-WG 3/2 (22 December 2010)

The Need for Refinement of the Energy Efficiency Design Index (EEDI), submitted by the International Chamber of Shipping (ICS), MEPC 59th Session, Agenda Item 4, IMO Doc MEPC 59/4/13 (9 April 2009)

Nordic Council's International Conference on Pollution of the Seas, Greenpeace 18, Annex 2 (1990)

Operational Energy Efficiency of New and Existing Ships, submitted by the International Chamber of Shipping (ICS), MEPC 64th Session, Agenda Item 5, IMO Doc MEPC 64/5/11 (27 July 2012)

Paris MOU, *New Targeting Lists Paris MOU* (16 June 2014) <https://www.parismou.org/new-targeting-lists-paris-mou>

Partnerships in Environmental Management for the Seas of East Asia, *Korea Maritime Institute (KMI)* <http://www.pemsea.org/organization/kmi>

Phase-in Implementation of the Energy Efficiency Design Index for Standard and Complex Ship Types, submitted by the Community of European Shipyards' Associations (CESA), MEPC 59th Session, Agenda Item 4, IMO Doc MEPC 59/4/38 (20 May 2009)

Possible Incompatibility between the WTO Rules and Market-based Measures for International Shipping, submitted by India and Saudi Arabia, MEPC 64th Session, Agenda Item 5, IMO Doc MEPC 64/5/3 (29 June 2012)

Possible Incompatibility between WTO Rules and A Market-based Measure for International Shipping, submitted by India, MEPC 62nd Session, Agenda Item 5, IMO Doc MEPC 62/5/27 (20 May 2011)

Procedures for Port State Control, IMO Doc Res A. 787(19) (23 November 1995)

Project to Develop a SEEMP Using a Structured Methodology and the Resulting Improvement in Energy Efficiency, submitted by Oil Companies International Marine Forum (OCIMF), MEPC 62nd Session, Agenda Item 5, IMO Doc MEPC 62/INF.12 (6 May 2011)

Promotion of Technical Co-operation and Transfer of Technology relating to the Improvement of Energy Efficiency of Ships, IMO Doc Res MEPC.229(65) (17 May 2013)

Promotion of Technical Cooperation and Transfer of Technology Relating to the Improvement of Energy Efficiency of Ships, submitted by Angola, China, Jamaica, South Africa and Venezuela, MEPC 64th Session, Agenda Item 4, IMO Doc MEPC 64/4/24 (27 July 2012)

Proposal for Amendments to the 2012 Guidelines on the Method of Calculation of the Attained Energy Efficiency Design Index (EEDI) for New Ships for Inclusion of A

Correction Factor to Account For Shallow Water Restrictions, submitted by Greece, MEPC 65th Session, Agenda Item 4, IMO Doc MEPC 65/4/17 (8 March 2013)

Proposal for an Energy Efficiency Design Index Verification Process, submitted by International Towing Tank Conference (ITTC), MEPC 60th Session, Agenda Item 4, IMO Doc MEPC 60/4/45 (25 January 2010)

Proposal for New Concept Approach to EEDI for New Ships, submitted by the Republic of Korea, Intersessional Meeting of the Greenhouse Gas Working Group 2nd Session, Agenda Item 2, IMO Doc GHG-WG 2/2/12 (6 February 2009)

Proposal of the United States to Enhance Energy Efficiency in International Shipping, submitted by the United States, MEPC 65th Session, Agenda Item 4, IMO Doc MEPC 65/4/19 (8 March 2013)

Proposal Submitted on behalf of the Group 77, UN Doc.A/CONF.151/PC/WG.III/L.20/REV.1 (1992)

Protection of Global Climate for Present and Future Generations of Mankind, GA/Res 43/53, 43rd sess, 70th plen mtg, UN Doc A/RES/43/53 (6 December 1988)

Protection of Global Climate for Present and Future Generations of Mankind, UNGA. A/RES/44/207 (22 December 1989)

Protection of Global Climate for Present and Future Generations of Mankind, UNGA. A/RES/45/212 (21 December 1990)

A Rebate Mechanism for A Market-based Instrument for International Shipping, submitted by the International Union for Conservation of Nature (IUCN), MEPC 60th Session, Agenda Item 4, IMO Doc MEPC 60/4/55 (29 January 2010)

Recommendation of the Council on Guiding Principles concerning International Economic Aspects of Environmental Policies (26 May 1972) OECD Recommendation C (72) 128

Reducing Global Ship Emissions Using a Speed-Related GHG or Compensation Fund, submitted by Clean Shipping Coalition (CSC), MEPC 64th Session, Agenda Item 5, IMO Doc MEPC 64/5/8 (27 July 2012)

Refinements to the "Draft Guidelines on the Method of Calculation of the Energy Efficiency Design Index for New Ships" for Conventional Passenger Ships, submitted by Cruise Lines International Association (CLIA), MEPC 59th Session, Agenda Item 4, IMO Doc MEPC 59/4/12 (9 April 2009)

Resolution on Clean Air Programmes for Ports, submitted by the International Association of Ports and Harbors (IAPH), MEPC 56th Session, Agenda Item 4, IMO Doc MEPC 56/INF.14 (11 May 2007)

Response to the Outcome of the First Intersessional Meeting of the Working Group on Greenhouse Gas Emissions from Ships, submitted by the International Bunker Industry Association (IBIA), MEPC 58th Session, Agenda Item 4, IMO Doc MEPC 58/4/19 (1 August 2008)

Results of Data Gathering Exercise for the Assessment of the Energy Efficiency Design Index (EEDI) for Ships Carrying Liquefied Gases in Bulk, submitted by the Society of International Gas Tanker and Terminal Operators (SIGTTO), MEPC 60th Session, Agenda Item 4, IMO Doc MEPC 60/4/44 (22 January 2010)

Results of the Sixth Seoul International Maritime Forum, submitted by the Republic of Korea, MEPC 65th Session, Agenda Item 5, IMO Doc MEPC 65/INF.6 (4 February 2013)

The Revision of MARPOL Annex VI, submitted by the International Chamber of Shipping (ICS), MEPC 57th Session, Agenda Item 4, IMO Doc MEPC 57/4/28 (13 February 2008)

Rio Declaration on Environment and Development, 31 ILM 874 (14 June 1992)

Scientific Study on International Shipping and Market-Based Instruments, IMO Doc. MEPC 60/INF.21 (15 January 2010)

Ship Efficiency Management Plan, submitted by ICS, BIMCO, Intercargo, Intertanko and OCIMF, MEPC 58th Session, Agenda Item 4, IMO Doc MEPC 58/INF.7 (28 July 2008)

Speed Reduction—the Key to the Fast and Efficient Reduction of Greenhouse Gas Emissions from Ships, submitted by the Clean Shipping Coalition (CSC), MEPC 61st Session, Agenda Item 5, IMO Doc MEPC 61/5/10 (23 July 2010)

Stockholm Declaration on the Human Environment, 11 ILM 1416 (16 June 1972)

Technical Evaluation of Market-based Instruments, submitted by Oil Companies International Marine Forum (OCIMF), MEPC 59th Session, Agenda Item 4, IMO Doc MEPC 59/4/17 (7 May 2009)

The Evaluation on the Relative Strengths and Weaknesses of the Reduction Mechanisms Employed by the MBM Proposals, submitted by the Republic of Korea, Intersessional Meeting of the Working Group on GHG Emissions from Ships 3rd Session, Agenda Item 3, IMO Doc GHG-WG 3/3/1 (25 February 2011)

Uncertainties and Problems in Market-based Measures, submitted by China and India, MEPC 61st Session, Agenda Item 5, IMO Doc MEPC 61/5/24 (5 August 2010)

Updated Technical Evaluation of Proposed GHG Reduction Measures, submitted by Oil Companies International Marine Forum (OCIMF), MEPC 60th Session, Agenda Item 4, IMO Doc MEPC 60/4/50 (29 January 2010)

World Trade Organization's Views on Document MEPC 64/5/4 Submitted by India and Saudi Arabia, note by the Secretary-General, MEPC 65th Session, Agenda Item 5, IMO Doc MEPC 65/INF.18 (21 February 2013)

Abbott, Tony, *The Coalition's Plan to Abolish the Carbon Tax* (2013) <https://www.liberal.org.au/our-plan-abolish-carbon-tax>

American Mathematical Society, *Developing Countries List* <http://www.ams.org/membership/individual/types/mem-develop>

Ares, Elena, *EU ETS and Aviation* (23 May 2012) <www.parliament.uk/briefing-papers/SN05533.pdf>

Asia Shipbuilding & Offshore Information Service (ASIASIS), *Korea Welcomes EU Environmental Regulations* (17 November 2010) <http://www.simic.net.cn/news_show.php?lan=en&id=80211>

Asian Shipowners' Forum (ASF), *The 18th ASF Joint Statement* (27 May 2009) <http://www.jsanet.or.jp/e/pressrelease_e/2009/pdf/20090529e_2.pdf>

Asian Shipowners' Forum (ASF), *Note of Understanding* (2 November 2011) <http://www.jsanet.or.jp/e/pressrelease_e/2011/pdf/a20111108e.pdf>

Asian Shipowners' Forum (ASF), *Press Release* (15 September 2008) <http://www.jsanet.or.jp/e/pressrelease_e/2008/pdf/20080930.pdf>

Australian-German Climate and Energy College, *INDC FactSheets* (22 April 2016) <http://www.climate-energy-college.net/indc-factsheets>

Australian Government: Ministry for Foreign Affairs, *List of Developing Countries as Declared by the Minister for Foreign Affairs* (2013) <http://www.ausaid.gov.au/ngos/Documents/list-developing-countries.pdf>

Australian Maritime Safety Authority, *Vessels* (2013) <https://www.amsa.gov.au/vessels/>

Bhattacharya, Suranjana Roy, 'Chinese Shipping Industry Is Big but Not Powful', *Gulf News* 19 July 2010 <http://gulfnews.com/business/opinion/chinese-shipping-industry-is-big-but-not-powerful-1.656076>

BIMCO, *About BIMCO* <https://www.bimco.org/About/About_BIMCO.aspx>

Bodansky, Daniel and Lavanya Rajamani, *Key Legal Issues in the 2015 Climate Negotiations* (June 2015) <http://www.c2es.org/docUploads/legal-issues-brief-06-2015.pdf >

Boehm, Peter, *Global Warning: Devastation of an Atoll* The Independent <http://www.commondreams.org/headlines06/0830-07.htm>

Bultheel, Clément, Romain Morel, et al., *COP21: Success at 'the End of the Beginning'* (18 December 2015) < http://www.i4ce.org/wp-core/wp-content/uploads/2015/12/15-12-18-I4CE-Climate-Brief-38-COP211.pdf>

Centre for Climate and Energy Solutions, *New Jersey Classifies Carbon Dioxide as Air Contaminant* (18 October 2005) <http://www.c2es.org/us-states-regions/news/2005/new-jersey-classifies-carbon-dioxide-air-contaminant>

China's Xinhua News Agency, *China Sets Conditions on Binding Climate Change Commitment after 2020* (6 December 2011) <http://english.peopledaily.com.cn/90883/7667257.html>

China Classification Society (CCS), 《绿色船型认可*EEDI*》 [*Green Ships Recognition EEDI*] (2012) <http://www.ccs.org.cn/ccswz/font/fontAction!moudleIndex.do?moudleId=61>

Clarkson, Perter, *Indian Yards Ask Government Action* (6 April 2009) <http://www.glob-maritime.com/news/shipbuilding-a-repair/4401-indian-yards-ask-government-action>

Cooper, Kevin, *Reducing Shipping Emissions: An Overview of Recent International Initiatives* (25 September 2012) <http://www.safety4sea.com/analysis/89/134/reducing-shipping-emissions->

Delingpole, James, *Meet the Man Who has Exposed the Great Climate Change Con Trick* (11 July 2009) <http://www.mannkal.org/downloads/environment/meettheman-whohasexposedthegreatclimatechangecontrick.pdf>

Devli, Anil, *Overview of the Shipping Sector in India* (1 April 2011) <http://www.ahlers.com/images/news/2011/overview%20of%20the%20shipping%20sector%20in%20india%20%28mr%20anil%20devli%29.pdf>

European Commission, *Emissions Trading System (EU ETS)* <http://ec.europa.eu/clima/policies/ets/index_en.htm>

European Federation for Transport and Environment, *Bunker Fuels and the Kyoto Protocol: How ICAO and the IMO Failed the Climate Change Test* <http://www.transportenvironment.org/sites/te/files/media/2009_06_aviation_shipping_icao_imo_history.pdf>

Gao, Weijie, *Development Strategy of Chinese Shipping Company under the Multilateral Framework of WTO* (2003) <http://www.cosco.com/en/pic/forum/654923323232.pdf>

Government of Canada, *Explanatory Notes--Asian Newly Industrialized Countries (NICs)* (14 June 2012) <http://www.ic.gc.ca/eic/site/tdo-dcd.nsf/eng/00042.html>

Gratsos, George A., *Green and More Profitable Shipping* (13 November 2012) <http://www.nee.gr/downloads/183NEWSFRONT%20NAFTILIAKI%2013-11-12.pdf>

Harvey, Fiona, *UK Shipping Industry Rejects EU's Carbon Reduction Programme* (9 August 2011) <http://www.guardian.co.uk/environment/2011/aug/09/shipping-industry-rejects-carbon-trading>

IACS, *Classification Societies: What, Why and How?* (March 2011) <http://www.iacs.org.uk/document/public/explained/class_whatwhy&how.pdf>

IMO, *IMO Takes Further Action on Climate Change* (22 April 2016) <http://www.imo.org/en/MediaCentre/PressBriefings/Pages/11-data-collection-.aspx >

IMO, *Member States, IGOs and NGOs* (2013) <http://www.imo.org/About/Membership/Pages/Default.aspx>

IMO, *Republic of Korea to Assist IMO in Building Capacity in Developing Countries to Address Greenhouse Gas Emissions from Ships* (21 April 2011) <http://www.imo.org/MediaCentre/PressBriefings/Pages/Home.aspx>

IMO, *Status of Conventions* (25 June 2014) <http://www.imo.org/About/Conventions/StatusOfConventions/Pages/Default.aspx>

IMO, *Summary of Status of Conventions* (2 December 2013) <http://www.imo.org/About/Conventions/StatusOfConventions/Pages/Default.aspx>

INTERCARGO, *About INTERCARGO* <http://www.intercargo.org/about/78-about-intercargo.html>

INTERFERRY, *Interferry Welcomes EEDI Decision* (12 October 2012) <http://www.marinelink.com/news/interferry-welcomes348466.aspx>

Intergovernmental Panel on Climate Change (IPCC), *Organization* <http://www.ipcc.ch/organization/organization.shtml>

Intergovernmental Panel on Climate Change (IPCC), *Principles and Procedures* <http://www.ipcc.ch/organization/organization_procedures.shtml>

Intergovernmental Panel on Climate Change (IPCC), *Structure* <http://www.ipcc.ch/organization/organization_structure.shtml>

Intergovernmental Panel on Climate Change (IPCC), *Understanding Climate Change: 22 Years of IPCC Assessment* <http://www.ipcc.ch>

International Association of Ports and Harbors (IAPH), *History/NGO Consultative Status* (2011) <http://www.iaphworldports.org/AboutIAPH/HistoryNGOConsultativeStatus.aspx>

International Maritime Organisation (IMO), *Market-based Measures* <http://www.imo.org/OurWork/Environment/PollutionPrevention/AirPollution/Pages/Market-Based-Measures.aspx>

International Maritime Organisation (IMO), *Status of Conventions: List of IMO Conventions and Their Amendments* (16 May 2013) <http://www.imo.org/About/Conventions/StatusOfConventions/Pages/Default.aspx>

International Maritime Organization (IMO), *Introduction to IMO* <http://www.imo.org/About/Pages/Default.aspx>

International Maritime Organization (IMO), *Mandatory Energy Efficiency Measures for International Shipping Adopted at IMO Environment Meeting* (15 July 2011) <http://www.imo.org/MediaCentre/PressBriefings/Pages/42-mepc-ghg.aspx>

International Maritime Organization (IMO), *Member States, IGOs and NGOs* <http://www.imo.org/About/Membership/Pages/Default.aspx>

International Maritime Organization (IMO), *Port State Control* <http://www.imo.org/blast/mainframe.asp?topic_id=159>

International Union of Marine Insurance (IUMI), *Membership* <http://www.iumi.com/about-iumi/membership>

INTERTANKO, *About Us* <http://www.intertanko.com/About-Us/>

IPCC, *Organization* <http://www.ipcc.ch/organization/organization.shtml>

JECKU, *CESS 2012 Press Release* (23 August 2012) <http://www.sajn.or.jp/e/press/CESS_AM2012_en,pdf>

JECKU, *CESS Annual Meeting 2009 Press Release* (2 September 2009) <http://www.sajn.or.jp/e/press/CESSPressRelease2009.pdf>

JECKU, *Chairman's Note (18th JECKU Top Executive Meeting October 29, 2009, Berlin)* <http://www.sajn.or.jp/e/press/Press_Berlin_TEM.pdf>

Khan, Munjurul Hannan, *Opinion: Paris Agreement – Opportunities and Challenges for Developing Countries* (23 February 2016) < http://cdkn.org/2016/02/opinion-paris-agreement-opportunities-and-challenges/?loclang=en_gb>

Korea Offshore & Shipbuilding Association (KOSHIPA), *Korean Shipbuilders Fall Behind Chinese Rivals* (2013) <http://www.koshipa.or.kr/eng/koshipa/koshipa3/news_view.jsp?kind=eng_n&idx=150&s_section=&s_keyword=>

Korean Register of Shipping (KRS), *KR Expands into Green Growth Business* (September 2009) <http://www.krs.co.kr/kor/html/webzine/sub/SubRead_V2_e.asp?chasu=20&idx=7&gubun=4&lan=E>

Kretschmer, Bianka, Felix Fallasch, *From Paris to Songdo: How the Green Climate Fund's New Strategic Vision Supports the Paris Agreement* (7 April 2016) < http://climate-analytics.org/blog/2016/from-paris-to-songdo>

Kurdziel, Marie, Thomas day, et al., *Challenges and Lessons Learned in the Preparation of Intended Nationally Determined Contributions (INDCs)* (1 March 2016) <http://mitigationpartnership.net/sites/default/files/challenges_lessons_indcs.pdf>

Lee, Alex, *South Korean Shipbuilding Faces Hard Times, Hyundai Heavy Reflects* (2 February 2012) <http://gcaptain.com/south-korean-shipbuilding-faces/>

Lloyd's Register, *World Shipbuilding Deliveries* (11 October 2010) <http://shipbuilding-history.com/today/statistics/world.htm>

Loris, Nicolas, *EPA Formally Declares CO_2 a Dangerous Pollutant* (7 December 2009) <http://blog.heritage.org/2009/12/07/epa-formally-declares-co2-a-dangerous-pollutant/>

Maritime Journal, *Maritime Industry Supports UK* (21 Dec 2012) <http://www.maritime-journal.com/news101/industry-news/maritime-industry-supports-uk>

Merchant, Brian, *First Official Climate Change Refugees Evacuate Their Island Homes for Good* (8 May 2009) <http://www.treehugger.com/corporate-responsibility/first-official-climate-change-refugees-evacuate-their-island-homes-for-good.html>

Nichols, Will, *EU Launches Attempt to Deliver Shipping Emissions Trading Scheme* (24 January 2012) <http://www.businessgreen.com/bg/news/2140997/eu-launches-attempt-deliver-shipping-emissions-trading-scheme>

OECD, *Environmental Externalities* (4 March 2003) <http://stats.oecd.org/glossary/detail.asp?ID=824>

OECD, *List of Developing Countries in Alphabetical Order* <http://www.icml9.org/public/documents/pdf/es/OECD.pdf>

OECD, *Market-based Instruments* (23 July 2007) <http://stats.oecd.org/glossary/detail.asp?ID=7214>

Oil Companies International Marine Forum (OCIMF), *Introduction* (2010) <http://www.ocimf.com/Organisation/Introduction>

Oxford Economics, 'The Economic Impact of the UK Maritime Services Sector: Shipping' (May 2015) <http://www.alinkfor.me/CoS/02/Shipping_The_economic_impact_of_the_UK_Maritime_Services_Sector.pdf>

Ozone Secretariat, UNEP, *Exemption Information* <http://ozone.unep.org/Exemption_Information/>

Pacific Islands Forum Secretariat, *Smaller Island States Leaders Statement on Climate Change* (6 September 2011) <http://www.forumsec.org.fj/resources/uploads/attachments/documents/SIS_Leaders_Outcomes_2011.pdf>

Paris MOU, *List of Paris MOU Deficiency Codes* (2013) <https://www.parismou.org/content/publishedmedia/2da7e155-8f2b-4afc-af3a-ea51a51a7aea/list%20of%20paris%20mou%20deficiency%20codes.pdf>

Paris MOU, *Paris Memorandum of Understanding on Port State Control* (23 May 2013) <http://www.guardiacostiera.it/servizi/documents/sicurezzanavigazione/MOU,%20incl.%2029th%20Amendment.pdf>

Paris MOU, *Paris MOU Reward Quality Flags* (2 November 2010) <https://www.parismou.org/Content/PublishedMedia/1039b41c-e715-4b08-8f17-080dd9c7238d/Paris%20MoU%20rewards%20quality%20flags%20%282%20November%29.pdf>

Parker, Dorothy, *Paul Sheehan, Ian Plimer, Heaven and Earth, and Is It Safe?* <http://themichaelduffyfiles.blogspot.com/2009/04/paul-sheehan-ian-plimer-heaven-and.html>

Parliament of Australia, *Carbon Pollution Reduction Scheme* (22 October 2010) <http://www.aph.gov.au/About_Parliament/Parliamentary_Departments/Parliamentary_Library/Browse_by_Topic/ClimateChange/Governance/Domestic/national/cprs>

PPMC, *Transport @ COP 21 Paris* (6 December 2015) < http://www.kas.de/wf/doc/kas_43927-1522-2-30.pdf?160119103248 >

Secretariat of the Pacific Community (SPC), *Pacific Island Populations: Estimates and Projections of Demographic Indicators for Selected Years* (2011) <http://www.spc.int/sdd/>

Shipping Australia Limited, *Shipping Australia Supports Studies into Ship Emissions in the Australian Region* (31 August 2012) <http://shippingaustralia.com.au/wp-content/uploads/2012/01/20123108ShippingAustraliasupportsstudiesintoshipemissionsintheAustralianregion.pdf>

Tayor, Lenore, *Australia Kills Off Carbon Tax* (17 July 2014) <http://www.theguardian.com/environment/2014/jul/17/australia-kills-off-carbon-tax>

The International Association of Port and Harbors (IAPH), *IAPH Tool Box for Greenhouse Gases* (2008) <http://iaphtoolbox.wpci.nl/DRAFT%20IAPH%20GHG%20TOOL%20BOX%20dea.pdf>

The International Association of Port and Harbors (IAPH), *IAPH Tool Box for Port Clean Air Programs* (2008) <http://iaphtoolbox.wpci.nl/DRAFT%20IAPH%20TOOL%20BOX%20%20dea.pdf>

The International Association of Port and Harbors (IAPH), *Member Ports of the IAPH Welcome the International Maritime Organization's New Mandatory Measures to Reduce Greenhouse Gas Emissions from International Shipping* (22 July 2011) <http://www.iaphworldports.org/IAPHPressReleases.aspx>

The Parliament of the Commonwealth of Australia, *Rebuilding Australia's Coastal Shipping Industry: Inquiry into Coastal Shipping Policy and Regulation* (October 2008) <http://www.aph.gov.au/Parliamentary_Business/Committees/House_of_Representatives_Committees?url=itrdlg/coastalshipping/report.htm>

The Shipbuilders and Shiprepairers Association (SSA), *Maersk Line's Focus on Energy Efficiency Pays Off* (8 March 2013) <http://www.ssa.org.uk/news/industry/maersk-lines-focus-on-energy-efficiency-pays-off>

The Shipping Corporation of Indian Ltd. (SCI), 'Global Compact Initiative' (2012) <http://www.shipindia.com/pdf/MediaSpeak/COPfor2012-5c80e7.pdf>

The United States of America, *U.S. Submission on Elements of the 2015 Agreement* (12 February 2014) <https://unfccc.int/files/documentation/submissions_from_parties/adp/application/pdf/u.s._submission_on_elements_of_the_2105_agreement.pdf>

The World Bank, *Country and Leading Groups* (2012) <http://data.worldbank.org/about/country-classifications/country-and-lending-groups>

Tokyo MOU, *List of Tokyo MOU Deficiency Codes* (20 August 2013) <http://www.tokyo-mou.org/doc/Tokyo%20MOU%20deficiency%20codes%20%2820%20August%202013%29.pdf>

Tokyo MOU, *Memorandum of Understanding on Port State Control in the Asia-Pacific Region* (29 October 2013) <http://www.tokyo-mou.org/doc/memorand.pdf>

Tokyo MOU, *Tokyo MOU Will Introduce A New Inspection Regime (NIR) from 1st January 2014* (15 February 2013) <http://www.tokyo-mou.org/doc/PRESS-NIR.pdf>

UK Chamber of Shipping, *UK Chamber of Shipping Refutes Conclusion of Energy and Climate Change Committee That International Solution on Reducing Carbon Is A 'Delaying Tactic'* (27 Jan 2012) <http://www.ukchamberofshipping.com/news/2012/01/27/uk-chamber-shipping-refutes-conclusion-energy-and-climate-change-committee-international-solution-reducing-carbon-delaying-tactic/>

UNFCCC, *Bodies* (2014) <http://unfccc.int/bodies/items/6241.php>

UNFCCC, *A Brief Overview of Decisions* <http://unfccc.int/documentation/decisions/items/2964.php>

UNFCCC, *Doha Amendment* (12 June 2014) <https://unfccc.int/kyoto_protocol/doha_amendment/items/7362.php>

UNFCCC, *The Doha Climate Gateway* (2012) <http://unfccc.int/key_steps/doha_climate_gateway/items/7389.php>

UNFCCC, *Durban: Towards Full Implementation of the UN Climate Change Convention* (2011) <http://unfccc.int/key_steps/durban_outcomes/items/6825.php>

UNFCCC, *International Bunker Fuels under the AWG-LCA* <http://unfccc.int/methods/emissions_from_intl_transport/items/6141.php>

UNFCCC, *Meetings* <http://unfccc.int/meetings/items/6240.php>

UNFCCC, *Record Support for Advancing Paris Climate Agreement Entry into Force* (22 April 2016) <http://newsroom.unfccc.int/paris-agreement/closing-paris-agreement-signing-press-release/>

UNFCCC, *Warsaw Outcomes* (2013) <http://unfccc.int/key_steps/warsaw_outcomes/items/8006.php>

Union of Greek Shipowners, *MBIs for Reduction of CO_2 Emissions from Tramp Shipping* (2010) <http://www.shippingandco2.org/UGSMBIsForReductionOfCO2EmissionsFromTrampShipping%20.pdf>

Union of Greek Shipowners, *Prevention of Environmental Pollution by Ships: Regulation and Compensation Regimes and Industry Standards* (2011) <http://www.nee.gr/default.asp?t=anakoinoseisDetails&id=13>

United Nations Economic Commission for Europe (UNECE), *The 1979 Geneva Convention on Long-rang Transboundary Air Pollution* <http://www.unece.org/env/lrtap/lrtap_h1.html>

United Nations Environment Programme (UNEP), *About UNEP* <http://www.unep.org>

United Nations Environment Programme (UNEP), *Climate Change* <http://www.unep.org/gc/gc26/factsheet/pdfs/Climate_change.pdf>

United Nations Framework Convention on Climate Change (UNFCCC), *Status of Ratification of the Convention* <http://unfccc.int/essential_background/convention/status_of_ratification/items/2631.php>

United States Coast Guard (USCG), *Adding International Energy Efficiency (IEE) Certificate to List of Certificates A Recognised Classification Society May Issue* (10 December 2012) <https://www.federalregister.gov/articles/2012/12/10/2012-29749/adding-international-energy-efficiency-iee-certificate-to-list-of-certificates-a-recognized>

United States Coast Guard (USCG), *MARPOL Annex VI International Energy Efficiency (IEE) Certificate Implementation Guidance* (5 May 2013) <http://www.regulations.gov/#!documentDetail;D=USCG-2012–1095–0002>

United States Coast Guard (USCG), *Overview of the United States Coast Guard* (19 September 2013) <http://www.uscg.mil/top/about/>

Vidal, John, *Maritime Countries Agree First Ever Shipping Emissions Regulation* <http://www.guardian.co.uk/environment/2011/jul/18/maritime-countries-shipping-emissions-regulation>

World Meteorological Organization (WMO), *Intergovernmental Panel on Climate Change (IPCC)* <http://www.wmo.int/pages/themes/climate/international_ipcc.php>

World Meteorological Organization (WMO), *WMO and Climate Change* <http://www.wmo.int/pages/themes/WMO_climatechange_en.html>

World Meteorological Organization (WMO), *WMO at UNFCCC/COP Sessions* <http://www.wmo.int/pages/prog/wcp/cop17/background_en.html>

World Ports Climate Initiative, *World Ports Climate Declaration* (July 2008) <http://www.wpci.nl/docs/Declaration.pdf>

World Trade Organization (WTO), *Chronological List of Disputes Cases* <http://www.wto.org/english/tratop_e/dispu_e/dispu_status_e.htm>

大公报[Takungpao], 《太平洋造船集团新船首获能效设计指数证书》[*New Ship Built by SinoPacific Shipbuilding Group First Obtained the EEDI Certificate*] (24 February 2012) <http://www.csoa.cn/xuliebd/jienengjp/201202/t20120224_1198135.html>

王尔德[Wang Erde], '《减排谈判首次共识，中国减排将始于 2019 年》[Agreement Achieved through Reduction Negotiations and China Is to Start Its Reduction in 2019]', *21st Century Business Herald* (Beijing), 26 July 2011 <http://stock.sohu.com/20110726/n314586469.shtml>

肜新春[Rong Xinchun], 《试论新中国海运事业的发展和变迁(*1949–2010*)》[*Development and Evolution of China's Shipping Industry (1949–2010)*] (31 October 2012) <http://economy.guoxue.com/?p=7575>

谢宗惠[Xie Zonghui], '《合力打造绿色航运产业链》Jointly Establish A Green Industry Chain for the Shipping Industry', 中国水运报 *China Waterborne Transportation Newspaper* 22 September 2011 <http://www.zgsyzz.com/Article/ShowInfo.asp?ID=5318>

Index

www.ingramcontent.com/pod-product-compliance
Lightning Source LLC
LaVergne TN
LVHW010623300926
849162LV00001B/37